Conversion tables: Metric and English systems.*

	ENGLISH TO METRIC						METRIC TO ENGLISH		
lb/A	equals	lb/ha	equals	kg/ha	equals	g/m²	kg/ha	equals	lb/A
0.25		0.62		0.28		0.028	0.25		0.22
0.33		0.82		0.35		0.035	0.50		0.44
0.50		1.24		0.56		0.056	0.75		0.67
0.67		1.63		0.69		0.069	1.00		0.89
0.75		1.85		0.84		0.084	1.50		1.33
1.00		2.47		1.12		0.112	1.75		1.55
2.00		4.95		2.24		0.224	2.00		1.78
3.00		7.42		3.37		0.337	2.50		2.23
4.00		9.89		4.49		0.448	3.00		2.67
5.00		12.36		5.61		0.561	3.50		3.11
6.00		14.84		6.73		0.673	4.00		3.56
7.00		17.31		7.85		0.785	4.50		4.01
8.00		19.78		8.98		0.898	5.00		4.45
9.00		22.26		10.10		1.010	5.50		4.90
10.00		24.70		11.22		1.122	6.00		5.34
15.00		37.06		16.83		1.683	6.50		5.79
20.00		49.50		22.44		2.244	7.00		6.23
25.00		61.86		28.05		2.805	7.50		6.68
							8.00		7.12
							8.50		7.57
							9.00		8.01
							9.50		8.46
							10.00		8.90
							15.00		13.36
							20.00		17.80
							25.00		22.26
							30.00		26.72
							35.00		31.15
							40.00		35.60
							45.00		40.05
							50.00		44.50
							60.00		53.44
							70.00		62.30
							80.00		71.20
							90.00		80.10
							100.00		89.00

*To convert lb/A to kg/ha, divide lb/A by 0.893.
To convert kg/ha to lb/A, multiply kg/ha by 0.893.

WEED SCIENCE

Principles
and
Applications

THIRD EDITION

WEED SCIENCE

Principles and Applications

THIRD EDITION

WOOD POWELL ANDERSON

Professor Emeritus of Agronomy
New Mexico State University
Las Cruces, New Mexico

WEST PUBLISHING COMPANY
Minneapolis/St. Paul   New York
Los Angeles   San Francisco

WEST'S COMMITMENT TO THE ENVIRONMENT

In 1906, West Publishing Company began recycling materials left over from the production of books. This began a tradition of efficient and responsible use of resources. Today, up to 95 percent of our legal books and 70 percent of our college and school texts are printed on recycled, acid-free stock. West also recycles nearly 22 million pounds of scrap paper annually—the equivalent of 181,717 trees. Since the 1960s, West has devised ways to capture and recycle waste inks, solvents, oils, and vapors created in the printing process. We also recycle plastics of all kinds, wood, glass, corrugated cardboard, and batteries, and have eliminated the use of Styrofoam book packaging. We at West are proud of the longevity and the scope of our commitment to the environment.

Production, Prepress, Printing and Binding by West Publishing Company.

PRODUCTION CREDITS

Copyeditor and Proofreader: Luana Richards

Interior and Cover Design: Roslyn Stendahl, Dapper Design

Illustrations: Carlyn Iverson, Absolute Science
Radiant Illustration and Design

Composition: Carlisle Communications, Ltd.

The main text typeface is New Caledonia. The display typeface is Frutiger.

 TEXT IS PRINTED ON 10% POST CONSUMER RECYCLED PAPER

 PRINTED WITH SOY INK™

British Library Cataloging-in-Publication Data. A catalogue record for this book is available from the British Library.

Library of Congress Cataloging-in-Publication Data

Anderson, Wood Powell.
    Weed science : principles and applications / Wood Powell
      Anderson.
    —3rd ed.
        p.   cm.
    Includes bibliographical references (p.   ) and index.
    ISBN 0-314-04627-5 (hardcover : alk. paper)
      1. Weeds—Control. 2. Weeds. 3. Herbicides. I. Title.
SB611.A52 1996
632'.58—dc20
                                    95-30901
                                      CIP

BRIEF CONTENTS

CONTENTS

PREFACE

This third edition of *Weed Science: Principles and Applications* expands and updates selected topics presented in earlier editions and introduces new topics gaining prominence in weed science. The title of this edition has been expanded to include "and Applications" in recognition of newly added chapters concerned with weeds and weed control in selected crops. The subject matter is organized into three sections; they are Weed-Control Principles (Chapters 1–11), Herbicide Families (Chapters 12–30), and Weed Control in Selected Crops: An Introduction (Chapters 31–46). Although the review chapters on soils, botany, and organic chemistry included in the earlier editions were well received by students, they have been deleted from this edition at the request of manuscript reviewers. Herbicides no longer in use have also been deleted.

New chapters include Weed Ecology, Conservation Tillage Systems, and Weed Resistance and Transgenic Crop Tolerance, and 15 chapters concerned with weed control in selected crops, turfgrass, and pastures and rangelands. New herbicide families include the aryloxyphenoxypropionates, cyclohexanediones, imidazolinones, and sulfonylureas. Chapter 7, Modes and Sites of Action of Herbicides, reflects the great advances made within the past decade in how and where the activity of individual and families of herbicides takes place within the plant. The characteristics of 128 individual active ingredients (herbicides), grouped in 18 families and 1 nonfamily category, are highlighted as to their respective chemical structure, principal use, mode and site of action, translocation, and selectivity.

Proficiency in weed identification is vital to effective weed control. However, due to the wide diversity of weed species among crops, states, and regions, it was deemed impractical to cover this topic in this text. What may be an important weed to one is of no importance to another. This is apparent from reviewing the lists of weeds included in Chapters 32–46 concerning weed control in selected crops. There are a multitude of excellent illustrated books, manuals, and pamphlets available that are devoted just to weed identification—for example, *Arizona Weeds,* 338 pp. (Univ. Arizona Press); *Weeds of the West,* 630 pp. (Univ. Wyoming); *Weeds and Poisonous Plants of Wyoming and Utah,* 280 pp. (Univ. Wyoming); *Weeds of the North Central States,* 303 pp. (Univ. Illinois); *Weeds of Colorado,* 218 pp. (Colorado State Univ.); *California Growers Weed Identification Handbook* (231 weeds) (Univ. Calif., Berkeley); *Weeds of Southern Turfgrasses,* 208 pp. (Univ. Georgia); *Weed Identification Guide* (about 400 weeds) (Southern Weed Sci. Soc.); *Selected Weeds of the United States,* 463 pp. (USDA); *The Biology of Canadian Weeds,* 2 Vol., 794 pp. *(Canadian J. Plant Sci.).*

Although the author has drawn heavily on the works of others, especially original research papers and scientific reviews, in developing his subject, he has not cited these works, with some exceptions, in the text itself. However, the principal references upon which the content of each chapter is based (with the exception of those chapters concerned with characteristics of individual herbicides, sprayer calibration, and dosage calculations) are included at the end of their respective chapters. This practice should prove less distracting to the student.

■ ACKNOWLEDGEMENTS

The author wishes to express his appreciation to Drs. Arden A. Baltensperger and Donald J. Cotter, Professors Emeriti, Departments of Agronomy and Horticulture, respectively, New Mexico State University, and to José Alba "Buddy" Viramontes, Sales Representative, DuPont Agricultural Products for their support and encouragement in my endeavors to complete this third edition. Very special thanks are extended to my wife Nancy, who has borne the brunt of the author's enthusiasms, frustrations, and loss of companionship during the preparation of the manuscript.

Christopher R. Conty and Elizabeth Riedel, Editor and Developmental Editor, respectively, West Educational Publishing, Amesbury, Massachusetts, provided valued encouragement and direction in the preparation of the manuscript from its conception. Appreciation is extended to Steven Yaeger, Production Editor, West Publishing Company, Eagan, Minnesota, for his competence and constructive suggestions in finalizing the manuscript and for his pleasant manner during our many phone conversations. Steve has been a pleasure to work with. Special thanks are expressed to Luana Richards, the manuscript copyeditor, for a superb job of editing. Appreciation is also extended to the following individuals who constructively reviewed all or portions of the manuscript:

Carol Bubar   *Olds College*

Brent Pearce   *Iowa State University*

Tom Peeper   *Oklahoma State University*

Glen Wehtje   *Auburn University*

Jim Kells   *Michigan State University*

Chris Boerboom   *Washington State University*

John Masiunas   *University of Illinois*

Pat Delwich   *California State University, Chico*

Clarence Swanton   *University of Guelph*

Albert E. Feldman   *Mesa Community College (Ariz.)*

Gordon Harvey   *University of Wisconsin*

WEED-CONTROL PRINCIPLES

1

WEEDS

■ INTRODUCTION

A weed may be defined simply as "any plant growing where it is not wanted." Weeds are familiar plants in our environment that are seen, consciously and unconsciously, infesting lawns, sidewalks, roadsides, fencerows, ditches and ditch banks, ponds and waterways, gardens, croplands, rangelands, and forests. Weeds are a part of the agriculturalist's everyday life. In general, weeds adversely affect the use, economic value, and aesthetic aspect of the land and waters they infest.

Weeds are controversial plants that are neither all bad nor all good, depending on one's outlook. Weeds can be enjoyed for their showy flowers and seedpods. Weeds prevent or reduce wind and water erosion of the land. They appear and grow quickly to cover unsightly scars in the landscape caused by man and nature. Some weeds, such as downy brome and Russian thistle, provide excellent forage for livestock during certain periods of the year, especially in the spring. Weeds provide shelter and food for birds, rabbits, small rodents, and other wildlife. Weeds provide employment for local and transient labor hired to control or remove them. They provide the physician and veterinarian with patients afflicted with allergies, dermititis, and poisoning resulting from exposure to the pollens and irritating or poisonous chemicals that they produce. Weeds serve as hosts for insects and disease organisms that move on to parasitize nearby crop and ornamental plants.

Weeds are familiar objects, yet they are not always easy to define. Some plants are readily recognized as weeds because of their general undesirableness—for example, common chickweed, field bindweed, puncturevine, purple nutsedge, quackgrass, sandbur, and Russian thistle. Other plants, however, are not so readily accepted as weeds; this depends entirely on how the plants affect the interests of an individual or community as to whether or not the plants are categorized as weeds. For example, annual morningglories are considered despicable weeds by cotton farmers, but they may be favored ornamental plants of many home gardeners, grown for their screening effect and showy flowers. Johnsongrass, canarygrass, and bermudagrass are desirable forage plants for livestock in some parts of the United States, yet in some locales they are ranked among the worst weeds. Volunteer plants from a previous crop, arising from seed shattering prior to harvest or dropped during harvest, may also become weeds in fallow land, as in the case of small grains, or in a succeeding crop—for example, barley in wheat, corn in soybean, and sorghum in cotton, or even a different variety or strain of wheat growing in a wheat field intended for pure, certified seed.

3

TABLE 1.1 ■ Crops, forages, and ornamentals that have become weeds in the United States.

WEED/CROP	INTENDED USE
Sorghum-almum	Forage
Bermudagrass	Forage
Crabgrass	Forage
Johnsongrass	Forage
Shattercane	Cultivated sorghum
Wild proso millet	Cultivated millet
Artichoke thistle	Cultivated artichoke
Common lambsquarters	Food (leaf salad)
Jimsonweed	Ornamental
Kochia	Ornamental
Showy crotalaria	Green manure crop

SOURCE: Adapted from D. Bronsten and B. Simmonds. 1989. Crops gone wild. *Agrichemical Age* 33(5):28.

Examples of plants introduced as potential crops, forages, and ornamentals that have become weeds are given in Table 1.1.

In general, weeds are plants that follow civilization, and few weeds are found where humankind has not taken up abode. Weeds are plants characteristic of lands where we have replaced the native vegetation with a controlled system of cropping and management and, on other lands, where we have forcibly altered the vegetation for other purposes. Our intervention in the natural ecology of an area is radically disruptive of the balanced and mutually dependent system of interrelated plant and animal life in the area. Often, the original plant species disappear and are replaced by other plant species that are better adapted to the new environment. A corresponding change in the animal population also occurs, with certain of the original animal species replaced by species that are better adapted to the changed environment.

Weeds comprise the first stage of plant succession on lands where the native vegetation has been disturbed. *Weeds are the pioneer plants of disturbed soils.* Common pokeweed is a pioneer weed species common to eastern North America, colonizing most places where the soil has been disturbed such as raw river alluvium, landslides, diggings, and abandoned cultivated fields. Pokeweed is, within limits, indifferent to shade, temperature, moisture, and soil composition and structure, an indifference that greatly contributes to its pioneer status.

In central Oklahoma, weeds comprise the first stage in plant succession in the return of abandoned farmland to characteristic climax prairie vegetation, followed by stages comprised of annual grasses, perennial bunchgrasses, and true prairie vegetation. Of these stages, the weed stage lasts for 2–4 years; the annual grass stage, dominated by prairie threeawn, lasts from 9 to 13 years; the perennial bunchgrass stage, dominated by little bluestem, may still be present 30 years after abandonment of the land; and the true prairie stage may not be reached until more than 40 years after abandonment. The true prairie in central Oklahoma is characterized by little bluestem, big bluestem, switchgrass, and Indian grass. In central Kansas, abandoned farmland requires more than 33 years to attain a climax cover.

A list of the worst weeds of U.S. croplands in 1991 is given in Table 1.2. One must not, however, be misled as to the finality of such lists; they are based on broad, general areas, and they are, by necessity, somewhat arbitrary. The reader may be familiar with a particularly troublesome weed that is not included in the above lists, such as pitted morningglory and common sunflower in corn and soybeans in the Corn Belt; dogfennel in pasture and rangelands in

TABLE 1.2 ■ The most prevalent problem weeds in croplands of the United States, 1991.

GRASS SPECIES	BROADLEAF SPECIES
Annuals	*Annuals*
Barnyardgrass	Annual morningglories
Broadleaf signalgrass	Common cocklebur
Crabgrass, large	Common lambsquarters
Crabgrass, smooth	Kochia
Downy brome	Morningglory, ivyleaf
Panicum, fall	Morningglory, tall
Panicum, Texas	Pennsylvania smartweed
Foxtail, giant	Pigweed, prostrate
Foxtail, green	Pigweed, redroot
Foxtail, yellow	Pigweed, smooth
Shattercane	Prickly sida
Wild oat	Russian thistle
Perennials	Sicklepod
Bermudagrass	Tansymustard
Johnsongrass	Velvetleaf
Quackgrass	Wild buckwheat
Sedges	Wild mustard
Purple nutsedge	*Perennials*
Yellow nutsedge	Canada thistle
	Field bindweed
	Leafy spurge

SOURCE: *Meister's 1992 weed control manual,* p. 363. Willoughby, Ohio: Meister Publishing.

the Southeast; jointed goatgrass in wheat grown in the West; or wild poinsettia in southern soybeans. Also not included are sicklepod and tropic croton in the southeast; yellow rocket in forage crops in southeastern Canada and the northern tier of states of the United States; common chickweed in alfalfa in the northeastern United States; red rice in cultivated rice grown in California and the southern United States; and annual bluegrass in lawns and turf throughout most of the United States. Thus a list of the worst weeds is all in the eye of the beholder, but these official lists do serve the purpose of focusing attention on the most difficult weed-control problems on a regional, national, and international basis.

■ LOSSES DUE TO WEEDS

In 1991, the estimated average annual monetary loss caused by weeds in 46 crops grown in the United States was $4.1 billion. Of this loss, about 82% occurred in field crops, 5% in noncitrus fruit crops, 3% in citrus crops, 1% in nut crops, and 9% in vegetable crops (Bridges, 1992). If herbicides had not been used in 1991, the estimated annual loss would have been $19.6 billion.

In Canada, the estimated average annual monetary loss caused by weeds in 58 crops was $984 million, with losses of $372 million in eastern Canada and $612 million in western Canada. Of the estimated loss in eastern Canada, about 50% occurred in hay crops, 33% in field crops, and 17% in fruit and vegetable crops; while in western Canada, 84% of the loss occurred in field crops, 12% in hay crops, 3% in fruits and vegetables, and 1% in tree fruits (Swanson et al., 1993).

■ ADVERSE EFFECTS OF WEEDS

In nonagricultural lands, weeds are a potential hazard to humans. Weed pollen may cause hay fever or other allergies, and toxic chemicals present in their sap or on their leaves may cause skin irritations or rashes when brushed against, as in the case of persons allergic to poison ivy, poison oak, or poison sumac. Some substances produced by weeds are deadly to man or animals when ingested.

Dense, moderately tall weeds obstruct visibility at roadway intersections, conceal warning signs and markers, and induce small animals and deer to forage along roadways, providing them with cover and a false sense of security. Weeds tend to hide tools and equipment, switches and valves, irrigation gates, and even holes in the ground.

Dense, moisture-holding weed growth aids in the deterioration of wooden structures and the rusting of metal fences, buildings, and immobile machinery. Dry, dead weeds constitute a fire hazard, subject to ignition by a spark from train wheels, a carelessly tossed cigarette, or even a piece of glass magnifying or reflecting sunlight. Weeds reduce the enjoyment of recreation areas. They provide protection for mosquitoes, spiders, chiggers, and other pests that attack humans and livestock. They impede the flow of water in drain ditches and waterways. They impede the movement of boats on infested waters.

In agricultural lands, weeds increase the time and costs involved in crop production, interfere with crop harvesting, and reduce crop yields and quality. In addition, weeds create problems in range and livestock management. Some weeds are undesirable in hay, pasture, and rangelands because of the injuries they inflict on livestock. Woody stems, thorns, and stiff seed awns cause injury to the mouth and digestive tracts of livestock, and the hairs and fibers of some plants tend to ball up and obstruct the intestines, especially in horses, causing serious problems. Ingested by milk cows, some weeds, such as ragweeds, wild garlic, and mustards, impart a distinctly distasteful odor or flavor to milk and butter. Barbed seed-dispersal units may become so entangled in the wool of sheep as to greatly diminish its market value. Parasitic plants, such as dodder, broomrape, and witchweed, rob their host plants of organic foods.

■ WEEDS AS HOST PLANTS

Weeds can serve as hosts for plant pests. Pepperweed and tansymustard, for example, are hosts to the turnip aphid and green peach aphid. Several weed species of the nightshade family (*Solanaceae*) are hosts to insects that commonly attack eggplant, pepper, potato, and tomato; for example, horsenettle is a host of the Colorado potato beetle, and black nightshade is a host of the cabbage looper. Annual morningglories are important hosts of insects attacking sweet potato, especially the highly destructive sweet potato weevil. Common ragweed serves as a host for many kinds of insects. Hoary cress is the main source of beet western yellows virus, but it also develops in plantains, shepherd's-purse, and pigweeds. Field bindweed and, to a lesser extent, Canada thistle are principal weeds on which young larvae of redbacked cutworms, a pest on asparagus, develop in April and May. European barberry is an essential host of the wheat stem rust in the northern United States. Goosegrass and purple nutsedge are hosts of barley yellow dwarf virus. Currants and gooseberries are hosts for white pine blister rust. Waterlettuce is the preferred host for *Mansonia* mosquitoes and also an insect vector for the human diseases encephalitis and rural filariasis.

■ HAY FEVER AND DERMATITIS

Hay fever is an affliction of millions of people in the United States each year. It is caused by the adverse effects of protein in the pollen of various plants on the respiratory system of susceptible people. Ragweeds are the principal plants causing hay fever east of the Rocky Mountains, although other plants such as goldenrod contribute to a lesser extent. West of the Rocky Mountains, hay fever is caused by a wide variety of plants, including trees, shrubs, grasses, and broadleaf plants. Many people are particularly allergic to the pollen of bermudagrass, an unfortunate situation for those who live in areas of the South and Southwest where bermudagrass is grown for turf and forage.

Dermatitis is an inflammation of the skin. Many people are troubled with a dermatitis resulting from contact with poison ivy, poison oak, and poison sumac. These plants produce and store a substance identified as *urushiol,* a mixture of several derivatives of the chemical *catechol* having 15-carbon side chains that, upon hydrogenation, yield *3-pentadecylcatechol,* the actual skin irritant. The chemical structures of catechol and 3-pentadecylcatechol are as follows:

catechol   3-pentadecylcatechol

Upon contact with the skin, urushiol induces intense itching, inflammation, and blistering, and in some cases marked swelling and puffiness. People vary greatly in susceptibility; some appear immune, while others are highly susceptible. A small amount of urushiol can induce a severe, spreading skin rash in susceptible persons. In some cases, highly susceptible people have been hospitalized or have died as a result of the effects induced by this plant-produced chemical.

Urushiol occurs in great abundance in the sap of all parts (roots, stems, leaves, fruits) of these plants. Urushiol is usually transferred from the plant to one's skin by direct contact. It may also be transferred from one object to another, and it may be carried in the smoke of burning plants. Urushiol-contaminated clothing and the hair of dogs and cats are potential means by which urushiol can be indirectly transferred.

■ CLASSIFICATION OF WEEDS

Weed classification is achieved by "grouping together those weed species whose similarities are greater than their differences." For convenience, weeds are commonly classified in a variety of ways. They are grouped in categories such as terrestrial and aquatic, woody and herbaceous, or simply as trees, shrubs, broadleaved, grasses, sedges, and ferns. For preciseness, weeds are grouped botanically by families, genera, species, and variety.

Weeds are also commonly grouped according to similar life cycles. On this basis, weeds are grouped as *annuals, biennials,* and *perennials.*

Annuals complete their life cycles in one growing season (within 12 months). Annuals that complete their life cycles during the period from spring to fall are referred to as *summer annuals,* or simply as *annuals;* those that complete their life cycles during the period from fall to spring or early summer are called *winter annuals.* The majority of annual weeds are summer annuals. However, some very troublesome weeds are found among winter annuals—for example, annual bluegrass, downy brome, field pennycress, London rocket, and tansymustard. Some annual weeds have indeterminate life cycles; that is, germination and growth may occur at any time of the year, but their life cycles are completed within a 12-month period. Common chickweed is an example of a weed with an indeterminate life cycle.

Biennials require two growing seasons in which to complete their life cycles. They usually form rosettes (radial clusters of leaves lying close to the ground) the first season and, during the second season, they send up flower stalks (referred to as *bolting),* set seed, and die. Biennial weeds are not as common as annual weeds. Examples include mullein, burdock, wild carrot or Queen Anne's lace, sweetclover, and wild parsnip.

Perennials live for three or more years. Depending on the species, perennial weeds may not flower the first year and, with some species, it is common for a portion of these early-produced seeds to be nonviable. Perennials may be subgrouped as *simple perennials* and *creeping perennials.*

Simple perennials propagate and spread primarily by seeds—for example, curly dock, common milkweed, and dandelion. Simple perennials possess a root crown, topping a fleshy taproot or a mass of fibrous roots, that produces one or more new shoots each year, replacing the previous year's shoots that matured and died.

Creeping perennials propagate by seeds and asexual means; examples include common bermudagrass, field bindweed, johnsongrass, purple and yellow nutsedges, quackgrass, and silverleaf nightshade. Asexual reproduction is by underground roots and rhizomes and by aboveground runners or stolons. The ability of creeping perennial weeds to reproduce asexually is what makes them so difficult to control. Some of our worst weeds are found in this group.

Botanical Families of Troublesome Weeds

The following is a list of botanical families in which many of the most troublesome weeds are members, with representative weeds in each family identified by common name. The newer family names are used. Where applicable, the older name is in parentheses.

Amaranthaceae	Pigweeds, khakiweed (creeping chaffweed).
Apiaceae (Umbelliferae)	Cow parsnip, poison hemlock, wild carrot.
Asclepiadaceae	Milkweeds.
Asteraceae (Compositae)	Asters, broom snakeweed, Canada thistle, knapweeds, ragweeds, sunflowers, sagebrushes, Texas blueweed.
Brassicaceae (Cruciferae)	Flixweed, mustards, pepperweeds, shepherd's-purse, yellow rocket.
Chenopodiaceae	Goosefoot, lambsquarters, halogeton, kochia, Russian thistle, saltbush.
Convolvulaceae	Dodder, field bindweed, annual morningglories.
Cyperaceae	Purple nutsedge, yellow nutsedge.
Euphorbiaceae	Leafy spurge, prostrate spurge, spotted spurge.
Fabaceae (Leguminosae)	Black medic, locoweeds, mesquites, sicklepod.
Poaceae (Gramineae)	Annual bluegrass, barnyardgrass, bermudagrass, crabgrasses, foxtails, johnsongrass, jointed goatgrass, panicums, quackgrass, wild oat.
Polygonaceae	Curly dock, knotweeds, smartweeds, sorrels, wild buckwheat.
Portulacaceae	Common purslane.
Rosaceae	Chokecherry, cinquefoils, wild rose, multiflora rose.
Solanaceae	Buffalobur, groundcherries, jimsonweed, nightshades.
Urticaceae	Stinging nettle.

■ WEED DISSEMINATION

The distribution of weeds throughout the world has been directly associated with humankind's explorations and colonizations. As we have moved about the surface of the world, we have carried with us the familiar food plants and seeds upon which we sustain ourselves and our livestock. Inadvertently, through ignorance or oversight, we have also carried with us the seeds of weeds common to the regions from which we came. These weed seeds have been "stowaways" among our goods and foodstuffs, not too different from the "seeds" of human diseases, such as diptheria, measles, smallpox, and tuberculosis, that we have also carried with us. Weed seeds were present in ships' ballasts (subsequently unloaded at innumerable foreign ports and inlets), and in the hay and bedding of livestock, between the ticking of mattresses, and intermixed as contaminants in crop seed to be planted at "trail's end." These seed stowaways were also nestled in vegetative matter used for packing and between boards of boxes and crates, or fastened to clothing and the hair and wool of domestic animals.

Many of the most troublesome weeds now present in the United States were introduced from Europe during the colonization and early settlement of North America. Such weeds include johnsongrass, Russian thistle, field bindweed, Canada thistle, giant foxtail, and St. Johnswort or Klamath weed, among others.

Weeds spread within a locale or from place to place by the dispersal of their reproductive parts, sexual (seeds) and asexual (aerial bublets, bulbs, corms, roots, rhizomes, stolons, tubers). *Annual weeds* are spread near and far only by the dissemination of their seeds. In some instances, annual weeds may root at a node touching moist soil or they may reroot after being uprooted, but these are of little importance to their general distribution. *Perennial weeds* usually invade distant places by the distribution of their seed, rarely by vegetative parts without the aid of man. Perennial weeds spread locally by sexual and/or asexual reproductive parts, with ever-spreading clusters, or colonies, evolving from vegetative parts spreading and radiating from the parent plant. The spread of perennial weeds in cultivated land is aided and enhanced by the movement of tillage equipment through the field, cutting and dragging asexual plant parts to new areas where they quickly become reestablished. Harvesting equipment distributes weed seeds, harvested along with the crop grain or forage, from one location to another as it moves within a field or from field to field. The combine harvester has been described as the perfect weed seed dispersal device for late-maturing weed species.

Weed seeds are dispersed by *natural* and *artificial* means. The natural means of seed dispersal include

(1) wind, (2) water, (3) animals, and (4) forceful dehiscence (rupturing) of seedpods or capsules. Artificial means of seed dispersal are largely associated with humankind and our agricultural pursuits.

Natural Dissemination

Wind

Wind is the most common of the natural means of weed seed dispersal. Some weed seeds, such as seeds of witch-weed, are so tiny and light that they are readily wind-borne

and carried from place to place. Other seeds are too heavy to travel on light breezes without special appendages to improve their aerodynamics, such as "parachutes," wings, and tufts of hair. Such appendages are usually not part of the seed itself but are instead part of the structure of the fruit that encases the seed. The *seed-dispersal unit* is composed of the seed, seed capsule, and appendages. Various seed-dispersal units that enhance dissemination of weed seeds are illustrated in Figure 1.1.

Weed species with seed-dispersal units equipped with a feathery pappus that serves as a "sail" or "parachute" include dandelion, prickly lettuce, common salsify, and

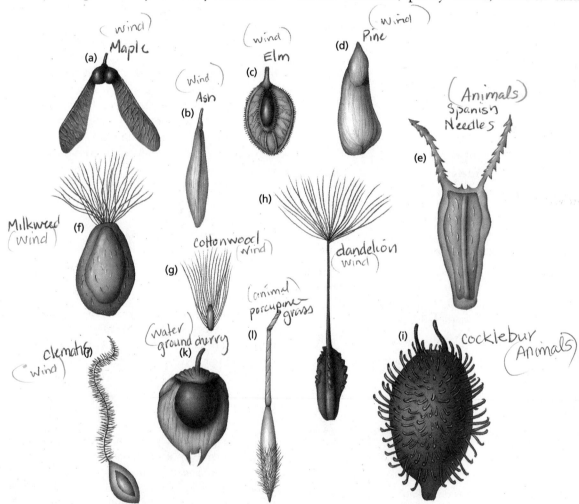

FIGURE 1.1 ■ Various seeds and fruits showing devices for dissemination.   (a) Winged fruit of maple (*Acer*). (b) Winged fruit of ash (*Fraxinus*). (c) Winged fruit of elm (*Ulmus*). (d) Winged seed of pine (*Pinus*). (e) Barbed fruit of spanish-needles (*Bidens*). (f) Milkweed (*Asclepias*) seed with tuft of hair. (g) Hairy seed of cottonwood (*Populus*). (h) Fruit of dandelion (*Taraxacum*) with parachutelike tuft of hairs. (i) Hooked fruit of cocklebur (*Xanthium*). (j) Fruit of *Clematis* with long plumelike style. (k) Fruit of groundcherry (*Physalis*), with membranous envelope. (l) Bearded fruit of porcupinegrass (*Stipa*).

SOURCE: R. M. Holman and W. W. Robbins. 1939. *General botany,* p. 294. New York: John Wiley & Sons, Inc.

annual sowthistle. Seed-dispersal units with tufts of hair at one end, enabling the seed to become airborne, are common to aster, dogbane, and milkweed. Winglike appendages are more characteristic of the taller woody plants, such as trees, tall shrubs, and climbing woody vines, than of herbaceous plants. Such appendages are common to maple, pines, and spruce, among others. Certain herbaceous weeds, however, do possess winged seed-dispersal units— for example, wild carrot, halogeton, yucca, and dock.

Some weed seeds are spread by the tumbling action of the dead mature plant about the landscape as the plant is uprooted or broken off and blown about by the wind. Such plants are often referred to as "tumbleweeds," and two examples are tumbling pigweed and Russian thistle. Tumbleweeds are common to the prairies and semidesert areas of the western United States and Canada.

Seeds of many weed species, such as ragweeds and mustards, may be rolled long distances on the surface of crusted snow during the winter by the action of wind. Light breezes may transport small, light seed for some distance, while strong winds may carry rather heavy seeds for miles, regardless of the structure of their seed-dispersal units.

Flowing Water

Flowing water is capable of transporting almost any type of seed, depending on the volume and swiftness of the water, but it is the light, buoyant seeds or seed-dispersal units that are carried the furthest. The movement of large volumes of water over the ground is generally confined to channels such as rivers, streams, canals, ditches, and furrows, with occasional uncontrolled flooding, especially after heavy rains. Weed seeds may be introduced into flowing water by sheet erosion (as from flooding or heavy rains) of the soil and by wind, but the most common means is by the seeds dropping into the water from mature plants growing on the channel banks.

Some seed-dispersal units are especially adapted for dispersal by water, just as some are adapted for dispersal by wind; such dispersal units are "corky" in structure or are air-filled "bladders" or pods. Seed-dispersal units possessing a light, corky structure include those of the parsley family (*Apiaceae*), some of the cucumber family (*Curcubitaceae*), and jimsonweed, among others. The seed-dispersal units of the docks (*Rumex* spp.) are characterized by air-filled bladders.

Over 19.7 million A (8 million ha) of land in the western United States is irrigated with surface irrigation water. The dissemination of weed seeds by irrigation water onto croplands is a common occurrence throughout this area. The monthly distribution of certain weed seeds in irrigation water in California in 1960 is shown in Figure 1.2.

In an Alberta, Canada, study, 1.7 million seeds of 13 weed species were carried in the upper 1.6-in. (4-cm) layer of water flowing past a given point in a 10-ft-wide irrigation canal during a 24-hour period, September 2, 1925. A study in Colorado in 1918 and 1919 showed similar results.

A study conducted in Washington during 1973 and 1974 of weed seeds being carried by the Columbia River revealed the presence of 77 weed species in the river water. Land irrigated directly from the Columbia River showed 203 million seeds (501 million/ha) in the upper 12-in. (31-cm) layer of irrigated soil. Adjacent, nonirrigated land had 51 million seeds/A (126 million/ha) in the upper 12-in. soil layer. Of the seeds present in the irrigated soil, 90% were seeds of barnyardgrass, common lambsquarters, and redroot pigweed. In contrast, none of these species' seeds were found in the nonirrigated soil. Seeds of these three weed species ranked among the 12 most numerous in the Columbia River.

A comparison of the number of weed seeds carried by the Columbia River with the number in each of two irrigation canals supplied with water from the river showed that, during a single season, one canal carried twice the number of weed seeds than did the river and the second carried seven times as many weed seeds as did the river.

In a Nebraska study during 1977 and 1978, irrigation water disseminated 20,000 weed seeds/A (50,000/ha) onto the sampled field. The majority of the weed seeds in the irrigation water were floating on the water surface. Weed seeds were collected from the North Platte River and from water flowing in two irrigation canals supplied with water from the river. The canals were found to carry two to five times more weed seeds than the river. Seeds of 77 weed species were identified from waters sampled from the canals and the river.

It is interesting to note similarities when comparing the weed species' seed found in the irrigation waters of Nebraska and Washington. Although both studies identified a total of 77 species, there were differences in the species and numbers of seeds present. In the Nebraska study, the 8 most prevalent weed species were (arranged in descending order) redroot pigweed, involute-leaved sedge, barnyardgrass, pale smartweed, curly dock, redtop bentgrass, common lambsquarters, and smooth groundcherry. In the Washington study, the 13 most prevalent weed species were (arranged in descending order) pale smartweed, prostrate verbain, tumble mustard, common lambsquarters, ladysthumb, prickly lettuce, redroot pigweed, horseweed, barnyardgrass, Russian thistle, common chickweed, tall hedge mustard, and slimleaf lambsquarters.

Kind of seed*	Month											
	JAN	FEB	MAR	APR	MAY	JUN	JUL	AUG	SEP	OCT	NOV	DEC
Agrostis spp. bentgrass					■		■	■	■			
Amaranthus spp. pigweed							■	■		■	■	
Chenopodium spp. goosefoot							■	■		■		
Compositae						■	■	■	■			■
Cynodon dactylon bermudagrass	■	■	■	■	■	■	■	■	■			
Cyperus spp.	■	■	■						■	■	■	■
Distichlis spicata saltgrass		■	■									
Echinochloa crus-galli barnyardgrass		■	■		■	■		■	■	■	■	■
Eclipta prostrata (eclipta) false daisy	■	■					■	■				
Helianthus annuus common sunflower						■	■					
Heliotropium curassavicum salt heliotrope									■	■	■	
Hordeum pusillum little barley				■	■	■	■					
Leptochloa uninervia	■					■	■	■		■		■
Lactuca scariola prickly lettuce			■	■	■		■	■	■	■		
Physalis lanceifolia lanceleaf groundcherry			■	■				■	■			
Polygonum argyrocoleon silversheath knotweed	■	■	■		■	■	■	■		■	■	
Polygonum hydropiperoides water smartweed	■	■	■		■		■	■	■	■	■	■
Potamogeton pectinatus pondweed		■	■	■					■			
Rumex spp. docks		■	■	■			■	■				
Scirpus paludosus bayonetgrass	■						■	■	■	■	■	■

*State of California Seed Laboratory Identification

FIGURE 1.2 ■ Kind and distribution of weed seeds in irrigation water in California in 1960. A shaded section indicates that seeds of the given species were present in the irrigation water during the given month.

SOURCE: H. Shull. 1962. Weed seed in irrigation water. *Weeds* 10:248–249.

Animals

Animals are often unwitting carriers of weed seeds whose dispersal units are equipped with barbs, bristles, and hairs that catch in the animals' fur or hair, or of seeds that have sticky or mucilaginous seed coats that adhere to passing animals. Wild animals and livestock may carry attached weed seeds long distances before they are shaken, scraped, or shed from the fur or hair to which they were attached.

Seed-dispersal units possessing hooks or bristles include common cocklebur, field sandbur, spanishneedles, and tall beggarticks. The seed-dispersal unit for puncturevine has a sturdy, sharp projection that can become imbedded in the feet of animals or the soft soles of people's shoes.

Seeds with sticky or mucilaginous seed coats include the plantains and small mistletoe. Seeds with viscid or gluey seed coats may be dispersed by adhering to dry leaves or other vegetation being blown about by the wind.

Animals carry viable seed from one place to another in their digestive tracts, excreting the seed in their feces during their meanderings as they graze or move along migratory trails. Seeds of mesquite have been distributed by cattle in this manner over millions of acres of grazing land in the southwestern United States. Migratory birds, which fly long distances in a single day, are responsible for distributing weed seeds to distant areas. Birds roosting along fence lines excrete viable seeds in their droppings. Other animals, and even insects, may carry weed seeds short distances.

Forceful Dehiscence

Forceful dehiscence of seedpods or capsules disperse seed by ejecting or propelling the seeds for distances of a few feet to as much as 20 ft (6 m) or more (Figure 1.3). Species of *Impatiens* and *Oxalis* disperse their seeds in this manner, as do many other weed species. Woodsorrel is a common weed of greenhouses and, when all is quiet, a person working alone can often be startled by the forceful dehiscence of its seedpods and the subsequent rattle of the seeds as they fall about the greenhouse.

Artificial Dissemination

We humans distribute weed seeds by means of our transport system, which includes wagons, trucks, trains, airplanes, and by farm machinery, especially harvesting equipment, moving within a field and from field to field. However, *the principal means by which we introduce weeds into our croplands is by planting crop seed contaminated with weed seeds.*

In Utah, a survey of 1232 samples of grain seed collected directly from drill boxes on the seeder at planting time showed that over half of the samples contained weed seeds and that seeds of wild oat were present in 36% of the contaminated samples. The worst sample collected contained seed of five noxious weeds, with a total of 167,000 noxious weed seeds/100 lb (45 kg) of grain seed. In this instance, an average of 54 noxious weed seeds were being planted in each 10.8 ft^2 (1 m^2) of the grain field, assuming uniform distribution.

Weed seeds and vegetative propagules of perennial weeds may be carried in the mud or soil adhering to the wheels or bodies of farm equipment. It is not uncommon to observe weed species new to an area springing up along roadsides where their seeds have fallen or bounced from the beds of trucks hauling hay, feed, or livestock. Weed seeds are often carried from place to place in trouser cuffs or attached to socks or other clothing.

Perennial weeds are spread near and far by their vegetative propagules hidden in the soil of potted or balled ornamental plants shipped to new areas for planting. Perennial weeds, such as bermudagrass, johnsongrass, purple nutsedge, and quackgrass, are often dispersed in this manner.

Spread of Perennial Weeds

Perennial weeds can spread and propagate in cultivated lands by seeds or asexual means or by both. The dragging of asexual plant parts (roots, rhizomes, tubers) by tillage equipment is a common means by which perennial weeds are spread within a field, as well as from field to field.

Roots of a single field bindweed can spread laterally about 2.5 ft (0.76 m) and penetrate to a soil depth of about 4 ft (1.2 m) during the first season of growth. By the end of the third season of growth, a single plant of field bindweed is capable of spreading throughout an area of about 18 ft in diameter and penetrating to a depth of 18–20 ft (5–6 m). Roots of a single plant of Canada thistle may spread laterally throughout an area 20 ft in diameter during the first growing season. Although roots of Canada thistle may penetrate as deep as 20 ft in some soils, most are found in the upper 15-in. layer of soil. Both field bindweed and Canada thistle develop many new shoots from adventitious buds originating on their extensive root systems, thereby giving rise to dense stands or colonies of these weeds.

Purple nutsedge and yellow nutsedge propagate asexually from tubers. Yellow nutsedge, but not purple nutsedge, may also propagate from seed. Planted in soil at 12-in. (38-cm) intervals in Mississippi, tubers of purple nutsedge

FIGURE 1.3 ■ **Forceful dehiscence of the seedpod of the squirting cucumber.**

SOURCE: R. M. Holman and W. W. Robbins. 1939. *General botany,* p. 295. New York: John Wiley & Sons, Inc.

produced the equivalent of 3.1 million plants/A (7.7 million/ha) and 4.4 million tubers and bulbs/A (10.9 million/ha) during the first season of growth. A single tuber of yellow nutsedge produced 146 tubers and bulbs within a 14-week period after planting, infesting an area 6.5 ft (2 m) in diameter, equivalent to a yield of 8.9 tons/A (22 tons/ha) of new tubers. For comparison, the average yield of potatoes in the United States is about 12 tons/A (30 tons/ha). During the first year of growth, this single tuber of yellow nutsedge was ultimately responsible for the development of 1918 plants and 6864 tubers. None of the tubers was located deeper than 12 in., and most were in the upper 8-in. soil layer.

While many perennial weeds spread primarily by asexual means, others spread primarily by seed—for example,

dandelion and curly dock. Some perennial weeds spread by both asexual and sexual means. Johnsongrass is an excellent example of a perennial weed that spreads by both means. A single 14-week-old plant of johnsongrass may already have produced as much as 85 ft (26 m) of rhizomes, with a total capacity of developing 200–300 ft (60–90 m) of rhizomes during a full growing season. A single johnsongrass plant may produce as many as 80,000 seeds, many of which are viable, during a full growing season. Bermudagrass is another perennial weed that spreads by both asexual and sexual means. It spreads asexually by the rapid growth of both rhizomes and stolons. A stand of bermudagrass may produce 4–6 million seeds/A (10–15 million/ha). Bermudagrass seeds have good viability and can lie dormant in soil. They survive immersion in water for as long as 50 days.

■ WEED–CROP COMPETITION

To the agriculturalist, the economic aspect of weeds growing among crop plants is of primary importance. Weeds growing in croplands are, like the crop plants themselves, merely trying to grow and perpetuate themselves, drawing upon the soil and air for these needed essentials. Unfortunately for the farmer, weeds obtain these essentials at the expense of the adjacent crop plants.

Competition occurs between two or more neighboring plants when the supply of one or more factors essential to growth and development falls below the combined demands of the plants. Successful competition between plants occurs with the disproportionate acquisition of one or more growth factors by one plant that proves detrimental to another's growth.

Some weed and crop species enhance their competitiveness by the production of phytotoxic or growth-inhibiting substances that adversely affect the growth and development of other plants. These biochemicals are released into the soil as root exudates or as leachates of the living or dead plants. The resulting biochemical interaction between plants is called *allelopathy.*

The term *interference,* widely used in the weed science literature today, is an all-inclusive term that denotes all the direct effects that one plant might impose upon another, such as competition, allelopathy, parasitism, and indirect effects (usually unknown) without referring to any one effect in particular. The vast majority of weed research reported in the literature is concerned with weed–crop competition and allelopathy, even though the term "interference" appears in the title of the research paper. There has been little or no research on parasitism and on the "indirect" (unknown) effects of interference.

Since both weeds and crops are plants, they basically have the same requirements for normal growth, development, and reproduction. They require and compete for an adequate supply of the same nutrients, water, light, heat energy (temperature), carbon dioxide, and growing space.

In an intermixed community of weed and crop plants, the more aggressive plants usually dominate. Aggressiveness is favored by greater root elongation and branching, resulting in a vigorous, rapidly spreading root system that absorbs water, nutrients, and oxygen from the soil at the expense of adjacent plants. Aggressiveness is also favored by taller plant species that grow more quickly than adjacent plants or by plants that climb their neighbors as vines, producing foliar canopies that shade shorter or slower growing plants in the community. Thus weeds successfully compete with crop plants by (1) being more aggressive in

growth habit; (2) obtaining and utilizing the essentials of growth, development, and reproduction at the expense of the crop plants; and (3) in some cases, secreting chemicals that adversely affect the growth and development of crop plants.

Reduction in crop yields is an accepted parameter for determining weed–crop competition. *Crop quality* is most adversely affected when green, moist vegetation and the reproductive parts of weeds are harvested along with the crop.

Key factors in crop-yield reduction resulting from weed–crop competition are as follows: weed species present; weed–crop emergence; competition duration; weed life cycle and growth habit; density of weed and crop plants, crop species and cultivars and their life cycles and growth habits; crop planting date, depth, and row spacing; and edaphic, climatic, and other environmental factors.

The Critical Weed-Free Period

The *critical weed-free period* is that period during crop production in which weeds are most likely to reduce crop growth and yield, a time during which weed-control efforts must be maintained to prevent loss in crop yield. To prevent this reduction, the crop must be kept weed-free during this period of development. The significance of the critical weed-free period is that, by the end of this period, the crop can successfully compete with later-emerging weeds, primarily by shading the soil surface.

Weed competition tends to have the greatest adverse effect on *crop yields* of summer annual crops when weeds are allowed to compete with the crop during the first 4–6 weeks or so after crop planting.

Weed emergence in relation to crop emergence is an important factor in weed–crop competition. Weeds that emerge prior to crop emergence may reduce crop yields more than weeds that emerge later. For example, white mustard seedlings emerging 3 days *before* field peas and allowed to compete season-long reduced the fresh weight of pea vines by 54%; when they emerged 4 days *after* the peas and were allowed to compete season-long, vine weights were reduced 17%. The adverse effect of white mustard on the growth of field peas was attributed to competition for moisture and nutrients.

Weeds that emerge along with crop plants also have an adverse effect on crop yields. For example, giant foxtail had no adverse effect on yields of corn or soybeans when it emerged 3 weeks after the crop, but it caused a grain-yield reduction of 13 and 27%, respectively, following season-long competition. Sunflower, grown as a seed crop, produced maximum yields when kept weed-free during the first 4–6 weeks after planting. Cotton seed yields were not affected by barnyardgrass when kept weed-free for 9 weeks after planting; kept weed-free for 6 weeks, cotton seed yields were reduced 15% by subsequent competition with barnyardgrass. Common purslane is most competitive with beans and table beets during the first 2 weeks following crop emergence. Shattercane did not reduce soybean yields if the young plants were removed by 6 weeks after crop emergence. Johnsongrass significantly reduced corn grain yield when allowed to grow with the corn during the first 6 weeks after planting. Wild poinsettia is most competitive with peanuts during the first 10 weeks after crop emergence.

In general, soybeans kept weed-free for the first 4 weeks after planting showed little yield loss from later-emerging weeds. Weeds that emerge after soybeans cause less competition, since established soybeans compete well with weeds. The critical duration of giant ragweed competition in soybeans was between 4 and 6 weeks after crop emergence. Full-season competition of giant ragweed at densities of 117,400 plants/A (290,000/ha) resulted in almost complete soybean yield loss. Soybeans kept free of sicklepod competition for 4 weeks after emergence resulted in yields equivalent to season-long control. The critical johnsongrass-free period for soybeans was approximately 4 weeks after planting.

In some instances, weeds emerging along with the crop plants have little or no adverse effect on crop yields unless they are allowed to remain in the crop longer than the first 2–3 weeks after planting. For example, peanut yields were not reduced by weeds growing along with the crop plants during the first 3 weeks after planting, but yields were reduced if weeds were not removed from the crop prior to the fourth week and the crop kept weed-free thereafter. Yields of grain sorghum were reduced when pigweed competed with the sorghum during the third and fourth weeks after crop emergence, but grain yields were not reduced if the pigweed was present only during the first 2 weeks after crop emergence or did not emerge until 4 weeks after crop emergence.

Weed Competitiveness

In general, among annual weeds, broadleaved weeds are more competitive than are grass weeds. For example, velvetleaf was twice as competitive in soybeans as were either green foxtail or yellow foxtail, compared on the basis of the weight of mature plants. Redroot pigweed was more competitive in sugar beets than was green foxtail. White mustard was nine times more competitive in field peas than

was foxtail millet; after season-long competition, 3 white mustard plants/ft^2 of cropland reduced pea yields by 58%—the same reduction caused by 27 foxtail millet plants/ft^2.

Competitiveness varies among grass species. For example, among foxtail species, giant foxtail was more competitive with soybeans than were either green or yellow foxtails. This difference in competitiveness was attributed to differences in the characteristic size among species, with giant foxtail plants much taller than those of green or yellow foxtails.

Competitiveness also varies among broadleaved weed species. In Saskatchewan, Canada, wild mustard competed strongly with both wheat and cow cockle, reducing wheat grain yields by 38% and the dry weight and seed production of cow cockle by about 50%. Tall morningglory was more competitive in cotton than was sicklepod. Sicklepod at densities of 4, 8, 16, and 32 plants/15 m of row reduced cotton seed yields by 21, 23, 42, and 55%, respectively; tall morningglory at the same densities reduced cotton seed yields by 19, 41, 64, and 88%, respectively.

Weed Densities

In general, crop yield decreases as weed density increases. For example, the 3-year averages of soybean yields were reduced 10, 28, 43, and 52% by full-season competition with common cocklebur at densities of 3300, 6600, 13,000, and 26,000 plants/ha, respectively. Giant ragweed at a density of 2 plants/9 m of row reduced soybean seed yields by 48%. Corn cockle at a density of 340 plants/m^2 reduced wheat grain yields by 60%. One kochia plant/7.6 m of row reduced the average yield of sugar beet roots by 5.9 metric tons and sucrose yields by 1.07 metric tons/ha. Broadleaf signalgrass at densities of 8, 16, and 1050 plants/10 m of row reduced peanut seed yields by 14, 28, and 69%, respectively. Perennial sowthistle, at a density of about 90 shoots/m^2 reduced dry edible bean (kidney bean) and soybean yields by 83 and 87%, respectively. Spring wheat yield was not reduced by wild oat densities of 64 or 118 plants/m^2 until the wild oat reached the 5- and 7-leaf stage, respectively. Removing wild oat plants prior to the 7-leaf stage or the 5-leaf stage at densities of 64 or 118 plants/m^2, respectively, did not increase wheat culm or fresh weight production. The influence of density of selected weed populations on crop yields is shown in Tables 1.3 and 1.4.

Root Elongation

Differences in root elongation, accompanied by an increase in potential for water and nutrient absorption, contribute to

TABLE 1.3 ■ The competitive effect of giant foxtail on yields of corn and soybean following season-long competition.

NUMBER OF FOXTAIL PLANTS PER FOOT OF ROW	REDUCTIONS IN GRAIN YIELDS (%)*	
	Corn	Soybean
Weed-free	0	0
0.5	3	5
1	7	5
3	10	7
6	13	10
12	17	13
Band of foxtail[†]	24	28

*Based on yields of 94 bu/A for corn and 39 bu/A for soybeans when crops were weed-free.

[†]Band 4 in. wide, unthinned, in crop row: averaging 54 foxtail plants per foot of row.

SOURCE: E. L. Knake and F. W. Slife. 1962. Competition of *Setaria faberii* with corn and soybeans. *Weeds* 10:26–29. (Table altered to show plants per foot of row and percent reduction of yields.)

TABLE 1.4 ■ The competitive effect of pigweeds on yield of soybean following season-long competition.

PIGWEED PLANTS PER 8 FEET OF ROW	SOYBEAN YIELD (BU/A)	PIGWEED (LB/A; DRY WT.)
0	50	0
1	35	1,704
2	25	4,065
4	22	4,882
8	27	3,395
16	22	4,580
32	14	6,485
Natural pigweed population, unthinned	10	8,072

SOURCE: J. Asberry and C. Harvey. 1969. The effect of pigweeds on the yield of soybeans. *Proc. Southern Weed Sci. Soc.* 22:96.

successful competition of one species over another in mixed plant populations. Of eight weed species studied, the rate of root elongation was greatest for common cocklebur and least for tumblegrass, with the roots of cocklebur elongating 33 in. (83 cm) in 15 days while those of tumblegrass elongated 11 in. (28 cm). During the same 15-day period, the roots of Russian thistle elongated 15 in. and those of sorghum (var. RS 610) 20 in. (50.8 cm). For each species studied, the rate of root elongation always exceeded the increase in shoot height. However, comparing plant species, the rate by which plant height increased for the more vigorous species exceeded the rate of root elongation of the less vigorous species. The moisture-extraction profiles for the root systems of sorghum and the eight weed species studied are shown in Figure 1.4. The spread in the moisture-extraction profiles varied from 20 ft^2 for kochia to 44 ft^2 for cocklebur.

Competition for Water

Weeds, like other plants, consume large quantities of water, and most of it is lost by transpiration to the atmosphere. Under field conditions, the water requirements for various weed species vary from about 330 to 1900 lb of water per pound of dry matter produced (150–1900 kg per kilogram of dry matter produced). Of those weeds studied, kochia, puncturevine, and Russian thistle required the least amount of water, and buffalobur required the most water, per kilogram of dry matter produced. It has been estimated that, by the time they are 6 in. (15 cm) tall, a single plant of Russian thistle, kochia, Palmer amaranth (carelessweed), and green foxtail will have removed from the soil the equivalent of 0.5, 0.6, 0.9, and 1.4 in. (1.3, 1.5, 2.3, and 3.6 cm) of moisture, respectively. It has also been estimated that one Russian thistle plant consumes about the same amount of water as do three sorghum plants; one mustard plant as much as four oat plants, and one sunflower plant as much as two and a half corn plants. In a conservation tillage system with cotton, coastal bermudagrass significantly reduced soil water to a depth of 6 in., with a 25% reduction in seed cotton yield.

Competition for Nutrients

Nitrogen is often the first nutrient to come into short supply as the result of weed–crop competition. Blue mustard, a winter annual, competes successfully with winter wheat, even when the wheat emerges before the weed does. The competitive effect of blue mustard for nitrogen during the winter months accounts for more than 50% of the total reduction in grain yield caused by season-long competition with winter wheat; the remainder of the reduction in yield is apparently due to the weed–crop competition for light during the latter part of the season. In Nebraska, increasing nitrogen rates to winter wheat decreased annual grass weed population and weed biomass, but excessive nitrogen (above 30 lb/A or 34 kg/ha N) reduced wheat yields.

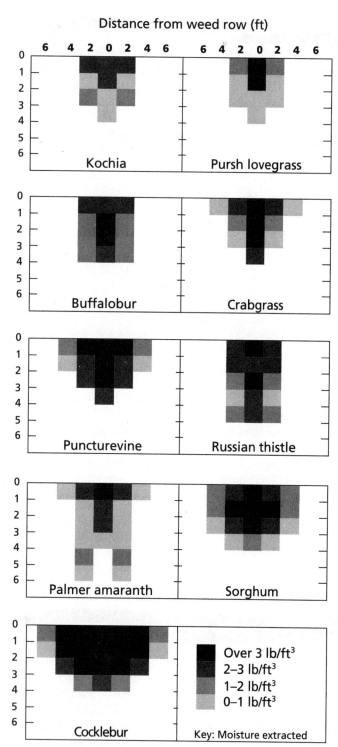

FIGURE 1.4 ■ Moisture extraction for the root systems of nine plants.

SOURCE: R. G. Davis, A. F. Weise, and J. L. Pafford. 1965. Root moisture extraction profiles of various weeds. *Weeds* 13:98–102.

Reduction in corn yields caused by competition between corn and foxtail plants results from competition for nitrogen; added nitrogen fertilizer partially overcomes such yield reductions. The addition of phosphate fertilizer enhanced the growth of corn growing alone and also of corn and weeds growing together, but the grain yields from the corn grown weed-free were significantly greater than those of corn grown in competition with weeds.

Comparison of the nutrient content of corn with that of five weeds common to this crop revealed that each weed species contained a higher percentage of nutrients than did the corn (Figure 1.5). On a dry weight basis, the weeds contained about twice as much nitrogen, 1.5 times as much phosphorus, 3.5 times more potassium, 7.5 times more calcium, and over 3 times more manganese as did the corn.

It has been advanced from time to time that clean cultivation is not necessary in row crops if heavier rates of fertilizer are applied so as to meet the needs of both the crop and weed plants. Champions of "no cultivation" (i.e., no effort to control weed growth) point out that weeds supply additional organic matter to the soil, that they tend to control soil erosion, and that the additional cost of the extra fertilizer is more than offset by savings gained by

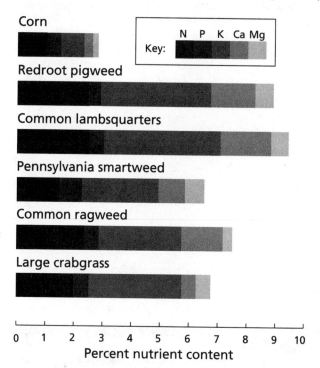

FIGURE 1.5 ■ Percent of nutrients in corn and five weeds common to this crop, based on dry weight of the whole plant.

SOURCE: H. A. L. Greer. 1966. Weeds: Costly competitors for nutrients. *Plant Food Rev.* 12(1):17.

eliminating cultivations. However, investigations concerning no-cultivation practices in row crops indicate that the presence of weeds in the crops is more detrimental than beneficial to the interests of the farmer. Conservation tillage practices today recognize the need to use herbicides for weed control in reduced- or no-tillage systems.

In general, a given parcel of land is capable of producing just so much total vegetation, and it makes little difference whether this vegetation is composed of a crop–weed mixture or a pure stand of one or the other kind of plants (Tables 1.5 and 1.6). When the maximum potential production of total vegetation is hindered by an insufficient supply of one or more nutrients, the addition of these nutrients to overcome the shortage will result in a general increase, within limits, in total vegetation produced. However, unlimited production is held in check by space limitations and by other limiting factors such as available moisture and light. Increased fertilization enhances the growth of both the weed and crop plants, and competition between plants occupying the land continues as before, with the more aggressive plants dominating the plant community.

Competition for Light

Under the stress of plant competition, where water and nutrients are adequate, low light intensity, such as from shading, is a major factor in limiting plant growth. Plants compete for light by positioning their leaves to intercept available light more favorably than their neighbors.

TABLE 1.5 ■ **Dry-matter production from corn and giant foxtail grown in competition (3-year average).**

NUMBER OF GIANT FOXTAIL PLANTS PER FOOT OF CROP ROW	POUNDS PER ACRE (DRY MATTER)				
	Corn			Foxtail Vegetation	Total
	Grain	Cobs	Stalks		
0	4,430	940	2,940	0	8,310
0.5	4,280	920	2,860	90	8,150
1	4,090	900	2,810	240	8,040
3	4,020	880	2,600	600	8,100
6	3,880	860	2,690	820	8,250
12	3,710	850	2,750	1,020	8,330
Band*	3,340	730	2,320	1,520	7,910

*Band of weeds 4 in. wide, unthinned, directly in the crop row, averaging 54 giant foxtail plants per foot of row.

SOURCE: E. L. Knake and F. W. Slife. 1962. Competition of *Setaria faberii* with corn and soybeans. Table copied from *Weeds* 10:26–29. Permission to reproduce table was granted by the editor of *Weed Science* and the authors.

TABLE 1.6 ■ **Dry-matter production from soybeans and giant foxtail grown in competition (3-year average).**

NUMBER OF GIANT FOXTAIL PLANTS PER FOOT OF CROP ROW	POUNDS PER ACRE (DRY MATTER)			
	Soybean Grain	Vegetation		Total
		Soybean	Foxtail	
0	2,030	2,160	0	4,190
0.5	1,960	2,180	70	4,210
1	1,940	2,100	120	4,160
3	1,910	2,040	240	4,190
6	1,830	2,000	380	4,210
12	1,680	1,910	520	4,110
Band*	1,460	1,720	1,170	4,350

*Band of weeds 4 in. wide, unthinned, directly in the crop row, averaging 54 giant foxtail plants per foot of row.

SOURCE: E. L. Knake and F. W. Slife. 1962. Competition of *Setaria faberii* with corn and soybeans. Table copied from *Weeds* 10:26–29. Permission to reproduce table granted by the editor of *Weed Science* and the authors.

Plants that are most successful in competing for light are those that are able to position their leaves for greater light interception by starting growth earlier in the season and by growing taller than neighboring plants. Broadleaved weeds have a distinct advantage over narrow-leaved (grass) plants in competing for light.

The overshadowing of shorter plants by taller plants is a principal means by which plants successfully compete for light. Weed density and morphology affect both the light distribution in the crop canopy and the light absorption by the crop. For example, climbing vine weeds, such as the morningglories and field bindweed, manage to position their leaves in favorable positions for light interception by ascending neighboring plants and forming a tangled mat over these plants. Other weeds, such as redroot pigweed, common lambsquarters, barnyardgrass, and johnsongrass, reach heights of as much as 6.7 ft (2 m) when mature, shading smaller crop plants such as cotton and soybean and causing yield reductions. Tomato fruit yields were reduced 54% when competing with the taller plants of eastern black nightshade at a density of 0.5 plant/ft^2 (4.8/m^2), whereas the shorter plants of black nightshade did not affect total fruit yield.

Kochia competes with sugar beets for light. It is a tall, troublesome annual broadleaf weed in sugar beets and other western crops. A severely infested sugar beet field may contain as many as 238,000 kochia plants/A (588,000/ha), compared to about 24,000 sugar beet plants/A (59,000/ha). Kochia seedlings emerge along with sugar beet seedlings in early May, and new seedlings continue to emerge until about the end of June; few kochia seedlings emerge after the last week of June. By midsummer, kochia plants that emerged along with the sugar beets have attained a height of more than three times that of the sugar beets, resulting in the sugar beet plants being in shade on the sunless side of the kochia plants. The shading effect of kochia begins to adversely affect beet growth and sugar content in about the fifth week of competition. An uncontrolled population of kochia competing season-long may reduce beet yields by 95%.

Kochia plants can compete with sugar beets during the first 4 weeks after crop emergence without significantly reducing beet root or sugar yields. However, to avoid the adverse shading effect of the older kochia plants, the kochia plants must be killed or removed from the beet crop by the fifth week and the crop kept kochia-free thereafter.

A population of one kochia plant/25 ft (7.6 m) of crop row (equivalent to 950 plants/A or 2347/ha), competing season-long, is sufficient to reduce yields of beet roots by about 2.6 tons/A (6.4 metric tons/ha) and of sucrose by about 950 lb/A (1067 kg/ha). An average yield of sugar beet

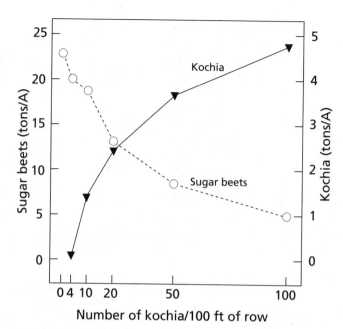

FIGURE 1.6 ■ Yield of sugar beet roots and kochia (*Kochia scoparia* L.) (dry weight of tops) where live densities of kochia competed with sugar beets all season in 1967 and 1968.

SOURCE: D. M. Weatherspoon and E. E. Schweizer. 1971. Competition between sugar beets and five densities of kochia. *Weed Sci.* 19:125–128.

roots in Colorado is about 22 tons/A (54 metric tons/ha), with a sucrose yield of 18% of the beet yield. The competitive effect of kochia on sugar beet roots is illustrated in Figures 1.6 and 1.7.

Weeds are not the only successful competitors for light. Tall, dense crops, such as corn, cotton, and soybeans, successfully compete with shorter plants and seedlings for light, especially those that emerge later in the season when the crop plants are well established. When weeds are controlled in such crops during the early part of the season, the shading effect of the crop plants largely prevents or reduces weed competition during the remainder of the season. Corn maintained free of johnsongrass for 4 weeks after planting resulted in 86% reduction in johnsongrass at the end of the growing season; in corn maintained johnsongrass-free for less than 4 weeks, johnsongrass reduction was 59%.

■ ALLELOPATHY

Plants are chemical factories, and some plants produce, store, and secrete chemicals that are harmful to other plants. These chemicals may be located in the leaves, stems, and roots of plants. They may be secreted from roots into the soil environment, leached or washed from the

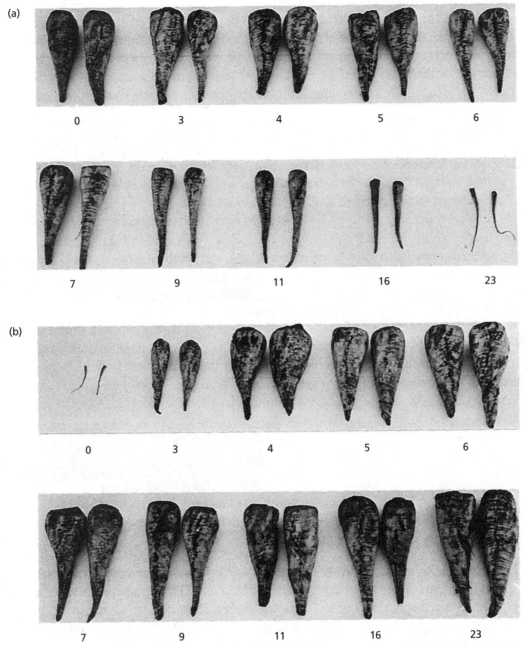

FIGURE 1.7 ■ The competitive effect of kochia on sugar beet roots.   (a) Sugar beet roots harvested from plots where kochia competed for the number of weeks indicated, after emergence. (b) Sugar beet roots harvested from plots kept free of kochia for the number of weeks indicated, after emergence.

SOURCE: D. M. Weatherspoon and E. E. Schweizer. 1969. Competition between sugar beets and kochia. *Weed Sci.* 17:464–467.

surface of aerial plant parts, leached from dead plant parts lying on the soil surface, or released from decaying vegetation incorporated in the soil.

Allelopathy is defined as the biochemical interaction between plants. The most obvious and probably the most

significant consequence of allelopathy is the control and modification of population densities in plant communities.

Competition between plants is enhanced by allelopathic interactions. This occurs when a growth-inhibitory chemical is released into the environment by one species and absorbed by another sensitive species, resulting in an adverse effect on the normal growth pattern of the sensitive species.

Biochemical interactions between weed and crop plants are generally more inhibitory than stimulatory. Such interactions may result in inhibition of crop seed germination, prevention or reduction of root elongation, cellular disorganization in roots, formation of abnormal crop seedlings, and crop-yield reductions, among others.

An example of allelopathic interaction is that of the bitter sneezeweed. Aqueous extracts of the leaves of this weed reduced alfalfa and Italian ryegrass seedling growth by as much as 50% at a concentration of 0.5% w/v. Leaf extracts were more phytotoxic than either stem or root extracts. Bitter sneezeweed tissue mixed into potting soil at concentrations as low as 0.3% w/w reduced alfalfa seedling numbers by 43%, height by 26%, and production of foliar dry matter by 54%, compared to an untreated control. The potential concentration of bitter sneezeweed leaf material in soil in a pasture ecosystem was 0.5% in the liquid phase and 0.2% in the solid phase.

Numerous researchers have reported inhibition of plant growth after plant residues were soil-incorporated. Plant residues of redroot pigweed were the most phytotoxic of several plant residues in western Texas, and redroot pigweed residues reduced soybean yields in Wisconsin. Soil-incorporated residues of Palmer amaranth inhibited the growth of carrot by 49% and of onion by 68%. Palmer amaranth, grown at population densities of 300 plants/10.8 ft^2 (300 plants/m^2), attained heights of nearly 6.5 ft (2 m) and biomass of 0.4–0.6 lb/ft^2 (2–3 kg/m^2) within 10 weeks after planting. After soil incorporation, phytotoxicity of this biomass to carrots and cabbage plantings persisted for 11 weeks.

Water extracts of small everlasting (a weed) inhibited vegetative development of leafy spurge. Soil taken from the immediate vicinity of the roots of small everlasting prevented seed germination and radicle elongation of various weed species seeded in this soil. In the field, small everlasting strongly competes with leafy spurge, and its successful competition is attributed to its allelopathic effect on the growth of leafy spurge. The scarcity of other broadleaf species growing among the leafy spurge colonies suggests that leafy spurge also exerts allelopathic effects on some plant species.

Dried plants and leachates of cluster tarweed contain a substance that inhibits germination of intermediate wheatgrass and induces abnormal seedling development. Where it is found growing in the field, the average amount of dry vegetation of cluster tarweed is about 108 g/m^2 of ground

area, and the leachate from this amount of dry matter was sufficient to significantly reduce seed germination of intermediate wheatgrass. Seed germination of intermediate wheatgrass was also inhibited by leachates of Douglas knotweed.

When grown together in agar in a test tube, seedlings of jimsonweed inhibited cell elongation in cabbage roots and induced cellular disorganization of its root tissue. Similarly grown green foxtail seedlings disrupted the normal growth of cabbage roots, and velvetleaf and wild mustard seedlings caused an increase in the number and size of parenchyma cells in the cortex of the roots of cabbage seedlings.

Terpenes released from the leaves of whiteleaf sage greatly inhibited cell division and the longitudinal elongation of parenchyma cells in roots and hypocotyls of cucumber seedlings.

Black walnut contains a phytotoxic chemical that kills or injures plants, such as alfalfa, apple, and tomato, growing in close proximity to its roots. Apparently only those plants with roots in close contact with the roots of black walnut are affected, as the toxicant in the soil is localized within the vicinity of the roots of black walnut. The toxicant occurs in all parts of the plant. It has been extracted, purified, and identified as 5-hydroxy-α-naphthaquinone (juglone).

The phenolic acids and associated compounds derived through the shikimic acid pathway are the most common of the many types of allelopathic chemicals produced by plants. Examples of allelopathic chemicals include caffeic acid, chlorogenic acid, t-cinnamic acid, p-coumaric acid, ferulic acid, gallic acid, p-hydroxybenzaldehyde, 5-sulfosalicylic acid, vanillic acid, and vanillin.

Although a considerable amount of information relative to allelopathy has accumulated in the scientific literature, its significance to agriculture and in particular to weed science has yet to be fully recognized and appreciated.

Toxic Chemicals Accumulated in Plants

Many weeds accumulate nitrate (NO_3) at such high levels that they are poisonous to livestock. Examples of weeds that tend to accumulate high levels of nitrate include lambsquarters and goosefoot species, mustards, nightshades, pigweeds, and members of the composite family. In some cases, treating plants with herbicides enhances nitrate accumulation, as well as the palatability of the plants—an unfortunate combination contributing to livestock poisoning. Nitrate concentrations of 1.5% (as KNO_3, dry weight) in forage is generally lethal to livestock. The high nitrate content persists in plants after harvest, even through the ensiling process or the curing of hay. Ruminants (cattle, sheep, goats) are more susceptible to nitrate poisoning than are animals with simple stomachs such as horses and swine. Cattle are poisoned more frequently by high nitrate levels than are other kinds of livestock.

Certain range plants absorb and accumulate toxic minerals, such as cadmium, copper, fluorine, lead, manganese, molybdenum, and selenium, from the soil. Selenium probably accounts for most of the instances of livestock poisoning. Plants that absorb and store large amounts of selenium include locoweeds, two-grooved milkvetch, saltbushes, and asters.

Induced Fetal Abnormalities

Some poisonous plants contain substances that induce abortion in livestock and cause abnormalities in the developing fetus. Mulvihill (1972) and MacLean and Davidson (1970) provide interesting discussions on this subject.

Photoactive Skin Irritants

Certain weeds produce chemicals that are photoactive (energized by light) and, when ingested or absorbed by the skin, induce injury similar to that of severe sunburn.

St. Johnswort, or Klamath weed, produces a photoactive pigment (hypericin) that, when ingested, passes through the intestinal wall into the bloodstream and is transported to the epidermal areas of the skin. In areas of light-colored or white skin not protected from light, the pigment is temporarily raised to high energy levels by light, resulting in injury similar to sunburn to nearby tissues. The mouths of cattle and sheep can be so sensitized (painful) as to discourage them from grazing.

Greater ammi produces a photoactive chemical that photosensitizes the skin of animals including cattle, sheep, and geese. Symptoms develop following the animals' contact with the plant during grazing or bedding, as well as from ingesting the plant. Symptoms are similar to severe sunburn, with redness followed by blistering, loss of skin, and sores on body parts such as the muzzle, ears, and udder; both dark- and light-colored animals are affected. Calves are orphaned when a cow's teats and udder become too sore for nursing. Greater ammi injury is a major problem for the cattle industry along the Gulf Coast of Texas, from the Rio Grande to the Sabine River. Although it has been a problem since about 1913, greater ammi was not associated with the disease until 1977. Humans are also susceptible to the photoactive chemical produced by greater ammi, with symptoms of severe sunburn developing after contact with the plant.

Spring parsley produces two photoactive furocoumarines (xanthotoxin and bergapten) that, following ingestion, can so photosensitize the udders of ewes (female sheep) that the ewes refuse to let their lambs nurse.

■ POISONOUS PLANTS

It is well known that many plants contain poisonous substances that, when ingested, cause death in humans and animals. Most of us are familiar with stories of poisoning associated with poison hemlock, especially the classic one concerning the poisoning of Socrates in 399 B.C. Most of us, however, are not familiar with the poisonous nature of many plants grown in our own homes or in our flower and vegetable gardens.

Poisonous Horticultural Plants

Dumbcane is an evergreen foliage plant commonly grown as an ornamental in many homes and office foyers. Most people are, however, unaware that it can be deadly to the person, child or adult, who bites its stem or leaf. Dumbcane contains large amounts of two harmful substances: the enzyme *asparagine* and *calcium oxalate*, which is present as sharp, needlelike crystals. Of these two substances, asparagine is the more harmful, causing severe burning and irritation of the mouth. When dumbcane is bitten into, the mouth parts are severely burned and the tongue may swell to such a proportion as to impede speech for days (hence the name "dumbcane") or, in extreme cases, block the air passage to the lungs causing death by asphyxiation.

Oleander is an ornamental shrub grown extensively in the southern and southwestern United States for its showy flowers. Unfortunately, this plant contains two highly poisonous cardiac glycosides (heart stimulants): *nerioside* and *oleandroside*. There is enough poison in a single leaf of oleander to kill a child or adult if ingested, and people have been killed by eating meat that had been skewered on a woody oleander stem and roasted over an open fire. Horses may be poisoned from nibbling on the plant while briefly tethered within reach.

Castor bean is another plant commonly grown as an ornamental. It is easy to grow and grows quickly, providing dense screening with deep-green, interestingly shaped leaves. The seeds of castor bean contain *ricin*, one of the most deadly poisons known; as little as 10 µg of ricin has been known to kill a 200-lb man. To be effective, ricin must enter the bloodstream, which it may do through abrasions or cuts in the skin or in the mucous membrane of the intestinal tract. Its action resembles that of snake venom, although it is considerably more deadly. Home remedies are inadequate for treating ricin poisoning. All classes of livestock are susceptible to ricin poisoning. Horses are most susceptible, with cattle, sheep, and hogs less so, and poultry the least susceptible. Seeds of castor bean present as contaminants in feed grain are a source of ricin poisoning in livestock.

Rhubarb is a common vegetable plant grown in many home gardens for its tasty, tart petioles or "stems." The leaf blade (the large, broad, flat parts of the plants) contain *oxalic acid* and *soluble oxalates* that, when ingested, crystalize in the kidneys, causing severe damage, and occasionally, death. The pleasant acidity of the petioles is derived from the presence of malic acid rather than oxalic acid.

Jimsonweed seeds and leaves are toxic and sometimes fatal when ingested by humans and livestock; as little as 4–5 g of the leaves or seeds has caused fatalities in children. In the 1980s, a man was reported to have died from drinking a tea brewed with jimsonweed seed.

Livestock Poisoning

Each year poisonous plants exact their toll on the livestock and dairy interests throughout the United States. A plant that is toxic or poisonous to livestock is one that causes biochemical or physiological changes when consumed by the livestock. The toxic effects vary from mild sickness to death. They may interfere with reproduction, normal growth, milk production, wool production, or animal maintenance. More than 200 species of range plants indigenous to the grazing lands of the United States and numerous introduced species produce chemicals toxic to livestock, some identified and others unidentified.

Livestock poisoning is localized, sporadic, and seasonal; it is most likely to occur in spring and winter. These are the times when grass is sparse and livestock are forced to feed promiscuously, eating plants that they would ordinarily avoid. Poisoning during summer and fall is largely associated with conditions that restrict grass production such as drought, infertile soil, and poorly managed pastures and rangelands—conditions that encourage indiscriminate browsing. Poisonous plants growing along fencerows and ditches and in wet places in woodland pastures and rangelands pose a hazard to livestock in the fall and summer months.

Plant toxicants may adversely affect the following body systems of animals:

1. **Reproductive system**—breeding and fetus development.

2. **Cardiopulmonary system**—heart, lungs, blood, and associated blood vessels (arteries, veins, and capillaries) necessary for oxygen transport.
3. **Nervous system**—brain and central nervous system (spinal cord and its many branches).
4. **Gastrointestinal system**—stomach and intestines.
5. **Hepatic system**—liver. [Photosensitization may be associated with some types of liver injury. However, some plants produce photodynamic chemicals that induce photosensitization without liver injury—for example, desert parsley (*Cymopterus watsonii*).]
6. **Renal system**—kidneys and associated structures.
7. **Musculoskeletal system**—muscle and bone.
8. **Dermal system**—skin.

Certain agronomic crops may be poisonous to livestock when the plants are ingested at certain stages of growth or when the plants are subjected to adverse growing conditions. For example, young plants of sorghum, sudangrass, johnsongrass, and sorghum almum contain dhurrin, a cyanogenetic glycoside that, when ingested, is enzymatically converted to the toxic *hydrocyanic acid* (prussic acid) in the rumen of cattle, sheep, and goats and within the young plants themselves following injury from freezing, wilting, or crushing. Small plants, young branches, and tillers of these plant species are high in hydrocyanic acid content, which decreases as the plants approach maturity. The leaves of these plants contain 3–25 times more hydrocyanic acid than corresponding portions of their stalks. Sorghum and its relatives are unsafe for pasturing until the plants are mature and no new growth is present. When death occurs from hydrocyanic acid, it is actually due to cyanide poisoning. Certain conditions, such as wilting, frosting, and an abundance of second-growth shoots or suckers, generally tend to increase the toxic properties of these plants.

■ FOREIGN PLANTS INTRODUCED INTO THE UNITED STATES

Williams (1980) presents an interesting insight into the advantages and disadvantages associated with foreign plants introduced into the United States, reproduced here with permission of the author and the Weed Science Society of America.

The introduction of plant germplasm from abroad has provided the United States with a remarkable diversity of useful plants. These introductions find use as food, fiber, forage, decorative ground cover, erosion control, and ornamentals. Moreover, introduced germplasm is used in plant breeding to increase yields and quality of crops and to improve resistance to disease and insects. Some of the introduced plants find use in medicine. Modern agriculture and the abundance we enjoy would not be possible without these introductions. Continued acquisition of useful germplasm is mandatory if we are to maintain our preeminence in world agriculture.

Unfortunately, some purposely introduced plants have become major weeds, costing the United States millions of dollars annually by reducing yields of crops and livestock, clogging navigable and recreational waterways, reducing the productivity and value of grazing lands, poisoning humans and livestock, and in rising control costs. One introduction, marijuana, has passed from a useful crop (fiber, medicine) to one whose possession, sale, or use can result in a fine and/or imprisonment.

The dollar loss from introduced plant pests is difficult to estimate, yet it accounts for a sizable and growing part of the 7.5 billion dollars lost annually to weeds. The following examples illustrate the seriousness of the problem.

Hydrilla, a submerged vascular hydrophyte, was introduced into Florida from South America in the late 1950s for use in aquariums. The plant is thought to have been started in several places to provide a crop for the trade. Hydrilla now infests approximately 284,000 ha [701,800 A] in Florida, 1600 ha [4000 A] in Louisiana, and 2400 ha [5900 A] in Texas and has been found in Georgia, Alabama, Mississippi, California, and Iowa. Unofficial reports suggest that the plant has spread to Colorado and Wisconsin. Between 6 and 8 million dollars was spent in Florida on hydrilla control in 1976.

Waterhyacinth, considered one of the 10 worst weeds of the world, was introduced as an ornamental [in] about 1884 as part of a horticultural exhibit. Within 13 years of [its] introduction, waterhyacinth sufficiently clogged some waterways so as to interfere with navigation. By 1956, losses were collectively estimated at 45 million dollars in Florida, Alabama, Mississippi, and Louisiana. Nearly 15 million dollars were spent in 1976 to control waterhyacinth in Florida and Louisiana.

Macartney rose was introduced as an ornamental from China. Infestations now occur from Florida and Virginia to Texas. Macartney rose infested 16,000 ha [39,520 A] of Texas grassland in 1948. By 1966, it had infested over 200,000 ha [500,000 A] in 40 southeastern Texas counties. Mature rose hips are eaten by birds, wildlife, and livestock. Seeds remain viable after passing through the digestive tract.

Brazilian peppertree was introduced into Florida from Brazil, and melaleuca or cajeput was introduced from Australia. Both produce pollen that is a respiratory irritant and sap that can cause painful blisters and skin rash. They are notorious invaders that crowd out natural vegetation. The fruit of Brazilian peppertree is known to kill racoons and horses and has

produced illness in children. Melaleuca, introduced because of its swamp-drying properties, is drying up hectares of surface water and may present a threat to areas that should remain swamps.

Halogeton, although inadvertently introduced, is a classic example of a failure to intercept a poisonous new introduction and the disastrous consequence that ensued.

Halogeton was discovered along railroad tracks near Wells, Nevada, about 1935. The first plants probably grew from seed that had fallen from a train carrying products from an area where halogeton was indigenous. The plant was thought to be innocuous because the literature provided no hint of its poisonous properties. Like purposefully introduced plants, interest centered about possible uses for this plant [which] was drought and salt tolerant and [which] showed promise for firebreaks and forage.

The failure to discover the poisonous properties of halogeton in time to eradicate the plant was due in part to the timing of its introduction. Almost no trained, full-time weed scientists existed in the 1930s; the nation was in a depression so that little money was available for research; and the following war years tied up funds and manpower through 1945. By then, halogeton was spreading rapidly throughout the Intermountain Region.

By the time sheep losses were traced to halogeton, the plant was beyond control, much less eradication. Within 25 years after its discovery, halogeton had engulfed between 4.0 and 4.5 million ha [10 and 11 million A] of desert range in Utah, Nevada, Idaho, Montana, and Wyoming, and had invaded Colorado, Oregon, and California. Thousands of sheep were killed by halogeton, sometimes 500 to 1700 in a single day. Even though the plant is widely recognized and largely avoided today, single losses of several hundred sheep are often reported. Moreover, halogeton was found to host the beet leafhopper and the curly top virus. Halogeton has been a costly lesson in terms of livestock losses, greatly reduced range productivity, and control costs.

Other purposeful introductions that have become serious pests include Dalmatian toadflax, which is spreading rapidly in the western United States, and johnsongrass, a pest in warm sections of the United States. Dalmation toadflax was introduced as an ornamental. Johnsongrass, used principally as forage, is a serious weed in crops and is also classed as one of the 10 worst weeds in the world. [Other purposefully introduced plants that have become pests are listed in Tables 1.7 and 1.8.]

Preventive weed control research should concentrate on locating problem plants before they are seeded in the United States. Suspect plants should be tested under rigid control that will prevent escape during evaluation and permit rapid eradication if the plants prove undesirable. Each undesirable plant intercepted before it became a problem would prevent millions of dollars in losses and control costs and be counted as dollars saved for the American farmer and consumer.

TABLE 1.7 ■ Purposeful foreign plant introductions that have become problem weeds (other than poisonous) in the United States.

SPECIES	PURPOSE OF INTRODUCTION	WEED TYPE	PROBLEM AREAS
Bermudagrass *Cynodon dactylon* (L.) Pers.	Forage	Perennial	Pastures
Dalmatian toadflax *Linaria genistifolia* (L.) Mill. subsp. *dalmatica* (L.) Maire & Petitmengin	Ornamental	Perennial	Rangeland
Dyers woad *Isatis tinctoria* L.	Dyes	Biennial	Rangeland, crops
Hydrilla *Hydrilla verticillata* (L.f.) Royle	Aquarium trade	Perennial	Lakes, reservoirs, waterways
Cogongrass *Imperata cylindrica* (L.) Beauv.	Forage	Perennial	Southern farms
Japanese honeysuckle *Lonicera japonica* Thunb.	Ornamental	Perennial	Wooded areas, pastures
Japanese knotweed *Polygonum cuspidatum* Sieb. & Zucc.	Ornamental	Perennial	Lowlands, homesites
Kochia *Kochia scoparia* (L.) Schrad.	Forage, ornamental	Annual	Widespread
Kudzu *Pueraria lobata* (Willd.) Ohwi	Ornamental, erosion control, forage	Perennial	Forests, rights of way, field borders
Macartney rose *Rosa bracteata* Wendl.	Ornamental	Perennial	Pastures
Multiflora rose *Rosa multiflora* Thunb. ex Murr.	Windbreaks, cover plantings	Perennial	Pastures
Reed canarygrass *Phalaris arundinacea* L.	Forage	Perennial	Canal and ditch banks
French tamarisk *Tamarix gallica* L.	Ornamental	Perennial	Pastures, food plains, waterways
Strangier vine *Morrenia odorata* (Hook. & Arn.) Lindl.	Ornamental	Perennial	Citrus
Waterhyacinth *Eichhornia crassipes* (Mart.) Solms	Ornamental	Perennial	Canals, lakes, waterways
Wild melon *Cucumis melo* L.	For observation	Annual	Imperial Valley, cropland
Yellow toadflax *Linaria vulgaris* Mill.	Ornamental	Perennial	Rangelands

SOURCE: M. C. Williams. 1980. Purposefully introduced plants that have become noxious or poisonous weeds. *Weed Sci.* 28:300–305. Reproduced with permission of the Weed Science Society of America and the author.

TABLE 1.8 ■ **Purposeful foreign plant introductions that have become poisonous problem weeds in the United States.**

SPECIES	PURPOSE OF INTRODUCTION	WEED TYPE	PROBLEM AREAS	TOXIC PRINCIPLE
Belladonna *Atropa bella-donna* L.	Herb	Annual	Roadsides, waste areas	Atropine *Alcaloid*
Bouncingbet *Saponaria officinalis* L.	Ornamental	Perennial	Pastures	Saponins
Brazilian peppertree *Schinus terebinthifolius* Raddi	Ornamental	Perennial	Parks, forests, yards	Irritants
Buckthorn *Rhamnus* spp.	Ornamental	Perennial	Grazing areas	Anthraquinone
Chinaberry *Melia azedarach* L.	Ornamental	Perennial	Grazing areas	Unknown
Corn cockle *Agrostemma githago* L.	Ornamental	Annual	Wheat, grasslands	Saponins
Crotalaria *Crotalaria spectabilis* Roth *Crotalaria retusa* L.	Green manure, hay	Annual to perennial	Ranges, waste areas, soybeans	Monocrotaline
Foxglove *Digitalis purpurea* L.	Ornamental	Biennial	Pastures, waste areas	Aglycones
Goatsrue *Galega officinalis* L.	Forage	Perennial	Pastures, canal banks	Galegin
Hemp *Cannabis sativa* L.	Fiber, medicine	Annual	Pastures, waste areas	Tetrahydrocannabinol
Henbane *Hyoscyamus niger* L.	Medicine	Annual or biennial	Roadsides, waste areas	Atropine, scopolamine, hyoscyamine
Jimsonweed *Datura stramonium* L.	Ornamental	Annual	Pastures, cropland	Solanaceous alkaloids
Johnsongrass *Sorghum halepense* (L.) Pers.	Forage	Perennial	Cropland, pastures	Cyanogenetic glycosides
Lantana *Lantana camara* L.	Ornamental	Perennial	Fence rows, ditch banks, fields	Lantadene
Melaleuca *Melaleuca leucadendron* (L.) L.	Ornamental	Perennial	Swamps, cities	Respiratory and skin irritants
Precatory bean *Abrus precatorius* L.	Ornamental	Perennial	Fence rows, roadsides	Abrin
Sicklepod milkvetch *Astragalus falcatus* Lam.	Forage	Perennial	Rangeland	Nitro compounds
Tansy *Tanacetum vulgare* L.	Herb	Perennial	Old gardens, roadsides	Unknown

SOURCE: M. C. Williams. 1980. Purposefully introduced plants that have become noxious or poisonous weeds. *Weed Sci.* 28:300–305. Reproduced with permission of the Weed Science Society of America and the author.

■ SELECTED REFERENCES

Anderson, W. P. 1987. Weed science as it relates to crop production. In *CRC handbook of plant science in agriculture,* Vol. 2, pp. 99–116. B. R. Christie, ed. Boca Raton, Fl.: CRC Press.

Anonymous. 1993. Regulatory file. In *Farm chemicals handbook,* pp. D-1 to D-64. Willoughby, Oh.: Meister Publishing.

Anonymous. 1993. Environmental and safety section. In *Farm chemicals handbook,* pp. E-1 to E-41. Willoughby, Oh.: Meister Publishing.

Anonymous. 1992. *Meister's 1992 weed control manual.* Willoughby, Oh.: Meister Publishing.

Baysinger, J. A., and B. D. Sims. 1991. Giant ragweed (*Ambrosia trifida*) interference in soybeans (*Glycine max*). *Weed Sci.* 39:358–362.

Bridges, D. C., ed. 1992. *Crop losses due to weeds in the United States.* Champaign, Ill.: Weed Science Society of America.

Bridges, D. C., B. J. Brecke, and J. C. Barbour. 1992. Wild poinsettia (*Euphorbia heterophylla*) interference with peanut (*Arachis hypogaea*). *Weed Sci.* 40:37–42.

Bronsten, D., and B. Simmonds. 1989. Crops gone wild. *Agrichemical Age* 33(5):6, 7, 26, 28.

Buchanan, G. A., and R. D. McLaughlin. 1975. Influence of nitrogen on weed competition in cotton. *Weed Sci.* 23:324–328.

Burnside, O. C. 1979. Soybean (*Glycine max*) growth as affected by weed removal, cultivar, and row spacing. *Weed Sci.* 27:562–565.

Burnside, O. C., et al. 1993. Alternative weed management systems for the production of kidney beans (*Phaseolus vulgaris*). *Weed Technol.* 7:949–945.

Donald, C. M. 1963. Competition among crop and pasture plants. *Adv. Agron.* 15:1–11.

Fellows, G. M., and F. W. Roeth. 1992. Shattercane (*Sorghum bicolor*) interference in soybean (*Glycine max*). *Weed Sci.* 40:68–73.

Forcella, F., and S. J. Harvey. 1988. Patterns of weed migration in northwestern U.S.A. *Weed Sci.* 36:194–201.

Hall, M. R., C. J. Swanton, and G. W. Anderson. 1992. The critical period of weed control in grain corn (*Zea mays*). *Weed Sci.* 40:441–447.

Keeley, P. E., and R. J. Thullen. 1991. Growth and interaction of barnyardgrass (*Echinochloa crus-galli*) with cotton (*Gossypium hirsutum*). *Weed Sci.* 39:369–375.

Kirkland, K. J. 1993. Spring wheat (*Triticum aestivum*) growth and yield as influenced by duration of wild oat (*Avena fatua*) competition. *Weed Technol.* 7:890–893.

MacLean, G. J., and J. H. Davidson. 1970. Poisonous plants: A major source of livestock disorders. *Down To Earth* 26(2):5–11.

McGiffen, M. E., Jr., J. B. Masiunas, and J. D. Hesketh. 1992. Competition for light between tomatoes and nightshades (*Solanum nigrum* or *S. ptycanthum*). *Weed Sci.* 40:220–226.

Mulvihill, J. J. 1972. Congenital and genetic disease in domestic animals. *Science* 176:132–137.

Olivier, A., and G. D. Leroux. 1992. Root development and production of witchweed (*Striga* spp.) germination stimulant in sorghum (*Sorghum bicolor*) cultivars. *Weed Sci.* 40:542–545.

Perry, K. M., R. Evans, and L. S. Jeffery. 1983. Competition between johnsongrass (*Sorghum halepense*) and corn (*Zea mays*). *Proc. Southern Weed Sci. Soc.* Vol. 36, p. 345. Champaign, Ill.: Southern Weed Science Society.

Putnam, A. R. and C-S. Tang, eds. 1986. *The science of allelopathy.* New York: John Wiley & Sons, Inc.

Rice, E. L. 1988. *Allelopathy,* 2nd ed. Academic Press. New York.

Shull, H. 1962. Weed seed in irrigation water. *Weeds* 10:248–249.

Swanton, C. J., K. N. Harker, and R. L. Anderson. 1993. Crop losses due to weeds in Canada. *Weed Technol.* 7:537–542.

Thompson, A. C. 1985. *The chemistry of allelopathy.* ACS Symposium Series No. 268. Washington, D.C.: American Chemical Society.

Valenti, S. A., and G. A. Wicks. 1992. Influence of nitrogen rates and wheat (*Triticum aestivum*) cultivars on weed control. *Weed Sci.* 40:115–121.

Vencill, W. K., L. J. Giraudo, and G. W. Langdale. 1992. Response of cotton (*Gossypium hirsutum*) to coastal bermudagrass (*Cynodon dactylon*) density in a no-tillage system. *Weed Sci.* 40:455–459.

Williams, M. C. 1980. Purposely introduced plants that have become noxious or poisonous weeds. *Weed Sci.* 28:300–305.

Zimdahl, R. L. 1980. *Weed–crop competition.* Corvallis, Ore.: International Plant Protection Center, Oregon State University.

Zollinger, R. K., and J. J. Kells. 1993. Perennial sowthistle (*Sonchus arvensis*) interference in soybean (*Glycine max*) and dry edible bean (*Phaseolus vulgaris*). *Weed Technol.* 7:52–57.

2 WEED ECOLOGY

■ INTRODUCTION

Weed ecology is concerned with the interrelationships between weed species and the environment. In this sense, the environment is the summation of all living and nonliving factors that surround and potentially affect weed species. There are numerous facets to the interaction of weed species with the environment such as geography, climate, soil components, agronomic practices, plant characteristics, photoperiodism, seed germination, and allelopathy. The facets of weed ecology presented in this chapter focus on the weed seed bank and shifts in weed populations due to cultural changes.

With respect to weed ecology, there would be no weeds if (1) plants would not grow where they were not wanted; (2) people did not cultivate plants as crops, in the process leaving ecological niches that are readily filled by other plants that we designate as "weeds"; and (3) there were no seeds or other propagules of undesired plants present in land being cropped.

Weeds commonly arise and infest croplands from seeds and asexual propagules already present in or on the soil when the crop was planted. If weed propagules are not present when the crop is planted and none are subse-

quently introduced, weeds will not infest the crop. Similarly, plants of a specific weed species will not be found among the weed population of a field if none of its propagules are present or introduced into the soil.

Prolific seed production, longevity of the seeds in soil, and vegetative regeneration of perennials are important characteristics of successful weeds.

■ WEED SEEDS

Seeds are the principal means by which weed species perpetuate themselves and invade new areas. Seeds are tiny, encapsulated embryonic plants. They enable embryonic life to exist in suspended animation for days or years and to later be revived as replicas or near replicas of their parents, even after their parents are dead and gone. To the plant, seeds are a way to encapsulate life—to preserve, protect, and ensure survival of the species: seeds are the vehicle of dissemination. To the farmer, seeds are vital to the propagation of most crops, but weed seeds spell "trouble"—they also are the means by which weeds are introduced and spread on croplands.

■ WEED SEED BANK

The weed seed bank consists of the viable seeds on the soil surface and in the soil, usually considered to the plowshare depth (about 8 in. or 20 cm). The seed bank contains an enormous reservoir of dormant and nondormant seeds. It is comprised of many age groups of seeds, from ones freshly deposited to others deposited many years before. Seeds in the weed seed bank may be depleted by death, germination, parasitism, or predation. Seed mortality is the factor most responsible for reducing the numbers of seed in the seed bank. Weed seeds buried deeply in the soil decay without germinating, or they fail to emerge if germination does occur. Germination occurs more often at or just below the soil surface, varying with species and depth of burial. Freshly deposited seeds are more likely to germinate than older seeds in any given year. Indications are that only about 5–10% of the viable seed population in the seed bank germinate and emerge at any one time, while the remaining seeds stay dormant.

Weed problems originate with the seed bank, and they seldom end even though attempts are made to prevent the last weed in the field from producing seed; there always seems to be one more viable seed left. In Nebraska, a 5-year study to determine the extent of weed seed mortality in soil, during which seed production by the weed flora was prevented, found that the population of viable seeds in the soil declined by 95%. However, during the next 5 years, when weed seed production was allowed, the population in the seed bank recovered to within 90% of its original level.

The size and composition of the seed bank, as well as the aboveground weed flora, reflect past, present, and future weed problems. A single plant of redroot pigweed can produce over 100,000 seeds, and a single plant of giant foxtail can produce 10,000 seeds. Each weed seed falling onto or mixed into the soil becomes a deposit in the seed bank.

Crop-rotation sequence, tillage, and herbicides are three primary factors that impact seed bank composition and population. Over a several-year period, cropping sequence is the most important factor influencing seed bank composition and weed flora in cropland soil. Intensive tillage during annual crop rotations favors annual weeds, while minimal tillage favors perennial weeds. Cropping sequences dictate the time and type of tillage and also patterns of herbicide usage.

■ WEED SEED PRODUCTION

Weed seeds usually infest cropland soils in the amount of millions per acre. A single weed plant may produce vast numbers of seeds—tens to hundreds of thousands—and these seeds easily escape detection when scattered on or in the soil, making their presence known only as they germinate and become seedling plants.

The seeds of weed species that do not possess some natural means of dispersal drop to the ground around the parent plant, and many of these seeds may germinate simultaneously, while others remain dormant for extended periods. For example, field pennycress has no natural means of scattering its seed, and a single plant of this weed species produces about 7000 seeds, which fall about the base of the parent plant when ripe. In a 2-year study of field pennycress, counts of its seedlings arising from seed dropped about the base of mature plants averaged 341 and 220 seedlings/yd^2 (408 and 263/m^2) in respective years. In turn, these seedlings grew into plants that produced a total of 156,000 and 107,400 seeds in their respective year of study.

An average stand of diffuse knapweed produced some 32 million seeds/A (79 million/ha), or about 275 seeds/plant. Curly dock can produce 40,000 seeds/plant. Seed production varies greatly with species. Among species of goosefoot, the number of seeds produced per plant ranged from 1136 seeds for pigweed goosefoot to 446,082 seeds for Jerusalem-oak goosefoot. Field studies with velvetleaf and common sunflower indicate that they will produce about 4300 and 7750 seeds/plant, respectively. A single plant of witchweed will produce from 50,000 to 500,000 seeds, each measuring about 0.2 mm. A single plant of biennial wormwood produced 1,075,000 seeds weighing a total of 0.15 lb (70.6 g.).

The number of seeds produced per plant is apparently related to the total weight of the seeds produced, and the lighter and smaller the seeds, the more seeds produced per plant. In one study of a number of weed species, burdock produced the greatest weight of seed, with a total of 31,000 seeds/plant, weighing 0.52 lb (237 g.).

Where natural vegetation is disturbed, the first plants to appear are weeds, and most of these plants arise from seed lying dormant in the soil. In cultivated lands, annual weeds also arise from seeds lying buried in the soil, a legacy of preceding plants. Cultivated lands become reservoirs for vast numbers of viable weed seeds, ranging in numbers from a few million to well over a hundred million per acre.

■ SEED DORMANCY

Seed dormancy is the failure of seeds to germinate because of factors associated with their *embryo, seed coat,* and/or *environment.* Many seeds are dormant even though they are, or appear to be, mature. The failure of seeds to germinate does not necessarily mean that they are dormant; such failure may be due to unfavorable environmental conditions rather than to dormancy *per se*. It is often difficult to determine under natural conditions whether seeds are actually dormant or merely quiescent (inactive). Seedling emergence is a tangible expression of soil and climatic factors interacting favorably on previously dormant seed.

Seed dormancy is a survival mechanism, ensuring germination when conditions favor seedling survival. It tends to regulate germination of weed seed populations, ensuring a reservoir of ungerminated but viable seed for later seasons. The dormancy of annual weed seeds is the principal factor contributing to their success as problem weeds.

Dormancy can be divided into three categories: (1) *innate,* (2) *induced,* and (3) *enforced. Innate dormancy* is inherent in the mature seed as it falls from its parent to the soil surface. Innate dormancy precludes germination in a hostile environment and preserves seed for burial in the safety of the soil. Innate dormancy may require a particular stimulus (such as light on imbibed seed at a favorable temperature); chilling (variously termed stratification, low-temperature after-ripening, or prechilling); or alternating temperatures (the magnitude of change is usually 34°F or 10°C). The requirements are often complex and involve multiple stimuli. *Induced dormancy* is the condition of a seed that has failed to germinate following loss of its innate dormancy and is recycled into the dormant condition by factors (such as high temperature or excess moisture) within the soil. Induced dormancy is also called "secondary dormancy." *Enforced dormancy,* or *quiescence,* occurs when seed germination is prevented by external environmental conditions. The end result of prolonged enforced dormancy is reversion to induced or secondary dormancy.

Almost without exception, seeds of weed species possess one or more of the factors contributing to seed dormancy. Such factors have, however, been removed from the seeds of crop plants by selective breeding to facilitate rapid, uniform germination—a desired quality.

In temperate climates where seasons change regularly, some seeds in the weed seed bank cycle between dormancy and nondormancy. This cycling is strongly influenced by soil temperatures. Rainfall modifies the time of seed germination and seedling emergence within the seasons as defined by temperatures. Dormant seeds of summer annuals break dormancy during winter, and they reach a state of maximum germinability in the spring and early summer. Seeds that fail to germinate may return to the dormant state as the summer progresses.

Dormancy of vegetative propagules of perennial weeds is as important as dormancy of annual weed seeds with respect to survival and persistence. Vegetative propagule dormancy is an integral part of the entire life history of perennial weeds. The following characteristics contribute to dormancy:

1. Seed coat impermeable to water and oxygen or both
2. Hard seed coat
3. Immature embryo
4. Embryos that require an "after-ripening" period
5. Endogenous chemical germination inhibitors

In its simplest form, seed dormancy is the result of the *impermeability* of the seed coat. As a result, the factors required for germination (water and oxygen) are blocked from entering the seed, and the vital first step in the process of seed germination is blocked. Examples of weed species whose seeds have impermeable seed coats are velvetleaf and annual morningglories.

The effect of *hard seed coats* on seed germination differs from *impermeable seed coats* in that hard seed coats are readily permeable to water and oxygen (within limits), but the seed coat is too strong to be split or ruptured by the force of the swelling embryo as it imbibes water—a necessary event to allow the developing embryo to escape confinement. Seed dormancy due to impermeable or hard seed coats may be overcome by cutting, removing, or puncturing the seed coats; by exposing them to strong acid treatment or to attack by soil microorganisms; or by scouring or abrading them, as occurs when the seed is tumbled with soil particles during tillage. Once a hard seed coat has absorbed water and then dries, it tends to rupture and crack. If the seed coat remains wet continually, without alternately drying, it is not weakened and germination is inhibited. After the mechanical obstruction of the impermeable or hard seed coat has been overcome and conditions are favorable, seed germination begins immediately upon imbibition of water. Examples of weed species with hard seed coats are the pigweeds, mustards, pepperweed, shepherd's-purse, and spurred anoda.

Some weed species drop or shed their seeds while their embryos are still *immature*. In such cases, the seed is dormant until the embryo completes its morphological maturity. In contrast, some weed species drop or shed their seeds after their embryos are *fully formed*, but the embryos require more time in which to complete their physiological development, and the seeds remain dormant until development is complete. This process is called *after-ripening.*

Chemical Germination Inhibitors

Chemical germination inhibitors have been found in all parts of the seed, including the embryo, the endosperm, nucellus, seed coat (testa), and pericarp. In general, these chemicals are water-soluble, and they may be removed from the seed by leaching, a natural phenomenon in soils. Water-soluble substances that inhibit seed germination have also been found in other plant parts such as leaves, buds, and roots.

Water-soluble germination inhibitors present in seed-dispersal units act as "chemical rain gauges." The amount of inhibitor in the dispersal unit is apparently adjusted so that the quantity of rainfall needed to leach it from the dispersal unit will at the same time moisten the soil sufficiently to ensure subsequent growth of the seedling. Such chemical germination inhibitors are of high survival value to the seedling, delaying germination until enough soil moisture has accumulated for seedling establishment.

Germination of *Atriplex dimorphostegia,* a desert plant, is rain-dependent under natural conditions. Its seed-dispersal unit is made up of a single-seeded fruit having a thin pericarp and enclosed between a pair of distinctly shaped fruit bracts. The bracts contain a water-soluble germination inhibitor. Under light rainfall conditions, the inhibitor is leached from the bracts, absorbed by the enclosed seed, and prevents germination. Under heavier rainfall conditions, corresponding to conditions of sufficient soil moisture to ensure seedling survival, the inhibitor is leached from both the fruit bracts and the seed, allowing germination to proceed if other conditions are favorable.

The embryo of cocklebur seed contains two water-soluble germination inhibitors. Because the seed has an impermeable seed coat, these inhibitors cannot be leached from the embryo while the seed coat remains intact. Once the seed coat is ruptured, leaching of the embryo removes both inhibitors, and germination takes place when other conditions are favorable.

■ SEED GERMINATION AND THE ENVIRONMENT

Seed germination may be defined as the emergence of the radicle or, in some cases, another embryonic organ, through the seed envelope or enclosing structures. Weed seed germination is dependent on favorable conditions *internal* and *external* to the seed. The germination of weed seeds in soil follows a cyclic pattern influenced by such factors as seed dormancy and the environment. External conditions are environmental factors favoring seed germination. Environmental factors that favor seed germination are as follows:

1. Water.
2. Oxygen.
3. Temperature.
4. Light, its presence or absence.
5. A chemical germination stimulant (usually excreted by the host plant, as with seed of the parasitic plants witchweed and dodder) is sometimes required.

Water imbibition (absorption) by the seed is the first step in germination. Without an adequate supply of available water, seeds will not germinate. In the early stages of germination, the seed swells, increasing its size by 25–200%, due primarily to water uptake.

Oxygen is essential for the respiratory (energy-producing) reactions associated with seed germination. If excess water reduces available oxygen in the soil, the seed is unable to respire normally and is likely to rot, the result of attack by anaerobic microorganisms.

Temperature requirements for seed germination usually coincide with those associated with active plant growth. Optimal temperatures vary widely with species. In general, seeds of cool-season species tend to germinate at lower temperatures than do those of warm-season species. The extremes of temperatures favoring seed germination are generally 32°F (0°C) and 120°F (49°C). At temperatures above 120°F, many seeds are killed or forced into "secondary dormancy." Death is due to enzyme destruction and coagulation. The exact cause of the induced secondary dormancy is unclear.

Seeds of Florida pusley germinated almost 100% when exposed to a constant temperature of 86°F (30°C), or at alternating temperatures of 68/86°F (20/30°C). Seeds of goatweed germinated 69% when exposed to day/night alternating temperatures of 77/68°F (25/20°C) and 86/77°F (30/25°C) for 7 days. These temperature regimes were optimal for goatweed seed germination.

Exposure of creeping woodsorrel seed for 10 minutes to any temperature up to 212°F (100°C) did not alter the expected 100% germination, perhaps because of a stimulatory mechanism. Germination was 97% after exposure of the seed to 30 minutes at 185°F (85°C). Thus soil sterilization by solar heating or steam sterilization may not destroy seeds of creeping woodsorrel. Steam sterilization of soil usually involves heating soil to around 180°F (82°C) for 30 minutes.

Light, its presence or absence, may favor or inhibit germination of weed seeds. For example, seeds of Virginia pepperweed require exposure to light for germination; they do not germinate in darkness. In contrast, seeds of henbit germinate only in darkness; exposure to light inhibits germination. Seeds larger than a few millimeters, or with impermeable seed coats, seldom display germination photosensitivity.

Seeds of broom snakeweed, threadleaf snakeweed, and common goldenweed readily germinate if placed on the soil surface or partially pressed into the soil. Germination is reduced by covering their seed with as little as 1 mm of soil, and few seedlings emerge from a soil depth of 0.8 in. (2 cm). Seed burial probably provides a form of enforced dormancy in the small, long-lived seed of broom and threadleaf snakeweeds.

Seeds of hairy galinsoga require exposure to light for germination. Covering the seeds with as little as 5 mm of soil reduces germination to near zero. Soil cultivation repeatedly exposes buried hairy galinsoga seed to light, resulting in their subsequent germination. Seeds of Florida pusley do not germinate in continuous darkness, but they do germinate if exposed to light. Continuous exposure to light for 12 hours or more resulted in 95% or greater seed germination. Seeds of goatweed require light for germination with continuous exposure for 10 hours, which results in 35% germination when incubated at 86/77°F (30/25°C) day/night temperatures for 7 days. Increasing light intensity also increased the germination rate. However, germination of goatweed seed at low light intensity confirms the observation that goatweed is well adapted to grow under plant canopies and in open, well-exposed areas such as between crop rows.

Seeds of creeping woodsorrel require exposure to light for germination, with a very low level of light intensity required for 100% germination. Seeds of this plant will not germinate in darkness. Fresh seeds are nearly 100% viable, regardless of the season of production. Germination of seeds stored as long as 1 year exhibited 83% viability, compared to fresh seeds. In greenhouses, creeping woodsorrel control is a year-round problem, and their seedlings need to be killed or removed within 5 weeks after germination to prevent seed dispersal within the greenhouse. An average of 50 viable seeds from each pod on a mature plant may be dehisced (thrown) to distances of up to 13 ft (4 m).

Seeds of the garden flower salvia require light for germination. Salvia seeds do not germinate in darkness or when covered with as little as ⅛ in. (0.3 cm) of soil. The author has used salvia seed in a classroom exercise to demonstrate light requirements for seed germination.

Light Characteristics

Light influences seed germination in a promotive or inhibitory manner. *Red light* (580–720 nm) promotes germination, whereas *far-red light,* referred to in Europe as *near*

infrared (peaking at 735 nm, just beyond visible light), inhibits germination. A nanometer (nm) denotes one-billionth of a meter. The influence of light on seed germination occurs only with seeds that have already imbibed water; dry seeds do not respond to exposure to light. Interestingly, red light *inhibits* flowering in short-day plants but *promotes* flowering in long-day plants.

A plant pigment, *phytochrome,* has been identified as the receptor of the light stimulus. So far, phytochrome is the only light receptor known to be involved in the control of seed germination. Phytochrome exists in two interconvertible forms: (1) an inactive form (P_r) and (2) an active form (P_{fr}). Exposure to red light transforms P_r to P_{fr} and seed germination will occur. If seed containing P_{fr} are irradiated with far-red light, P_{fr} is transformed to P_r and germination will not occur. These transformations, which take place rapidly, can dramatically switch germination on and off, with sustained germination dependent on the final form of phytochrome left after the sequence of transformations has ended. Following light absorption, phytochrome must react with another cellular component as yet unknown; indications favor a cell membrane.

Longevity

The longevity of weed seeds in the soil and in water is due primarily to seed dormancy, most likely to (1) the physical and chemical makeup of the seed coat, (2) the substrate preference of microorganisms, and (3) the inherent metabolic activity of the seed. Prior to incorporation of weed seeds into the soil, some seeds will be lost to predation by insects, arthropods, rodents, and birds. Arthropod and rodent predation of seed of some weed species can be as high as 70%. Predation of freshly dispersed seed of velvetleaf and common sunflower was estimated to be 40 and 75%, respectively. Following entry into the weed seed bank, the annual death rates of velvetleaf and common sunflower were estimated to be 20 and 40%, respectively.

Seeds of some weed species remain viable in the soil for many years. Seeds with hard seed coats comprise a high percentage of the longer-lived seeds. Seeds of spurred anoda failed to germinate the first year after burial, but simply piercing their hard seed coats increased germination to 99%. Seed burial studies show velvetleaf seed to be viable after lying undisturbed in soil for 39 years, and under the same conditions, seeds of moth mullein germinated at a rate of 42% after 100 years. In Mississippi, after 5.5 years of soil burial, the viability was 48% for johnsongrass seed; 36% for velvetleaf; 33% for purple moonflower; 30% for spurred anoda; and 18% for hemp sesbania. In Canada, seed of green foxtail were viable for up to 17 years after soil burial.

In Alaska, seeds of 11 of 17 weed species studied had a greater than 6% viability after 4.7 years of soil burial.

Seeds of tall larkspur and western false hellebore either germinate or disintegrate during the first year in the soil. In contrast, seeds of witchweed remain viable in field soils for 20 years or longer. Of the seeds of 20 weed species buried in soil in 1879 and sampled periodically since then, it was found that the seeds of 3 species—common evening primrose, moth mullein, and curly dock—were still viable 80 years later.

Velvetleaf seeds exhibit a high degree of dormancy, primarily because of their impermeable seed coat, and its seed may remain viable in soil for 40 years or more. The longevity of velvetleaf seed in soil may be greatly reduced by cultivation, primarily as a result of moving the seed closer to the soil surface where conditions are more favorable for seed germination or disintegration. In Minnesota, 4 years of intensive cultivation of fallow land reduced the velvetleaf seed population by 90%, but the remaining 10% still numbered 520,000 seeds/A (1.28 million/ha) in the upper 9-in. (23-cm) soil layer. In the same study, 56% of the original velvetleaf seed population remained viable after 4 years in an undisturbed stand in alfalfa, and 37% remained viable in undisturbed land kept weed-free by continuous chemical fallow. Thus, longevity of velvetleaf seed is greater in undisturbed soils than in cultivated soils. The seed population of velvetleaf in cultivated croplands can be immense. In Nebraska, a study of six fields devoted to continuous corn revealed a mean of 20.6 million velvetleaf seeds/A (50.9 million/ha) in the top 8-in. (20-cm) soil layer.

The viability of weed seeds transported by irrigation waters onto irrigated lands is generally unaffected during the relatively short time that the seeds are in the water. Even seeds that disintegrate comparatively rapidly in water such as barnyardgrass remain viable long enough to survive the journey from their source along the banks of the water channels onto irrigated fields. When stored in an open, freshwater canal, relatively high percentages of seeds of Canada thistle, annual morningglories, povertyweed, and Russian knapweed remained viable for more than 22 months. Depending on the species, the longevity of weed seeds held in flowing fresh water may range from a few months to 5 years or more.

In general, weed seeds buried in soil or immersed in flowing fresh water either disintegrate quickly (in days or months if conditions are unfavorable for their germination), or they remain in a viable but dormant condition, for many years if necessary, until conditions are favorable for germination. Weed species characterized by seeds remaining viable for long periods while buried in soil or immersed in irrigation water are potential problem weeds in croplands even after years of clean tillage.

■ DISTRIBUTION OF SEEDS IN SOIL PROFILE

The soil profile, with regard to weed seed distribution, is considered to be of plowshare depth—that is, the depth to which the moldboard plow works the soil, about 8 in. (20 cm). The exact distribution of weed seeds in the soil profile is generally not known, due largely to the fineness of the seed and the difficulties encountered in trying to separate the seed from the soil particles. Some weed seeds are buried deeply in the soil, especially where the moldboard plow has been used, whereas others lie on the soil surface or in the litter layer just above the soil surface, as with limited- or no-tillage practices. Deep burial of weed seeds in soil discourages germination and favors dormancy; shallow burial favors the opposite.

In pastures, some 64–99.6% of all weed seeds were found in the upper 4-in. (10-cm) layer of soil, with greater

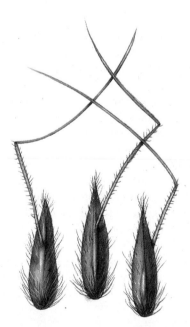

FIGURE 2.1 ■ Seed-dispersal unit of wild oat (*Avena fatua* L.).

SOURCE: U.S. Dept. of Agriculture. 1970. *Selected weeds of the United States.* Agri. Hndbk. No. 366, p. 38. Washington, D.C.: U.S. Printing Office.

numbers in the 1- to 4-in. layer than in the surface to 1-in. layer. In prairie soils of Alberta, Canada, the number of viable weed seeds in the upper 1-in. (2.5-cm) soil layer ranged from 204 to 12,432 seeds/m^2, with a general average of 780 seeds/m^2 or 7.8 million seeds/ha (3.2 million/A). The tiny seeds of witchweed have been found as deep as 60 in. (150 cm) in Lakeland sand in North Carolina, with the greatest number located in the upper 12-in. (30.5-cm) soil layer.

Tillage is the principal means by which weed seeds become buried in soil. The natural means of soil burial is by the seeds dropping or being washed into cracks, crevices, and pores of the soil, and by the burial action of earthworms and insects. Some weed seeds bury themselves in loose soil by the twisting and turning (corkscrewing) action of special appendages on their seed-dispersal units. The wedge-shaped seed-dispersal units of wild oat and filaree have long, bent, tightly spiraled, tail-like appendages that are moisture-sensitive (Figures 2.1 and 2.2). These appendages twist and turn like a corkscrew in response to changes in their moisture content, unwinding when moist and rewinding when dry. The appendages are so moisture-sensitive that even changes in humidity may induce them to twist and turn. If held in the palm of the hand, the spiraled appendage will unwind in response to the moisture obtained from the palm, causing the seed to wiggle across the palm. These seed-dispersal units move across the soil surface in a similar manner until their wedge-shaped seeds get caught in a soil opening or loose soil. The corkscrewing appendage then moves the seed deeper into the soil, and it continues to flail as the soil environment alternates between dry and moist, until the seed meets an obstacle that impedes its progress or the surrounding soil moisture remains constant.

Russian thistle seeds burrow into loose soil during germination as a result of the uncoiling action of their embryos (Figure 2.3). As the embryo uncoils, the radicle of the embryo is forced downward into loose pliable soil through cellular elongation. Germinating seeds of Russian thistle are unable to penetrate firm or hard soil surfaces; thus plants of Russian thistle are not usually found in undisturbed natural terrain. When Russian thistle seeds germinate on hard soil surfaces, their embryos dehydrate and die. Seeds of Russian thistle germinate at temperatures above 52°F (11°C) when in contact with moist soil. At temperatures above 80°F (27°C), they germinate within a matter of minutes if sufficient moisture is present. Rapid germination takes advantage of adequate soil moisture and, under favorable temperature conditions, aids in quick root penetration in moist, loose soil.

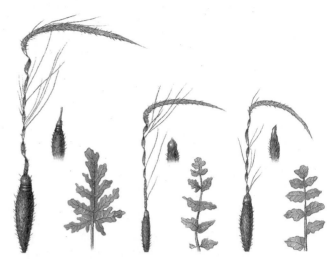

FIGURE 2.2 ■ **Fruits and leaves of _Erodium_ species.**
(a) Broadway filaree. (b) Redstem filaree. (c) Whitestem filaree.

SOURCE: W. W. Robbins, M. K. Bellue, and W. S. Ball. No publication date. _Weeds of California_, 2nd ed., p. 275. Sacramento, Calif.: California State Department of Agriculture.

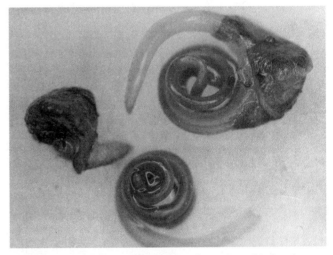

FIGURE 2.3 ■ **Germinating seeds of Russian thistle; the actual seed size is 1.0–1.3 mm.**

SOURCE: W. A. Rhoads, E. F. Frolich, and A. Wallace. 1967. Russian thistle seeds. _Calif. Agr._ 21(7):2.

■ SOIL ENVIRONMENT

The components of the soil environment that have the greatest impact on weed seed germination are (1) temperature, (2) moisture, (3) aeration, and (4) microorganisms. Temperature influences the time of weed seed germination, and many weed species, such as summer and winter annuals, typically germinate and emerge during certain times of the year. Rainfall and irrigation provide high levels of soil moisture and thereby modify the time of seed germination and seedling emergence within the seasons defined by temperature.

■ IMPACT OF SOIL TILLAGE

The germination of weed seeds in and on the soil is greatly favored by tumbling the soil, as occurs during tillage, provided that soil temperature and moisture are also favorable. Many observant farmers and gardeners are aware that cultivation of the soil results in the emergence of a batch of weed seedlings soon after tillage, when adequate moisture for germination occurs. The first weed seedlings to emerge on tilled land usually originate from viable seeds lying in or on the upper layer of soil, generally those in the upper 1-in. layer but occasionally as deep as 4 in. (10 cm) or more (Table 2.1 and Figure 2.4).

The enhancement of weed seed germination by soil tillage can be explained in part by exposure of the seeds to light, even just a flash of light, and in part to repositioning the seeds in the soil profile where the soil environment is more conducive to their germination and seedling emer-

TABLE 2.1 ■ Emergence of morningglory (*Ipomoea* spp.) seedlings as affected by depth of seeding in the field.

SEEDING DEPTH (IN.)	EMERGENCE (%)	
	Sandy Loam	Silt Loam
0	20	7
0.5	47	47
1	76	66
2	61	80
3	19	69
4	10	50
5	1	40
6	1	3
7	0	0

SOURCE: H. P. Wilson and R. H. Cole. 1966. Morningglory competition in soybeans. *Weeds* 14:49–51.

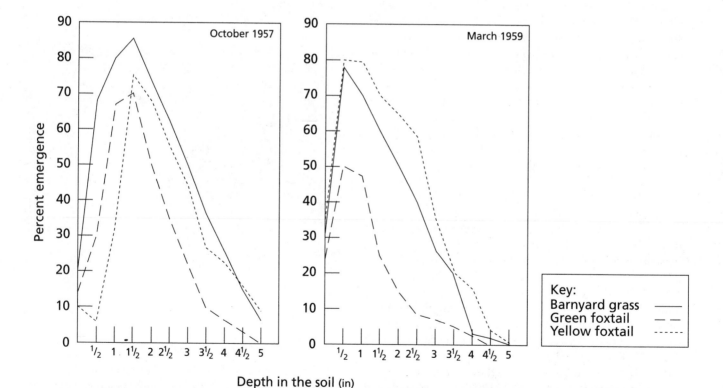

FIGURE 2.4 ■ Percent emergence of barnyardgrass [*Echinochloa crus-galli* (L.) Beauv.], green foxtail [*Setaria viridis* (L.) Beauv.], and yellow foxtail [*Setaria glauca* (L.) Beauv.] from seeds planted at various depths in the soil.

SOURCE: J. H. Dawson, and V. F. Bruns. 1962. Emergence of barnyardgrass, green foxtail, and yellow foxtail seedlings from various soil depths. *Weeds* 10:136–139.

gence. However, tillage may move other weed seeds deeper in the soil, where conditions are less favorable for germination. Thus soil tillage can influence weed seed germination within limits defined by temperature, soil moisture, exposure to light, and depth in the soil.

Seedbed preparation promotes weed seed germination because (1) positioning weed seeds on or near the soil surface favors germination; (2) providing larger voids between soil particles in the upper soil layer allows for faster drying and warming of the soil; (3) oxygen availability is increased in this soil layer; and (4) the tumbling action of the soil particles causes the seed coats to abrade.

■ REDUCTION OF THE WEED SEED POPULATION

Over the past 60 years, studies have been directed toward limiting the buildup of weed seeds in soil. These studies have included (1) mechanical and/or chemical fallowing; (2) monoculture and associated tillage practices; (3) monoculture and herbicide combinations; (4) crop rotations; (5) crop rotation and herbicide combinations; (6) herbicide combinations; and (7) conservation tillage and herbicide combinations.

In a 5-year study in Colorado on the monoculture of irrigated corn, the estimated average number of weed seeds initially present in the upper 10-in. (25-cm) soil layer was 512 million/A (1.27 billion/ha) with seeds of redroot pigweed accounting for 82% and common lambsquarters for 12% of this number. After 1 crop-year, where atrazine was applied preemergence in combination with other typical cultural practices, the number of seeds of redroot pigweed and common lambsquarters declined by 53 and 35%, respectively. After the sixth crop-year, the overall decline in the total number of redroot pigweed and common lambsquarters seeds was 99 and 94%, respectively. Where the use of atrazine was discontinued at the beginning of the fourth crop-year, there were still an estimated 130 million redroot pigweed and 28 million common lambsquarters seeds/A (321 and 69 million/ha) present in the soil. As a result of not using atrazine for just 1 crop-year, the seed bank of redroot pigweed increased by over 48 million seeds/A (118.6 million/ha), and after 3 years of not using atrazine, the number of redroot pigweed seeds had built up to over 239.4 million seeds/A (591.5 million/ha).

In Minnesota, a 7-year study was conducted in an attempt to eradicate seeds of wild mustard from the weed seed bank. Nine cultural or chemical practices were used, and the land selected for the study had been infested with wild mustard for many years. During the study, wild mustard plants emerged but were not allowed to produce seed at any time during the year in any of the treatments. The wild mustard seed population at the beginning of the study averaged 57.9 million seeds/A (143 million/ha) for all treatment locations, ranging from 48 to 81.5 million seeds/A (118.6 to 201.4 million/ha).

After seven growing seasons, 50% of the original mustard seeds was still present where the soil had not been disturbed (continuous grass sod or continuous chemical fallow). In contrast, less than 3% of the original wild mustard seeds remained where the soil received moldboard plowing and additional tillage throughout the growing season. However, this 3% was equivalent to 0.9 million seeds/A (2.2 million/ha). It was concluded that eradication of wild mustard from an infested field is impractical with currently available techniques.

■ SHIFTS IN WEED POPULATIONS

Weed species common to agricultural lands of a given region change over a period of time owing to changes in agricultural practices. A shift in the dominant weed species occurs because species differ in growth habits, survival mechanisms, germination requirements, response to environmental factors, competitive efficiency, and herbicide resistance. Examples of changing agricultural practices that result in shifts in the dominant weed species are (1) monoculture to rotational cropping, (2) tillage alone to tillage plus herbicides, (3) wide row-spacing to narrow row-spacing, (4) conventional to conservation tillage, and (5) changing herbicides.

In some instances, the establishment of one or more new weed species can occur by the introduction of as little as one seed of a particular weed species by wind, irrigation water, a passing bird or animal, tillage or harvesting equipment, a truck or train, or as a crop seed contaminant. Later, as one or a few plants of the introduced species goes unnoticed or ignored, the plant grows and matures, sets seed, and makes a deposit of hundreds or thousands of its seed to the seed bank as a new "account." It then perpetuates itself from plants arising from these seeds, which in turn produce and drop more seed to the soil in subsequent years.

Certain weed species are commonly associated with a particular crop because the weed and crop plants have similar growth habits and thrive under the same cultural practices. For example, downy brome, jointed goatgrass, cereal rye, and volunteer wheat have the same germination and growth requirements as that of winter wheat, while those of prickly sida and spurred anoda are similar to cotton. Velvetleaf thrives under the same cultural practices used in soybean production.

Some weed species escape early detection because their appearance in the seedling or young plant stage is similar to the crop. For example, spurred anoda seedlings are similar in appearance to young cotton seedlings to a casual observer. Red rice seedlings are similar to those of cultivated rice, requiring a selective herbicide to differentiate between the two species.

Effect of Herbicides on Weed Seed Populations

Prior to the widespread use of herbicides in agricultural lands, changes in the dominant weed species of a region occurred slowly, as changes in agricultural patterns did not vary greatly from year to year.

With the advent of herbicides, changes in the dominant weed species in agricultural areas have been relatively rapid and dramatic. The two classic examples whereby herbicides caused a shift in the dominant weed species are those of 2,4-D in corn and winter wheat and trifluralin in cotton.

2,4-D

Within the decade of 1945–1955, extensive postemergence use of the herbicide 2,4-D in corn grown in the Corn Belt, and in winter wheat grown in the western United States, removed broadleaf weeds as the major problem weeds. With the removal of these competitors in the Corn Belt, the way was opened for 2,4-D–resistant grass species, such as barnyardgrass, foxtail species, and panicum species, to move into the vacuum and become the dominant weed species. Similarly, downy brome, jointed goatgrass, cereal rye, and volunteer wheat replaced broadleaf weeds as the dominant weed species in winter wheat.

Trifluralin

From 1965 to 1970, the extensive use of the herbicide trifluralin, accompanied by reduced tillage, in cotton throughout the Cotton Belt of the United States resulted in a change in the dominant weed species. Trifluralin controlled most annual grass species and many broadleaf weed species. However, the removal of these weeds opened the way for trifluralin-resistant broadleaved weeds, such as prickly sida and spurred anoda, to multiply and grow unchecked until they became the dominant problem weeds not only in cotton, but also in crops rotated with cotton. These weed species were present, but of little importance, in cotton until released from competition with more aggressive weeds, first by trifluralin and later by other dinitroaniline herbicides.

Other herbicides

Herbicide use can in itself bring about changes in the composition of weed populations. For example, the postemergence herbicides dicamba, 2,4-DB, bromoxynil, and bentazon control broadleaf weeds but not grasses. In addition, dicamba and 2,4-DB translocate and are effective in controlling both perennial and annual species, while bromoxynil and bentazon do not translocate and are effective against annual, but not perennial, broadleaf weeds. In contrast, the postemergence grass herbicides diclofop-methyl, sethoxydim, and fluazifop-P-butyl, control grasses but not broadleaf weeds. Diclofop-methyl does not translocate and is effective on annual, but not perennial, grasses, while sethoxydim and fluazifop-P-butyl do translocate and are effective against both annual and perennial grass weeds.

The use of nonresidual herbicides, such as MSMA and glyphosate, to control johnsongrass infestations on irrigation ditch banks opened the way for other perennials, such as bermudagrass, field bindweed, quackgrass, and nutsedge species, to invade the area. With the possible exception of bermudagrass, these weeds are often more undesirable on the ditch banks than is johnsongrass. A benefit provided by perennial weeds such as bermudagrass and johnsongrass growing on earthen ditch banks is that their extensive root/rhizome systems tend to stabilize the banks against erosion. Where the ditches are concrete-lined, there are, of course, no benefits.

Many weeds that were previously controlled by specific herbicides have rapidly evolved biotypes with increased levels of resistance. Resistance has occurred with many different herbicides including 2,4-D, chlorimuron, fluazifop, imazaquin, linuron, paraquat, pendimethalin, propanil, triazines, and trifluralin. Over a period of years, the herbicide-resistant weeds become the dominant strain. By 1991, there were confirmed reports from around the world of triazine resistance in 57 weed species, with resistant biotypes found in 33 states in the United States, 4 Canadian provinces, and 18 other countries. Herbicide-resistant biotypes include atrazine-resistant biotypes of redroot pigweed, Powell pigweed, common lambsquarters, common ragweed, and common groundsel; trifluralin-resistant biotypes of goosegrass; sulfonylurea-resistant biotypes of kochia; and diclofop-resistant biotypes of Italian ryegrass.

It has often been said that "change is the way of life," and this old adage is especially appropriate when applied to changes in the dominant weed species comprising the weed populations of agricultural lands.

■ ECOLOGY AND WEED CONTROL

A knowledge of the major mechanisms of weed survival—seed and propagule production and seed longevity—is needed to develop effective methods of weed control. In addition, factors governing seed germination, sprouting of vegetative propagules, and plant development must also be addressed in planning an effective control program. Dormancy allows weed seeds to escape the effects of direct control measures such as tillage and herbicides and provides a mechanism for prolonged seed survival in soil, so it too must be taken into consideration.

Weed-control methods are based on the knowledge that seed dormancy will terminate and germination will follow. Correctly estimating when maximum numbers of weed seed will germinate allows the agriculturalist to apply preemergence herbicides prior to seed germination. It also provides an estimate of the time that lethal levels of the preemergence herbicides need to be maintained in the soil. Preemergence herbicides do not kill nongerminated seeds; thus there is no need to maintain lethal levels of premergence herbicides in the soil when weed seeds are not germinating.

■ SELECTED REFERENCES

Anderson, W. P. 1987. Weed science as it relates to crop production. In *Handbook of plant science in agriculture*, B. R. Christie, ed. Vol. 2, pp. 99–113. Boca Raton, Fla.: CRC Press.

Ball, D. A. 1992. Weed seed bank response to tillage, herbicides, and crop rotation sequence. *Weed Sci.* 40:654–659.

Ball, D. A., and S. D. Miller. 1989. A comparison of techniques for estimation of arable soil seed banks and their relationship to weed flora. *Weed Res.* 29:365–372.

Bayer, D. E. 1985. Mechanisms for weed seed survival. Anaheim, Calif.: Proceedings of 37th Calif. Weed Conference, pp. 50–52.

Benoit, D. L., D. A. Derksen, and B. Panneton. 1992. Innovative approaches to seed bank studies. *Weed Sci.* 40:660–669.

Bickford, E. D., and S. Dunn. 1972. *Lighting for plant growth*. Kent, Oh.: Kent State University Press.

Biswas, P. K., et al. 1975. Germination behavior of Florida pusley seeds. I. Effects of storage, light, temperature, and planting depth on seed germination. *Weed Sci.* 23:400–403.

Brenchley, W. E., and K. Warrington. 1930. The weed seed population of arable soil. *J. Ecol.* 18:235–272.

Brosten, D., and B. Simmonds. 1989. Crops gone wild. *Agrichem. Age.* 33(5):6, 7, 26.

Buhler, D. D., and T. C. Mester. 1991. Effect of tillage systems on the emergence depth of giant foxtail (*Setaria faberii*) and green foxtail (*Setaria viridis*). *Weed Sci.* 39:200–203.

Burnside, O. C. 1979. Soybean (*Glycine max*) growth as affected by weed removal, cultivar, and row spacing. *Weed Sci.* 27:562–565.

Burnside, O. C., et al. 1981. Germination of exhumed weed seed in Nebraska. *Weed Sci.* 29:577–586.

Burnside, O. C., et al. 1986. Weed seed demise in weed-free corn (*Zea mays*) production across Nebraska. *Weed Sci.* 34:248–251.

Cardina, J., E. Regnier, and K. Harrison. 1991. Long-term effects on seed banks in three Ohio soils. *Weed Sci.* 39:186–194.

Cavers, P. B., M. Kane, and J. J. O'Toole. 1992. Importance of seed banks for establishment of newly introduced weeds: A case study of proso millet (*Panicum miliaceum*). *Weed Sci.* 40:630–635.

Conn, J. S. 1990. Seed viability and dormancy of 17 weed species after burial for 4.7 years in Alaska. *Weed Sci.* 38:134–138.

Derksen, D. A., et al. 1993. Impact of agronomic practices on weed communities: Tillage systems. *Weed Sci.* 41:409–417.

Egley, G. H., and J. M. Chandler. 1983. Longevity of weed seeds after 5.5 years in the Stoneville 50-year buried seed study. *Weed Sci.* 26:230–239.

Egley, G. H., and R. D. Williams. 1991. Emergence periodicity of six summer annual weed species. *Weed Sci.* 39:595–600.

Fellows, G. M., and F. W. Roeth. 1992. Factors influencing shattercane (*Sorghum bicolor*) seed survival. *Weed Sci.* 40:434–440.

Forcella, F., et al. 1992. Weed seed banks of the U.S. Corn Belt: Magnitude, variation, emergence, and application. *Weed Sci.* 40:636–644.

Horng, L. C., and L. S. Leu. 1978. The effects of depth and duration of burial on the germination of ten annual weed seeds. *Weed Sci.* 26:4–10.

Jain, R., and M. Singh. 1989. Factors affecting goatweed (*Scoparia dulcis*) seed germination. *Weed Sci.* 37:766–770.

Johansen, C. 1984. How weed control affects insect and disease problems. *Ag Consultant and Fieldman.* January, pp. 15–16.

Kapusta, G., and R. F. Krausz. 1993. Weed control and yield are equal in conventional, reduced- and no-tillage soybeans (*Glycine max*) after 11 years. *Weed Technol.* 7:443–451.

Khedir, K. D., and F. W. Roeth. 1981. Velvetleaf (*Abutilon theophrasti*) seed population in six continuous-corn (*Zea mays*) fields. *Weed Sci.* 29:485–490.

Kivilaan, A., and R. S. Bandurski. 1981. The one-hundred-year period for Dr. Beal's seed viability experiment. *Am J. Bot.* 68:1290–1292.

Leck, M. A., V. T. Parker, and R. L. Simpson. 1989. *Ecology of soil seed banks.* San Diego, Calif.: Academic Press.

Lueschen, W. E., and R. N. Andersen. 1980. Longevity of velvetleaf (*Abutilon theophrasti*) seeds in soil under agricultural practices. *Weed Sci.* 28:341–346.

Lueschen, W. E., et al. 1993. Seventeen years of cropping systems and tillage affect velvetleaf (*Abutilon theophrast*) seed longevity. *Weed Sci.* 41:82–86.

MacDonald, G. E., B. J. Brecke, and D. G. Shilling. 1992. Factors affecting germination of dogfennel (*Eupatorium capillifolium*) and yankeeweed (*Eupatorium compositifolium*). *Weed Sci.* 40:424–428.

Majek, B. A., and S. S. Johnston. 1986. Diethatyl for hairy galinsoga (*Galinsoga ciliata*) control in peppers (*Capsicum annuum*) integrated with cultural practices for phytophthora blight control. *Weed Sci.* 34:569–571.

Maurizo, S., G. Zanin, and A. Berti. 1992. Case history of weed competition/population ecology: Velvetleaf (*Abutilon theophrasti*) in corn (*Zea mays*). *Weed Technol.* 6:213–219.

Mayeux, H. S., Jr. 1983. Effects of soil texture and seed placement on emergence of four subshrubs. *Weed Sci.* 31:380–384.

McCarty, M. K. 1982. Musk thistle (*Carduus thoermeri*) seed production. *Weed Sci.* 30:441–445.

Minotti, P. L., and R. D. Sweet. 1981. Role of crop competition in limiting losses from weeds. In *CRC handbook of pest management in agriculture,* D. Dimental, ed. Vol. 2, pp. 351–367. Boca Raton, Fla.: CRC Press.

Pike, D. R., M. D. McGlamery, and E. L. Knake. 1991. A case study of herbicide use. *Weed Technol.* 5:639–646.

Reddy, K. N., and M. Singh. 1992. Germination and emergence of hairy galinsoga (*Bidens pilosa*). *Weed Sci.* 40:195–199.

Roberts, H. A. 1968. The changing population of viable weed seeds in an arable soil. *Weed Res.* 8:253–256.

Roberts, H. A., and J. E. Neilson. 1981. Changes in the soil seed bank of four long-term crop/herbicide experiments. *J. Appl. Ecol.* 18:661–668.

Schreiber, M. M. 1992. Influence of tillage, crop rotation, and weed management on giant foxtail (*Setaria faberii*) population dynamics and corn yield. *Weed Sci.* 40:645–653.

Schweizer, E. E., and R. L. Zimdahl. 1984. Weed seed decline in irrigated soil after six years of continuous corn (*Zea mays*). *Weed Sci.* 32:76–83.

Schweizer, E. E., R. L. Zimdahl, and R. H. Mickleson. 1989. Weed control in corn (*Zea mays*) as affected by till-plant systems and herbicides. *Weed Technol.* 3:162–165.

Singh, M., and N. G. Achhireddy. 1984. Germination ecology of milkweedvine (*Morrenia odorata*). *Weed Sci.* 32:781–785.

Standifer, L. C., P. W. Wilson, and R. Porche-Sorbet. 1984. Effects of solarization on soil weed seed populations. *Weed Sci.* 32:569–573.

Staniforth, D. W., and A. F. Wiese. 1985. Weed biology and its relationship to weed control in limited-tillage systems. In *Weed control in limited-tillage systems,* A. F. Wiese, ed. Monograph No. 2, pp. 15–25. Champaign, Ill.: Weed Science Society of America.

Stevens, O. A. 1932. The number and weight of seeds produced by weeds. *Am. J. Bot.* 19:784–794.

Swanton, C. J., D. R. Clements, and D. A. Derksen. 1993. Weed succession under conservation tillage: A hierarchical framework for research and management. *Weed Technol.* 7:286–297.

Taylorson, R. B. 1987. Environmental and chemical manipulation of weed seed dormancy. *Rev. Weed Sci.* 3:135–154.

Thomas, A. G., J. D. Banting, and G. Bowers. 1986. Longevity of green foxtail seeds in a Canadian prairie soil. *Can. J. Plant Sci.* 66:189–192.

Toole, E. H., V. K. Toole, H. A. Borthwick, and S. B. Hendricks. 1956. Photocontrol of *Lepidium* seed germination. *Plant Physiol.* 30:15–21.

Triplett, G. B., Jr., and G. D. Lytle. 1972. Control and ecology of weeds in continuous corn grown without tillage. *Weed Sci.* 20:453–457.

Warnes, D. D., and R. N. Andersen. 1984. Decline of wild mustard (*Brassica kaber*) seeds in soil under various cultural and chemical practices. *Weed Sci.* 32:214–217.

Williams, M. C. 1980. Purposefully introduced plants that have become noxious or poisonous weeds. *Weed Sci.* 28:300–305.

Wruckle, M. A., and W. E. Arnold. 1985. Weed species distribution as influenced by tillage and herbicides. *Weed Sci.* 33:853–856.

Yenish, J. P., J. D. Doll, and D. D. Buhler. 1992. Effects of tillage on vertical distribution and viability of weed seed in soil. *Weed Sci.* 40:429–433.

Young, J. A., and R. A. Evans. 1976. Responses of weed populations to human manipulations of the natural environment. *Weed Sci.* 24:186–190.

Zamore, D. L., D. C. Thill, and R. E. Eplee. 1989. An eradication plan for plant invasions. *Weed Technol.* 3:2–12.

3

METHODS OF WEED CONTROL

■ INTRODUCTION

Weed control encompasses those practices whereby weed infestations are reduced but not necessarily eliminated. Weed control is a matter of degree, ranging from poor to excellent. The degree of weed control obtained is dependent on (1) the characteristics of the weed species involved and (2) the effectiveness of the control practice(s) used.

Eradication is the *ideal* of weed control but is rarely achieved. Eradication implies that a given weed species, including its seed and vegetative reproductive parts, has been killed or completely removed from a given area and that the weed will not reappear unless reintroduced into the area. Owing to its difficulty and high cost, eradication usually is attempted in relatively small areas (a few acres, a few thousand square feet, or less). Eradication is practiced through soil sterilization and other means by proficient nursery operators (especially with container-grown plants), greenhouse operators (in and under growing benches and in potting soil), and in certain high-value areas (ornamental plant beds and certain high-value vegetable crops, among others). The greatest obstacles to achieving weed eradication are the presence in the soil of (1) long-lived weed seeds and (2) vegetative reproductive plant parts.

Weed-control programs that employ combinations of weed-control practices are called *integrated weed-control systems* (IWCS), or just *integrated weed control* (IWC). Another accepted and frequently used term introduced in the early 1970s is *integrated weed management systems* (IWMS), or just *integrated weed management* (IWM). It has been argued that IWMS or IWM is simply new terminology for a system that was previously called "weed control" (Buchanan, 1976). When reading the weed science literature, the author has found the term "weed management" vague and frustrating and, for clarity, mentally substitutes the term "weed control" in its place.

Integrated pest-control programs directed toward a coordinated attack against two or more kinds of pests (such as weeds, insects, plant pathogens) are termed *integrated pest management* (IPM). Each of these programs, IWC and IPM, is discussed more fully at the end of this chapter.

Weed-control practices can be grouped into five general categories:

1. Preventive
2. Cultural
3. Mechanical (physical)

4. Biological
5. Chemical

■ PREVENTIVE WEED CONTROL

Preventive weed control is concerned with measures taken to prevent the introduction, establishment, and/or spread of specified weed species in areas not currently infested with these plant species. These areas may be national, state, or local. At the national and state levels, prevention is practiced primarily through laws and their enforcement. In 1968, Public Law 90-583 was enacted "To provide for the control of noxious plants on land under the control or jurisdiction of the Federal Government." In 1974, the Federal Noxious Weed Act (Public Law 93-629) was enacted "To provide for control and eradication of noxious weeds and potential carriers thereof, and for other purposes." Each state has its own seed laws that designate certain weeds as noxious and that regulate their movement in interstate commerce. However, federal and state legislative authority, regulations, and current programs do not provide adequate protection against the importation or interstate movement of noxious weeds, weed seeds or other weed propagules, or of materials so infested, such as crop seeds, feed grains, hay, straw, sod, soil, manure, nursery stock, livestock, and agricultural machinery (Shaw, 1982).

Prevention at the local level is largely the responsibility of individuals or groups of people with a common desire to prevent the introduction or spread of one or more noxious weed species. It may be practiced through "weed districts," county, city, and village maintenance departments, and through the efforts of individual farmers, gardeners, and nursery operators.

Responsible people at the national, state, and local levels need to be familiar with the importance (economic, health, and welfare repercussions) of weeds and the regulations and technology for their control. *People are the key to preventive weed control.*

Preventive weed control embraces all practices that control weeds, with the objective of preventing the introduction or propagation of weeds in a specified area. Weed-control practices that could be designated as preventive practices include (among others) the use of (1) weed-free crop seed, (2) weed-free manure and hay, (3) clean (weed-free) harvesting and tillage equipment, and (4) the elimination of weed infestations in areas bordering croplands and irrigation ditches and canals. A farmer hand-pulling a few scattered weeds in an alfalfa, corn, or onion crop before the weeds mature and drop their seeds is practicing preventive weed control, as is the gardener hand-pulling weeds in a flower bed prior to seed-drop. The effectiveness of preventive practices is improved through the following kinds of programs: (1) *regulatory,* (2) *research,* and (3) *education.*

Weed-Free Crop Seed

One of the most common means by which farmers introduce weed seeds into their croplands is by planting crop seed contaminated with weed seeds. Examples of this practice is illustrated in the following surveys of crop-seed drill boxes.

Jensen (1962) reported on a drill box survey in Utah in which 1232 samples of seed grain were collected from grain drills in the field at planting time and analyzed for the presence of weed seeds. Over half of the samples contained weed seeds, and wild oat seed was the most prevalent weed seed present. Tracing the source of grain seed surveyed showed that crop seed harvested from the farmers' own crop, and not cleaned, contained the most weed seeds. It was further determined that cleaned, tested, and tagged crop seed had fewer weed seeds than uncleaned and untested crop seed, with 99% of the uncleaned crop seed and 44% of the cleaned crop seed containing weed seeds. The worst sample contained seed of five noxious weed species, with a total of 167,000 noxious weed seeds/100 lb or 45 kg of crop seed. This farmer was planting an average of 54 noxious weed seeds/m^2 of cropland.

A 1969 spring wheat drill box survey of 109 boxes in South Dakota indicated that seeds of the primary noxious weeds field bindweed, Canada thistle, and quackgrass were present in 1.8, 0.9, and 31.2%, respectively, of the crop seed sampled, and that seeds of the secondary noxious weeds field pennycress, wild mustard, and wild oat were present in 11.0, 27.5, and 34.9%, respectively, of the crop seed sampled.

A seedbox survey in North Dakota in 1980-1981 found that about 70% of the farmers surveyed planted their own uncertified wheat seed and only about 7% purchased certified seed; 31% of the seedbox samples contained no weed seeds. Of the 325 seedboxes of hard red spring wheat sampled, 44% were contaminated with weed seeds, as were 35% of 139 seedboxes of durum wheat sampled.

A wheat drill box survey in Oklahoma in 1981 revealed that 69% of the farmers in 16 major wheat-producing counties planted seed they raised and 23% purchased seed from another farmer. Only 7% purchased seed from a dealer, but 9% of the planted seed was reportedly certified. However, over half of the samples labeled as a certified class of seed failed to pass the minimum field standards for certified wheat.

The concept of planting weed-free crop seed is as applicable today as it was 100 years ago. As an aid to the user, the kind and percentage of weed seed present in a bag of certified crop seed are listed on the seedbag tag, based on a representative sample of that particular crop seed.

Noxious Weed Defined

The Federal Seed Act of 1939 defines *noxious weed* somewhat differently than the Federal Noxious Weed Act of 1974. The Federal Seed Act defines noxious weed as "any weed or plant that is so declared by an authoritative group, with the legal power to make such a declaration, to be harmful or possess noxious characteristics." The Federal Noxious Weed Act defines a noxious weed as "any living stage (including, but not limited to, seeds and reproductive parts) of any parasitic or other plant, of a kind or subdivision of a kind, which is of foreign origin, is new or not widely prevalent in the United States, and can directly or indirectly injure crops, other useful plants, livestock, poultry, or other interests of agriculture, including irrigation, or navigation, or the fish, or wildlife resources of the United States, or the public health."

Federal Noxious Weed Act

The Federal Noxious Weed Act was signed into law on January 4, 1974. The responsibility for implementation of this law lies with the Animal and Plant Health Inspection Service (APHIS) of the U.S. Department of Agriculture (USDA). However, it was not funded by the federal government until 1979. This is the first law that regulates the importation of weed species into the United States, with the exception of the earlier Federal Plant Pest Act, which prohibited the importation (and interstate movement) of parasitic plants and seeds. Thirty-three plant species, representing 24 genera and all species of *Striga* (witchweed) have been designated as noxious weeds (Table 3.1); additional foreign plant species are being proposed for inclusion in this list of noxious weeds.

The Federal Noxious Weed Act restricts federal government involvement to those weeds that are of foreign origin and not widely dispersed within the United States. It authorizes inspections at ports of entry, weed surveys within the United States, and necessary eradication and quarantine measures to prevent the spread of a noxious weed.

In spite of legislation to regulate entry of noxious weeds into the United States, snafus do occur, as illustrated by the

TABLE 3.1 ■ Plants designated as noxious weeds by the U.S. Department of Agriculture under the Federal Noxious Weed Act.

TERRESTRIAL WEEDS	
Avena ludoviciana Dur.	*M. micrantha* Kunth
Carthamus oxycantha M. B.	*Mimosa invisa* Martius (great sensitive plant)
Commelina benghalensis L. (tropical spiderwort)	*Oryza longistaminata* A. Chev. & Roehr.
Crupina vulgaris Cassini (bearded creeper)	*O. punctata* Kotschy ex Steud.
Digitaria scalarum (Schweinf.) Chiov. (blue couch)	*O. rufipogen* Griff (red rice)
Drymaria arenarioides H.B.K. (alfombrilla)	*Pennisetum clandestinum* Hochst. (kikuyu grass)
Galega officinalis L. (goatsrue)	*P. pedicellatum* Trinius (kyasuma grass)
Imperata brasiliensis Trin.	*P. polystachion* (L.) Schult. (mission grass)
Ischaemum rugosum Salisb. (saromacca grass)	*Prosopis ruscifolia* Griseb. (mesquite)
Leptochloa chinensis (L.) Ness	*Rottboellia* exalata L. F. (itchgrass, raoulgrass)
Mikania cordata (Burm. F.) B. L. Robinson (mile-a-minute)	*Saccharum spontaneum* L. (wild sugarcane)

PARASITIC WEEDS	AQUATIC WEEDS
Orobanche aegyptiaca Persoon	*Eichhornia azurea* (Sw.) Kunth (anchored water hyacinth)
O. cernua Loeft.	*Hydrilla verticillata* F. Muell. (hydrilla)
O. lutea Baumg.	*Monochoria hastata* (L.) Solms
O. major L.	*M. vaginalis* (Burm. f.) Presi.
Striga spp. (witchweed)	*Salvinia molesta* D. S. Mitchell
	Sparganium erectum L.
	Stratiotes aloides L.

SOURCE: R.G. Westbrooks. 1981. Introduction of foreign noxious plants into the United States. *Weeds Today* 12(3): 16–17. Reproduced with permission of *Weeds Today* and the author.

following example. In 1983, serrated tussock (*Nassella trichotoma*) was designated as a Federal Noxious Weed in the United States. The following is excerpted by permission from an article by Westbrook and Cross (1993):

Serrated tussock was introduced into the United States in late 1988 as a contaminant of eight shipments of tall fescue

(*Festuca arundinaceae*) seeds from Argentina. The contaminant was detected during port-of-entry inspections by the U.S. Department of Agriculture (USDA). Detections were made by Plant Protection and Quarantine (PPQ) personnel with the Animal and Plant Health Inspection Service (APHIS) in Jacksonville, FL, Houston, TX, and Portland, OR.

Following positive identification of the contaminants, the APHIS PPQ staff in Hyattsville, Maryland, determined that the shipments could not be regulated under the Federal Noxious Weed Act (FNWA). This decision was based on a clause in Section 12 of the FNWA which prohibits regulation of Federal Seed Act shipments (seeds for planting) under the FNWA. With no perceived authority to regulate the entry of the shipments, PPQ released them to the importers and informed affected states of the decision. Upon release by U.S. Customs to the importers, the contaminated seeds were transported to inland locations for sale and distribution. In early 1989, the USDA Office of General Counsel reevaluated the FNWA and issued a legal opinion that the contaminated shipments could be regulated under Section 6 of the Act.

A majority of the approximately 294,450+ kg of contaminated seed was recovered and either cleaned, exported, or destroyed. In December 1989, 91,818 kg of the seeds were buried in a landfill in Springfield, Missouri. In January 1990, 5,000 kg of the seeds were burned in Kentucky. However, by the time a recall order was issued in early 1989, over 24,000 kg of the contaminated seeds had already been sold to retail customers in at least 49 counties in five states. Affected states included Illinois, Kentucky, Missouri, North Carolina, and South Carolina. Commercial outlets that sold the contaminated seeds have been documented. However, the majority of sales were anonymous. APHIS PPQ and the regulatory agencies in affected states are active in efforts to detect incipient infestations.

Educational efforts have been initiated by APHIS PPQ and the regulatory agencies in affected states to create public

TABLE 3.2 ■ Prohibited noxious weeds listed in the seed laws of five or more states of the United States in 1987.

COMMON NAME	SCIENTIFIC NAME	NUMBER OF STATES IN WHICH LISTED AS NOXIOUS
Balloonvine	*Cardiospermum halicacabum*	8
Bindweed, field	*Convolvulus arvensis*	42
Bindweed, hedge	*Convolvulus sepium*	13
Bursage, skeletonleaf	*Franseria discolor*	6
Camelthorn	*Alhagi pseudalhagi*	8
Cress, hoary	*Cardaria draba*	19
Crotalaria, longbeaked	*Crotalaria longisrostrata*	5
Crotalaria, showy	*Crotalaria spectabilis*	6
Fieldcress, Austrian	*Rorippa austriaca*	8
Halogeton	*Halogeton glomeratus*	6
Horsenettle	*Solanum carolinense*	11
Horsenettle, silverleaf	*Solanum elaeagnifolium*	8
Johnsongrass	*Sorghum halepense*	19
Knapweed, Russian	*Centaurea repens*	24
Nutsedge, purple	*Cyperus rotundus*	14
Nutsedge, yellow	*Cyperus esculentus*	10
Pepperweed, perennial	*Lepidium latifolium*	5
Quackgrass	*Agropyron repens*	31
Sorghum almum	*Sorghum almum*	14
Sorghum, perennial	*Sorghum* spp.	8
Sowthistle, perennial	*Sonchus arvensis*	21
St. Johnswort, common	*Hypericum perforatum*	6
Thistle, Canada	*Cirsium arvense*	40
Thistle, musk	*Carduus nutans*	7
Whitetop, hairy	*Cardaria pubescens*	13

SOURCE: Unpublished compilation by G. Hoxworth and W. P. Anderson, Agricultural Experiment Station, New Mexico State University, Las Cruces, N. Mex.

awareness of the potential hazards posed by serrated tussock to the United States.

Federal Seed Act

The Federal Seed Act, approved in 1939, *regulates* interstate and foreign commerce in seeds. Its *purpose* is to protect purchasers from mislabeled or contaminated crop seeds. The Federal Seed Act is administered by the USDA, which has cooperative agreements with all 50 states.

The Federal Seed Act requires, in part, that the following information be provided on seed labels in interstate commerce:

1. Percentage of pure seed of the named crop
2. Percentage of other crop seeds
3. Percentage of weed seeds
4. The names of noxious weed seeds present and the rate of their occurrence

Seeds of noxious weeds prohibited by the Federal Seed Act from entering the United States as contaminants of imported seeds are as follows: Canada thistle (*Cirsium arvensis*), dodder (*Cuscuta* spp.), field bindweed (*Convolvulus arvense*), johnsongrass (*Sorghum halepense*), leafy spurge (*Euphorbia esula*), perennial sowthistle (*Sonchus arvensis*), quackgrass (*Agropyron repens*), Russian knapweed (*Centauria repens*), and whitetop (*Cardaria draba*). Crop seeds containing in excess of 2% weed seeds of all kinds are also prohibited entry into the United States.

The Federal Seed Act supplements the state seed laws in interstate commerce. It requires labeling of agricultural seed in accordance with, and prohibits rates of occurrence in excess of, the requirements of the state into which the seed is shipped. It imposes no restriction on total percentage of weed seeds in interstate commerce.

State Seed Laws

Each state has it own seed laws in which certain weed species are designated as *noxious*. There are two general classifications of noxious weeds: (1) *prohibited* and (2) *restricted* or *secondary.* Crop seeds cannot be sold for seeding purposes in most states if they contain prohibited noxious weed seeds, excessive amounts of restricted weed seeds, or more than 2.5% by weight of all weed seeds. State seed laws are relatively uniform, although there is sufficient variation to create problems in interstate movement of seeds. *Prohibited noxious* and *restricted noxious* weed species in the seed laws of five or more states are listed in

Tables 3.2 (opposite) and 3.3 (pages 44 and 45), respectively. In 1991, North Carolina listed serrated tussock as a noxious weed.

Most state laws do not regulate crop seed that is sold on the grower's premises and not advertised in the public media. Thus the sale of crop seeds and the movement of contaminants in these seeds, from farm to farm or from farm to seed company by the grower, is not regulated by law.

■ CULTURAL WEED CONTROL

Cultural control of weeds utilizes practices common to good land, crop, and water management. These practices include manipulation of crop row-spacing, crop cultivars, and crop populations, maintaining critical weed-free periods, and using crop rotations and smother crops.

Crop Row-Spacing

Competitiveness between crops and weeds is affected by row-spacing, with narrow row-spacings favoring the crop plants. Planting a crop, such as corn, cotton, or soybeans, in narrow rows allows for a more rapid shading of the soil surface by the foliar canopy. Once the soil surface is shaded, further germination and emergence of weeds is essentially stopped. Shading also suppresses growth of established weeds. For example, the shading effect of corn planted in rows 30 in. (76 cm) apart reduced the vegetation of fall panicum by 260% when compared to that produced with row-spacings of 40 in. (102 cm). When cotton was planted in rows 21 in. (53 cm) apart, maximum yields were obtained with less weed maintenance than when the rows were spaced at 31 and 42 in. (79 and 106 cm). Soybeans planted in rows spaced 8 in. (20 cm) apart outyielded those spaced at 16 and 31 in. (40 and 80 cm) when sicklepod competed with the soybeans all season.

The general acceptance of narrow row-spacing is hampered by lack of available tractors and cultivating equipment designed for narrow rows in the various row crops.

Crop Cultivars

Crop cultivar selection, in conjunction with row-spacing, affects weed competition by determining the rapidity of foliar-canopy development. A corn cultivar, characterized by being shorter, early-maturing, and with horizontal leaves, planted in rows 36 in. (91 cm) apart reduced weed

TABLE 3.3 ■ Restricted noxious weeds listed in the seed laws of five or more states of the United States in 1987.

COMMON NAME	SCIENTIFIC NAME	NUMBER OF STATES IN WHICH LISTED AS RESTRICTED NOXIOUS
Bermudagrass, common	Cynodon dactylon	13
Bindweed, field	Convolulus arvensis	7
Bindweed, hedge	Convolulus sepium	7
Bluegrass, annual	Poa annua	10
Blueweed, Texas	Helianthus, ciliaris	7
Carrot, wild	Daucus carota	11
Cheat	Bromus secalinus	14
Chess, hairy	Bromus commutatus	11
Cockleburs	Xanthium spp.	6
Cocklebur, common	X. strumarium (formerly X. pensylvanicum)	9
Corncockle	Agrostemma githago	26
Daisy, oxeye	Chrysanthemum leucanthemum	9
Darnel	Lolium temulentum	11
Docks	Rumex spp.	16
Dock, broadleaf	R. obtusifolius	17
Dock, cluster	R. conglomeratus	12
Dock, curly	R. crispus	26
Dock, pale	R. altissinus	17
Dock, veiny	R. venosus	16
Dodder	Cuscuta spp.	47
Foxtail, giant	Setaria faberi	19
Garlic, wild	Allium vineale	26
Horsenettle	Solanum carolinense	25
Horsenettle, silverleaf	Solanum elaeagnifolium	11
Johnsongrass	Sorghum halepense	13
Knapweed, Russian	Centaurea repense	5
Lettuce, blue	Lactuca pulchella	6
Moonflower, purple	Calonyction muricatum (also Ipomoea turbinata)	7
Morningglories	Ipomoea spp.	7 *Continued*

growth and weed seed production more than did a corn cultivar, characterized as taller, full-season, and upright, planted in rows spaced at 30 in. (76 cm). Soybean cultivars differ in their rate of foliar-canopy development and thus in their ability to quickly shade the soil surface and suppress weeds.

Crop Population

Crop plant population and row-spacing can affect weed competition and crop yields. Using snap beans as an example, decreasing the crop row-spacing from 35 to 5 in. (89 to 12.7 cm) increased the rate of snap bean canopy closure and decreased the fresh weight of competing weeds by 50–87%. However, where the same in-row plant-spacing was used, the plant population (or plant density) increased sevenfold as row-spacing was decreased from 35 to 5 in. The commercially acceptable plant density for snap beans is about 4 plants/ft^2 (43/m^2). If plant density is held constant at this commercially acceptable level, then the spacing of the crop plants in the row is increased, as row-spacing is decreased. In a study conducted with snap bean row-spacings of 6, 10, 14, 18, and 36 in., all seeded at a constant density of 4 plants/ft^2 and weeds allowed to emerge with the crop, row-spacings of 6–14 in. suppressed weed growth by 18%, compared to the conventional 36-in.

TABLE 3.3 ■ Restricted noxious weeds listed in the seed laws of five or more states of the United States in 1987, *continued.*

COMMON NAME	SCIENTIFIC NAME	NUMBER OF STATES IN WHICH LISTED AS RESTRICTED NOXIOUS
Morningglory, purple	*I. turbinata*	7
Morningglory, ivyleaf	*I. hederacea*	7
Morningglory, tall	*I. purpurea*	7
Morningglory, wooly	*I. hirsutula*	7
Mustards	*Brassica* spp.	24
Mustard, birdsrape	*B. rapa*	24
Mustard, black	*B. nigra*	27
Mustard, Indian	*B. juncea*	29
Mustard, white	*Sinapis alba*	24
Mustard, wild	*S. arvensis*	34
Oat, wild	*Avena fatua*	15
Onion, wild	*Allium canadense*	26
Pennycress, field	*Thlaspi arvense*	10
Plantain, bracted	*Plantago aristata*	9
Plantain, buckhorn	*Plantago lanceolata*	34
Puncturevine	*Tribulus terrestris*	10
Quackgrass	*Agropyron repens*	19
Radish, wild	*Raphanus raphanistrum*	16
Rice, red	*Oryza sativa*	7
Rocket, yellow	*Barbarea vulgaris*	5
Sandbur, field	*Cenchrus pauciflorus*	7
Sorghum almum	*Sorghum almum*	8
Sorghum, perennial	*Sorghum* spp.	6
Sorrel, red	*Rumex acetosella*	18
Sowthistle, perennial	*Sonchus arvensis*	5
Starthistle, yellow	*Centaurea solstitialis*	7
Sumpweed, poverty	*Iva axillaris*	7
Thistle, blessed	*Cnicus benedictus*	7
Thistle, Canada	*Cirsium arvense*	9

SOURCE: Unpublished compilation by G. Hoxworth and W. P. Anderson, Agricultural Experiment Station, New Mexico State University, Las Cruces, N. Mex.

spacing. When weeds were controlled for the first half of the season, row-spacings of 6–14 in. suppressed weed growth by 82%, compared to the 36-in. spacing. The effect of the 18-in. spacing was variable. Weed suppression by narrow rows was explained by a higher rate of canopy closure in narrow rows as compared to wider rows. Snap bean yields in row-spacings of 6–18 in. produced similar yields and were about 23% greater than yields from snap beans in rows spaced 36 in. (Teasdale and Frank, 1983).

Crop row-spacing, morphology, and population regulate the amount and quality of light reaching the soil surface, which in turn influences weed seed germination, emergence, and growth.

When the weeds were allowed to grow in cotton for more than 2–4 weeks after planting, yield reductions occurred. In this case, the weed-free period was not influenced by row-spacing. However, row-spacing may affect the weed-free period as when cotton planted in rows spaced 21 and 31 in. (53 and 79 cm) apart required a 6-week weed-free period for yields equivalent to cotton maintained weed-free all season, while cotton in rows spaced at 42 in. (106 cm) required an 8-week weed-free period for equivalent yields. In another case, cotton in rows spaced 21 in. had a critical weed-free period of 6 weeks for maximum yield, while cotton in rows spaced at 31 and 42 in. required a weed-free period of 10 and 14 weeks, respectively.

Crop Rotation

Crop rotation is practiced as a means of weed control to prevent or reduce the buildup of high populations of certain weeds common to a particular crop. Any given crop, cultivated or not, is plagued by certain weed species that possess similar growth habits and thrive under the same cultural conditions provided for the crop. The particular weed species involved may vary from area to area, but they are generally characteristic to the crop.

When the same cultural practices are followed year after year on the same land, crop-associated weeds tend to multiply rapidly and successfully compete with the crop plants, reducing yields, quality, and economic return.

When weed control is a prime factor, the choice of crop plants to include in a crop-rotation sequence and the sequence itself should give preference to those crop plants whose growth and cultural characteristics are in sharp contrast to those of the previous crop and the predominant problem weeds.

Crop-rotation sequences commonly involve crops planted in sequence with noncultivated crops. Of necessity, perennial crops such as alfalfa and fruits are not rotated as often as annual crops, but they are commonly intermixed in a crop-rotation sequence.

In addition to weed control, crop rotations often result in improved crop yields and quality, improved soil conditions, and control of insects and plant diseases.

Smother Crops

Smother crops are those crops that are especially highly competitive for light, nutrients, and moisture with the weed species infesting the area. Crop plants used for this purpose include the small and large grains (barley, ensilage corn, millet, rye, sorghum), the legumes (alfalfa, clovers, cowpeas, sweetclovers), the grasses (crested wheatgrass, sudangrass), and others (e.g., buckwheat, sesbania, sunflower). Any highly competitive crop that is well adapted to an area may be suitable for use as a smother crop. Care must be taken not to use a crop that may in itself develop into a problem weed. Smother crops may be grown in rotation with less competitive crops.

■ MECHANICAL (PHYSICAL) WEED CONTROL

Mechanical methods of weed control include such practices as hand-pulling, hoeing, mowing, water management, smothering with nonliving material, artificially high temperatures, burning, and machine tillage.

Hand-Pulling

Hand-pulling is an effective practice for the control of weed seedlings and established annual and biennial weeds. It is of minor value in the control of established perennial weeds, as underground vegetative reproductive parts are usually not disturbed. Hand-pulling is best adapted to small garden and lawn areas, roguing a few scattered weeds in a crop, and removal of weeds growing near or between crop plants in the row where it is difficult to reach with a hoe or cultivator.

Hoeing

Hoeing is a highly effective means of weed control, and the hoe would remain one of the principal tools for weed control were it not for people's desire to spend their time more productively. With the competitive demands for labor and the availability of more economical tools resulting from our advancing technology, the "hoe-hand" is not as available nor as economical to use as in years past. However, hoeing is still used to supplement other weed-control practices in certain crops such as cotton, soybeans, vegetables, ornamentals, and small fruits. The hoe, like hand-pulling, is most effective against weed seedlings and annual and biennial weeds. It may be used to advantage against perennial weeds if practiced at intervals of 1–2 weeks during the growing season.

Mowing

Mowing has limited value as a means of weed control. It is primarily used to reduce seed production and to restrict unsightly or rank weed growth. Mowing is commonly used for these purposes in meadows and pastures, along roadsides, and in unimproved (waste) places. It is not uncommon for some weed species to flower and set seed below the height of the cutter bar in spite of repeated mowings; this may occur with species that normally grow tall enough to mow such as redroot pigweed, London rocket, and Russian thistle.

Water Management

Flooding is an effective means of weed control under certain conditions. Flooding is effective only when the roots and/or shoots of the weeds are completely covered by water for a sufficiently long period of time. Its use is limited by soil type (one that allows ponding of water) and the availability of sufficient water to provide flooding to the desired depth. The success of flooding is primarily confined to the control of terrestrial plants, especially certain peren-

nial weeds. Flooding kills weeds by depriving the plants of air: they die of "suffocation." More precisely, flooding displaces air from the soil, thereby preventing root absorption of oxygen that is vital to respiratory processes in the roots and, when the shoots are covered, it prevents leaf absorption of carbon dioxide from the atmosphere, which is essential to photosynthesis.

In arid regions, weed control can be achieved by withholding water at a crucial stage in the growth of weeds. Under such conditions, water may be withheld for a few months or a year, depending on the intended use of the land. Care must be taken to deprive the weeds of water before they set seed, or viable seed will be produced even though the parent plants die due to the drought conditions. Even under these conditions, adapted perennial weeds are difficult to eliminate because of the prolonged survival of underground asexual reproductive parts, and weed seeds lying dormant in the soil will rest peacefully until conditions favor their germination.

Smothering with Nonliving Material

Smothering with nonliving material is an effective means of weed control under certain conditions. The objective is to completely exclude light from the growing weed plants, thereby preventing photosynthesis and further growth. Materials used for this purpose include hay, manure, grass clippings, straw, sawdust, wood chips, rice hulls, paper, and plastic film. Black paper and black plastic film have been used successfully in various horticultural and agronomic crops. The cost of such materials may not be as economical, however, as some other methods of weed control. In Hawaii, the use of black plastic film has proved effective and economical for weed control in nonirrigated sugarcane and pineapple crops.

Artificially High Temperatures

Clear plastic tarps placed over seedbeds prior to crop planting and exposed to strong sunlight have effectively controlled weeds by raising the air and soil temperatures under the tarps to such high levels that weed seeds and seedlings are killed. The high temperatures from confined steam has been used for years by horticulturalists to sterilize greenhouse soils, raising the soil temperature to levels toxic to soil-borne weed seeds, insects, and fungi.

Burning

Burning or *flaming* has been practiced for many years as a means of general weed control in noncropped areas, such as railroad rights-of-way, irrigation canals, drainage ditches, and roadsides. The practical application of flaming for selective weed control in croplands (i.e., without significant injury to the crop plants) had its beginning in the early 1940s.

Selective flaming operates on the general principle that the crop plants are sufficiently tall that the flame, directed toward the ground near the base of the plants, will not strike their leaves or other tender parts; that their stems are woody and resistant to the flame's intense heat; and that the weeds are small, succulent, and susceptible to the heat.

The greatest use, by far, of selective flaming has been in cotton crops, although it has been used successfully in soybean and sugarcane. In cotton, flaming is normally used when the stems of the cotton plants are about 3/16 in. (0.5 cm) in diameter at ground level. At this stage of development, the cotton plants are about 8 in. (20 cm) tall.

The theory behind flaming is that the heat intensity (fuel rate) and exposure time (forward speed) are synchronized so as to cause the cell sap to expand, resulting in rupture of the cell walls, but not to cause actual combustion of the plant vegetation. The heat from flaming may also cause coagulation of cell protoplasm and inactivation of enzymes. The effect of flaming may not be apparent until several hours after application.

To be effective, flaming for selective weed control in croplands must be utilized when the weeds are not more than 1 or 2 in. (2.5–5 cm) tall. Therefore, it is imperative that other means of weed control, such as cultivation or chemical, be used until the crop plants reach a height safe for flaming.

The flaming apparatus is generally tractor mounted and consists of a fuel tank, pipelines, vaporizer, pressure regulator, and burners. The fuel is either butane or propane, or a mixture of the two, stored under pressure.

Machine Tillage

Machine tillage has two functions—namely (1) seedbed preparation, referred to as "primary tillage," and (2) weed control, referred to as "secondary tillage" or "cultivation."

Primary tillage equipment is used to break and loosen the soil to a depth of 6–36 in. Primary tillage tools include the moldboard, disk, chisel, subsoil, and power-driven rotary plows. The moldboard plow inverts the soil and is well suited for turning under and covering crop residues. The disk plow also inverts the soil and is adapted to conditions where the moldboard plow is unsuited such as in heavy, sticky, peaty, leaf-mold soils where the moldboard plow will not turn the slice; in dry, hard ground that the moldboard plow cannot penetrate; and in stony and rooty

FIGURE 3.1 ■ A two-way 3-bottom tractor-mounted moldboard plow.

SOURCE: Photo by the author.

FIGURE 3.2 ■ A tractor-mounted chisel plow equipped with six chisels and sweeps mounted on the chisels.

SOURCE: Photo by the author.

FIGURE 3.3 ■ Depth-controlled, spring-tooth, row-crop cultivator.

SOURCE: Courtesy of Dickey Machine Works, Pine Bluff, Arkansas.

FIGURE 3.4 ■ Cultivator equipped with shields, soil-firming plates, and spray system.

SOURCE: Courtesy of Dickey Machine Works, Pine Bluff, Arkansas.

ground where the disk plow will ride over these obstacles. The chisel plow stirs the soil more or less in place; it does not invert and pulverize the soil to the extent that moldboard and disk plows do. Subsoil plows are used to penetrate the soil to depths of 20–36 in. The power-driven rotary plow has L-shaped cutting knives driven by an auxiliary engine or the tractor power-takeoff. The various plows come in many shapes and sizes. The energy for plowing is derived from diesel, propane, or gasoline engines and, at times, animals. A two-way 3-bottom moldboard plow is shown in Figure 3.1, and a chisel plow with sweeps is shown in Figure 3.2.

Weed control is seldom the objective of primary tillage. However, moldboard and disk plowing invert the soil and bury weed seeds lying on or just below the soil surface to the depth of the worked soil. Short-lived seed such as that of downy brome, a troublesome weed in wheat, may not survive such deep burial. Following chisel plowing, weed seeds on or near the soil surface remain more or less in place, usually under conditions favoring germination.

Secondary tillage generally takes place after the seedbed has been prepared and the crop seed planted, but it may be utilized shortly before planting to control emerged weeds and, if necessary, before final seedbed preparation. Secondary tillage equipment works the soil to depths less than 6 in. Secondary tillage tools include harrows (disk, spike-tooth, spring-tooth), shovels and sweeps (large and small), chisels, rotary hoe or Lilliston, and rod weeders. Examples of machine tillage tools are shown in Figures 3.3 through 3.11.

The advantages of secondary tillage are (1) a wide selection of cultivating tools may be used, and (2) large areas may be weeded rapidly and economically. The prin-

FIGURE 3.5 ■ Gangs of rotary hoes (Lillistons) attached to pipe beams with swivel joints and mounted on 3-point hitch.

SOURCE: Photo by the author.

FIGURE 3.6 ■ Gangs of rotary hoes working a raised cotton seedbed.

SOURCE: Photo by the author.

FIGURE 3.7 ■ A rotary hoe showing the series of pronged wheels mounted on a common axis.

SOURCE: Courtesy of International Harvester Company.

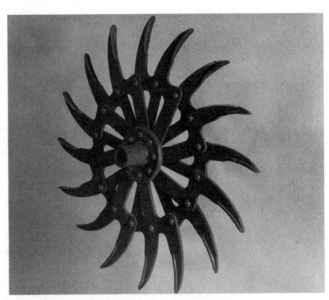

FIGURE 3.8 ■ General design of a single unit of a rotary hoe.

SOURCE: Courtesy of International Harvester Company.

cipal disadvantage of machine tillage is failure, or difficulty, in controlling weeds growing close to, or between, crop plants in the seed row. Weeds in check-rowed or cross-blocked crops can be controlled reasonably well by cross-cultivation, but weeds within the drilled row often require hand-hoeing or pulling or the use of specialized equipment such as flexible finger-weeders.

Weed control by cultivation is achieved primarily by (1) burial of seedlings and small annual weeds by the soil thrown over them by the action of the tillage tools, and (2) uprooting and/or root severance, resulting in death by dessication. Care must be taken in the tillage operation so as not to injure the roots or aboveground parts of the crop plants.

The principal reason for cultivation is weed control; other reasons include breaking soil-surface crusts to aid water penetration and emergence of crop seedlings and, in areas of furrow irrigation, to prepare the land for irrigation.

The two principal means by which weeds are controlled are through the use of cultivation and herbicides, usually in conjunction with one another. Today, the use of herbicides is by far the more important.

Cultivation tends to create favorable conditions for bacterial activity in soils, resulting in the rapid destruction of organic matter and a corresponding loss of exchange capacity in the worked layer of soil. Cultivation tends to shift the location of weed seeds in the soil profile, moving a portion of the ungerminated weed seeds closer to the soil surface where conditions are more favorable for germination. At the same time, it may also bury weed seeds deeper in the soil profile where conditions do not favor germination.

FIGURE 3.9 ■ A rotary hoe in action. This rotary hoe is equipped with two pneumatic-tired wheels (shown raised) that are lowered and used when the hoe is turned sideways when passing through gateways or other narrow openings.

SOURCE: Courtesy of International Harvester Company.

FIGURE 3.10 ■ A hydraulically adjustable tandem disk in action.

SOURCE: Courtesy of International Harvester Company.

FIGURE 3.11 ■ A hydraulically adjustable tandem disk.

SOURCE: Courtesy of International Harvester Company.

Shallow cultivation tends to create favorable conditions for germination of weed seed near the soil surface.

Conservation Tillage Systems

Conservation tillage systems are systems of managing crop residue on the soil surface with minimal or no tillage. The objective of these systems is threefold: (1) to leave enough plant residue on the soil surface at all times for wind and water erosion control, (2) to reduce energy use, and (3) to conserve soil and water. Demands for improved water quality in streams, rivers, and groundwater have also stimulated the practice of leaving crop residues on the soil surface. Weed control is not an objective of conservation tillage systems. However, weed control is vital to the success of such systems, and herbicides are an integral part of the overall program. Conservation tillage systems are frequently referred to as stubble mulching, ecofallow, limited tillage, reduced tillage, minimum tillage, no tillage, and direct drill. These systems are used throughout the United States and the world, and they can be applied to all kinds of crop residue in many cropping systems (Unger and McCalla, 1981).

■ BIOLOGICAL WEED CONTROL

Biological weed control involves the utilization of natural enemies for the control of specific weed species. The objective of biological control is not eradication of the target weed species, but the reduction of its population and crop competitiveness to an acceptable level under the conditions involved. This may be achieved by *direct* or *indirect* action of the biotic agent. The *direct action* of such a biotic agent includes (1) boring into the plant and weakening its structure so that it collapses, and (2) the consumption or destruction of "vital" plant parts. *Indirect action* is attributed to (1) the biotic agent cancelling the competitive advantage of the weed by reducing its vigor of growth and reproductive capability, and (2) the enhancement of conditions favorable to the biotic agent. To be effective, a biotic agent need not kill the weed but only reverse its competitive advantage over the desired plants.

Biotic Agents of Weed Control

Phytophagous insects and *pathogenic fungi* are the principal biotic agents used for the control of terrestrial weeds. *Herbivorous fish* have been the principal biotic agent for the control of aquatic weeds. In either case, various organisms have been successful agents of weed control, and new candidates are continually under investigation for this purpose. Such agents include insects, microorganisms, competitive and parasitic higher plants, sheep, goats, geese, snails, birds, and the nonherbivorous mud carp.

Selection of Biotic Agents

The general principles involved in the selection of a biotic agent for weed control are as follows:

1. To ensure that the agent will attack only the targeted weed species. It is illogical, however, to deny the introduction of an insect as an agent of biological control because, at some stage in its life cycle, it is capable of feeding on some plant of economic importance when such feeding is unlikely to be more than trivial.

2. To obtain biotic agents from areas that are climatically similar to those in which they are to be introduced.

3. To reduce competition between two or more introduced agents by selecting agents with different feeding habits—for example, insects that attack different plant parts such as roots, stems, leaves, flowers, or seeds, or insects that are gall-formers.

Host Specificity of Insects

As our knowledge of phytophagous insects increases, it becomes more apparent that the fear of an introduced species attacking desired plants has been greatly exaggerated, provided that adequate precautions are taken in its screening and final selection. Apparently, instances in which a phytophagous insect has been reported moving from a weed to a crop plant are glaring examples of our lack of knowledge concerning the habits of that particular insect.

Research indicates that insects maintain a relatively rigid habit of host specificity, rather than promiscuous or changeable feeding habits, and that there is no greater chance that an introduced phytophagous insect will change its diet—that is, become a pest of desired plants—than there is for the genetic mutation of insects that are presently innocuous in this respect. Humans are so accustomed to their own varied vegetable menu and to the somewhat varied diet of large grazing animals that they are often startled to learn of the "fastidiousness," or high degree of host specificity, exhibited by many phytophagous insects. The restriction of certain insects to a single species of food plant or a series of related species is by no means rare. Phytophagous insects that do not exhibit some such host specificity are the exception rather than the rule. The real risk associated with the introduction of phytophagous insects, as well as other biotic agents, lies not with the insects themselves but with our ignorance of their natural habits or tendencies.

The preference of phytophagous insects for particular food plants is largely influenced by the presence of *phagostimulants* (feeding stimulants) in or on the plant. Other features of the host plant, such as nutritional balance, surface hairiness, and other physical conditions, also influence selection. Botanical groups of plants usually possess characteristic phagostimulants, and the feeding habits of phytophagous insects are largely determined by these characteristic stimulants. When the same phagostimulant appears sporadically in unrelated plants, these plants may also be selected as food plants by the insects. It is not uncommon for a phagostimulant to be present in plants that are botanically widely separated. Also, the specific phagostimulant present in a plant may vary with the time of day, the growth stage of the plant, the tissue, the season of the year, and climatic and soil conditions. Thus the presence and kind of phagostimulants are central to the selection of food plants by phytophagous insects.

The manner in which a phytophagous insect kills an individual weed is not of prime importance; what is of importance is that it kills existing *stands* of the weed, by direct or indirect action. Insects that attack the seeds or fruits of perennial weeds are generally of minor importance to their control. On the other hand, such insects could prove highly effective in the control of annual weeds, as these plants are dependent on their seed for propagation and spread. Insects that are leaf-eaters may be as effective in controlling a weed as are those that bore into the plant and destroy "vital" parts from within. Individual plants may survive unless attacked by large numbers of insects, or they may succumb to the action of only one or a few insects.

Conflicts of Interest

The use of biological methods of weed control has been hindered by conflicts of interest in the particular plant involved. A plant that may be considered a problem weed in one situation may be considered beneficial in another. For example, yellow starthistle is a problem weed in grain and seed crops and rangelands in California. At the same time, it is considered to be a key plant in the bee industry in California, essential to the maintenance of the bee population at a level necessary for the adequate pollination of the state's fruit and seed crops. Russian thistle serves as an alternate host in the transmission of the curly top virus to tomatoes, sugar beets, and certain other crops *via* the beet leaf hopper. Ranchers consider this plant, in young stages of growth, a desirable forage plant on rangelands, while ecologists consider it beneficial to soil conservation. The annual grass, downy brome, is a major pest in wheatlands of the western United States, but it is a desirable forage plant on rangelands during spring. Thus, prior to initiating a program for biological control of a specific weed, it is mandatory that an accurate appraisal be made of the net economic position of the weed over the entire region in which it is found, taking into consideration future as well as present values. Successful biological control, once started, is a continuing process, not bound by economic borders or by time.

Because of conflicts of interest, biological control has been practiced primarily on lands too inaccessible or of such low economic value as to warrant other methods of control, or in critical situations where other methods have failed. Biological weed control has therefore been directed predominantly against perennial weeds in uncultivated areas, and these have included some of the world's worst noxious weeds. In view of this, it is not surprising to note that the major successes of biological control have been with perennial weeds rather than annuals. Progress has been made in recent years in using fungi for control of specific weeds in croplands.

Plant Pathogens for Weed Control

Fungi are the principal microbial biotic agent used in the most recent work with plant pathogens to suppress or eliminate weeds. The potential use of fungi as biotic agents for weed control is attractive because these organisms are ubiquitous, highly host specific, destructive to the host plant, persistent, and can be readily produced *in vitro*. The use of fungi for weed control employs two approaches: (1) an initial innoculation of the weed with self-sustaining pathogenic fungi (this is referred to as the *classical* approach), and (2) an annual application of endemic or foreign pathogenic fungi in a manner similar to the use of herbicides (this has been termed the *mycoherbicide* approach). It is of interest to point out that where fungi are used for weed control, the fungi produce chemicals (toxicants) that are herbicidal. Thus the use of fungi for weed control is in a real sense a facet of chemical weed control.

The *classical* approach to the use of fungi for weed control is better suited to situations where annual applications of a pathogen are not economically feasible such as the control of perennial weeds in aquatic, pasture, rangeland, and waste areas. In the classical approach, persistence and efficient natural dispersal of the pathogen (as demonstrated by rust fungi) are vital to the successful suppression or control of weeds. Two examples of weed control by the classical approach are (1) control of skeletonweed in Australia by a fungal rust (*Puccinia chondrillina*), and (2) suppression of curly dock in Europe with another fungal rust (*Uromyces rumicis*).

Two *mycoherbicides* (Collego and Devine) have been developed from specific fungal pathogens of weeds. Collego contains spores of the fungus *Colletotrichum gloeosporioides* f. spp. *aeschynomene* and is recommended for control of northern jointvetch in rice and soybean. Devine contains spores of the fungus *Phytophthora palmivora* and is recommended for control of honeyvine milkweed in perennial crops. These two mycoherbicides are living products that control specific weeds in agricultural crops as effectively as herbicides. They are applied as aqueous sprays to the weed foliage in the same manner as postemergence herbicides.

Biological Control of Terrestrial Weeds

With few exceptions, the biological control of terrestrial weeds has been achieved by the action of *introduced* (imported) natural enemies, primarily phytophagous insects. The introduction of such natural enemies is based on the fact that many, if not most, of the serious problem weeds have been introduced into an area from a foreign source, whereas their indigenous natural enemies were left behind. This does not preclude, however, the introduction of insects or other agents for the control of indigenous weeds, nor does it preclude the transfer of indigenous biotic agents within a region or country for the control of indigenous or introduced weeds. In general, natural enemies are sought in the natural habitat of the weed. Experience has demonstrated that effective biotic agents may also be found in areas other than the weed's native habitat, often as the natural enemies of related plant species. Examples of biological control of terrestrial weeds are presented here.

Prickly Pear Cactus

The classic example of biological weed control is the control of two *Opuntia* species, common prickly pear and spiny prickly pear, in Australia by the phytophagous insect *Cactoblastis cactorum*, introduced from Argentina for this purpose (Figures 3.12 and 3.13, see pages 54–56). Prior to 1900, about six species of *Opuntia* had become established in Australia and all were introduced, as *Opuntia* is native only to the Western Hemisphere.

In 1839, common prickly pear was introduced to Australia in a flower pot, and from this single plant, cuttings and plants were transplanted to pastoral areas where they were grown as hedges around homesteads. About 1870, the spread of common prickly pear by natural means had gotten out of hand. By 1925, about 60 million A (24 million ha) were infested with *Opuntia* species, about half so densely that the land was useless. Of these species, common prickly pear was by far the most prevalent and serious pest, covering about 50 million A (20 million ha) in Queensland and New South Wales. Spiny prickly pear was the next most prevalent and troublesome species, infesting several million hectares in central Queensland. Of the whole, about 80% of the infested land was in Queensland with the remainder in New South Wales. Chemical and mechanical methods of controlling the prickly pears were effective but uneconomical on such a broad scale.

In 1925, eggs of the insect *Cactoblastis cactorum* were imported to Australia from Argentina, and initial releases were made the following year. By 1930, the insect population had increased to such great numbers that huge areas of prickly pear, rather than individual plants, were killed. By 1930–1932, most of the original stands of common prickly pear and spiny prickly pear had been killed. In the next year, millions and millions of starving caterpillars went in search of food, and during this time, they were not observed feeding on any plant other than the two prickly pear species, proving host specificity. Owing to the tremendous depletion of their food supply, the population of *C. cactorum* was greatly reduced, and there was much regrowth of the prickly pears from seed and surviving plants from 1933 to 1935. By this time, the insect's population had recovered sufficiently to again effectively attack the prickly pears. Within the next few years, virtually complete control of these two species in Queensland and northern New South Wales occurred. In southern New South Wales, in areas below latitude 32° south, *C. cactorum* was ineffective in controlling prickly pear. In areas where the two species were controlled, a formerly minor species of prickly pear (*O. aurantiaca*) became widely distributed in Queensland and developed into a major pest. Satisfactory control of this species has been obtained with cochineal insects (*Dactylopius* spp.). A number of cochineal insects are known, and all are restricted to feeding on prickly pear and other cacti plants.

The destructive form of *Cactoblastis cactorum* is its larva or caterpillar. The adult is a plain brown moth with a wing expanse of slightly over 1 in. (2.5 cm). The moths live but a few days, during which they mate and lay their eggs. In Australia, there are two generations annually, with the moths emerging from the cocoon stage in September–October and January–February. The moths are free flying and have been known to fly at least 10 mi (16 km) before depositing eggs, but most lay their eggs in the immediate vicinity of their emergence. The eggs are laid in chains, called sticks, preferably on a prickly pear spine or spinelike projection. A female moth may lay 3–4 such sticks, with an average of about 75 eggs per stick. Thus great numbers of eggs, even to the extent of 1 million/A (2.5 million/ha), may be laid in a limited area. The eggs incubate in 3–6 weeks, at the end of which the larvae (caterpillars) are hatched. The caterpillars are orange to orange-red, with black cross stripes; they grow to about 1 in. long. Following hatching, the caterpillars bore into the prickly pear plants and pass their existence within the plant, rarely emerging except in search of additional food plants. Within the plant, they tunnel from joint to joint, eating out the interiors, and penetrate into the underground bulbs and even into the roots. In young segments of the plant, the whole inside is eaten,

(a)

(b)

FIGURE 3.12 ■ Biological control of prickly pear cactus (*Opuntia* spp.) in Queensland, Australia, by *Cactoblastis cactorum* Berg. (a) Roadway bordered by cactus in 1938. (b) Same scene 18 months later. (Courtesy of A. P. Dodd, 1940; H. L. Sweetman, 1958)

SOURCE: The Alan Fletcher Research Station, Department of Lands, Queensland, Australia.

leaving only a thin papery cuticle. Older segments are not destroyed entirely, but "wet rot," caused by various fungi and bacteria, becomes established, aiding in the destruction of the plant. When full grown, the caterpillars spin loose white silken cocoons under bark or fallen debris or between the destroyed pear joints.

(a)

(b)

FIGURE 3.13 ■ Prickly pear cactus (*Opuntia* spp.) control by *Cactoblastis cactorum* Berg. in Queensland, Australia. (a) Dense stand of prickly pear cactus in 1926. (b) Same scene in 1929, following cactus destruction by *Cactoblastis cactorum* Berg.

Klamath Weed (St. Johnswort)

The classic example of biological weed control in the United States is the control of Klamath weed (since re-named common St. Johnswort) by the leaf-feeding beetle *Chrysolina quadrigemina* (also named *C. gemellata*) in the mid-1940s (Figure 3.14, see page 57).

(c)

(d)

FIGURE 3.13 ▪ *continued.*
(c) Same area after trees have been cut and area burned over, 1930. (d) Same area in 1931 showing prolific growth of Rhodes grass (*Chloris*) and excellent pasture for dairy cattle. (Courtesy of A. P. Dodd, 1940; H. L. Sweetman, 1958)

SOURCE: The Alan Fletcher Research Station, Department of Lands, Queensland, Australia.

Klamath weed is a perennial weed native to Europe. It was first introduced into the eastern United States shortly before 1800, and it was first reported in the western United States about 1900 in the vicinity of Klamath River in northern California. It is known in Europe and the eastern United States as St. Johnswort and in the western United States and Canada as Klamath weed or as goatweed. Although the weed is not outright poisonous, when ingested it photosensitizes white-skin areas of grazing animals, reducing appetite, and it also causes unthriftiness. It is a threat to agriculture primarily on rangelands in the western United States and western Canada, where soil moisture is ample from winter to summer but deficient later in the year. Overgrazing fosters its spread, enabling its deep root system to overcome even the sturdiest grass competition. Its minute seeds readily adhere to the oily hair and hide of grazing animals, and its spread is associated with the movement of sheep and cattle and with hay and seeds of other plants. The plant grows 1–5 ft (0.3–1.5 m) tall, more or less woody at the base but herbaceous above. Its aboveground parts are killed in the fall by cold weather and replaced by new growth the next year. In addition to

FIGURE 3.14 ■ St. Johnswort or Klamath weed (*Hypericum perforatum*) control by the beetle *Chrysolina quadrigemina* at Blocksburg, California, in 1948. The left foreground shows the weed in heavy flower while the remainder of the field has been killed by the beetle. (Photo courtesy of C. B. Huffacker, University of California, Berkeley)

SOURCE: C. B. Huffaker. 1957. Fundamentals of biological control of weeds. *Hilgardia* 27:101–157.

erect stems, it has lateral runners or horizontal stems growing at or just below the surface of the ground. The root system is extensive and highly branched.

In 1945–1946, *Chrysolina quadrigemina* was introduced into northern California and released for the control of Klamath weed. This insect has been highly effective in controlling this weed, reducing established stands by 99%, and it continues to hold the weed in check. Its effectiveness is attributed to the insect's host specificity for this weed and to the perfectly synchronized relationship between the life cycle of the insect and the growth stages of the weed. The adult beetles emerge from their pupal cells, located just below the soil surface at or near the base of Klamath weed plants, in April and early May. They feed voraciously on the leaves of the plants during May and June, a time coinciding with flowering of the plants. By late June and early July, the beetles have completed preparations for their summer hibernation in debris, under stones, and in crevices in the soil. The beetles spend 4–6 months in this inactive condition without food or water. While the beetles are resting, the weed also enters a relatively dormant phase at which time it drops most of its leaves, becomes hard and woody, and develops and ripens its seed. The rains in the fall and early winter reactivate both the weed and the beetle. The beetles mate and deposit many eggs on the new leafy growth of the weed, which is sending out vigorous, prostrate, leafy shoots in rosettes at the base of the flowering stalks. Egg laying

(oviposition) begins in the fall and extends into the spring. The insect is thus present as the adult, larva, and egg during these months, and all three stages survive the winter conditions. The larvae actively feed in the warm periods in winter and spring, and their intensive feeding keeps the plants stripped of leaves over a long, critical period when its food reserves are low. The fully mature larvae enter the soil to pupate at about the time the plant begins to develop shoots that become flower-bearing stalks. The appearance of the adult beetle completes the insect's life cycle, which covers 1 year. Thus the weed is under continual attack by the insect's larval and adult stages and is unable to manufacture needed food; its root system becomes greatly reduced; it can no longer compete successfully with other plants; and it eventually dies during the hot dry summer months.

Local dispersal of the insect is normally by adult beetles crawling from plant to plant or from area to area, less commonly by flight. About 3 years are required for local control of a Klamath weed infestation in a situation where only a few thousand beetles are present. By the third reproductive year, the beetle population reaches a level that can exert a controlling effect on the weed. By 1952, the beetle had dispersed over large areas of northern California and effected general control of Klamath weed over an area hundreds of square miles in size. This insect has now been introduced into other western states and British Columbia where Klamath weed is also a problem.

FIGURE 3.15 ■ The grass carp is an effective biotic agent for aquatic weed control when used judiciously.

SOURCE: © Tom McHugh, The National Audubon Society Collection/Photo Researchers.

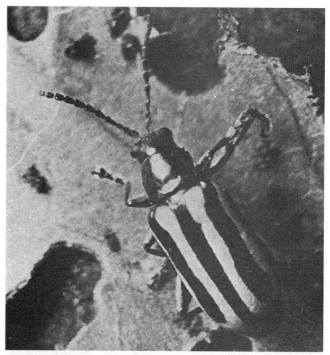

(a)

(b)

FIGURE 3.16 ■ Effective biotic agents for weed control.   (a) Flea beetle (*Agasicles connexa*). (b) Tropical freshwater snail (*Marisa cornuarietis*).

SOURCE: U.S. Department of Agriculture. 1968. *The yearbook of agriculture*, pp. 229–234. Washington, D.C.

Biological Control of Aquatic Weeds

Although *herbivorous fish* have been the principal biotic agent for the control of aquatic weeds, other biotic agents are of considerable value. The reader is referred to a recent paper presenting the history and development of aquatic weed control in the United States (Gallagher and Haller, 1990). Examples of some effective biotic agents for control of certain aquatic weeds follow.

The susceptibility of most herbivorous fish to low temperatures is the principal factor limiting their use in the United States. An exception is the grass carp (formerly known as the white amur, *Ctenopharyngodon idella*, Figure 3.15). The grass carp has a voracious appetite for many aquatic plants. It grows at a rate of 9–11 lb (4–5 kg) per year and may, in time, reach a weight of 100 lb (45 kg). It can tolerate low water temperatures and other water-quality extremes. However, once introduced into a waterway, the grass carp is nearly impossible to eliminate, and by 1987, introduction of the diploid (fertile) grass carp had become illegal in 28 states. Its voracious appetite for aquatic vegetation is extremely detrimental to other aquatic life, especially game fish. In 1984, a practical method was devised to mass-produce triploid (sterile) grass carp, and in 1988, Arizona legalized the use of the triploid grass carp in golf course ponds and streams. The use of the grass carp seems best suited to relatively small, closed ponds, such as farm and golf course ponds, where the fish are confined.

The mud carp (*Cyprinus carpio*), a nonherbivorous fish, is a mud-bottom feeder. Its control of submerged aquatic weeds is apparently due to the uprooting of aquatic plants and breaking up of mats of algae in its search for food. The scarcity of vegetation in many bodies of water in Wisconsin has been associated with the presence of the mud carp.

Herbivorous fish, such as the Congo tilapia (*Tilapia melanopleura*) and Java tilapia (*Tilapia mossambica*), have effectively controlled algae (*Chara* and *Nitelia*) and saw weeds (*Najas*) in Africa. These herbivorous species have shown promise in the southern United States for algae control, but they generally succumb to cool winter temperatures.

A flea beetle (*Agasicles connexa*) (Figure 3.16a) was introduced from Argentina for possible control of alligator-weed in the United States. In 1965, 266 adults were released in extensive alligatorweed beds on the Ortega River in Duval County, Florida. A flea beetle population explosion followed this introduction, and within 18 months after releasing the adult flea beetles, dying and decaying alligatorweed islands were breaking away from the shore-line and carrying the flea beetles to other alligatorweed infestations downriver. The flea beetle proved so successful in control of alligatorweed, that 2 years after its introduction, it was found in essentially all significant infestations in Florida and, subsequently, introduced into other states from Florida. The alligator flea beetle has effectively controlled alligatorweed in the warmer regions of its distribution in the United States (Figure 3.17, see next page).

The large, tropical, freshwater snail (*Marisa cornuarietis*) (Figure 3.16b) feeds voraciously on submerged weeds such as common coontail, Illinois pondweed, and southern naiad. The snail prefers submersed weeds to floating or emersed weeds. However, it does feed on salinia, a floating weed, nearly as readily as on the submerged weeds noted above. The snail, native to Colombia and Venezuela, is dark brown, sometimes with dark brown stripes, and it attains a maximum size of 0.75 by 2.5 in. (1.9 by 6.4 cm). As a biotic agent for weed control, it has two disadvantages: (1) it eats rice plants that are less than 4 or 5 weeks old as well as watercress and waterchestnut, and (2) it cannot survive at temperatures below 48°F (9°C).

■ CHEMICAL WEED CONTROL

Chemicals that kill plants are called "phytotoxic chemicals." Phytotoxic chemicals that are used for weed control are termed *herbicides*. A herbicide has been defined as "any chemical substance or cultured biological organism used to kill or suppress plant growth." In this text, the term "herbicide" will signify a chemical substance, while the term "mycoherbicide" will signify a cultured biological microorganism. Herbicides are the most widely used pesticide in agriculture today (Table 3.4, see page 61).

The use of phytotoxic chemicals to control weeds has been practiced to some extent since about 1900. However, the development of this practice along scientific lines has largely occurred since 1944, following the discovery of the herbicides 2,4-D and MCPA. The importance of herbicides and mycoherbicides to agriculture is that they can selectively kill undesired plant species without significant injury to desired species. The utilization of herbicides has become so extensive in agriculture and allied fields that specialties have developed in their use such as in agronomic crops, horticultural crops, turfgrass, pastures and rangelands, forests, aquatics, and noncrop areas. The scientific literature relative to herbicides and their use is extensive.

In 1994, there were more than 125 different active ingredients and 100 premixes of these ingredients marketed in the United States for use as herbicides. They are offered to the consumer in an almost bewildering array of trade-named products, each varying from the other as to active ingredient, salt and/or ester forms of active ingredients, concentration of active ingredient, mixtures of active ingredients, and formulation constituents, although sometimes only the trade name is different. In general, the active ingredients in these products are of relatively low toxicity to man and animals, often less toxic than aspirin. However, all herbicides should be handled with respect and, to avoid possible injury, treated as though each was poisonous. If ingested, it may be the formulation constituents, rather than the active ingredient, that is toxic to man or animals. One must bear in mind that "dosage determines toxicity"; too much of anything can prove injurious or fatal. The comparative toxicity of many herbicidal active ingredients is listed in the Appendix, Table A.4.

Herbicides are effective and economic tools for use in our eternal struggle with weeds, but *herbicides demand respect*. Properly used, herbicides can effectively and safely accomplish their objective; misused, they can cause severe economic loss. The misuse of herbicides is usually due to (1) ignorance of their characteristic activity, and/or (2) carelessness in application.

Characteristics of Herbicides

To be effective, some herbicides must be applied to the soil *prior* to weed seed germination and seedling emergence from the soil; others must be applied after emergence. Some herbicide are root-absorbed while others are not. Some are absorbed from the soil via the emerging shoot. Some herbicides are only absorbed by aerial plant parts while others enter the plant only via underground plant parts. Some herbicides are inactivated by the soil while others remain relatively unaffected for various periods of time. Some herbicides are readily leached in soils; others are not. Some herbicides may be used safely in croplands; others may not. Some herbicides persist in the soil for more

(a)

(b)

FIGURE 3.17 ■ A weed-control success story. (a) This canal is heavily infested with alligator-weed; growth is so heavy that canal is not navigable. (ARS, USDA; PN-1467) (b) Same canal after flea beetle had been released to feed on alligator-weed. (ARS, USDA; PN-1468)

SOURCE: Anonymous. 1968. Insects destroy weeds. *Agri. Res.* 15(8):10.

TABLE 3.4 ■ Comparative use of pesticides in selected crops in the United States in 1992.

CROP	ACRES PLANTED (× 1000)	PERCENT ACREAGE TREATED		
		Herbicides	Insecticides	Fungicides
Field crops				
Corn	71,375	96	29	<1
Cotton	10,115	88	65	7
Fall potatoes	1,068	81	90	72
Rice	2,030	97	11	21
Soybeans	53,050	91	1	<1
Winter wheat	36,390	33	5	2
Durum wheat	2,200	93	<1	1
Other spring wheat	15,350	87	1	4
Vegetables				
Asparagus	89.3	86	64	28
Beans, lima (fresh)	4.2	82	83	69
Beans, snap:				
Fresh	65.7	52	77	62
Processing	158.5	95	68	55
Broccoli	118.6	58	95	31
Cabbage	64.0	49	96	53
Carrots	100.3	67	37	79
Celery	34.0	82	100	98
Corn, sweet:				
Fresh	154.1	75	84	41
Processing	486.3	92	75	19
Lettuce, head	206.0	68	97	76
Onions:				
Dry bulb	113.6	86	79	83
Peas, green:				
Processing	329.4	91	49	1
Peppers, bell	61.8	65	85	66
Tomatoes:				
Fresh	105.1	75	95	56
Processing	252.3	90	80	92

SOURCE: *Agricultural chemical usage: 1992 field crop summary,* and *Agricultural chemical usage: Vegetables, 1992 summary,* United States Department of Agriculture, Government Printing Office, Washington, D.C.

than one crop-year; others are inactivated within a few days, weeks, or months. Some herbicides are volatile, others are not.

Herbicide Application

Herbicides may be applied in croplands in more than a dozen different ways (Figure 3.18, see page 62). They are commonly applied directly to the soil and soil-incorporated or not. Soil application may be made prior to seeding the crop, or after seeding but prior to crop emergence. The term *preplant surface* (PP) is used when the herbicide is applied prior to planting the crop and it is not incorporated. The term *preplant-incorporated* (PPI) is used when the application is prior to planting and the herbicide is incorporated. The term *preemergence* (PRE or Pre) is used when the herbicide is applied after planting but prior to crop emergence. Unless otherwise designated, preemergence denotes a surface application that is not incorporated; if shallowly incorporated, the term is PREI or simply

Growth stage of crop	Preemergence to weeds	Postemergence to weeds
Preplant		
Preemergence		
At emergence		
Broadcast postemergence		
Directed postemergence		
Band preemergence		Key:

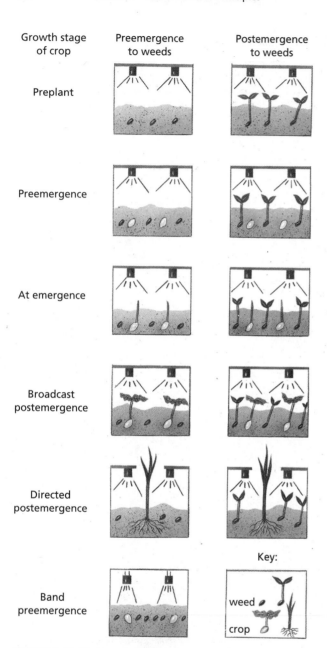

FIGURE 3.18 ■ Stages of growth of weed and crop plants during which herbicides may be applied.

SOURCE: T. D. Fryer and S. A. Evans, eds. 1968. *Weed control handbook, Vol. 1, Principles,* p.63. Oxford, England: Blackwell Scientific Publishers.

preemergence-incorporated. A preemergence-incorporated herbicide application denotes shallow incorporation so as not to displace the crop seed.

The term *postemergence* (POST or Post) usually implies after crop emergence, but it is just as appropriate to use the term to denote after weed emergence. Therefore, to avoid confusion, it is desirable to specify whether one is referring to the crop plants, the weeds, or both. If used postemergence to the crop, the herbicide may be applied to the soil surface prior to weed emergence and soil-incorporated between the crop rows, or it may be applied to the soil surface and not incorporated. If used postemergence to the crop, the herbicide may be applied as a *broadcast* spray, covering both the crop and weed foliage (a true test of herbicide selectivity), or it may be applied as a basal-directed spray to the crop plants, contacting the weed foliage but avoiding contact with the crop foliage. It is not uncommon to apply a herbicide to emerged weeds in a seedbed prior to crop planting or after planting but prior to crop emergence.

With certain crops, other descriptive terms may be used to designate a herbicide treatment, such as "at-cracking," a time just prior to soil emergence of peanut seedlings; "at-emergence," when the crop seedling has just emerged through the soil surface; or "spike stage," when the emerging corn seedling is still in the coleoptilar stage as it emerges from the soil.

Herbicides may be applied as sprays, in water or oil, or as dry granules or pellets. Their activity may be altered by surfactants, crop oil concentrates, or liquid nitrogen fertilizers added to the spray mixture, or by the kind of formulation used.

Carelessness in herbicide application can produce unsatisfactory results. It includes such factors as (1) applying an improper dosage, too high or too low (the old adage that "a little more won't hurt" is not true when using selective herbicides); (2) applying the wrong herbicide; (3) failing to properly calibrate and adjust the application equipment; (4) improper soil incorporation; (5) timeliness of application with respect to growth stage of the crop and weed plants.

Applied in light to strong winds, the mist from herbicide sprays may drift from one field to another, as may the vapors of highly volatile herbicides. *Herbicides that unintentionally cross fence lines are just as active, dose for dose, as are those intentionally applied.* Thus the drifting of herbicide vapors and sprays present a real hazard to nearby susceptible crop plants, as their movement is uncontrolled. For example, the vapor and spray drift of the herbicides 2,4-D and clomazone (Command) pose a hazard to nearby crops. In the case of 2,4-D, drift injury of 2,4-D on grapes and tree fruits has occurred as much as 10 mi (16 km) or more from the site of aerial application. In 1986, widespread use of Command on soybeans resulted in major vapor-drift injury to neighboring field crops, lawns, vegetable gardens, and fruit and ornamental plants.

Phytotoxicity

Herbicides are either *selective* or *nonselective* with respect to the kinds of plants that they kill. In either case, they kill the plants on contact or they kill by more subtle means following plant absorption. Herbicides that kill upon contact usually do not translocate in the plant, killing that portion of the plant with which they come into direct contact. Thus best results are obtained with complete plant-spray coverage, including the meristems (buds). Herbicides that kill by more subtle means do so by disrupting one or more physiological processes vital to the life of the plant. They may exert their toxic effect (1) near their site of entry into the plant, or (2) at some point remote from the site of entry. In the first case, the herbicide is said not to translocate, although some short-distance transport across cell walls and membranes or plant organ does occur. In the second case, the herbicide must translocate within the plant to reach its site(s) of activity.

Grouping of Herbicides

Herbicides may be grouped in a variety of ways, all dependent on some similar characteristic such as *chemical similarity, mode of action, application, weed species–controlled,* or *crop selectivity.* With few exceptions, herbicides are organic compounds. The grouping and discussion of herbicides in this text is, where possible, by *similarity of molecular structure,* a grouping hereafter referred to as *herbicide families.* One of the attributes of grouping by herbicide families is that differences, sometimes seemingly slight, in molecular structure can be compared with differences in herbicidal activity. The molecular structures of some herbicides are unique, not meeting the requirements for a particular herbicide family, and the characteristics of each of these herbicides are discussed separately.

Herbicide Mixtures

The extensive use of herbicides has brought about changes in weed problems, and it is evident that no single herbicide can be expected to cope with these changes. To control new problem weeds that appear in a field, herbicide mixtures are used to control a broader spectrum of weeds. When so labeled, farmers and others have mixed two or more herbicidal products in their spray tanks to combat these new weed infestations. Such mixtures usually involve mixing two compatible broadleaf herbicides, mixing a broadleaf herbicide with a compatible grass herbicide, or mixing two compatible herbicides that control one or more different grass and broadleaf weeds. Care must be taken in making such mixtures, as the formulations may not necessarily be compatible and may form a sludge or other unsprayable mixture, or the active ingredients may antagonize the phytotoxicity of one or the other resulting in reduced weed control of some weed species.

Premixed products, containing two or more herbicides, are now on the market. These premixes offer convenience to the user such as ease of handling, eliminating compatibility problems, and sometimes a cost savings. At the same time, they can have a negative side: a ratio of active ingredients may not fit the user's requirements for crop tolerance, or there may not be a sufficient dosage of an active ingredient key to controlling specific weeds. Thus the user is compelled to adjust the ratio by adding additional amounts of one or more of the active ingredients. Also, the proliferation of available premixes has created an inventory nightmare for the distributor. Advertisements for the premixes image them as "new," whereas in reality they are usually mixtures of known, established herbicides offered under a new trade name. Premix trade names do not give a clue as to the individual active ingredients in the mixtures.

In addition to the appearance of new weed species in a field or region, herbicide-resistant strains of some weed species have evolved through the continual use of one or more herbicides with the same modes of action. To control these resistant strains, herbicides with different modes of action are used in alternate years or in mixtures with ones to which a weed species has developed tolerance.

■ HERBICIDE REGULATION

Before a pesticide can be marketed for use on food crops, it must be registered under the Federal Insecticide, Fungicide, and Rodenticide Act of 1947 (Public Law 86-139; 73 Stat. 286; June 25, 1947), referred to as FIFRA. However, pesticide tolerances (maximum permitted residue levels) are established under Section 409 of the Federal Food, Drug, and Cosmetic Act (FFDCA). Pesticides found in processed foods are treated as food additives.

The Delaney Clause

In 1958, the Delaney Clause (named after its original sponsor, Representative James Delaney of New York) was added to Section 409 of FFDCA to address concerns about the risk of cancer from food additives. The Delaney Clause states that a food additive shall not be permitted "if it is found to induce cancer when ingested in man or animal or if it is found, after tests which are appropriate for evaluation of the safety of food additives, to induce cancer in man or animal." *The Delaney Clause permits "zero residue" of a carcinogen in processed foods,* in spite of today's advanced technology capable of detecting pesticides in amounts as minute as parts per billion and, in some cases, parts per trillion.

Even products considered to pose virtually no risk in the human diet because of very low levels of residues in food are denied tolerances under the Delaney Clause. In February 1993, the U.S. 9th District Court and the Circuit Court of Appeals overturned an earlier EPA interpretation of the Delaney Clause that would have allowed pesticide residues in processed foods provided that the residue amounts did not exceed levels at which the risk of human cancer was proved to be negligible. Thus, if the level of pesticide residue was determined by research not to cause cancer in more than one case in 1 million people over a 70-year life span, it had been deemed, by the EPA, a "negligible risk" and therefore safe to consume. This is no longer allowed under the 1993 ruling.

The Delaney Clause relates only to tolerances for processed food and not for unprocessed food (raw agricultural commodities) such as fruits and vegetables eaten as "fresh" commodities. Tolerances for raw agricultural commodities are established under Section 408 of the FFDCA, which allows tolerances to be established for chemicals with carcinogenic tendencies if the tolerance is set low enough to mitigate risks to the public.

The Environmental Protection Agency

In December 1970, the Environmental Protection Agency (EPA) was established under FIFRA and given the responsibility for administering the regulations governing pesticides. This was formerly the responsibility of the U.S. Department of Agriculture (USDA).

The FIFRA, through the EPA, grants the states significant latitude in regulating the distribution, sale, and use of pesticides; many of the federal laws have comparable state laws. A pesticide must have EPA registration and approved product labeling prior to being sold or distributed in intrastate or interstate commerce. An EPA registration is often referred to as a FIFRA Section 3 registration. It is imperative to note that *a pesticide must be registered with the appropriate state agency in each state before being marketed or used in the state.*

Section 18 of FIFRA

Section 18 of FIFRA authorizes the administrator to exempt state and federal agencies from any provision of the Act, if it is determined that emergency conditions exist that re-

quire an exemption. The term *emergency condition* means an urgent, nonroutine situation that requires the use of a pesticide(s) and shall be deemed to exist when:

(1) No effective pesticides are available under the Act that have labeled uses registered for control of the pest under the conditions of the pest.

(2) No economically or environmentally feasible alternative practices that provide adequate control are available.

(3) The situation:

    (i) Involves the introduction or dissemination of a pest new to or not theretofore known to be widely prevalent or distributed within or throughout the United States and its territories, or

    (ii) Will present significant risks to human health, or

    (iii) Will present significant risks to threatened or endangered species, beneficial organisms, or the environment, or

    (iv) Will cause significant economic loss due to:

        (A) An outbreak or an expected outbreak of a pest, or

        (B) A change in plant growth or development caused by unusual environmental conditions where such change can be rectified by the use of a pesticide(s).

There are four types of emergency exemptions that may be authorized: *specific, quarantine, public health,* and *crisis exemptions.*

A *specific exemption* may be authorized in an emergency condition to avert:

(1) A significant economic loss, or

(2) A significant risk to:

    (i) Endangered species,

    (ii) Threatened species,

    (iii) Beneficial organisms, or

    (iv) The environment.

A *quarantine exemption* may be authorized in an emergency condition to control the introduction and spread of any pest new to or not theretofore known to be widely prevalent or distributed within and throughout the United States and its territories.

A *public health exemption* may be authorized in an emergency condition to control a pest that will cause a significant risk to human health.

A *crisis exemption* may be utilized in an emergency condition when the time from discovery of the emergency to the time when the pesticide use is needed is insufficient to allow for the authorization of a specific, quarantine, or public health exemption.

The *duration of exemptions* is also subject to regulation. The EPA shall allow use of a pesticide under a *specific* or *public health* exemption for as long a period as is reasonably expected to be necessary but *in no case for longer than 1 year*. The EPA shall allow use of a pesticide under a *quarantine exemption* for as long a period as is deemed necessary but *in no case for longer than 3 years*. Quarantine exemptions may be renewed.

Section 24c of FIFRA

Section 24c of FIFRA authorizes each state to register a new end-use product for any use, or an additional use of a federally registered pesticide product, if the following conditions exist:

1. There is a special local need for the use within the State;
2. The use is covered by necessary tolerances, exemptions, or other clearances under the FFDCA, if the use is a food or feed use;
3. Registration for the same use has not previously been denied, disapproved, suspended, or cancelled by the Administrator, etc.
4. The registration is in accord with the purposes of FIFRA.

Section 2ee of FIFRA

Section 2ee of FIFRA authorizes the use of any registered pesticide in a manner inconsistent with its labeling, provided:

(1) The pesticide is applied at a dosage, concentration, or frequency *less* than that specified on the labeling, unless the labeling specifically prohibits deviation from the specified dosage, concentration, or frequency.

(2) The pesticide is applied against any target pest not specified on the labeling if the application is to the crop, animal, or site specified on the labeling, unless the Administrator has required that the labeling specifically state that the pesticide may be used only for the pests specified on the labeling after the Administrator has determined that the use of the pesticide against other pests would cause an unreasonable adverse effect on the environment.

(3) Application of the pesticide may employ any method of application not prohibited by labeling unless the labeling specifically states that the product may be applied only by the methods specified on the labeling.

(4) Mixing a pesticide or pesticides with a fertilizer when such mixture is not prohibited by the labeling.

(5) Any use of a pesticide in conformance with Section 5, 18, or 24 of FIFRA.

(6) Any use of a pesticide in a manner that the Administrator determines to be consistent with the purposes of FIFRA.

Herbicides are regulated in the United States through EPA-approved product *labeling*. Approval for such labeling is based on (1) the product's proven effectiveness, (2) its safety to humans and animals, and (3) its potential effect on the environment.

Sources of Weed-Control Guides

To aid farmers and others in the selection of the proper herbicide, agricultural experiment stations and extension services in many states (or groups of states such as New England and the Pacific Northwest) publish guides relative to available herbicides and their use in the principal crops grown in their respective states or regions. These guides are usually offered free of charge or for a nominal fee. In addition, weed-control guides may be available through private publishers such as the Meister Publishing Company in Willoughby, Ohio.

■ SELECTED REFERENCES

Andres, L. A. 1982. Integrating weed biological control into a pest-management program. *Weed Sci.* 30(Supplement 1):25–30.

Anonymous. 1988. Arizona legalizes sterile grass carp. *Landscape Management.* 27(9):6. (Sept. issue.)

Ball, W. S., A. Mann, and V. Anderson. 1982. North Dakota drill box seed survey 1980 and 1981. N. D. State Univ. Coop. Ext. Serv.

Buchanan, G. A. 1976. Weeds and weed management in cotton production. Proc. Beltwide Cotton Prod. Res. Conf., pp. 166–168.

Dodd, A. P. 1940. *The biological campaign against prickly pear.* Brisbane, Australia: Commonwealth Prickly Pear Board.

Donald, W. W., and J. D. Nalewaja. 1990. Northern Great Plains. In *Systems of weed control in wheat in North America.* W. W. Donald, ed. Monograph Series, No. 6, p. 105. Champaign, Ill.: Weed Science Society of America.

Gallagher, J. E., and W. T. Haller. 1990. History and development of aquatic weed control in the United States. *Rev. Weed Sci.* 5:115–192.

Hill, G. D. 1982. Herbicide technology for integrated weed management systems. *Weed Sci.* 30(Supplement 1):35–39.

Jensen, L. A. 1962. Are we planting weeds? Proc. Western Weed Control Conf., p. 52. Las Vegas, Nev.

Joye, G. F. 1990. Biological control of aquatic weeds with plant pathogens. In *Microbes and microbial products as herbicides,* pp. 155–173. R. E. Hoagland, ed. Symposium Series 439. American Chemical Society.

Lueschen, W. E., R. N. Andersen, T. R. Hoverstad, and B. K. Kanne. 1993. Seventeen years of cropping systems and tillage affect velvetleaf (*Abutilon theophrasti*) seed longevity. *Weed Sci.* 41:82–86.

Martin, A. M., et al. 1991. The economics of alternative tillage systems, crop rotations, and herbicide use on three representative east-central Corn Belt farms. *Weed Sci.* 39:299–307.

McFarland, J. 1993. Delaney background and update. In *WSSA Newsletter,* pp. 8, 19. Champaign, Ill.: Weed Science Society of America.

McWhorter, C. G., and W. C. Shaw. 1982. Research needs for integrated weed management systems. *Weed Sci.* 30:40–45.

Miller, J. H., L. M. Carter, and C. Carter. 1983. Weed management in cotton (*Gossypium hirsutum*) grown in two row spacings. *Weed Sci.* 31:236–241.

Moomaw, R. S. and A. R. Martin. 1984. Cultural practices affecting season-long weed control in irrigated corn (*Zea mays*). *Weed Sci.* 32:460–467.

Murdock, E. C., P. A. Banks, and J. E. Toler. 1986. Shade development effects on pitted morningglory (*Ipomoea lacunosa*) interference with soybeans (*Glycine max*). *Weed Sci.* 34:711–717.

Nalewaja, J. D. 1981. Integrated pest management for weed control in wheat. In *CRC handbook of pest management in agriculture.* Vol. 3, pp. 343–354. Boca Raton, Fla.: CRC Press.

Oraze, M. J., and A. A. Grigarick. 1992. Biological control of ducksalad (*Heteranthera limosa*) by the waterlilly aphid (*Rhopalosiphum numphaeae*) in rice (*Oryza sativa*). *Weed Sci.* 40:333–336.

Peeper, T. F., and A. F. Wiese. 1990. Southern Great Plains. In *Systems of weed control in wheat in North America.* W. W. Donald, ed. Monograph Series, No. 6, p. 173. Champaign, Ill.: Weed Science Society of America.

Quimby, P. C., chairman. 1986. Symposium: microbiological control of weeds. *Weed Sci.* 34(Supplement 1):1–53.

Quimby, P. C. and H. L. Walker. 1982. Pathogens as mechanisms for integrated weed management. *Weed Sci.* 30(Supplement 1):30–34.

Rogers, N. K., G. A. Buchanan, and W. C. Johnson. 1976. Influence of row spacing on weed competition with cotton. *Weed Sci.* 24:410–413.

Schweizer, E. E., D. W. Lybecker, and R. L. Zimdahl. 1991. Systems approach to weed management in irrigated crops. *Weed Sci.* 36:840–845.

Shaw, W. C. 1982. Integrated weed management systems technology for pest management. *Weed Sci.* 30(Supplement 1):2–12.

Shear, G. M. 1985. Introduction and history of limited tillage. In *Limited-tillage systems.* A. F. Wiese, ed. Monograph Series, No. 2, pp. 1–14. Champaign, Ill: Weed Science Society of America.

Staniforth, D. W., and A. F. Wiese. 1985. Weed biology and its relationship to weed control in limited-tillage systems. In *Weed control in limited tillage.* A. F. Wiese, ed. Monograph Series, No. 2, pp. 15–25. Champaign, Ill.: Weed Science Society of America.

Swanton, C. J., K. N. Harker, and R. L. Anderson. 1993. Crop losses due to weeds in Canada. *Weed Technol.* 7:537–542.

Sweetman, H. L. 1958. *The principles of biological control,* chap. 15. Dubuque, Iowa: Wm. C. Brown Co.

Teasdale, J. R., and R. Frank. 1983. Effect of row spacing on weed competition with snap beans (*Phaseolus vulgaris*). *Weed Sci.* 31:81–85.

Templeton, G. E. 1982. Biological herbicides: Discovery, development, deployment. *Weed Sci.* 30:430–433.

Templeton, G. E., R. J. Smith, Jr., and D. O. Tebeest. 1986. Progress and potential of weed control with mycoherbicides. *Rev. Weed Sci.* 2:1–14.

Templeton, G. E., and E. E. Trujillo. 1981. The use of plant pathogens in the biological control of weeds. In *CRC handbook of pest management in agriculture.* Vol. 2, pp. 345–350. Boca Raton, Fla.: CRC Press.

Thomas, T. W., M. A. Martin, and C. R. Edwards. 1988. The adoption of integrated pest management by Indiana farmers. *J. Prod. Agric.* 1:257–261.

Unger, P. W. and T. M. McCalla. 1981. Conservation tillage systems. *Adv. Agron.* 33:1–44.

Walker, H. W., and G. A. Buchanan. 1982. Crop manipulation in integrated weed management systems. *Weed Sci.* 30(Supplement 1):17–24.

Walker, H. W., et al. 1984. Effect of insecticide, weed-free period, and row spacing on soybean (*Glycine max*) and sicklepod (*Cassia obtusifolia*) growth. *Weed Sci.* 32:702–706.

Westbrook, R. G. and G. Cross. 1993. Serrated tussock (*Nassella trichotoma*) in the United States. *Weed Technol.* 7:525–529.

4 CONSERVATION TILLAGE SYSTEMS

■ INTRODUCTION

Conservation tillage systems manage crop residue on the soil surface with minimal tillage or no tillage. The crop residue may be shredded with rotary or flail-type shredders. Such systems encompass all tillage and planting practices that maintain at least 30% plant-residue ground cover to reduce soil erosion by water. Runoff water from cropland can contain sediment, dissolved nutrients, and pesticides, and a plant-residue ground cover of 30% can reduce surface erosion of pesticides by as much as 70%. Where wind erosion is of primary concern, such practices require a ground cover equivalent to at least 1000 lb/A (5450 kg/ha) of flat, small-grain residue throughout the critical wind erosion period. Terminology denoting conservation tillage systems is still evolving; terms currently in use include limited-tillage, minimum-tillage, mulch-till, no-till, reduced-tillage, and ridge-till. Chemical fallow, direct drill, double-crop, ecofallow, sleeping sod, sod planting, stale seedbed, and stubble-mulch are also part of the terminology. Favored terms vary with the region or with a variant in a conservation tillage practice; they are defined at the end of this chapter and in the Glossary in the Appendix.

The objectives of conservation tillage are threefold: (1) to control soil erosion by wind and water with plant residue left on the soil surface, (2) to conserve soil moisture, and (3) to conserve energy by reducing equipment use.

In the United States in 1993, conservation tillage was practiced on 97 million A (39 million ha), or 35% of the total land in crop production, compared to 26% in 1989. The acreages devoted to no-till, ridge-till, and mulch-till for five major agronomic crops grown in the United States in 1993 are shown in Table 4.1. In Ontario, Canada, in 1989, approximately 206,000 A (83,500 ha) of crops were grown under no-till. Stubble-mulching in small grain is a common practice in the prairie provinces of Canada since wind erosion can be severe.

Today, conservation tillage practices are widely used in many crops as growers see yields equal to, or better than, those achieved with conventional tillage. It is predicted that more than 90% of the acreage in cotton, corn, and soybeans will be grown under conservation tillage by the end of this decade. As of 1994, peanuts and rice were among the increasing number of crops taking up conservation tillage. It is predicted that by the year 2000 more than 50% of the

TABLE 4.1 ■ Acreages of five major crops grown under three conservation tillage systems in the United States in 1993.

CROP	TOTAL ACREAGE (MILLIONS)	ACRES IN CONSERVATION TILLEAGE (MILLIONS)				PERCENT OF CROP
		No-till	Ridge-till	Mulch-till	Total	
Cotton	13.6	0.4	0.1	0.9	1.4	10
Field corn (full-season)	73.3	12.4	2.3	17.1	31.8	43
Grain sorghum (full-season)	10.0	1.0	0.2	2.4	3.6	35
Soybean:						
Full-season	55.1	12.0	0.8	13.2	26.0	47
Double-crop	5.4	3.0	0.01	0.6	3.6	65
Small grains:						
Spring-seeded	35.9	1.5	0.01	8.7	10.2	28
Fall-seeded	51.4	2.9	0.03	12.5	15.4	30

SOURCE: Anonymous. 1993. *CTIC national crop residue management survey. Executive summary.* West Lafayette, Ind.: Conservation Technology Information Center.

crops grown in the United States will utilize one or more conservation tillage practices. Under some conditions, conservation tillage in combination with either terracing or contour cropping may prove more effective in soil and water conservation than either used alone.

The major factors delaying adoption of conservation tillage have been (1) breaking with the long-standing tradition of tilling the soil, especially of clean-tilling; (2) anticipated crop-yield reduction; (3) the inability of planters to handle the plant residue on the seedbed and provide consistent uniform stands; (4) disease and insect problems due to not burying crop residue; (5) slower soil warming in the upper Midwest; (6) incomplete and inconsistent weed control; and (7) shifting weed populations. The removal of crop residues from seedbeds has often been accomplished by burning (see Figure 4.1). Figures 4.2, 4.3, and 4.4 show various planters that have been developed to handle plant residue on seedbeds in crops where conservation tillage is practiced.

■ WEEDS

Problem weed species may change under conservation tillage systems. With a conservation tillage practice that does not disturb the soil, newly produced seed remain on or near the soil surface. No-till seems to favor small-seeded species, while large-seeded broadleaf species diminish as problems. Many small-seeded species such as common lambsquarters require light for germination and thus are

FIGURE 4.1 ■ The removal of wheat stubble from seedbed by burning in eastern New Mexico.

SOURCE: Photo by the author.

favored. However, weeds can be controlled with preemergence herbicides or controlled soon after emergence with postemergence herbicides. Without cultivation, more deeply buried weed seeds remain dormant and do not become problems. Under conservation tillage, especially those of reduced-tillage and no-till systems, perennial species increase. Volunteer crop plants such as corn and small grains can be problem weeds in reduced-tillage and no-till systems.

In an 11-year study in Illinois (Kapusta and Krausz, 1993), there was a distinct shift in weed populations in

(a)

(b)

(c)

(d)

FIGURE 4.2 ■ Yetter equipment mounted in front of planter to handle plant residue trash covering the seedrow. (a) Residue Manager Combo; (b) Residue Manager; (c) Trash Master; (d) Residue Brush to be mounted in front of planter to clear the seedrow.

SOURCE: Courtesy of Yetter Mfg. Company, Colchester, Ill..

(a)

(b)

FIGURE 4.3 ■ Tools for use in no-till conservation tillage systems.   (a) Finger-wheel/coulter system for handling plant residue over the seedrow; (b) finger-wheel row cleaner that mounts just in front of the planter to clear plant residue from the seedrow.

SOURCE: Courtesy of Dawn Equipment Company, Sycamore, Ill..

FIGURE 4.4 ■ Martin Row Cleaner clearing a path in heavy residue covering seedbed.   Note cleared seedrow shown just to right of large wheel. Equipment moving toward the left in photo.

SOURCE: Courtesy of Martin and Company, 169 Allegre Road, Elkton, Ky.

conventional, reduced, and no-till tillage systems when herbicides were not used. Horseweed, dandelion, gray goldenrod, prickly lettuce, and downy brome were observed only in no-till. In no-till, horseweed was the dominant spring-emerging species in the first years of the study. Following its initial appearance, gray goldenrod became the dominant species within 2 years and remained so during the last 4 years of the study. By the third year, the population of gray goldenrod was estimated at 100,000 plants/A (240,000/ha). Gray goldenrod rapidly develops a tall canopy that tends to preclude other species from becoming established. Gray goldenrod may have caused a rapid reduction in horseweed, dandelion, and prickly lettuce populations through adverse allelopathy. Of the annual grasses, giant foxtail occurred most frequently and in the greatest numbers in all tillage systems. It occurred in especially large numbers in conventional and reduced-tillage systems, with a maximum density of approximately

5.3 million plants/A (13 million/ha). Dominant populations of gray goldenrod and giant foxtail apparently resulted in the decline and disappearance of populations of large crabgrass, fall panicum, and yellow nutsedge.

■ WEED CONTROL

The use of cover crops in no-till systems can add organic matter to the soil, suppress or prevent annual weed growth, reduce the cost of weed control, reduce soil erosion, and minimize surface-water runoff. However, cover crops consume soil moisture that may be needed for the marketable crop. Various cereal crops have been evaluated for their use in no-till systems. Following the use of herbicides to kill emerged weeds, vegetative residues of sorghum, rye, or wheat provided up to 90% control of annual weeds for 30–60 days. A living cover of spring-planted winter rye reduced early-season biomass of common lambsquarters by 98%, large crabgrass by 42%, and common ragweed by 90%. Vegetative residues of fall-planted rye, spring-killed with a desiccant-type herbicide, reduced total weed biomass by 82%. Adverse allelopathic effects on weed seed germination and growth have been attributed to using residues of cereals as cover crops. Italian ryegrass, hairy vetch, or clovers have also been used as cover crops in conservation tillage systems.

The adoption of conservation tillage for a given situation may well be dependent on the availability of selective herbicides to control weeds, especially perennial weeds, without crop injury. In reduced-tillage systems, herbicides are substituted for tillage to control weeds; in conventional tillage systems, herbicides supplement tillage. Thus dependency on herbicides in the reduced-tillage systems is greater, and their effectiveness is more critical. Reduced-tillage systems may restrict or eliminate the use of soil-incorporated herbicides that are widely used in conventional tillage practices.

Results of an 11-year study in Illinois (Kapusta and Krausz, 1993) indicate that soybean population and yield are equal in conventional, reduced-tillage, and no-till systems if a high level of weed control is achieved. Soybean herbicides currently available can provide excellent control of most annual and perennial weeds in each of these three tillage systems. A combination of desiccant-type, preemergence, and postemergence herbicides may be required to achieve a high level of weed control in all tillage systems.

Weed-control programs in conservation tillage generally begin with a desiccant-type or translocated herbicide treatment, which is applied at or before planting. This is followed by a residual preemergence herbicide. A selective postemergence herbicide is applied broadcast or as a directed spray to row middles on an as-needed basis.

Gyphosate and paraquat are the predominant desiccant-type herbicides used in conservation tillage across all crops. Preemergence and postemergence herbicides vary according to the crop and weeds involved.

With preemergence herbicides, the effectiveness of weed control in no-till is directly related to timely rainfall or sprinkler irrigation to move the herbicides into the soil. The consistency of weed control in no-till in soybeans has improved with the development and use of chlorimuron, clomazone, and imazaquin.

Postemergence herbicides are applied to selectively control young, actively growing weeds. The particular herbicide or mixture used will depend on the weeds present and the tolerance of the crop.

For example, several herbicides, such as clethodim (Select), diclofop (Hoelon), fluazifop-P-butyl (Fusilade), and sethoxydim (Poast), selectively control grasses. These herbicides do not control broadleaf plants. They may be safely used postemergence in broadleaf crops such as soybeans, but they cannot be safely used postemergence in grass crops such as corn. Diclofop does not translocate in grasses; thus it does not control perennial grasses. Clethodim, fluazifop-P-butyl, and sethoxydim translocate in grasses and control both annual and perennial grasses.

Bentazon (Basagran, Galaxy) and 2,4-D (Weedar 64, Weedone LV4) are postemergence herbicides that control only broadleaf weeds. Bentazon may be safely used postemergence in soybeans, a broadleaf crop, but 2,4-D, applied similarly, severely injures or kills soybeans. Either herbicide may be safely used postemergence for broadleaf weed control in field corn, a grass crop. Bentazon does not translocate in broadleaf plants; thus it controls young annual broadleaf weeds but not established perennials. 2,4-D translocates in broadleaf plants and controls both annual and perennial broadleaf weeds.

A herbicide registered for use preemergence or postemergence in a specified crop will usually be suitable in either conventional or conservation tillage systems used with that crop. However, in limited-tillage systems, soil-incorporated herbicides are usually not recommended.

■ CONSERVATION TILLAGE TERMINOLOGY

No-till The soil is left undisturbed from harvest to planting, except for nutrient injection. Planting and drilling is accomplished in a narrow seedbed with coulters, row cleaners, disk openers, in-row chisels, or roto-tillers. Weed control is achieved primarily with herbicides, with cultivation used for emergency weed control. In a modified no-till system, herbicides may be applied in a band, thereby reducing the amount of herbicide applied per acre, accompanied by interrow cultivation.

Ridge-till   Ridge-till is characterized by permanent ridges that are left intact throughout the year. There is usually no preplant tillage or tillage between successive crops with ridge-till. At planting time, the top of the ridge is cleaned and an interrow cultivation is performed. Annual weeds are controlled with herbicides applied in a band over the row, with one or two interrow cultivations.

Mulch-till   The soil is disturbed prior to planting using tillage tools such as chisels, field cultivators, disks, or sweeps. The stubble of the preceding crop is partially buried.

Limited-tillage   Limited-tillage encompasses a wide range of practices within the extremes of no-till to reduction of only one postseeding cultivation in a conventional clean-tillage system. Its significant impact has been substitution of sweeps and the chisel plow for moldboard plowing.

Conventional tillage   A moldboard plow is used for primary tillage followed by implements such as disks and other harrows for seedbed preparation. Weed control is accomplished with cultivation and/or herbicides.

Double-crop   Two crops are harvested in one growing season.

Sleeping sod   A living cover crop is maintained while the marketable crop is growing. Growth of the cover crop is suppressed or partially killed with a sublethal application of a herbicide such as atrazine.

Stubble-mulch   The stubble of the preceding crop is left on the surface with or without soil disturbance, as with mulch-tillage.

Stale seedbed   No soil preparation is done within 2–3 months prior to spring planting. Weed control is accomplished with a "burndown" herbicide applied 30–45 days ahead of planting.

■ SELECTED REFERENCES

Anonymous. 1983. Southern no-till depends on new herbicides and cultivars. *Agrichem. Age.* Dec.: 20, 24, 26.

Anonymous. 1984. Groundwater: How you can help keep it clean. *Farm Chemicals.* June: 28, 31, 58.

Anonymous. 1993. *CTIC national survey of conservation tillage practices. Executive summary.* West Lafeyette, Ind.: Conservation Technology Information Center.

Anonymous. 1993. Raising residue. *Farm Journal.* Nov.: 14–15.

Anonymous. 1993. Less plowing cuts costs and erosion. *Progressive Farmer.* Dec.: 60.

Anonymous. 1993. *Groundwater bulletin, Illinois groundwater consortium.* Urbana, Ill.: University of Illinois.

Blackshaw, R. E. 1991. Control of downy brome (*Bromus tectorum*) in conservation fallow systems. *Weed Technol.* 5:557–562.

Buhler, D. D., and E. S. Oplinger. 1990. Influence of tillage systems on annual weed densities and control in solid-seeded soybean (*Glycine max*). *Weed Sci.* 38:158–165.

Colvin, D. L., et al. 1985. Weed management in minimum-tillage peanuts (*Arachis hypogaea*) as influenced by cultivar, row spacing, and herbicides. *Weed Sci.* 33:233–237.

Derksen, D. A., et al. 1993. Impact of agronomic practices on weed communities. *Weed Sci.* 41:409–417.

Eadie, A. G., et al. 1992. Banded herbicide applications and cultivation in a modified no-till corn (*Zea mays*) system. *Weed Technol.* 6:535–542.

Eadie, A. G., et al. 1992. Integration of cereal cover crops in ridge-tillage corn (*Zea mays*) production. *Weed Technol.* 6:553–560.

Faulkner, E. H. 1943. *Plowman's folly.* New York: Grossett and Dunlap.

Fee, R. 1993. Coulter choices keep growing. *Successful Farming.* Oct.: 42, 44.

Fink, C. 1994. Straight from the field (coulters). *Farm Journal*, Jan.: 10–16.

Gebhardt, M. R. et al. 1985. Conservation tillage. *Science* 230:625–630.

Gebhardt, M. R., and K. J. Fornstrom. 1985. Machinery for reduced tillage. In *Weed control in limited-tillage systems,* A. F. Wiese, ed. Monograph No. 2, pp. 242–254. Champaign, Ill.: Weed Science Society of America.

Johnson, G. A., M. S. Defelice, and Z. R. Helsel. 1993. Cover crop management and weed control in corn (*Zea mays*). *Weed Technol.* 7:425–430.

Kapusta, G., and R. F. Krausz. 1993. Weed control and yield are equal in conventional, reduced- and no-tillage soybean (*Glycine max*) after 11 years. *Weed Technol.* 7:443–451.

Kidwell, B. 1994. No-tillers take out the trash. *Progressive Farmer,* Jan.: 22–23.

Knake, E. L. 1992. *Weed control systems for lo-till and no-till.* Circular 1306. Urbana-Champaign, Ill.: University of Illinois, Cooperative Extension Service.

Lewis, W. M. 1985. Weed control in reduced-tillage soybean production. In *Weed control in limited-tillage systems,* A. F. Wiese, ed. Monograph No. 2, pp. 41–50. Champaign, Ill.: Weed Science Society of America.

Liebl, R., et al. 1992. Effect of rye (*Secale cereale*) mulch on weed control and soil moisture in soybean (*Glycine max*). *Weed Technol.* 6:838–846.

Lueschen, W. E., and T. R. Hoverstad. 1991. Imazethapyr for weed control in no-till soybean (*Glycine max*). *Weed Technol.* 5:845–851.

McWhorter, C. G., and T. N. Jordan. 1985. Limited tillage in cotton production. In *Weed control in limited-tillage systems,* A. F. Wiese, ed. Monograph No. 2, pp. 61–76. Champaign, Ill.: Weed Science Society of America.

Richardson, L. 1983. Surface herbicides and reduced-tillage residue. *Agrichem. Age,* Jan.: 40–44.

Shear, G. M. 1968. The development of the no-tillage system in the United States. *Outlook Agric.* 5:247–251.

Shear, G. M. 1985. Introduction and history of limited tillage. In *Weed control in limited-tillage systems,* A. F. Wiese, ed. Monograph No. 2, pp. 1–14. Champaign, Ill.: Weed Science Society of America.

Smith, E. D. 1993. Conservation tillage: A special report. *Farm J.* Nov.: C-1 to C-20.

Standifer, L. C., and C. E. Beste. 1985. Weed control methods in vegetable production with limited tillage. In *Weed control in limited-tillage systems,* A. F. Wiese, ed. Monograph No. 2, pp. 93–100. Champaign, Ill.: Weed Science Society of America.

Staniforth, D. W., and A. F. Wiese. 1985. Weed biology and its relationship to weed control in limited-tillage systems. In *Weed control in limited-tillage systems,* A. F. Wiese, ed. Monograph No. 2, pp. 15–25. Champaign, Ill.: Weed Science Society of America.

Triplett, G. B. 1985. Principles of weed control for reduced-tillage corn production. In *Weed control in limited-tillage Systems,* A. F. Wiese, ed. Monograph No. 2, pp. 26–40. Champaign, Ill.: Weed Science Society of America.

Triplett, G. B., et al. 1983. *Weed control for reduced tillage systems.* Washington, D.C.: USDA, Agricultural Extension Service.

Unger, P. W., and T. M. McCalla. 1981. Conservation tillage systems. *Adv. Agron.* 33:1–45.

Westbrooks, R. G. and G. Cross. 1993. Serrated tussock (*Nasella tritochoma*) in the United States. *Weed Technol.* 7:525–529.

Wicks, G. A. 1985. Weed control in conservation tillage systems—small grains. In *Weed control in limited-tillage systems,* A. F. Wiese, ed. Monograph No. 2, pp. 77–92. Champaign, Ill.: Weed Science Society of America.

Wiese, A. F., P. W. Unger, and R. R. Allen. 1985. Reduced tillage in sorghum. In *Weed control in limited-tillage systems,* A. F. Wiese, ed. Monograph No. 2, pp. 51–60. Champaign, Ill.: Weed Science Society of America.

Wrucke, M. A., and W. E. Arnold. 1985. Weed species distribution as influenced by tillage and herbicides. *Weed Sci.* 33:853–856.

5 HERBICIDES AND THE SOIL

■ INTRODUCTION

All soils share a number of common characteristics, namely: (1) they are four-phase systems (solid, liquid, gaseous, living); (2) they are open systems to which substances may be added or lost; (3) all have profiles, some with more distinct horizons than others; (4) the basic soil entities are three-dimensional, having length, width, and depth; and (5) local soils, as a rule, tend to merge with one another, with gradation the normal mode of change. A representative soil profile is illustrated in Figure 5.1.

To the weed scientist, soils are storehouses for (1) weed seeds and other plant propagules; (2) water for weed and crop seed germination and subsequent growth; (3) nutrients for plant growth; (3) organic and inorganic colloids that adsorb herbicide ions and molecules, making them available for ionic exchange and preventing their loss by leaching; and (4) microflora populations that interact with soil-applied herbicides, resulting in their degradation. Soils also serve as the medium in which the roots of weed and crop plants develop, providing support and anchorage for the plants and sites of uptake for water, nutrients, and herbicides from the soil environment. Water percolating through the soil may incor-

porate a herbicide into the weed seed germination zone or carry it deeper into the soil profile.

Some herbicides are applied directly to the soil to control weeds in the seedling stage or at more mature stages. Such a practice is dependent for success on absorption of the herbicides by the weed seeds, developing seedlings, and/or underground vegetative parts. Herbicides inadvertently reach the soil when applied as a postemergence spray to emerged weeds, with that portion of the spray not intercepted by the weed foliage falling onto the soil. When a herbicide comes in contact with the soil, it is immediately subjected to a variety of processes common to soils—ones that will be discussed in this chapter.

Examples of herbicides applied to the soil prior to weed seedling emergence are alachlor (Cropstar, Lasso), clomazone (Command), EPTC (Eptam, Eradicane), metolachlor (Dual), and trifluralin (Treflan, Trilin). These herbicides are used primarily for the control of germinating grass weeds. Similarly, linuron (Lorox), imazaquin (Scepter), imazethapyr (Pursuit), and norflurazon (Zorial) are soil-applied to control germinating broadleaf weeds. To be effective, these herbicides are incorporated into the weed seed germination zone by shallow tillage and/or by percolating water.

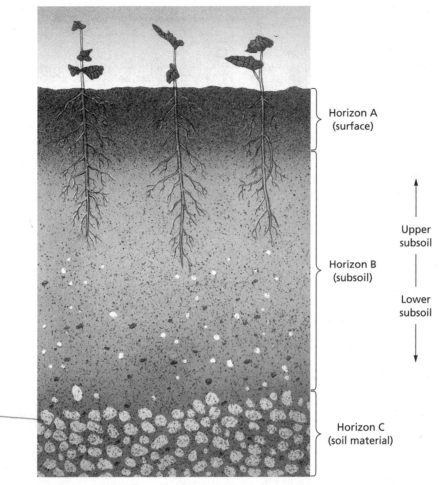

Horizon A
(surface)

Upper
subsoil

Horizon B
(subsoil)

Lower
subsoil

Horizon C
(soil material)

FIGURE 5.1 ■ A representative soil profile.   Horizon A is relatively high in organic matter and becomes the furrow slice when the land is plowed. Horizon B, while markedly weathered, usually contains very little organic matter. At variable depths, it *merges* gradually into the soil material designated as horizon C.

SOURCE: L. T. Lyon and H. O. Buckman. 1937. *The nature and properties of soils,* p. 3. New York: Macmillan. Copyright 1937 by Macmillan Publishing Co., Inc., renewed 1965 by Harry O. Buckman.

■ THE FATE OF HERBICIDES IN SOILS

The *fate of herbicides in soils* pertains to their interaction with the components of the soil. It is governed by three general processes that occur in soils: (1) *chemical,* (2) *physical,* and (3) *microbial.* These processes influence the availability and phytoxicity of soil-applied herbicides. Each of these encompasses specific interactions with soil components, as follows:

1. Chemical processes
   a. Adsorption
   b. Ionic exchange
   c. Photochemical decomposition
   d. Chemical reactions with soil constituents
   e. Absorption by plants and microorganisms

2. Physical processes
   a. Soil erosion by wind and water
   b. Leaching
   c. Volatility
   d. Vapor drift

3. Microbial decomposition

■ CHEMICAL PROCESSES

Adsorption

Adsorption in soils is the process whereby herbicide ions and molecules are held to the surface of soil colloids through the electrical attraction between themselves and the colloidal particles. It is a process similar to the attraction of iron filings to a magnet or lint to nylon fabric. *Adsorption is the most important factor by which herbicides become unavailable for uptake (absorption) by plants and microorganisms.* All soil-applied herbicides are adsorbed to some extent, and their herbicidal activity is reduced in direct proportion to the amount that is retained on the soil colloids by adsorption. Adsorbed herbicides are in a passive state and are unavailable to biological, physical, and chemical processes until desorption occurs.

The adsorption of herbicide ions and molecules occurs on the clay (inorganic) and humic (organic) colloidal fractions of soils. The cationic portion of herbicides that are salts is more strongly adsorbed to the clay colloids than to the humic colloids. Herbicides that are nonionic are more strongly adsorbed to humic colloids than to clay colloids.

In general, herbicides are more strongly adsorbed when applied to dry soils than to wet soils. Most adsorbed herbicides are readily displaced from their adsorptive sites on clay colloids by the competitive action of water molecules for these sites. They may also be displaced from these sites by ions or molecules of similar or stronger charge. Herbicides are not readily displaced by water from their adsorptive sites on humic colloids.

Humus (organic colloids) has a high adsorptive capacity for herbicides. Humus is considered to be the most important single factor contributing to herbicide adsorption in soils. In general, such adsorption is greatly enhanced by even small increases in the humic content of a soil.

Anionic herbicides are more strongly adsorbed in acid soils than in alkaline soils. In contrast, cationic herbicides are more strongly adsorbed in alkaline soils than in acid soils. However, this effect is only slight for nonionized herbicides in soils with a pH range of about 6–8 (the practical crop-production range).

In contrast to adsorption, the term *absorption* denotes the uptake, or surface penetration, of ions or molecules by, or into, any substance or organ such as the uptake of herbicides or nutrients by plant roots, or water by a sponge. *Desorption* refers to the release, or displacement, of adsorbed or absorbed ions or molecules from the substance to which they are adsorbed or absorbed. *Sorption* refers to either adsorption or absorption or both and does not specifically indicate one or the other.

Ionic Exchange

Ionic exchange accounts for the adsorption of herbicide ions or molecules (with an unsymmetrical electrical charge) to clay and humic colloids, which removes them from the soil solution. By the same token, it accounts for the release of these adsorbed ions and molecules from their adsorptive sites, thereby enabling them to move into the soil solution as free ions and molecules.

Photochemical Decomposition

Exposed to sunlight, some herbicides undergo molecular alterations that result in their deactivation; such alterations are caused by interaction with the ultraviolet portion of sunlight.

Under field conditions, photodecomposition is an important mechanism for deactivating certain herbicides left exposed on the soil surface (or upon plant foliage). Photochemical decomposition may be prevented by incorporating the herbicides into the soil soon after application, thereby shielding them from sunlight with a soil barrier.

Chemical Reactions with Soil Constituents

In contrast to biological reactions involving soil microorganisms in the deactivation of herbicides, reactions involving purely chemical systems in soils result in transformations that are not as well understood. This lack of understanding is due primarily to the difficulty in obtaining a sterile soil system (free of microorganisms) with which to work. However, it is known that herbicides do react with chemicals present in soils and that such reactions occur as the result of the following reactions:

1. Oxidation–reduction
2. Hydrolysis
3. Formation of water-insoluble salts
4. Formation of chemical complexes

Oxidation–reduction reactions involve the transfer of electrons from one reactant to another, resulting in the formation of electrically charged (ionized) compounds.

Hydrolysis involves the reaction of herbicide molecules with water, a reaction in which chemical bonds are broken and one or the other of the ions of water (H^+ or OH^-) becomes bonded to the herbicide molecule. Generally, in such a reaction, one or more atoms or groups of atoms on the herbicide are replaced by hydroxyl ions (OH^-) from the water. The resulting substitution(s) renders the molecules herbicidally inactive (nonphytotoxic).

Oxidation and hydrolytic reactions may occur separately or concurrently in soils. In soils of high calcium content, herbicides may react with the calcium ions to form *water-insoluble calcium salts*. Some herbicides tend to form *stable complexes* with metal ions in soils, ions such as cobalt, copper, iron, magnesium, and nickel.

Chemical reactions between herbicides and soil constituents tend to deactivate the herbicides.

Absorption by Plants and Microorganisms

The uptake of herbicides from soils by the underground parts of plants and by soil microorganisms results in their removal from the soil environment, at least temporarily. With certain herbicides (e.g., dicamba, and picloram), a portion of the absorbed herbicide's concentration may be circulated through the plant and returned (excreted) to the soil in an unchanged form by root exudation. The absorption of herbicides from soils by biological systems results in a reduction in their concentration in the soil. If the concentration is appreciably reduced in this manner, subsequent weed control will be poor.

■ PHYSICAL PROCESSES

The leaching and/or volatility of soil-applied herbicides result in their loss, to a greater or lesser extent, from soils. Wind and water erosion of soils also contribute to loss of the herbicides, chemically unchanged, from soils to which they were applied.

Leaching

Leaching is the movement of herbicides in soils as influenced by water flow. Leaching may occur in any direction in soils, depending on the direction of water movement. Due to the large volumes of percolating water in soils following heavy rains or irrigations, the most common direction in which herbicides are leached in soils is downward. It is the common assumption that when the leaching of a herbicide is referred to in an unqualified manner, *the direction of movement is downward*. In general, herbicides may be leached downward in soils to distances of less than 1 in. (2.5 cm) to more than 3 ft (0.9 m). Sidewise leaching is referred to as *lateral movement*, and upward leaching is referred to as *reverse leaching* or *upward movement*.

Leaching results in the physical movement of the ions and molecules of a herbicide dispersed in the soil solution; strongly adsorbed herbicide ions and molecules are not leached until displaced from their adsorptive sites and returned to the soil solution. The leaching of herbicides in soils may or may not be desirable.

Desirable attributes of leaching include (1) incorporation of a soil surface–applied herbicide, and (2) enhancing plant–herbicide interception.

Undesirable effects of leaching include (1) crop injury because the herbicide is transported into the absorption zone of susceptible crop plants; (2) poor weed control following partial or complete removal of a herbicide from the weed seed germination zone before seed germination occurs; (3) accumulation of a herbicide in the seed furrows in amounts toxic to otherwise tolerant crop plants; (4) contamination of groundwater; and (5) enhanced loss of a herbicide from soils by volatility following movement of the herbicide to the soil surface by reverse-leaching.

Factors Directly Influencing Leaching

The principal factors that influence herbicide leaching in soils are:

1. Soil texture
2. Soil permeability
3. Volume of water flow
4. Adsorption of herbicide to soil colloids
5. Water solubility of herbicide

Leaching is enhanced in loose-textured soils such as coarse, sandy loams. Reverse-leaching is enhanced by fine-textured soils such as clay loams. Leaching tends to increase as the volume of water moving past a given point increases. It is restricted in soils having water-impermeable layers, such as hardpan or caliche, by the location and continuity of such barriers. Water solubility of herbicides is only an indicator of their leachability. Herbicides that are water soluble and that remain dissolved in the soil solution are highly vulnerable to leaching. The leaching of water-soluble herbicides that are strongly adsorbed to soil colloids is greatly restricted, as are herbicides that react with chemicals in the soil to form water-insoluble compounds or complexes.

Herbicides that are relatively insoluble in water may be leached in soils. Molecules of such herbicides may remain suspended in the soil solution and be transported with the flow of water as suspended particles.

In general, adsorption is the most important factor influencing the leachability of herbicides in soils. Herbicides that are adsorbed to soil colloids do not leach unless the soil colloids themselves move with the flow of water. Adsorption may be overcome by desorptive processes such as ionic exchange. The desorbed herbicide ions or molecules move from their adsorptive sites into the soil solution, where they are vulnerable to leaching.

Factors Indirectly Influencing Leaching

Two factors in soils that indirectly influence herbicide leaching are (1) soil pH, and (2) soil colloids (inorganic and organic).

The effect of soil pH on herbicide ion and molecule leaching in soils lies primarily in its influence on ionic exchange and the adsorption of the herbicide ions and molecules to the soil colloids. Soil pH also affects chemical reactions between the herbicide ions and molecules and various soil constituents.

The influence of soil colloids on herbicide leaching in soils is primarily one of providing adsorptive sites for these chemicals, thereby removing them from the soil solution. Increases in the clay and humic colloidal content of a soil results in increased herbicide ion and molecule adsorption, accompanied by decreased leachability.

Volatility

Volatility is the process by which a substance passes from the solid or liquid state to the gaseous state.

The volatility of a herbicide becomes of practical importance when economic loss occurs. Economic loss from volatility occurs when (1) poor weed control results from significant reduction in herbicide concentration, and (2) severe crop injury results from the movement of herbicide vapors from the soil onto susceptible crop plants.

Herbicides that readily volatilize are lost much more rapidly from the surface of *moist soils* than from *dry soils.* Volatile herbicides present in the soil solution may be carried to the soil surface by reverse-leaching and lost to the atmosphere by volatilization. Examples of herbicides subject to volatility are butylate (Sutan+), EPTC (Eptam, Eradicane), and trifluralin (Treflan, Trilin).

Herbicide loss through volatility is greatly reduced (1) *by soil incorporation,* and (2) *by using special formulations such as granules.* When the volatile nature of a herbicide requires its immediate incorporation into the soil, it is best accomplished by mechanical means such as a power-driven tiller mounted behind, but a part of, the sprayer, or a separate tillage operation following immediately behind the sprayer.

Vapor Drift

The movement of herbicide vapors in the atmosphere is called *vapor drift.* The direction of this movement is governed by airflow moving in the same direction as the breezes or winds. In sufficient concentration, such vapors drifting onto susceptible plants may be highly injurious. Examples of such volatile herbicides are clomazone (Command) and 2,4-D esters. Vapor drift of these herbicides has resulted in severe economic loss. Cotton, grapes, and tomatoes are highly susceptible to vapor drift from 2,4-D

esters, and vapor drift of clomazone poses a hazard to many crop plants, other than a few tolerant crops such as soybeans, green peas, and pumpkins.

■ MICROBIAL DECOMPOSITION OF HERBICIDES

Microbial decomposition is the most important means by which herbicides are destroyed in soil. The agent of microbial decomposition is enzymatic regulation. Microbial decomposition occurs predominantly in the aerated upper 12-in. (30-cm) layer of the soil profile.

Organic herbicides are subject to attack and utilization by soil microorganisms in their search for nutrients and energy. The principal microorganisms in soils are *bacteria, fungi,* and *actinomycetes.* These organisms are capable of changing and breaking apart herbicide molecules, which results in their deactivation (nonphytotoxic).

Microbial decomposition is dependent on the catalytic action of specific enzymes produced by soil microorganisms. These enzymes govern herbicide degradation reactions. Such enzymes may be present within the microbes themselves, necessitating herbicide molecules to absorb the organisms, or the enzymes may be secreted or released from the microbes into the surrounding soil solution where they encounter and react with the herbicide molecules.

Enzymes are biochemical catalysts, complex compounds that accelerate specific biochemical reactions without themselves becoming part of the end products. Enzymatic reaction with herbicide molecules result in the following molecular alterations:

1. Dehalogenation
2. Dealkylation
3. Amide or ester hydrolysis
4. Beta-oxidation
5. Ring hydroxylation
6. Ring cleavage
7. Reduction of nitro groups (NO_2) under anaerobic conditions

Soil factors that affect microbial metabolism and multiplication also favor herbicide decomposition; such factors are:

1. Favorable moisture
2. Aeration
3. Mild temperature
4. Acidity–alkalinity (pH)
5. Organic-matter content

In general, the optimal conditions for microbial activity are (1) soil moisture at 50–100% of field capacity; (2) well-aerated soil; (3) temperatures of 80–90°F (27–32°C); (4) soil pH of 6.5–8; and (5) even slight increases in organic-matter content.

■ MICROBIAL POPULATIONS

Soil applications of herbicides result in an adjustment of the microbial population in the soil. When microorganisms capable of herbicide degradation are already present in the treated soil, there is an initial lag period of possibly a week or two in which the herbicide concentration is not appreciably affected by the microbes. During this period, however, the microbes are rapidly multiplying as they utilize the herbicide molecules as an energy source. When their population is sufficiently large, their attack significantly reduces the concentration of the herbicide in the soil, and the activity of the herbicide is impaired. When the concentration falls below that required for herbicidal activity, the herbicide, for all practical purposes, has been inactivated.

As this food source is depleted, the large, select microbial population dies back. In the event that additional applications of the herbicide are made while the microbial population is at its peak, the added herbicide will be decomposed much more rapidly than in the initial application. The resulting degree of weed control will be poor, unless a compensating higher dosage is applied. If an additional application is made after the microbial population has died back, there will again be a lag period before herbicide degradation becomes significant.

The importance of the microbial degradation of herbicides is especially apparent in a monoculture system such as corn following corn, where the same herbicides may be used repeatedly. For example, EPTC (Eradicane) is used preplant-incorporated in corn for control of annual grass and broadleaf weeds. Following application, the herbicide is attacked almost immediately by bacteria in the soil and, due to the now plentiful "food" source, the bacterial population increases rapidly, accompanied by a reduction of EPTC in the soil. To overcome this adverse effect, the amount of EPTC applied would have to be increased or, alternatively, its molecules protected from bacterial attack. The first choice offers a temporary solution, but the bacterial population would continue its rapid expansion while the food supply was plentiful. The second choice is more desirable and is now possible with the development of Eradicane Extra, which contains EPTC and the *herbicide extender* "dietholate." Dietholate protects EPTC from mi-

crobial attack in the soil long enough for the herbicide to carry out its intended purpose.

When microorganisms capable of decomposing a particular herbicide are not present in the herbicide-treated soil, microbial populations that can cause the herbicide's degradation apparently develop from mutants of one or more of the microbial species present. Such mutants result from genetic changes in the parent microbe; the progeny possess the characteristics of the mutant parent.

Decomposition of the herbicide molecules and the by-products of this initial degradation continues until they are eventually reduced to carbon dioxide, water, and basic elements, or become part of the humus.

■ HERBICIDE PERSISTENCE AND RESIDUES IN SOILS

Herbicides are applied to soils for the control of weeds under certain conditions such as prior to their emergence from the soil and for absorption by underground vegetative parts of established perennials. For best results, the herbicides must remain in the soil in an active (phytotoxic) and available form until this purpose is achieved. However, phytotoxicity is only desirable up to the time that the herbicides have accomplished their intended purpose; longer persistence in the soil poses a hazard to subsequent land use and is therefore undesirable.

Soil persistence or *soil-residual life* is defined as the length of time that a herbicide remains active in the soil. *Herbicide residue* refers to the amount of herbicide persisting in a soil after its mission has been accomplished, even though it may not be present in phytotoxic concentrations or be immediately available for plant absorption. In general, the decomposition by-products may be considered to be residues of the herbicide, and they may or may not pose a problem to subsequent land use.

The soil persistence of a herbicide may pose a hazard when:

1. Crop failure necessitates replanting.
2. A herbicide-susceptible crop follows a short-term crop treated with a long-term (persistent) herbicide.
3. A susceptible crop follows in rotation on land treated with a herbicide that persists longer than 1 year.
4. A persistent herbicide applied in the fall is followed by a susceptible crop the following year because of a market-induced change in plans.
5. Herbicide decomposition proceeds slowly due to adverse weather conditions that contribute to its persistence in the soil.

Herbicide residues in soils are undesirable primarily because (1) they may result in injury to sensitive crops in a crop-rotation system; (2) they may be absorbed by rotational crops and accumulated in unlawful amounts in plant parts harvested for food (seeds, fruit, vegetation); and (3) they may accumulate in excessive amounts in groundwater (water tables) and water supplies due to percolation and runoff.

Each year, injury occurs when susceptible crops are planted, or replanted, in land previously treated with a soil-applied herbicide recommended for an initial crop that failed due to hail, frost, or heavy rains. In other cases, crop injury occurs to plantings that follow the application of a persistent herbicide recommended for the previous crop but that is incompatible with the current crop. To minimize such losses, chemical companies have included suggestions as to which crops may be safely planted in soil treated with their herbicide and what delays are appropriate before such a planting can be safely made.

The soil-residual life of a herbicide may be minimized by:

1. Applying the lowest dosage of the herbicide consistent with obtaining the desired effect.

2. Applying a herbicide mixture resulting in the desired effect with a smaller amount of the persistent herbicide.

3. Applying the herbicide in a band directly over or between crop rows, rather than broadcasting it, to reduce the total amount of herbicide used.

4. Applying the herbicide as early as is feasible in the warm part of the growing season, thereby allowing more time for microbial decomposition.

5. Cultivating the land to encourage microbial decomposition of the herbicide.

6. Soil-incorporating large quantities of activated charcoal to adsorb the herbicide (however, this may only be feasible on small areas).

■ SELECTED REFERENCES

Helling, C. S., P. C. Kearney, and M. Alexander. 1971. Behavior of pesticides in soils. *Adv. Agron.* 23:147–240.

Loux, M. M., and K. D. Reese. 1992. Effect of soil pH on adsorption and persistence of imazaquin. *Weed Technol.* 40:490–496.

Nye, P. H. 1979. Diffusion of ions and uncharged solutes in soils and soil clays. *Adv. Agron.* 31:225–271.

Parfitt, R. L. 1978. Anion adsorption by soils and soil materials. *Adv. Agron.* 30:1–50.

Roeth, F. W. 1986. Enhanced herbicide degradation in soil with repeat applications. *Rev. Weed Sci.* 2:45–65.

Simon-Sylvestre, G., and J. C. Fournier. 1979. Effects of pesticides on the soil microflora. *Adv. Agron.* 31:1–93.

Stougaard, R. N., P. J. Shea, and A. R. Martin. 1990. Effect of soil type and pH on adsorption, mobility, and efficacy of imazaquin and imazathapyr. *Weed Sci.* 38:67–73.

Walker, A. 1987. Herbicide persistence in soil. *Rev. Weed Sci.* 3:1–17.

Wilson, J. A., and C. L. Foy. 1992. Influence of various soil properties on the adsorption and desorption of ICI-0051 in five soils. *Weed Technol.* 6:583–586.

6 ENTRY AND MOVEMENT OF HERBICIDES IN PLANTS

■ INTRODUCTION

Herbicides are applied to the soil and/or the foliage of plants. Thus herbicides enter plants via their underground and/or aerial vegetative parts. The manner in which a herbicide is applied is dependent on the characteristics of the herbicide and those of the plant species involved. Certain herbicides are only effective when absorbed by underground plant parts, others only when absorbed by aerial plant parts, and still others are effective regardless of site of entry.

The primary sites of entry for soil-applied herbicides are the roots and developing shoots of plants. Herbicides are also absorbed from the soil by seeds, rhizomes, bulbs, corms, and tubers. The primary site of entry for herbicides applied to aerial plant parts is the foliage.

An understanding of the possible sites of herbicide entry is vital to: (1) herbicide selection, (2) timely herbicide application, and (3) proper herbicide placement.

Following entry into plants, a herbicide must move from site of entry to site of activity (toxicity). The movement of herbicides in plants from their site of entry to other locations within the plants is called *translocation*. Herbicides that are translocated in plants are said to be *mobile*, whereas those that are not translocated are *immobile*.

This chapter is concerned with (1) the sites of herbicide entry in plants, and (2) the movement of herbicides from the exterior surface of plants to their sites of activity within the plants.

■ SOIL-APPLIED HERBICIDES

Roots

To be effective, some herbicides must be applied directly to the soil. They are subsequently moved into the upper 1- to 4-in. (2.5- to 10-cm) soil layer by cultivating tools or leaching. For such herbicides, the roots of most broad-leaved (dicot) plants are a major site of entry. However, not all plant species readily absorb a given herbicide via their roots, nor are all herbicides readily root-absorbed by a given plant species. With some soil-applied herbicides, root absorption by grass plants plays a minor role in herbicide effectiveness, a situation to be discussed later in this chapter.

Herbicides may penetrate the walls of epidermal and cortical cells of roots by mass flow, but absorption is not complete until these herbicides have penetrated the plasmalemma of these cells and been released in the cytoplasm.

In a sense, herbicides that have penetrated the "apparent free space" (i.e., the cell walls and intercellular spaces) of roots have "entered" the root, but entry only this far is of doubtful significance, except as an initial step in reaching the symplast.

Herbicides that enter the apparent free space of roots but fail to penetrate the plasmalemma are barred from further movement into the plant via the apoplast pathway (cell walls, intercellular spaces) by a waxy barrier (the Casparian strip) present in the cell walls of the *endodermis*. Such herbicides may accumulate in the apparent free space of the roots or freely move back into the soil solution. In either case, since they fail to reach the symplast, they do not alter the normal processes of the plants—unless they disrupt the organization or permeability of the plasmalemma itself. Thus the barriers to herbicide root absorption are (1) the Casparian strip in the endodermis, and (2) the plasmalemma of epidermal and corticular cells.

Examples of herbicides that are readily absorbed by roots of broadleaved plants include the growth regulator–type herbicides, dinitroanilines, imidazolinones, sulfonylureas, triazines, and ureas. All of these herbicides are readily translocated from the roots to the leaves except for the dinitroanilines, which are translocated little, if at all, and 2,4-D (a growth regulator–type), which is translocated slightly. When the acid form of 2,4-D is root-absorbed, it tends to react with sodium ions present in the cytoplasm of root cells to form the insoluble sodium salt of 2,4-D, an immobile salt that accumulates in the root cells.

In broadleaved plants, the roots are the principal site of absorption of soil-applied herbicides, and the hypocotyl and shoot are of much less importance. However, hypocotyl or shoot absorption may occur as these plant parts push their way toward the soil surface, resulting in death of the seedlings or in stunted or malformed plants. For example, the herbicide DCPA is not translocated in broadleaved seedlings following root absorption, but when absorbed by hypocotyls, translocation is primarily upward. In common purslane and *Galinsoga*, DCPA is freely translocated from the hypocotyls to the roots, as well as upward in the shoot.

Developing Grass Shoots

Prior to their emergence aboveground, the developing shoots of grass seedlings are important sites of absorption for certain soil-applied herbicides. The *coleoptilar node* and *crown node* of emerging grass seedlings are the most important sites of herbicide absorption, whereas the coleoptile, leaves and buds encased by the coleoptilar sheath, and the epicotyl are of lesser importance.

Depending on the grass species involved, the activity of shoot-absorbed herbicides may be much greater than that of the same herbicides that are root-absorbed. In such cases, the herbicides tend to remain mobile in the grass plants following shoot absorption, whereas those that are root-absorbed are immobilized in the roots. Thus *the difference between herbicide shoot absorption and root absorption by grass seedlings is that the herbicide bypasses translocation barriers present in the root.*

Examples of soil-applied herbicides that gain entry into grass seedlings via shoots moving toward the soil surface include DCPA, EPTC, and trifluralin. The absorption of DCPA via the developing shoot of grass seedlings is greater than its absorption via the roots when it is shoot-absorbed.

Seed Uptake

Soil-applied herbicides may be either adsorbed or absorbed or both by seeds present in the soil. This may occur prior to, or during, seed germination. When adsorbed, herbicide molecules remain on the outer surface of the seed coat; later they may be absorbed by the seedling as it emerges through the seed coat following germination. Absorbed herbicide molecules enter the seeds by mass flow (dispersed in imbibed water) or by diffusion.

Herbicide Interception

To be effective, soil-applied herbicides must come in contact with, and be absorbed by, underground plant parts. Such contact is made by the following processes: (1) *mass flow*, (2) *interception*, and (3) *diffusion*.

Mass flow is a passive process whereby the soil-applied herbicide molecules are transported in the soil solution as it moves toward or into underground plant parts.

Interception denotes contact between the soil-applied herbicide and underground plant parts (primarily the developing root tip) as the plant part grows through the soil.

Diffusion is a passive process in which the herbicide molecules move along a concentration gradient in the soil solution, or in the vapor phase via soil pores, to the surface of underground plant parts.

Of these three processes by which soil-applied herbicides come in contact with underground plant parts, *mass flow* is the most important for nonvolatile herbicides. *Diffusion* is perhaps the most important for volatile herbicides, especially in a dry soil. Once contact has been made, entry of the herbicide molecules into the cell walls and intercellular spaces is by *mass flow* or *passive diffusion* (nonmetabolic processes) and entry through the plasmalemma is by *active uptake* (a metabolic process).

■ FOLIAR-APPLIED HERBICIDES

Foliar-applied herbicides are commonly used to control weeds that have emerged from the soil. The leaves are the primary target, but stems and buds may also intercept the herbicides. Thus the sites of entry into plants for herbicides applied as foliar sprays include (1) *leaves,* (2) *stems,* and (3) *buds.*

Leaves

The *leaves* of plants are the most important site of entry for herbicides applied to aboveground plant parts. The upper leaf surface is the surface onto which herbicide sprays are commonly deposited. Due to a thinner cuticle and a greater density of stomates, the lower leaf surface is usually penetrated more rapidly by herbicides, but this surface is rarely targeted. To gain entry, herbicides deposited on leaf surfaces must

1. Penetrate the waxy cuticle covering epidermal cells.
2. Penetrate the cell wall of epidermal cells.
3. Penetrate the plasmalemma of leaf cells.
4. Be released into the cytoplasm within the cells.

Foliar absorption is not completed until the herbicide molecules or ions have been released into the cytoplasm following their passage through the plasmalemma. Foliar-absorbed herbicides penetrate the cuticle and cell walls of leaves by *diffusion.* They penetrate the plasmalemma and are released into the cytoplasm by active, energy-requiring processes.

Not all parts of the leaf surface serve as sites of herbicide absorption. *The primary sites of entry are associated with regions of high concentrations of ectodesmata such as the guard cells of the stomata, hairs and trichomes, and anticlinal epidermal cell walls.* Other important sites of entry include the modified epidermal cells overlying the leaf veins and openings through the leaf cuticle due to the cuticle being stretched by expanding cells below it or by cracks and tears in the cuticle (Figure 6.1).

The entry of herbicides into the leaf via the stomata opening is of minor importance. Both oil and aqueous solutions penetrate the cuticle slowly, each by separate pathways. Nonpolar molecules are thought to penetrate the cuticle more readily than polar molecules (oils are nonpolar; water and salts are polar).

Stems

With foliar sprays, herbicides generally come in contact with the *stems* of treated plants, and some herbicide penetration of the stems undoubtedly occurs. However, it is in the control of certain perennial plants, notably woody plants, that the stem becomes the primary target for treatment. Such treatments are commonly applied to the basal portions of the stems and trunks of woody plants and to their stumps following cutting. A suitable oil, diesel or burner oil, is often used as the carrier for the herbicide when treating woody plants in this manner. Many plants that are resistant to foliar applications of herbicides are susceptible to stem applications.

Buds

Active Cell division
Meristematic Tissue

The *buds* of plants are of primary importance as sites of entry for *contact-type* herbicides. When contact-type herbicides are applied as overall sprays to plants, it is essential that the buds be directly contacted by the herbicide; otherwise, the buds may continue to grow, flower, and produce seed if the stems of the plant survive the herbicide treatment. The buds are minimally important as sites of entry when applying a systemic (translocated) herbicide because the herbicide is translocated to the buds.

Factors Influencing Foliar Absorption

Herbicide foliar retention and subsequent penetration of the barriers to absorption in leaves are the most important factors influencing absorption of foliar-applied herbicides.

Herbicides that remain on the external surface of the leaf cuticle are of little or no practical value. Indeed, they may pose a chemical residue problem on harvested plant parts. Foliar retention of herbicides applied as foliar sprays to plants is affected by:

1. The inherent wettability of the leaf surface
2. The surface tension of the spray solution

The inherent wettability of the leaf surface varies among plant species; for example, the leaves of barley plants resist wetting with aqueous sprays, whereas those of mustard are readily wetted. The inherent wettability of leaf surfaces is affected by:

1. The waxiness and characteristic structure of the cuticle
2. The pubescence (hairiness) of the leaf surface

Leaf surfaces with thick, waxy cuticles, such as those of many mature leaves, and leaf surfaces that are densely pubescent tend to resist wetting by aqueous solutions. The surfaces of young leaves and those of moderately pubescent leaves tend to wet much more readily.

The surface tension of water is relatively high, tending to hold water droplets in the form of spheres on waxy

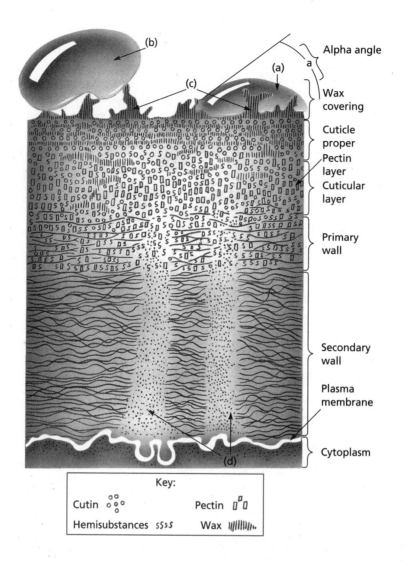

FIGURE 6.1 ■ Simplified scheme of the outer wall of an epidermal cell. (a) Water droplet with detergent, (b) without detergent, (c) wax rodlets, and (d) ectodesmata as nonplasmatic structures.

SOURCE: Redrawn from W. Franke. 1967. Mechanisms of foliar penetration of solutions. *Ann. Rev. Plant Physiol.* 18:284.

surfaces. Herbicides applied in aqueous solutions to plant foliage may not be retained by the foliage because of the high surface tension of the solution. The wettability of the aqueous solution can be greatly increased by reducing its surface tension; this is accomplished by adding an appropriate surfactant to the solution. Reduction of the surface tension results in more uniform spreading of the aqueous solution over the leaf surface and enhanced retention, or adherence, of the solution to this surface. The surface tension of oil solutions is relatively low, and oil sprays applied to plant foliage readily wet the leaf surfaces.

Surfactants (spray adjuvants), in addition to lowering the surface tension of aqueous solutions, enhance penetration of leaf surfaces. Enhanced penetration is the result of:

1. Elimination of air films between the spray solution and leaf surface

2. Increased area of contact between herbicide and leaf surface

3. Surfactants acting as cosolvents or stabilizing agents for the herbicide during cuticular penetration

4. Surfactants serving as humectants to retard the drying out of the aqueous spray solution on the leaf surface (A humectant promotes moisture retention, thereby increasing the length of time that a herbicide remains in solution and available for leaf absorption.)

5. Enhanced movement of herbicide via intercellular spaces

Environmental factors also influence foliar absorption of herbicides, and such factors include:

1. Light
2. Temperature
3. Humidity
4. Rainfall
5. Wind

Light enhances foliar absorption by actuating photosynthesis, which leads to a reduction in concentration of the leaf-absorbed herbicide within the leaf due to its molecules being transported from the leaf via the photosynthate stream, carried along with products of photosynthesis. However, some herbicides (e.g., the triazines and ureas) are not free to move in the photosynthate stream, and these remain entrapped in the leaves.

Atmospheric temperature may enhance or inhibit the effectiveness of foliar-applied herbicides. For example, temperatures of 60°F (16°C) or less tend to inactivate the herbicide bromoxynil when applied to the foliage of otherwise susceptible plants. Temperatures above 85° or 90°F (28° or 32°C) enhance the loss by volatility of the ester forms of certain foliar-applied herbicides, thereby reducing their effectiveness and increasing the hazard of vapor drift to nearby susceptible plants. High temperatures encourage the formation of a thicker, less permeable cuticle, whereas more moderate temperatures tend to promote foliar absorption. High temperatures may alter the nature of the cuticle, making it more permeable.

Low humidity adversely affects foliar absorption by causing aqueous spray droplets to dry (evaporate) rapidly, and it promotes water stress in the plant. *High humidity* favors foliar absorption of herbicides by delaying the evaporation of aqueous spray droplets and reducing water stress in the plant. Humidity affects hydration of the cuticle and polar absorption pathways.

Rain falling soon after, or during, foliar application of herbicides may wash residues from the foliage of treated plants. The importance of rainfall on foliar absorption varies among herbicides; some herbicides are rapidly absorbed, whereas others penetrate the leaf surface more slowly. In general, phytotoxic amounts of a herbicide are absorbed within the first hour following foliar application. The rate of entry of most foliar-applied herbicides is surprisingly fast. A time factor that pertains to rainfall and herbicide effectiveness is stated on some herbicide product labels. Herbicide formulation adjuvants—stickers, for example—reduce the loss of the herbicide from the foliar surface due to the washing effect of rainfall.

Wind adversely affects foliar absorption of herbicides by enhancing the evaporation of aqueous spray droplets and loss, by volatility, of herbicide residues from the leaf surface.

Foliar Absorption: The Process

Leaves absorb both organic and inorganic chemicals, which may have beneficial or harmful effects on the plant's welfare. Plant scientists have taken advantage of this ability by applying chemicals to the leaves of plants that enhance or modify plant growth to the benefit of the crop; such chemicals include plant nutrients, growth regulators, and herbicides. Air pollutants and spray drift of herbicides that induce abnormal plant growth or death are examples of the adverse effects of foliar-absorbed chemicals.

Organic chemicals, and certain inorganic ones, are excreted from leaf surfaces. Such excretions (exudates) may be viewed as "reverse absorption." These exudates may volatilize from leaf surfaces, or they may be washed from the surfaces by heavy dew, rainfall, or sprinkler irrigation. Leached into the soil, a portion of these exudates may be reabsorbed by roots of plants from which they originate or be absorbed by roots of neighboring plants. In plant communities where the foliage of the plants intermingle, the exudates may fall upon, and be absorbed by, the leaves of neighboring plants. Foliar exudates that are not lost from leaf surfaces are decomposed by microflora of the phyllosphere. Certain leaf exudates are involved in allelopathic interactions between plants.

The leaf presents barriers to the movement of chemicals inward from its outer surface, and not all chemicals can penetrate these barriers. Three such barriers are:

1. The cuticle
2. The cell walls
3. The plasmalemma

The general location of these barriers in the leaf is illustrated in Figure 6.1.

Cuticle

The cuticle is the first barrier to the absorption of a chemical resting on the outer surface of a leaf. The cuticle cannot be circumvented except via an occasional puncture or tear in its structure. Thus, for a foliar-applied herbicide to be absorbed, it must first penetrate the cuticle.

The cuticle is the outer waxy covering of the epidermal cells of the leaf. It is composed of units of 18 carbon atoms interconnected to form fatty acids and alcohols containing hydrophilic (—COOH and —OH) and lipophilic (—CH$_2$ and —CH$_3$) groups joined by ester bonds and peroxide bridges; together these form a coating of overlapping, pectin-cemented, lipoid platelets, insoluble in water and most organic solvents. Minute spaces exist throughout the cuticle, presumably large enough for the passage of small molecules.

The structure of the cuticle has been likened to a sponge in which the framework is of spongy cutin and the holes are filled with wax. The cutin forming the framework is elastic, and the waxy components between the cutin are inelastic. When a plant is under water stress, the cutin shrinks, pulling the waxy units closer together, thereby decreasing the permeability of the cuticle. In the case of plants that are turgid (not under water stress), the cutin imbibes water and swells, which results in spreading the waxy units further apart and increasing permeability of the cuticle to water-soluble chemicals. *Turgid plants absorb herbicides more readily than do plants that are under water stress.*

The thickness of the cuticle influences its penetration by chemicals. A thick cuticle is more resistant to penetration than a thinner cuticle. Mature leaves, having thicker cuticles, are more resistant to herbicide penetration than are younger leaves.

Foliar-absorbed chemicals move through the cuticle by diffusion. Diffusion is a physical process involving the movement of a chemical along a concentration gradient; the movement is from a higher to a lower concentration. Three general pathways along which chemicals may diffuse through the cuticle are as follows:

1. Penetration via intermolecular spaces
2. For water-soluble solutes, movement via water-filled and swollen pectin corridors between lipoid platelets
3. For oils and oil-soluble substances, movement directly through the waxy portions of the cuticle

Chemicals may also gain entry to the leaf through the cuticle via cracks, abrasions, punctures, or areas not completely covered by the waxy cuticle. However, such entry does not constitute passage through the cuticle *per se.*

Penetration of the cuticle by herbicides is influenced by their polarity—that is, the kind and strength of their electrical charge. All chemicals are either polar or nonpolar—either they do (polar), or do not (nonpolar) possess an electrical charge (negative or positive). Molecules with asymmetric force fields are polar, whereas those with symmetrical force fields are nonpolar. The degree of polarity between polar compounds varies depending on the strength of the asymmetric force fields. Ions are highly polar, as is water, whereas aliphatic (open-chain) hydrocarbons are nonpolar. The polarity of a molecule influences its solubility, as the solvents themselves vary in polarity. As a rule, "like dissolves like." Thus polar compounds are generally soluble in polar solvents, whereas nonpolar compounds tend to be soluble in nonpolar solvents. Ions are highly soluble in water but not in oils.

In general, herbicide molecules with a moderate degree of polarity penetrate the cuticle more readily than those that are either nonpolar or highly polar. A completely nonpolar molecule may accumulate in the waxy components of the cuticle and not pass through. A molecule that is too polar, with a high affinity for water and other polar substances, will not penetrate the cuticle readily. Undissociated molecules—that is, intact molecules—penetrate the cuticle more readily than do their ions, as penetration by the anionic (negatively charged) portion of the molecule, which is generally the herbicidally active portion, is impeded by the weakly acidic groups within the cuticle. As a rule, oils and oil-soluble herbicides readily penetrate the cuticle, whereas water and water-soluble compounds are absorbed more readily via the roots of plants than via the leaves.

Cell Wall

The cell wall is the second barrier to the absorption of chemicals by leaves. The cell wall is a finely interwoven network of cellulose strands of varying size and complexity, with interspaces ranging in size from 10 Å in the older primary walls up to 100 Å in the newer secondary walls. The interspaces within the cell wall commonly are filled with water, and the wall offers little resistance to the penetration of ions and molecules. Chemicals penetrate the cell wall by *diffusion,* the same process by which they penetrate the cuticle.

In general, cell walls have a high cation-exchange capacity. They readily imbibe water, and their hydrophilic properties are due to the hydroxyl (OH) groups of cellulose and the carboxyl (COOH) groups of pectin, and to both of these groups present on hemicellulose. These chemical groups facilitate the passage of water-soluble substances through the cell wall. The penetration of the cell wall by lipophilic substances may be more difficult than through the cuticle.

As with root cells, the cell walls and intercellular species in the leaf are considered *apparent free space* (solutes enter by free diffusion) and *Donnan free space* (ion exchange occurs against a concentration gradient without the expenditure of metabolic energy).

There is no sharp boundary between the cuticle and the outer cell wall of the epidermal cells of a leaf. Extensive movement of inorganic and organic chemicals through the cuticle and cell wall appears to occur in limited areas rather than over all, or most, of their surfaces. These areas of penetration are centered about very fine submicroscopic corridors extending through the walls of the epidermal cells up to, but not perforating, the cuticle. These submicroscopic channels are called *ectodesmata*. Under the electron microscope, the ectodesmata appear as fine strands (or channels) that may or may not be related to interfibrillar spaces within cell walls. They are different from plasmodesmata, although until recently they were thought to be the same. The ectodesmata appear to provide an almost direct pathway between the inner surface of the cuticle and the plasmalemma for chemicals moving through the cell wall (Figure 6.1).

Ectodesmata are found most abundantly in the walls of certain cells; these include the guard cells of the stomates; the epidermal cells surrounding trichomes and hairs on the leaf surface; the trichomes and hairs themselves; the anticlinal (slightly humped) cells of the epidermis; and the cells surrounding the vascular tissue (veins) within the leaf. In guard cells, the ectodesmata often show a typical distribution along the edges of the stomatal pores and the rear walls of the guard cells. *The sites in which ectodesmata are abundant are also the sites of extensive movement of chemicals through the cuticle and cell walls.*

Plasmalemma

The plasmalemma is the third and final barrier to the absorption of chemicals from the leaf surface. The plasmalemma is the outer plasma membrane of living cells, and it encloses the cytoplasm of cells. The plasmalemma is a semipermeable, bimolecular membrane about 75 Å thick composed of closely packed, globular lipoprotein molecules. The intermolecular spaces may be as large as 4 Å, and only water molecules (about 2 Å in size) may pass through without difficulty.

Ions and molecules that diffuse through the cuticle and cell wall are initially barred from reaching the cytoplasm within the cell by the plasmalemma barrier. However, some, if not all, of these ions and molecules may penetrate the plasmalemma by special energy-requiring processes, commonly referred to as active transport. The mechanisms of active transport appear to be the same for the cells of leaves as for those of roots. In general, the process of active absorption is as follows:

1. Ions and molecules are adsorbed to the outer surface of the plasmalemma.

2. Ions and molecules are transported through the plasmalemma by one, or more, of the following mechanisms:

   a. Change in membrane permeability due to enlargement of its intermolecular spaces.

   b. Invagination of the membrane, resulting in the formation of vesicles that are tied off and released into the cytoplasm—a process called *pinocytosis* (more easily envisioned as the inward bulging of the plasmalemma to form balloonlike vesicles enclosing the absorbed chemical); the vesicles seal off and separate from the membrane, leaving it intact, and move freely into the cytoplasm of the cell.

   c. Carrier molecules transport adsorbed ions or molecules through the plasmalemma, releasing them in the cytoplasm.

The mechanisms of active transport are not yet completely understood, but of the mechanisms proposed (Figure 6.2), the "carrier-transport" mechanism (Figure 6.3) appears most likely. The energy utilized in active transport in the leaf is derived from the following sources:

1. Oxidative phosphorylation in mitochondria
2. Photosynthetic phosphorylation in chloroplasts

The process of foliar absorption is complete when the chemical gaining entry has been released into the cytoplasm. Following absorption, this chemical may remain in the cytoplasm or in the vacuole of the cell into which it entered, or it may be translocated from the cell to other plant parts. Chemicals that are translocated cell to cell in plants move via cytoplasmic streaming in the symplast, an energy-consuming process.

Stomata

The guard cells surrounding the stomatal pores, rather than the stomatal cavities themselves, function as sites of herbicide absorption of foliar-applied herbicides. The importance of these sites to absorption is apparently due to the presence of numerous *ectodesmata* present in the walls of the guard cells, which enhance herbicide movement through the cell walls. However, the cuticular membrane covering the outer surface of the guard cells must first be penetrated.

Urea

The foliar absorption of *urea* is unique and worthy of special mention. Urea (NH_2CONH_2) is a common source

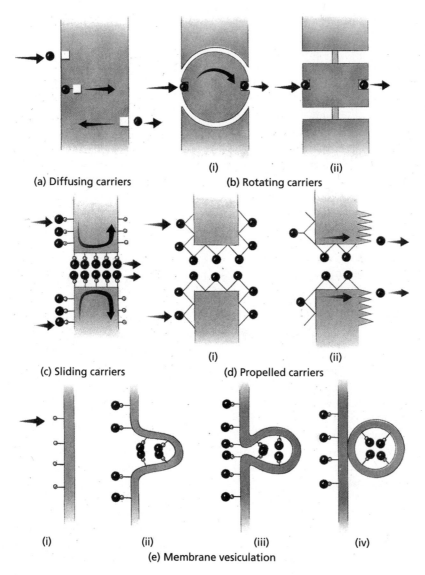

FIGURE 6.2 ■ Hypothetical mechanisms of ion transport across membranes. Ions are indicated by circles. (a) Diffusing carriers. (b) Rotating carriers. (c) Sliding carriers. (d) Propelled carriers, (e) Membrane vesiculation.

SOURCE: Redrawn from J. F. Sutcliffe. 1962. *Mineral salts absorption in plants,* p. 43. New York: Pergamon Press.

of nitrogen for plants, and when foliar-applied, it promotes the foliar absorption of other chemicals. The penetration of urea through leaf surfaces exceeds that of other nutrients by ten- to twentyfold, and it is independent of concentration. Apparently, penetration is due to some process other than simple diffusion. Urea enhances the absorption of other nutrients that are applied to the leaves at the same time. Enhancement of foliar absorption of anions by urea

occurs at the cuticular level, whereas that of cations occurs at both the cuticular and cellular levels.

The fertilizer ammonium sulfate enhances the absorption and activity of several postemergence-applied herbicides such as bentazon, dichlorprop, glyphosate, and sethoxydim. It has been proposed that (1) ammonium sulfate increases herbicide absorption by acting on plant membranes, and (2) the ammonium ion of ammonium sulfate acts to form the

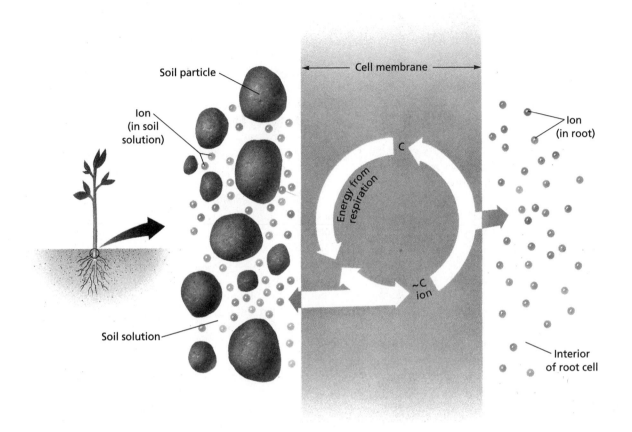

FIGURE 6.3 ■ **The carrier transport mechanism.** An illustration of how nutrient elements are thought to be transported from the soil solution into the root cell. An organic carrier C is specific for a given ion or group of ions. It can be energized through the process of respiration. In this state (~C), it will bind ions of a specific element and transport them across the membrane. At the inner surface of the membrane, the carrier releases the ion into the interior of the root cell.

SOURCE: Adapted from J. B. Hanson. 1967. Roots—selectors of plant nutrients. *Plant Food Review,* Spring 1967 Issue, p. 8.

ammonium salt of weak acid herbicides. The ammonium salt form of these herbicides is then readily absorbed by the plants.

Spray Adjuvants   *Surfactants*

The addition of adjuvants such as surfactants or crop oil concentrates to herbicide spray solutions increases foliar absorption and the activity of the herbicide. These adjuvants increase wetting and retention of the spray droplets on plant foliage. Surfactants that act as humectants increase herbicide penetration time by prolonging the drying period of spray droplets. In general, the maximum reduction in surface tension of a spray solution occurs in the surfactant concentration range of 0.01–0.1% v/v. Indications are that enhanced membrane permeability and herbicide toxicity take place at much higher concentrations than needed for wetting and retention. Thus, for field applications, the usual recommended rate for surfactants or crop oils added to aqueous spray mixtures is 0.25% v/v (1 qt/100 gal spray mixture) for nonionic surfactants and 1.0% v/v (4 qt/100 gal spray mixture) for crop oil concentrates. However, too high a surfactant concentration can kill the plant tissue underlying the cuticle, thereby blocking herbicide absorption and translocation and reducing herbicide activity. In addition, some surfactants may antagonize herbicide activity.

Carrier

The majority of herbicide applications use water as the carrier. Factors that influence the water carrier may also affect the performance of the applied herbicide. These factors are (1) *pH of the spray solution,* (2) *water hardness,* and (3) *spray droplet size.*

The pH of the spray solution primarily affects the leaf cuticle and dissociation of herbicide molecules that are weak acids. Herbicides that are weak acids and that are applied in spray solutions of low pH generally penetrate the cuticle best because their molecules are dissociated or protonated.

The activity of some herbicides may be adversely affected by the water hardness of the carrier due to interactions between the herbicide molecules and the inorganic ions present in the carrier. Hard water is alkaline in nature, and the activity of herbicides that are weak acids such as glyphosate may be reduced, apparently due to the formation of complexes between the herbicide molecules and the Ca^{2+}, Mg^{3+}, and Al^{3+} cations in the hard water.

In general, small droplet size of aqueous spray solutions increase spray retention and phytotoxicity of postemergence herbicides. However, leaf wettability varies among plant species. Applied to easy-to-wet leaves, droplet size has little or no effect on retention. Applied to hard-to-wet leaves, greater retention of the spray solution is obtained with small droplet sizes. However, *the smaller the spray droplet size, the greater the potential for spray drift.*

Oils as Carriers

Sometimes oils (diesel, paraffinic, and oils of plant origin) are used as carriers for postemergence herbicide applications such as low-volume applications and basal treatment of woody plants. Advantages attributed to the use of oil as the carrier for herbicide applications are (1) *increased foliar retention,* (2) *increased foliar absorption,* and (3) *reduced spray drift.* When using oil as the carrier, care must be taken to select a nonphytotoxic oil; otherwise absorption and activity of the herbicide may be blocked.

Liquid Fertilizers as Carriers

Liquid nitrogen fertilizers (28-0-0; 32-0-0) are recommended as the carrier for certain postemergence herbicides, and their use has resulted in enhanced phytotoxicity of the spray mixture.

■ TRANSLOCATION: THE MOVEMENT OF HERBICIDES IN PLANTS

Following their entry into plants, herbicide ions and molecules are immediately subjected to the influence of inter-actions with plant constituents and processes (physical, chemical, and biochemical) that may affect their movement in plants. In order to be phytotoxic, herbicide ions and molecules must reach their sites of toxicity within the plant following their absorption.

The movement in plants of herbicide ions and molecules from their sites of entry to other locations within plants is called *translocation.* Herbicide ions and molecules that are active (toxic) at sites near their point of entry undergo short-distance transport to reach these sites. Those that are active at sites more distant from their point of entry undergo long-distance transport.

Herbicide transport in plants is a dynamic system, not restricted to movement in only one transport pathway. A soil-applied herbicide (e.g., triazines and ureas) may be root-absorbed and translocated in the xylem via the transpiration stream to the leaves where it will then be retained. A foliar-applied herbicide (e.g., 2,4-D, dicamba, picloram, glyphosate) may be leaf-absorbed and translocated in the phloem via the photosynthate stream to underground plant parts. A root-absorbed herbicide (e.g., dicamba, picloram) may be transported in the xylem to the leaves, cross-over to the phloem, and be retranslocated to the roots. Conversely, a foliar-absorbed herbicide (e.g., dicamba, glyphosate, picloram) may be transported in the phloem to the roots, cross-over to the xylem, and be retranslocated to the leaves via the xylem. Thus some herbicides may actually circulate in the plant via the xylem and phloem conduits. Some herbicides (e.g., dicamba and glyphosate) that move in an apex-to-base direction may be secreted from the roots into the soil solution where they are then reabsorbed via the roots of the plants that secreted them or by adjacent plants and then translocated upward via the transpiration stream. Transport of a herbicide from a mature leaf to a young expanding leaf may occur exclusively in the phloem, or it may occur initially in the phloem, then continue in the xylem by transfer and translocation.

The translocation of herbicide ions and molecules in plants may be described as (1) *intracellular,* (2) *extracellular,* and (3) *intercellular.*

Intracellular translocation is the movement of the ions and molecules within an individual cell. It involves short-distance transport.

Extracellular transport is the movement of ions and molecules in the cuticle and the apoplast (cell walls and intercellular spaces). It primarily involves short-distance transport.

Intercellular translocation is the movement of ions and molecules from cell to cell via the symplast (the pathway of interconnected protoplasms of adjoining cells). This involves short-distance transport. *Intercellular transport* also takes place via the xylem and phloem tissues, conduits for the transpiration and photosynthate streams, respectively. This movement occurs between, but also within, cells. It involves long-distance transport.

Translocation and Sugar Movement

The translocation of herbicides in plants varies greatly among plant species, being influenced in part by growing conditions and plant age. The movement of some herbicides such as 2,4-D and MCPA is closely linked with the stage of leaf development and the movement of sugars from these leaves. In this regard, the following observations are of practical importance:

1. Very young leaves do not export sugars, and mobile herbicides absorbed by such leaves tend to remain in these leaves.

2. The movement of sugars from somewhat more mature leaves is toward the apex (tip) of the stem to which these leaves are attached, and mobile herbicides are carried along with the sugars from these leaves to active meristem regions.

3. The movement of sugars from lower, more mature leaves on the plant is predominantly downward toward the roots. From these leaves, herbicides are most likely to be transported with the sugars to the underground storage organs, an important factor in the control of perennial weeds.

4. From leaves of intermediate maturity, the movement of sugars is about equal in an upward and downward direction.

5. Sugars accumulate in developing fruit and seeds, and herbicides carried along with the sugars also accumulate in the fruit and seeds.

6. Depending on the stage of growth of the plant, the movement of sugars within the plant is at times predominantly upward into the growing and developing plant parts; at other times, movement is predominantly downward to the roots and other underground storage organs.

7. Whenever the transport of sugars within the plant is restricted, movement of herbicides via the photosynthate stream is likewise restricted. Such restriction of movement occurs in dormant plants and in plants exposed to low light intensity or darkness.

Transport Pathways

There are six solute-transport pathways by which herbicides may move in plants. In addition, the pathways may be categorized as *short-distance* and *long-distance* solute-transport pathways.

The short-distance transport pathways are:

- *Apoplast*—movement within cell walls and intercellular spaces
- *Symplast*—cell-to-cell cytoplasmic movement via plasmodesmata
- *Intracellular*—movement within an individual cell
- *Membranes*—penetration of plasmalemma, tonoplast, and organelle membranes

The long-distance transport pathways are:

- *Xylem conduits*—movement via xylem vessels and tracheids
- *Phloem conduits*—movement via phloem sieve tubes and companion cells

Short-Distance Transport Pathways

Apoplast

The *apoplast* is composed of (1) a network of connected (glued together) cells walls, and (2) any voids (intercellular spaces) between adjacent cell walls. The apoplast extends throughout the plant, forming a continuum from near the root tips to the upper extremities of the shoots, except for the blockage provided by the Casparian strip located in the walls of the endodermal cells separating the root cortex from the stele. The apoplast is a short-distance, nonliving, solute-transport pathway. Although nonliving, the cuticle and xylem conduits are not here considered part of the apoplast, but these plant components are in intimate contact with the apoplast. The apoplast is adjacent, but external, to the symplast. The external boundary of the apoplast is the inner surface of the cuticle, whereas its internal boundary is the outer surface of the plasmalemmas of the symplast.

The apoplast is normally filled with water, and it is permeable to water and its solutes. Solutes, including herbicide ions and molecules, that enter the symplast, xylem, and phloem transport pathways must first traverse the apoplast, even if only for the distance across one cell wall.

The basic mechanisms for solute transport in the apoplast are *diffusion* and, at times, *mass flow;* such movement is passive, requiring no metabolic energy.

Symplast

The *symplast* consists of the network of plasmodesmata-connected cytoplasms of living cells. The phloem conduits and their companion cells, although living, are not here considered part of the symplast transport pathway *per se.* The symplast is primarily a short-distance, living solute-transport pathway. The symplast extends throughout the plant, forming a continuum from the root tips to the shoot extremities. It is in intimate contact with, and surrounded by, the apoplast. The external boundary of the symplast is the inner surface of the plasmalemmas, and its internal boundary is the outer surface of the tonoplast.

The cell-to-cell movement of solutes follows the symplast transport pathway via the plasmodesmata-connected cytoplasms, and such movement is normally concerned with the distribution of solutes to assimilation sites and the subsequent movement of the assimilates to sites of utilization and storage, primarily over short distances.

The basic mechanisms responsible for solute transport within the symplast is *cytoplasmic streaming* and, to a lesser extent, *diffusion,* with the plasmodesmata serving as the cytoplasmic pathway through adjacent cell walls. Cytoplasmic streaming is an active process, requiring expenditure of metabolic energy. Diffusion is a passive process and does not utilize metabolic energy.

Intracellular

The *intracellular movement* of solutes occurs primarily within the cytoplasm of individual cells. It is a short-distance transport pathway. The cytoplasm is the "living" portion of cells, and it is in continuous motion, partly owing to forces that originate at or near the cell membranes and partly owing to brownian movement resulting from thermal agitation.

The intracellular transport of solutes involves:

1. Movement of transient solutes utilizing the cell's cytoplasm merely as a pathway, as part of the symplast, to more distant sites of utilization and storage

2. Movement of solutes that enter the cell's cytoplasm via the plasmodesmata or plasmalemma and move to an active site within the cell itself (Such sites of activity may be the chloroplasts, mitochondria, nucleus, or one or more of the various organelle membranes.)

3. Movement of solutes from the cytoplasm, across the tonoplast, into the vacuole

4. Movement of solutes from the vacuole, across the tonoplast, into the cytoplasm

5. Export from the cell of solutes, originating from sites of assimilation or storage within the cell, via the plasmodesmata or plasmalemma

The basic mechanisms responsible for intracellular solute transport are *cytoplasmic streaming, thermal diffusion,* and the complex *Golgi apparatus,* which controls the distribution of cellular material. Except for thermal diffusion, a passive process, intracellular transport of solutes involves active processes, requiring expenditure of metabolic energy. Solutes that remain indefinitely within a cell may be retained in the cytoplasm or stored in the vacuole.

Membranes

The cytoplasm is enclosed by the plasmalemma and each organelle within the cytoplasm is enclosed by its own membrane. Cellular membranes are about 75 Å thick; thus solute transport across these membranes involves short distances. The site of toxicity of herbicide ions and molecules may be within the membrane itself or within the organelle *per se.* For example, the chloroplast membrane is the site of photosynthesis, and the mitochondrial membrane is the site of the synthesis of adenosine triphosphate (ATP)—both are sites of herbicide activity.

Penetration of membranes occurs either by *diffusion* or by *active uptake* (carrier-mediated or facilitated transport). Almost all herbicides penetrate membranes by diffusion. The herbicides 2,4-D and glyphosate apparently penetrate the plasmalemma by active uptake, but they may also penetrate by diffusion.

Long-Distance Transport Pathways

There are two long-distance transport pathways in plants; they are (1) the *xylem* and (2) the *phloem.* Movement of solutes (including herbicide ions and molecules) from the roots or other underground plant parts to the leaves and upper extremities of the shoots of plants indicates xylem transport, whereas movement from the leaves and upper plant extremities toward the roots and other underground plant parts indicates phloem transport.

Xylem

The *xylem conduits* are the principal water-conducting tissue of vascular plants, and they are the "pipeline" for the transpiration stream. The xylem conduits comprise a long-distance, nonliving solute-transport pathway in plants. They consist of two kinds of conducting cells: (1) *vessel members,* and (2) *tracheids.* The vessel members and tracheids are competition. Sunflower, grown as a seed crop, produced maximum yields when kept weed-free during the first 4–6 weeks after planting. Cotton seed yields were not affected by barnyardgrass when kept weed-free for 9 weeks after planting; kept weed-free for 6 weeks, cotton seed yields were reduced 15% by subsequent competition with barnyardgrass. Common purslane is most competitive with beans and table beets during the first 2 weeks following crop emergence. Shattercane did not reduce soybean yields if the young plants were removed by 6 weeks after crop emergence. Johnsongrass significantly reduced corn grain yield when allowed to grow with the corn during the first 6 weeks after planting. Wild poinsettia is most competitive with peanuts during the first 10 weeks after crop emergence.

In general, soybeans kept weed-free for the first 4 weeks after planting showed little yield loss from later-emerging weeds. Weeds that emerge after soybeans cause less competition, since established soybeans compete well with weeds. The critical duration of giant ragweed competition in soybeans was between 4 and 6 weeks after crop emergence. Full-season competition of giant ragweed at densities of 117,400 plants/A (290,000/ha) resulted in almost complete soybean yield loss. Soybeans kept free of sicklepod competition for 4 weeks after emergence resulted in yields equivalent to season-long control. The critical johnsongrass-free period for soybeans was approximately 4 weeks after planting.

In some instances, weeds emerging along with the crop plants have little or no adverse effect on crop yields unless they are allowed to remain in the crop longer than the first 2–3 weeks after planting. For example, peanut yields were not reduced by weeds growing along with the crop plants during the first 3 weeks after planting, but yields were reduced if weeds were not removed from the crop prior to the fourth week and the crop kept weed-free thereafter. Yields of grain sorghum were reduced when pigweed competed with the sorghum during the third and fourth weeks after crop emergence, but grain yields were not reduced if the pigweed was present only during the first 2 weeks after crop emergence or did not emerge until 4 weeks after crop emergence.

Weed Competitiveness

In general, among annual weeds, broadleaved weeds are more competitive than are grass weeds. For example, velvetleaf was twice as competitive in soybeans as were either green foxtail or yellow foxtail, compared on the basis of the weight of mature plants. Redroot pigweed was more competitive in sugar beets than was green foxtail. White mustard was nine times more competitive in field peas than

was foxtail millet; after season-long competition, 3 white mustard plants/ft^2 of cropland reduced pea yields by 58%—the same reduction caused by 27 foxtail millet plants/ft^2.

Competitiveness varies among grass species. For example, among foxtail species, giant foxtail was more competitive with soybeans than were either green or yellow foxtails. This difference in competitiveness was attributed to differences in the characteristic size among species, with giant foxtail plants much taller than those of green or yellow foxtails.

Competitiveness also varies among broadleaved weed species. In Saskatchewan, Canada, wild mustard competed strongly with both wheat and cow cockle, reducing wheat grain yields by 38% and the dry weight and seed production of cow cockle by about 50%. Tall morningglory was more competitive in cotton than was sicklepod. Sicklepod at densities of 4, 8, 16, and 32 plants/15 m of row reduced cotton seed yields by 21, 23, 42, and 55%, respectively; tall morningglory at the same densities reduced cotton seed yields by 19, 41, 64, and 88%, respectively.

Weed Densities

In general, crop yield decreases as weed density increases. For example, the 3-year averages of soybean yields were reduced 10, 28, 43, and 52% by full-season competition with common cocklebur at densities of 3300, 6600, 13,000, and 26,000 plants/ha, respectively. Giant ragweed at a density of 2 plants/9 m of row reduced soybean seed yields by 48%. Corn cockle at a density of 340 plants/m^2 reduced wheat grain yields by 60%. One kochia plant/7.6 m of row reduced the average yield of sugar beet roots by 5.9 metric tons and sucrose yields by 1.07 metric tons/ha. Broadleaf signalgrass at densities of 8, 16, and 1050 plants/10 m of row reduced peanut seed yields by 14, 28, and 69%, respectively. Perennial sowthistle, at a density of about 90 shoots/m^2 reduced dry edible bean (kidney bean) and soybean yields by 83 and 87%, respectively. Spring wheat yield was not reduced by wild oat densities of 64 or 118 plants/m^2 until the wild oat reached the 5- and 7-leaf stage, respectively. Removing wild oat plants prior to the 7-leaf stage or the 5-leaf stage at densities of 64 or 118 plants/m^2, respectively, did not increase wheat culm or fresh weight production. The influence of density of selected weed populations on crop yields is shown in Tables 1.3 and 1.4.

Root Elongation

Differences in root elongation, accompanied by an increase in potential for water and nutrient absorption, contribute to

Animals

Animals are often unwitting carriers of weed seeds whose dispersal units are equipped with barbs, bristles, and hairs that catch in the animals' fur or hair, or of seeds that have sticky or mucilaginous seed coats that adhere to passing animals. Wild animals and livestock may carry attached weed seeds long distances before they are shaken, scraped, or shed from the fur or hair to which they were attached.

Seed-dispersal units possessing hooks or bristles include common cocklebur, field sandbur, spanishneedles, and tall beggarticks. The seed-dispersal unit for puncturevine has a sturdy, sharp projection that can become imbedded in the feet of animals or the soft soles of people's shoes.

Seeds with sticky or mucilaginous seed coats include the plantains and small mistletoe. Seeds with viscid or gluey seed coats may be dispersed by adhering to dry leaves or other vegetation being blown about by the wind.

Animals carry viable seed from one place to another in their digestive tracts, excreting the seed in their feces during their meanderings as they graze or move along migratory trails. Seeds of mesquite have been distributed by cattle in this manner over millions of acres of grazing land in the southwestern United States. Migratory birds, which fly long distances in a single day, are responsible for distributing weed seeds to distant areas. Birds roosting along fence lines excrete viable seeds in their droppings. Other animals, and even insects, may carry weed seeds short distances.

Forceful Dehiscence

Forceful dehiscence of seedpods or capsules disperse seed by ejecting or propelling the seeds for distances of a few feet to as much as 20 ft (6 m) or more (Figure 1.3). Species of *Impatiens* and *Oxalis* disperse their seeds in this manner, as do many other weed species. Woodsorrel is a common weed of greenhouses and, when all is quiet, a person working alone can often be startled by the forceful dehiscence of its seedpods and the subsequent rattle of the seeds as they fall about the greenhouse.

Artificial Dissemination

We humans distribute weed seeds by means of our transport system, which includes wagons, trucks, trains, airplanes, and by farm machinery, especially harvesting equipment, moving within a field and from field to field. However, *the principal means by which we introduce weeds into our croplands is by planting crop seed contaminated with weed seeds.*

In Utah, a survey of 1232 samples of grain seed collected directly from drill boxes on the seeder at planting time showed that over half of the samples contained weed seeds and that seeds of wild oat were present in 36% of the contaminated samples. The worst sample collected contained seed of five noxious weeds, with a total of 167,000 noxious weed seeds/100 lb (45 kg) of grain seed. In this instance, an average of 54 noxious weed seeds were being planted in each 10.8 ft^2 (1 m^2) of the grain field, assuming uniform distribution.

Weed seeds and vegetative propagules of perennial weeds may be carried in the mud or soil adhering to the wheels or bodies of farm equipment. It is not uncommon to observe weed species new to an area springing up along roadsides where their seeds have fallen or bounced from the beds of trucks hauling hay, feed, or livestock. Weed seeds are often carried from place to place in trouser cuffs or attached to socks or other clothing.

Perennial weeds are spread near and far by their vegetative propagules hidden in the soil of potted or balled ornamental plants shipped to new areas for planting. Perennial weeds, such as bermudagrass, johnsongrass, purple nutsedge, and quackgrass, are often dispersed in this manner.

Spread of Perennial Weeds

Perennial weeds can spread and propagate in cultivated lands by seeds or asexual means or by both. The dragging of asexual plant parts (roots, rhizomes, tubers) by tillage equipment is a common means by which perennial weeds are spread within a field, as well as from field to field.

Roots of a single field bindweed can spread laterally about 2.5 ft (0.76 m) and penetrate to a soil depth of about 4 ft (1.2 m) during the first season of growth. By the end of the third season of growth, a single plant of field bindweed is capable of spreading throughout an area of about 18 ft in diameter and penetrating to a depth of 18–20 ft (5–6 m). Roots of a single plant of Canada thistle may spread laterally throughout an area 20 ft in diameter during the first growing season. Although roots of Canada thistle may penetrate as deep as 20 ft in some soils, most are found in the upper 15-in. layer of soil. Both field bindweed and Canada thistle develop many new shoots from adventitious buds originating on their extensive root systems, thereby giving rise to dense stands or colonies of these weeds.

Purple nutsedge and yellow nutsedge propagate asexually from tubers. Yellow nutsedge, but not purple nutsedge, may also propagate from seed. Planted in soil at 12-in. (38-cm) intervals in Mississippi, tubers of purple nutsedge

FIGURE 1.3 ■ Forceful dehiscence of the seedpod of the squirting cucumber.

SOURCE: R. M. Holman and W. W. Robbins. 1939. *General botany*, p. 295. New York: John Wiley & Sons, Inc.

produced the equivalent of 3.1 million plants/A (7.7 million/ha) and 4.4 million tubers and bulbs/A (10.9 million/ha) during the first season of growth. A single tuber of yellow nutsedge produced 146 tubers and bulbs within a 14-week period after planting, infesting an area 6.5 ft (2 m) in diameter, equivalent to a yield of 8.9 tons/A (22 tons/ha) of new tubers. For comparison, the average yield of potatoes in the United States is about 12 tons/A (30 tons/ha). During the first year of growth, this single tuber of yellow nutsedge was ultimately responsible for the development of 1918 plants and 6864 tubers. None of the tubers was located deeper than 12 in., and most were in the upper 8-in. soil layer.

While many perennial weeds spread primarily by asexual means, others spread primarily by seed—for example, dandelion and curly dock. Some perennial weeds spread by both asexual and sexual means. Johnsongrass is an excellent example of a perennial weed that spreads by both means. A single 14-week-old plant of johnsongrass may already have produced as much as 85 ft (26 m) of rhizomes, with a total capacity of developing 200–300 ft (60–90 m) of rhizomes during a full growing season. A single johnsongrass plant may produce as many as 80,000 seeds, many of which are viable, during a full growing season. Bermudagrass is another perennial weed that spreads by both asexual and sexual means. It spreads asexually by the rapid growth of both rhizomes and stolons. A stand of bermudagrass may produce 4–6 million seeds/A (10–15 million/ha). Bermudagrass seeds have good viability and can lie dormant in soil. They survive immersion in water for as long as 50 days.

■ WEED–CROP COMPETITION

To the agriculturalist, the economic aspect of weeds growing among crop plants is of primary importance. Weeds growing in croplands are, like the crop plants themselves, merely trying to grow and perpetuate themselves, drawing upon the soil and air for these needed essentials. Unfortunately for the farmer, weeds obtain these essentials at the expense of the adjacent crop plants.

Competition occurs between two or more neighboring plants when the supply of one or more factors essential to growth and development falls below the combined demands of the plants. Successful competition between plants occurs with the disproportionate acquisition of one or more growth factors by one plant that proves detrimental to another's growth.

Some weed and crop species enhance their competitiveness by the production of phytotoxic or growth-inhibiting substances that adversely affect the growth and development of other plants. These biochemicals are released into the soil as root exudates or as leachates of the living or dead plants. The resulting biochemical interaction between plants is called *allelopathy*.

The term *interference,* widely used in the weed science literature today, is an all-inclusive term that denotes all the direct effects that one plant might impose upon another, such as competition, allelopathy, parasitism, and indirect effects (usually unknown) without referring to any one effect in particular. The vast majority of weed research reported in the literature is concerned with weed–crop competition and allelopathy, even though the term "interference" appears in the title of the research paper. There has been little or no research on parasitism and on the "indirect" (unknown) effects of interference.

Since both weeds and crops are plants, they basically have the same requirements for normal growth, development, and reproduction. They require and compete for an adequate supply of the same nutrients, water, light, heat energy (temperature), carbon dioxide, and growing space.

In an intermixed community of weed and crop plants, the more aggressive plants usually dominate. Aggressiveness is favored by greater root elongation and branching, resulting in a vigorous, rapidly spreading root system that absorbs water, nutrients, and oxygen from the soil at the expense of adjacent plants. Aggressiveness is also favored by taller plant species that grow more quickly than adjacent plants or by plants that climb their neighbors as vines, producing foliar canopies that shade shorter or slower growing plants in the community. Thus weeds successfully compete with crop plants by (1) being more aggressive in

growth habit; (2) obtaining and utilizing the essentials of growth, development, and reproduction at the expense of the crop plants; and (3) in some cases, secreting chemicals that adversely affect the growth and development of crop plants.

Reduction in crop yields is an accepted parameter for determining weed–crop competition. *Crop quality* is most adversely affected when green, moist vegetation and the reproductive parts of weeds are harvested along with the crop.

Key factors in crop-yield reduction resulting from weed–crop competition are as follows: weed species present; weed–crop emergence; competition duration; weed life cycle and growth habit; density of weed and crop plants, crop species and cultivars and their life cycles and growth habits; crop planting date, depth, and row spacing; and edaphic, climatic, and other environmental factors.

The Critical Weed-Free Period

The *critical weed-free period* is that period during crop production in which weeds are most likely to reduce crop growth and yield, a time during which weed-control efforts must be maintained to prevent loss in crop yield. To prevent this reduction, the crop must be kept weed-free during this period of development. The significance of the critical weed-free period is that, by the end of this period, the crop can successfully compete with later-emerging weeds, primarily by shading the soil surface.

Weed competition tends to have the greatest adverse effect on *crop yields* of summer annual crops when weeds are allowed to compete with the crop during the first 4–6 weeks or so after crop planting.

Weed emergence in relation to crop emergence is an important factor in weed–crop competition. Weeds that emerge prior to crop emergence may reduce crop yields more than weeds that emerge later. For example, white mustard seedlings emerging 3 days *before* field peas and allowed to compete season-long reduced the fresh weight of pea vines by 54%; when they emerged 4 days *after* the peas and were allowed to compete season-long, vine weights were reduced 17%. The adverse effect of white mustard on the growth of field peas was attributed to competition for moisture and nutrients.

Weeds that emerge along with crop plants also have an adverse effect on crop yields. For example, giant foxtail had no adverse effect on yields of corn or soybeans when it emerged 3 weeks after the crop, but it caused a grain-yield reduction of 13 and 27%, respectively, following season-long

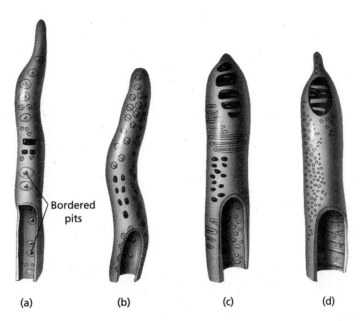

FIGURE 6.4 ■ **Tracheids and vessel members for secondary wood.**   (a) Tracheid from spring wood of white pine (*Pinus*). (b) Tip of tracheid from wood of oak (*Quercus*). (c) Tip of vessel member from wood of *Magnolia*. (d) Tip of vessel member from wood of basswood (*Tilia*).

SOURCE: W. W. Robbins, T. E. Weier, and C. R. Stocking. 1965. *Botany,* 3rd ed., p. 102. New York: John Wiley & Sons, Inc. Redrawn with permission of John Wiley & Sons, Inc.

highly specialized, and they are without cytoplasm at maturity. At maturity, they form dead, hollow, lignified, thick-walled, pipelike conduits through which large volumes of water and solutes move from the roots to upper plant parts, especially to the leaves.

The vessel members differ from the tracheids in that both their lateral and end walls are perforated with large, open pores, whereas the walls of tracheids contain only pits and pit membranes (Figure 6.4). The flow of water and solutes through the tracheids is more impeded than through the vessel elements. The highly evolved angiosperms contain both tracheids and vessel elements, whereas gymnosperms and lower vascular plants have only tracheids.

The xylem tissue is located in close proximity to the phloem tissue, and the two tissues together form "vascular bundles" and are referred to as the *vascular tissue*. The xylem conduction elements, upon maturity, have no direct continuity with the symplast or with the phloem sieve-tube elements. However, there is a limited exchange of solutes between the elements of the xylem conduits and the adjacent apoplast, symplast, and phloem sieve-tube elements. Transport of solutes in the xylem occurs only in a base-to-apex direction, with some lateral leakage from the system.

The basic mechanisms responsible for solute transport via xylem conduits are *mass flow* and *root pressure*. Mass flow is the mechanism by which solutes are transported via the transpiration stream. Under certain conditions, as when transpiration is not taking place or is greatly reduced and soil moisture is abundant, the humidity is high, and the light intensity is low, water is forced upward through the xylem conduits via *root pressure*, a phenomenon in roots resulting from an increase in the solute osmosis or electroosmosis, or both, near the inner face of the Casparian strip. *Guttation* results from conditions conducive to root-pressure flow. The process of guttation can be described as the upward flow of xylem sap (water and solutes) in the xylem conduits due to a positive, aqueous root pressure and the subsequent exudation of droplets of the sap from xylem conduit terminals at the leaf margins (the hydathodes).

Herbicide ions and molecules that move to the xylem are transported via the transpiration stream in the xylem conduits, moving from roots and other underground plant parts primarily to the leaves.

Phloem

The *phloem conduits* are the principal food-conducting tissue of vascular plants. They are the "pipeline" for the photosynthate stream. The phloem conduits provide

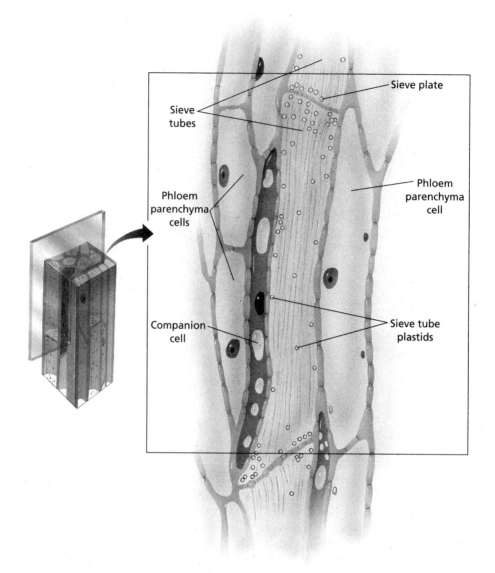

FIGURE 6.5 ■ Phloem tissue from the stem of *Nicotiana tabacum* showing sieve tubes, companion cells, and phloem parenchyma cells.

SOURCE: R. M. Holman and W. W. Robbins. 1945. *General botany,* p. 90. New York: John Wiley & Sons, Inc. Redrawn with permission of John Wiley & Sons, Inc.

a long-distance, living solute-transport pathway in plants. They are composed primarily of the sieve-tube elements, possibly with help from sieve-tube companion cells (Figure 6.5). The sieve-tube companion cells and adjacent phloem parenchyma cells apparently serve to channel solutes from the surrounding symplast, and possibly from the apoplast as well, into the cytoplasm of adjacent phloem sieve-tube elements via plasmodesmata. They also transfer solutes from the sieve-tube elements to the surrounding symplast.

Herbicide ions and molecules that move in the phloem are transported in the photosynthate stream, moving from leaves to regions of new growth (e.g., apical meristems in roots and stems; lateral buds in roots, rhizome, and stems; and fruits) and to storage areas (e.g., cortical cells in roots).

The terms *source* and *sink* occur in the scientific literature relative to translocation of herbicides in plants. Expanded leaves, mature stem tissue, and mature rhizomes, tubers, and storage roots function as *sources*. Meristem tissues and immature or developing leaves, developing seeds, tubers, rhizomes, and roots function as *sinks*. As a particular tissue develops, its designation as source or sink may change. Some tissues (e.g., roots of perennial weeds) may go through several transitions between source and sink during the plant's life cycle. Phloem transport moves in a source-to-sink direction.

Translocation of solutes in the phloem occurs in three major steps:

1. Loading of solutes (e.g., sugars and herbicide ions and molecules) into the phloem at the source
2. Transport of solutes via the photosynthate stream in the phloem conduits from the source to the sink
3. Unloading of solutes from the phloem into the sink tissue

Loading and unloading of solutes into and out of the phloem are under metabolic control, whereas solute transport in the phloem is by mass flow.

A prerequisite for translocation of herbicides in the phloem is that they enter the symplast and be retained long enough to be transferred to the phloem-conducting elements. Herbicides approach the symplast either by diffusion through the leaf cuticle and underlying cell walls or by a combination of diffusion and mass flow from the soil into the root and underground stem tissue. Penetration of the plasmalemma and entry into the symplast is primarily by active uptake. Herbicides that are more lipophilic penetrate the plasmalemma more rapidly than do less lipophilic herbicides.

Neutral, lipophilic herbicides readily penetrate the plasmalemma, but they do not accumulate in the cell unless they are bound or metabolized to more polar products. Such herbicides readily move out of the cells of the symplast. Consequently, translocation of these herbicides occur more in the xylem than in the phloem.

Herbicides that are weak acids can readily penetrate the plasmalemma into the symplast in the unassociated state; they then tend to deprotonate in the cytoplasm, due to the higher pH inside the cell. The dissociated weak acids are less able to recross the plasmalemma, and the herbicide accumulates in the cell. Other herbicides slowly penetrate the plasmalemma and are retained there as a result of the lower permeability of the plasmalemma to the absorbed form of the herbicide. One consequence of the cellular retention of herbicides that are weak acids or that exhibit intermediate permeability is that they are more likely to be transferred to the phloem and translocated through the plant with photosynthates.

Herbicides that have entered the symplast may be retained there by two mechanisms: (1) dissociation of weak acids in the cytoplasm (the dissociated forms are less lipophilic and less able to diffuse back through the plasmalemma than undissociated acids), and (2) the relatively low permeability of the plasmalemma to certain herbicides that do not behave as weak acids (such herbicides permeate the symplast slowly but leave it even more slowly). The conversion of ester forms of herbicides to their acidic forms results in accumulation of the anionic form in the symplast and, subsequently, in their translocation in the phloem. Thus dissociation of herbicides that are weak acids and increased lipophilicity of herbicides allow some herbicides to enter, accumulate, and/or be retained in the symplast; these characteristics also confer phloem mobility on the herbicides.

Herbicides that become less lipophilic when transferring from the apoplast to the symplast are more likely to be retained in the symplast and less likely to be translocated in the phloem. Those that become more lipophilic are more phloem-mobile.

The desired consequence of herbicide retention in the symplast is that the herbicide is located where it can be transferred to the phloem, along with photosynthates, and subsequently transported to other plant parts. However, retention in the symplast does not necessarily result in phloem mobility of a herbicide, since retention may be due to binding to cellular constituents (e.g., bromoxynil, a weak acid), or partitioning into lipid fractions or metabolism to more polar products (e.g., oryzalin and atrazine).

Herbicide activity that may disrupt phloem transport includes: (1) blocking photosynthesis directly or indirectly, thereby preventing production and subsequent movement of assimilates from *source* leaves; (2) disruption of plasmalemma structure or function; and (3) herbicide leakage from the phloem into the surrounding tissue, resulting in damage to the tissue and reduction in phloem transport.

Environmental Conditions

Humidity and Temperature

For foliar-applied herbicides, high relative humidity usually results in increased foliar absorption and greater translocation in the phloem. Increased translocation as influenced by high humidity may be due to increased CO_2 uptake and consequently increased assimilate production and subsequent translocation, a function independent of increased foliar absorption. Greater absorption and translocation of fluroxypyr (a pyridine-based herbicide structurally similar to picloram) occurred in leafy spurge plants exposed to high relative humidity (greater than 90%) compared to low humidity (less than 30%) for 6 hours or longer after treatment. Temperature had no effect on fluroxypyr absorption and translocation from leafy spurge leaves to the roots. Applied to pitted morningglory, foliar absorption of imazethapyr was greater when the relative humidity was 100% than when it was 40% (absorption rates of the herbicide were 88 and 47%, respectively), as was translocation (34 and 17%, respectively).

Greater herbicide translocation in the phloem conduits usually occurs at higher temperatures, whereas low temperatures usually reduce phloem translocation. Foliar absorption of picloram increased in leafy spurge by 1% for each 1°C increase in air temperature 24 hours *before* treatment. Conversely, foliar absorption decreased 1% by each 1°C decrease in air temperature *after* treatment. Translocation of picloram to the roots of leafy spurge declined as temperature increased. However, all plant species do not respond in the same way to high and low temperatures, resulting in variations in phloem translocation depending on plant species. Exposure to temperatures below 32°F (0°C) does not necessarily prevent translocation unless it results in lasting damage to the tissue.

Herbicide–Herbicide Interaction

Simultaneous application of two or more herbicides may result in reduced translocation of one of the herbicides compared to its translocation when applied alone. Glyphosate, applied alone, is rapidly foliar-absorbed and is translocated extensively in plants. However, absorption, translocation, and control of grasses (e.g., johnsongrass) by glyphosate was reduced by the addition of 2,4-D and dicamba to the spray solution. Haloxyfop-methyl toxicity to johnsongrass was reduced when applied in tank mix with 2,4-D. The reduced toxicity was due to reduced absorption, translocation, and greater metabolism of haloxyfop-methyl in the treated leaves. When haloxyfop-methyl was applied 24 hours after the application of 2,4-D, control of johnsongrass was equal to that from haloxyfop-methyl applied alone. When applied in tank mix, atrazine significantly reduced the translocation and activity of primisulfuron when applied to the leaves of velvetleaf and giant foxtail. Normally, primisulfuron is translocated upward in velvetleaf from treated leaves and downward in the leaves of giant foxtail.

■ SELECTED REFERENCES

Devine, M. D. 1989. Phloem translocation of herbicides. *Rev. Weed Sci.* 4:191–214.

Devine, M. D., H. D. Bestman, and W. H. Vanden Born. 1990. Physiological basis for the different phloem mobilities of chlorsulfuron and clopyralid. *Weed Sci.* 38:1–9.

Doohan, D. J., T. J. Monaco, and T. J. Sheets. 1992. Effect of field violet (*Viola arvensis*) growth stage on uptake, translocation, and metabolism of terbacil. *Weed Sci.* 40:180–183.

Fielding, R. J. and E. W. Stoller. 1990. Effects of additives on the efficacy, uptake, and translocation of the methyl ester of thifensulfuron. *Weed Sci.* 38:172–178.

Flint, J. L. and M. Barrett. 1989. Antagonism of glyphosate toxicity to johnsongrass (*Sorghum halepense*) by 2,4-D and dicamba. *Weed Sci.* 37:700–705.

Gerwick, B. C. 1988. Potential mechanisms for bentazon antagonism with haloxyfop. *Weed Sci.* 36:286–290.

Hart, S. E. and D. Penner. 1993. Atrazine reduces primisulfuron transport to meristems of giant foxtail (*Setaria faberi*) and velvetleaf (*Abutilon theophrasti*). *Weed Sci.* 41:28–33.

Hatzios, K. K. and D. Penner. 1985. Interactions of herbicides with other agrochemicals in higher plants. *Rev. Weed Sci.* 1:1–63.

Hull, H. M., D. G. Davis, and G. E. Stolzenberg. 1982. Action of adjuvants on plant surfaces. In *Adjuvants for herbicides*. Monograph, pp. 26–27. Champaign, Ill.: Weed Science Society of America.

Kent, L. M., G. D. Wills, and D. R. Shaw. 1991. Influence of ammonium sulfate, imazapyr, temperature, and relative humidity on the absorption and translocation of imazethapyr. *Weed Sci.* 39:412–416.

Lym, R. G. 1992. Fluroxypyr absorption and translocation in leafy spurge. *Weed Sci.* 40:101–105.

McWhorter, C. G. 1982. The use of adjuvants. In *Adjuvants for herbicides*. Monograph, pp. 10–25. Champaign, Ill.: Weed Science Society of America.

Mueller, T. C., M. Barrett, and W. W. Witt. 1990. A basis for the antagonistic effect of 2,4-D on haloxyfop-methyl toxicity to johnsongrass (*Sorghum halepense*). *Weed Sci.* 38:103–107.

Nishimoto, R. K., and G. F. Warren. 1971. Site of uptake, movement, and activity of DCPA. *Weed Sci.* 19:152–155.

Norris, R. F. 1982. Action and fate of adjuvants in plants. In *Adjuvants for herbicides*. Monograph, pp. 68–83. Champaign, Ill.: Weed Science Society of America.

Parr, J. F. 1982. Toxicity of adjuvants. In *Adjuvants for herbicides*. Monograph, pp. 93–114. Champaign, Ill.: Weed Science Society of America.

Smith, A. M. and W. H. Vanden Born. 1992. Ammonium sulfate increases efficacy of sethoxydim through increased absorption and translocation. *Weed Sci.* 40:351–358.

Wanamarta, G., and D. Penner. 1989. Foliar absorption of herbicides. *Rev. Weed Sci.* 4:215–231.

Wolf, T. M., et al. 1992. Effect of droplet size and herbicide concentration on absorption and translocation of ^{14}C-2,4-D in oriental mustard (*Sisymbrium orientale*). *Weed Sci.* 40:568–575.

7 MODES AND SITES OF ACTION OF HERBICIDES

■ INTRODUCTION

The mechanisms by which herbicides kill plants are called *modes of action*. The location at which herbicides exert their toxicity at the cellular level is called the *site of action*. In the last few years, great progress has been made in determining herbicide modes and sites of action, and this information is now known for many herbicides available today. However, the modes of action of some herbicides are still unknown—for example, the thiocarbamates, acid amides, and organic arsenicals.

A particular herbicide may interfere to a greater or lesser extent with more than one plant process, but the vital process first known to be blocked or disrupted, which results in the death of the plant, is considered the primary mode of action of that herbicide. One must not assume that all herbicides within a particular group possess the same mode of action, or that all plant species or varieties respond similarly to a given herbicide or group of herbicides.

Often all members of a herbicide family, due to similarities of molecular structure, exhibit the same mode and site of action as other members of the group. However, there are exceptions. For example, the mode of action of siduron differs from other urea herbicides; that of the phenylcarbamate herbicides differ from desmedipham and phenmedipham; and that of propanil and pronamide differ both from one another and from other acid amides. Different families of herbicides may exhibit the same mode of action as another family. For example, the ureas, triazines, and uracils block photosynthesis; the imidazolinones and sulfonylureas inhibit acetolactate synthase; and the aryloxyphenoxypropionates, thiocarbamates, and cyclohexanediones inhibit lipid formation. The bipyridiniums have the same mode of action (i.e., photosynthesis inhibition) as the triazines, ureas, and uracils, but their site of action is different.

At times, research redefines, or appears to redefine, the primary mode of action of a particular herbicide, as in the case of sethoxydim. Also, an increase in herbicide concentration above the minimum necessary to induce phytotoxicity of the herbicide under study may indicate a different mode of action, overshadowing the primary mode of action.

Remember, if all plants responded similarly to a given herbicide, its use in croplands would be of little practical value. It is this difference in response among plant species—called *selectivity*—that makes herbicides useful in agriculture.

■ MODES AND SITES OF ACTION

The known modes of action for herbicides in use today are as follows:

1. Acute toxicity
2. Chronic toxicity
3. Disruption of transport systems and interference with nucleic acid metabolism
4. Disruption of mitosis
5. Inhibition of carotenoid formation and inhibition of protoporphyrinogenoxidase
6. Inhibition of photosynthesis and inhibition of ATP formation
7. Inhibition of lipid formation
8. Inhibition of acetolactate synthase
9. Inhibition of the shikimic acid pathway and inhibition of δ-aminolevulinic acid synthesis
10. Inhibition of the biosynthesis of cellulose from glucose

Each of these modes of action will be discussed and their sites of activity noted in the following sections. At the end of each discussion, a list is provided of herbicides known to exhibit this particular mode of action as their primary mode.

Acute Toxicity

Contact-type herbicides kill by acute toxicity. Acute toxicity implies rapid kill, usually within minutes or a few hours after contact is made with the plant. In the case of some contact herbicides, such as sulfuric acid, acute toxicity may be described as *caustic* or *burning.* In the case of phytotoxic oils, the mode of action is the weakening and disorganization of cellular membranes accompanied by increased membrane permeability, which results in loss of cellular contents by leakage.

Herbicides of the contact type include the diphenyl ethers (acifluorfen, fomesafen, lactofen, oxyfluorfen), the bipyridiniums (diquat, paraquat), the methanearsonates (cacodylic acid, CAMA, DSMA, MSMA), sodium chlorate, sodium metaborate tetrahydrate, sulfuric acid, and phytotoxic oils.

Contact-type herbicides undergo short-distance transport, moving through the exterior waxy surface upon which they were deposited, through the cell walls to the plasmalemma, and, in some cases, through the plasmalemma and into the cytoplasm. Where the phytotoxicity of certain contact-type herbicides (e.g., diquat and paraquat) is dependent on light, such herbicides will translocate in the symplast if the treated plants are not exposed to light after treatment; after exposure to light, translocation ceases.

Certain herbicides may be referred to as "contact herbicides" on the basis of their rapid kill of treated plants, even though their primary mode of action is known to be different. For example, bromoxynil, diquat, and paraquat (which are inhibitors of photosynthesis), and the diphenyl ethers (which are inhibitors of protoporphyrinogen oxidase).

Chronic Toxicity

Systemic herbicides kill by chronic toxicity. Chronic toxicity implies "slow acting," with death of the plant occurring after a prolonged period of time (a few days or longer) following herbicide absorption, as with the aryloxyphenoxypropionate and cyclohexanedione herbicides. In some cases, the herbicidal effect may be apparent shortly (a few hours) after application, as with the growth regulator–type herbicides.

Systemic herbicides are translocated in plants from their site(s) of entry to their sites of phytotoxicity via (1) the *symplast,* (2) *transpiration stream,* (3) *photosynthate stream,* or by two or more of these pathways. Systemic herbicides may move from living cell to living cell, penetrating cell membranes and accumulating in toxic amounts at their sites of action without disrupting the living system while in transit.

Systemic herbicides may be absorbed, translocated, and their toxic effect induced in the same chemical form as when applied, or they may be absorbed in one molecular form, translocated in a modified form, while their phytotoxicity is induced in yet another molecular form. *On the molecular level, biological function is embedded in molecular structure.*

Disruption of Transport Systems and Interference with Nucleic Acid Metabolism

Disruption of Transport Systems

Herbicides that induce abnormal plant tissue development, such as leaf epinasty and malformations resulting from cell proliferation and enlargement, are generally referred to as *hormone-type* or *growth regulator–type herbicides.* The induced malformations caused by these herbicides can interfere with normal distribution of metabolites and nutrients within the plant. Growth regulator–type herbicides include dicamba, quinclorac, and members of the phenoxycarboxylic and pyridine herbicide families. The growth regulator–type herbicides are effective in controlling many broadleaf species. They do not effectively control grasses, but they may interfere with the normal root development of grasses.

As a group, the growth regulator–type herbicides are usually applied postemergence, but they may also be effective through the soil. They move systemically throughout the plant following absorption by roots and shoots. They induce similar symptoms in plants, but their precise modes and sites of action remain elusive. The following are general symptoms induced by these herbicides on susceptible broadleaf (dicot) plants, with those induced by the herbicide 2,4-D used as examples.

Following foliar application of 2,4-D, subsequent epinastic bending is rapid and is the first apparent symptom; mitosis ceases; cells stop elongating but continue radial expansion; stem and root tissues develop swollen, tumorous growths. Eventually, the xylem and phloem conduits become disoriented and the phloem conduits become clogged. Disorientation of the transport conduits leads to disruption of the movement of the transpiration and photosynthate streams, depriving the plant of assimilates needed for continued growth and development, which in turn results in the collapse and death of the plant.

Interference with Nucleic Acid Metabolism

In addition, 2,4-D appears to adversely affect numerous biochemical and metabolic processes in plants, many of which are undoubtedly of secondary and tertiary importance. However, the growth responses suggest that the adverse effects of 2,4-D on nucleic acid metabolism and metabolic aspects of cell wall plasticity are most relevant to the mode of action of 2,4-D. Although extensive research has been reported as to the activity of 2,4-D, its mode(s) and site(s) of action are still speculative. This is also true of the other hormone-type herbicides.

Of the herbicides included in this text, the following are grouped as growth regulator–type herbicides, with their exact modes and sites of action incompletely known:

Phenoxycarboxylics	Pyridines
2,4-D	Clopyralid
Dichlorprop	Picloram
2,4-DB	Triclopyr
MCPA	
MCPB	**No Subgroup**
Mecoprop (MCPP)	Dicamba
	Quinclorac

Disruption of Mitosis

Cell division is the process by which plant growth is made possible, and it is initiated and largely controlled by the nucleus of the cell. When cell division occurs, each daughter cell contains a duplicate of the varied structures present in the mother cell. The division of the nucleus itself is commonly called *mitosis*. However, the term *mitosis* is also used to denote the entire act of cell division, and it is used in this sense here.

Herbicides that disrupt mitosis as a primary mechanism of action can be grouped as those that

1. Cause arrested prometaphase configuration similar to colchicine
2. Disrupt spindle microtubule organization
3. Disrupt phragmoplast microtubule organization

Herbicides that arrest prometaphase configuration can be subgrouped as (1) those that result in complete microtubule depolymerization, such as the dinitroanilines, and (2) those with persistent, short kinetochore microtubule tufts, as with pronamide and dithiopyr. However, since the net effect of the herbicides is identical (i.e., no meaningful chromosome movement), they may be thought of as acting in the same manner. Herbicides in group 1 interact directly with tubulin (Vaughn and Lehnen, 1991).

Herbicides that affect mitosis as their primary mode of action appear to do so by interacting directly or indirectly with the *microtubules*. The various microtubules (kinetochore, spindle, phragmoplast) are the sites of action of the herbicides, depending on which site a specific herbicide interacts with.

Microtubules are contractile elements in the nucleus that control the movement of chromosomes during mitosis and meiosis. They attach to the *kinetochore* region of chromosomes and help pull them to the opposite poles during cell division. Microtubules are hollow cylinders with an external diameter of 15–25 nanometers (nm), a wall thickness of about 6 nm, and range in length up to 200 microns. They are primarily composed of the dimeric protein tubulin, which is composed of similar but distinct subunits of 55 kilodaltons[1] (kd) each. The microtubules of most species are made up of 13 protofilaments.

Microtubules act in groups rather than singularly. In higher plants, there are four arrays of microtubules: (1) cortical (interphase), (2) preprophase, (3) spindle, and (4) phragmoplast. Cortical microtubules are involved in cell shape, and spindle microtubules control chromosome movement to the poles. Microtubules apparently originate in the endoplasmic reticulum and the nuclear membrane. Microtubules and microfilaments are frequently found together in the cell, and they may work together in some processes such

---

1. The terms *dalton* and *molecular weight* are used interchangeably; for example, a 32-kd polypeptide has a molecular weight of 32,000.

as chromosome movement during mitosis. Microfilaments are thinner than microtubules, and they are involved in cytoplasmic streaming and possibly the positioning of organelles in the cell.

The dinitroanilines (e.g., trifluralin) are the largest group of herbicides that disrupt mitosis. They inhibit microtubule polymerization from free tubulin subunits, resulting in the loss of spindle and kinetochore microtubules. Because there are no spindle microtubules, chromosomes cannot move to the poles of the cell during mitosis. Although the exact molecular mechanism is unknown, it is assumed that the herbicides bind to the tubulin heterodimers in the cytoplasm, and as the herbicide–tubulin complex is added to the growing microtubule, further growth of the microtubule ceases. With depolymerization of the microtubules, the microtubules become shorter and shorter, eventually resulting in their complete loss. Without the cortical microtubules, cells cannot elongate, and they expand isodiametrically (square rather than rectangular in shape). Thus the swollen or club-shaped root tip results from the production of isodiametric cells following the loss of cortical microtubles. The prometaphase stage of mitosis is normal, but no metaphase or later mitotic stages occur.

Pronamide (an acid amide herbicide) and dithiopyr (a pyridine herbicide) also cause arrested or condensed prometaphase and root tip swelling, as do the dinitroanilines. However, pronamide treatment results in small tufts of microtubules at the kinetochore regions, rather than the complete absence of microtubules characteristic of the dinitroaniline herbicides. With these short kinetochore microtubules, no meaningful chromosome movement takes place, and the net effect is a cell arrested in prometaphase. Pronamide binds to tubulin and can prevent *in vitro* assembly of microtubules.

Dithiopyr causes much the same effects as pronamide, but it does not bind to tubulin. Rather, dithiopyr binds to a protein of about 65 kd, which is in the range of microtubule-associated proteins. Thus dithiopyr may interact with a microtubule-associated protein, resulting in shortened microtubules similar to those from pronamide treatment.

The discontinued phenylcarbamate herbicides barban, chloropropham, and chlorpham are examples of herbicides that disrupt mitosis in a mechanism different from herbicides that inhibit microtubule polymerization. These phenylcarbamate herbicides cause chromosome movement during anaphase toward three or more foci, rather than the two foci of a normal anaphase. After this multipolar division, the nuclear membranes re-form around the micronuclei, resulting in highly branched and oddly shaped phragmoplasts. The abnormal phragmoplasts organize ab-

normal and irregularly shaped cell walls. Apparently, these carbamate herbicides disrupt the spindle microtubule organizing centers, fragmenting them throughout the cell. The molecular site of action of these herbicides is not known.

The primary effect of DCPA (a phthalic acid herbicide) in disrupting mitosis seems to be as a disrupter of phragmoplast microtubule organization and production. The array of phragmoplast microtubules are abnormally oriented and are dispersed throughout the cytoplasm, rather than confined to the area set by the preprophase band as the division plane. The process blocked most effectively by DCPA is cell plate formation. In many cells, the newly formed cell plate is incomplete or misoriented so that the two nuclei are not separated at mitosis. The cell wall may contain loops or be abnormally thick in one part and thin in another. In the most severe cases, no wall is formed and tissue with multiple nuclei in a common cytoplasm are found. However, the precise mode of action of DCPA is not known.

Of the herbicides included in this text, the following ones disrupt mitosis as their primary mode of action:

Dinitroanilines	Phthalic Acids
Benefin	DCPA
Ethalfluralin	**Benzonitriles**
Oryzalin	Dichlobenil
Pendimethalin	**Acid Amides**
Prodiamine	Pronamide
Trifluralin	**Nonfamily**
	Asulam
	Dithiopyr

Inhibition of Carotenoid Formation and Protoporphyrinogen Oxidase

Carotenoids and chlorophyll are plant pigments and both are located in the chloroplasts. By definition, pigments are substances that strongly absorb visible light, and most pigments only absorb light in certain regions of the visible light spectrum. They transmit light of all other wavelengths and consequently they appear colored. Light passed through a chlorophyll solution appears green to the eye, but viewed from the same side as the directed (incident) light, the solution appears red, a phenomenon called *fluorescence*. Fluorescence is defined as the radiation of light of a different wavelength from that of the incident light.

Carotenoids are the yellow, ether-soluble pigments of plant cells, and they are almost always associated with chlorophyll. The carotenoids absorb light in the blue-wavelength portion of the visible spectrum, and they appear yellow (carotene), orange (xanthophyll), and red (lycopene) to the eye. Light is not necessary for carotenoid formation, as it is for chlorophyll formation, but the presence of light will increase the amount of carotenoids formed. Seedlings grown in darkness manufacture carotenoids but little or no chlorophyll.

Chlorophyll, a green pigment, is vital to the process of photosynthesis. It is present in plants as different isomers, designated *a, b, c,* and *d,* but the main isomers in most plants are chlorophyll *a* and chlorophyll *b.* Chlorophyll *a* is more abundant than chlorophyll *b* (in a ratio of 3:1), and chlorophyll *a* is capable of mediating photosynthesis in the absence of chlorophyll *b*.. Chlorophyll *a* and *b* complement one another in absorbing incident light. Light in the blue and red regions of the visible spectrum not absorbed by chorophyll *a* is absorbed by chlorophyll *b*.

Carotenoids and chlorophyll are localized in membranes (lamellae) within chloroplasts. The chloroplasts are located in the cytoplasm of green plant cells. They are made up of the following components:

1. An exterior, differentially permeable, double membrane
2. The *stroma,* a nongreen, liquid ground substance containing many of the enzymes involved in photosynthetic CO_2 fixation, and unstacked thylakoid membranes (or stroma lamellae)
3. Interior double membranes (lamellae) composed of
   a. Stroma lamellae: double membranes that are not stacked
   b. Grana lamellae: composed of stacked membranous discs called *thylakoids*

Cells of green plants that contain chloroplasts may have up to 80 chloroplasts per cell (each chloroplast is 2 by 5 microns in size). Each mature chloroplast may contain 40–60 grana, with an intricate thylakoid membrane system present in each grana.

Carotenoid and chlorophyll molecules are bonded to the same protein in the chloroplast structure, forming a complex known as *photosynthin.* The carotenoids are just as efficient harvesters of light as are the chlorophylls, and they play a part in photosynthesis by transferring "excited electrons" to chlorophyll *a.* In addition, they serve as "chemical buffers" to protect chlorophyll pigments and the chloroplasts themselves from photooxidation by singlet oxygen ($^1O_2$), a highly reactive species of oxygen.

Chlorophylls and carotenoids are formed by plant-specific pathways, and interruption of the biosynthesis of these pigments by herbicides results in "bleaching" of plant foliage and cells containing chloroplasts. Typically, bleaching occurs only in newly formed leaves; pigments present before herbicide treatment are not affected. The *site of action* of herbicides that inhibit carotenoid formation is the lamellae of chloroplasts.

Inhibition of Carotenoid Formation

The primary mode of action of herbicides that causes bleaching in plants is the inhibition of carotenoid formation and/or pigment destruction. (Sandmann et al., 1991, and Kunert and Boger, 1981, respectively). For example, inhibition of pigment biosynthesis dominates when using the phenyl pyridazinone herbicide norflurazon, while pigment destruction occurs when using the diphenyl ether herbicide oxyfluorfen. *Herbicides that inhibit carotenoid formation do so by inhibiting phytoene desaturase (PD),* a membrane-bound enzyme in the carotenogenic pathway (the site of action).

Diphenyl ether herbicides are multifunctional, with their modes of action dependent on substituents on their common molecular nucleus. They are light-dependent, fast-acting, contact, photobleaching herbicides. Their symptoms, rapid bleaching and dessication of green tissue, are similar to the effects of paraquat (a bipyridinium herbicide), and, although light is required for activity, photosynthesis is not (as with paraquat). Their most important activities are (1) inhibition of carotene biosynthesis and (2) inhibition of protoporphyrinogen oxidase.

Inhibition of Protoporphyrinogen Oxidase

The diphenyl ether herbicides acifluorfen, fomesafen, and lactofen inhibit protoporphyrinogen IX oxidase (Protox), the enzyme that converts protoporphyrinogen IX to protoporphyrin IX (Proto). This leads to uncontrolled autooxidation of the substrate, resulting in the accumulation of Proto. Proto is a potent photosensitizer that generates high levels of singlet oxygen in the presence of molecular oxygen and light, leading to light-induced oxidative breakdown of cell constituents. With acifluorfen, the herbicide itself is not the photodynamic dye nor the peroxidative agent. *The actual mode of action is the induction of the biosynthesis of the photodynamic compound Proto by the herbicide* (Duke et al., 1991).

At the ultrastructural level, the plasmalemma and tonoplast are generally the first membranes to break as a result of light peroxidation. The first detectable damage is cellular leakage, followed sequentially by inhibited photosynthesis, ethylene evolution, ethane and malondialdehyde evolution, and finally bleaching of chloroplast pigments. Protox is thought to be a membrane-bound enzyme, perhaps in, or on, the plastid envelope. Thus the cellular *site of action* of these herbicides may be the plastid envelope (Duke et al., 1991).

Of the herbicides included in this text, the following induce bleaching in plants, either by inhibition of carotenoids or by inhibition of protoporphyrinogen oxidase:

Diphenyl Ethers
Aciflurofen
Fomesafen
Lactofen
Oxyfluorfen

Nonfamily
Clomazone
Fluridone

Phenyl Pyridazinones
Norflurazon

Inhibition of Photosynthesis and Inhibition of ATP Formations

Photosynthesis is the process in plants by which light energy is utilized to shift energy, in the form of energized electrons, from one chemical (chlorophyll) to certain other chemicals. Ultimately, this energy (termed "biological energy") is stored in glucose in a form readily available to the plant and to all other living organisms to "drive" their metabolic reactions. Photosynthesis is unique to green plants and certain microorganisms.

Photosynthesis occurs via a sequence of reactions that may be grouped according to their dependence on light as:

1. Light-dependent phase (utilization of light energy to form biological energy)
   a. Photosystem II (PS II)
   b. Photosystem I (PS I)
   c. ATP formation (photo and cyclic phosphorylation)
2. Light-independent phase (atmospheric CO_2 fixation)
   a. Calvin-Benson cycle
   b. Hatch-Slack pathway

During the light-dependent phase, biological energy is derived from light energy, chlorophyll, and the dissociation (splitting) of water, and it is temporarily stored in the high-energy phosphate bonds of nicotinamide adenine dinucleotide phosphate (NADPH) and the two terminal, high-energy phosphate bonds of adenosine triphosphate (ATP). The biological energy derived from the light-dependent phase is utilized to convert carbon dioxide and water into carbohydrates, with the concomitant storage of biological energy in the carbon-to-carbon bonds of the carbohydrates.

In the photosynthetic process, photosystem II actually precedes photosystem I. The mechanisms of photosystem I were investigated prior to the discovery of photosystem II, and the designations I and II refer to the sequence of early investigations, rather than to their sequential occurrence in photosynthesis.

Sites of Photosynthesis

The light-dependent phase and the light-independent phase of photosynthesis take place in the chloroplasts, but they occur at different locations within this organelle. The light-dependent phase occurs within the membranous *thylakoids* (lamellae) of the chloroplasts, whereas the light-independent phase takes place in the *stroma*, the colorless proteinaceous matrix in which the chlorophyll-containing thylakoids are embedded.

Respiration

Respiration is the most important energy-releasing process in plants and animals. The primary function of respiration is to transfer biological energy from its storage in carbohydrates to ATP, a universal carrier for this energy. Biological energy, transferred to ATP, is then available for utilization by living organisms. ATP serves as a temporary storehouse and carrier of biological energy, an intermediate between more stable energy storehouses (carbohydrates such as sugars and starch) and energy utilization.

Inhibition of Photosynthesis

Only two biochemical mechanisms have been clearly demonstrated to be of primary importance in the inhibition of photosynthesis by herbicides; they are (1) inhibition of electron transport in photosystem II, and (2) diversion of electron transport through photosystem I. Photosynthetic electron transport in PS II and PS I occurs in chloroplast lamellae. Thus sites of action for herbicides inhibiting photosynthesis are in the chloroplast lamellae.

Herbicides that inhibit electron transport in PS II bind to the D1 protein of the PS II reaction center, blocking electron transfer to plastoquinone. Inhibition of PS II electron transport prevents the conversion of absorbed light energy into electrochemical energy, indirectly blocking the transfer of excitation energy from chlorophyll molecules to the PS II reaction center (Fuerst and Norman, 1991). A diagram of the electron transport chain in photosystem II and the sites of action of herbicides that interfere with electron transport in this chain is shown in Figure 7.1.

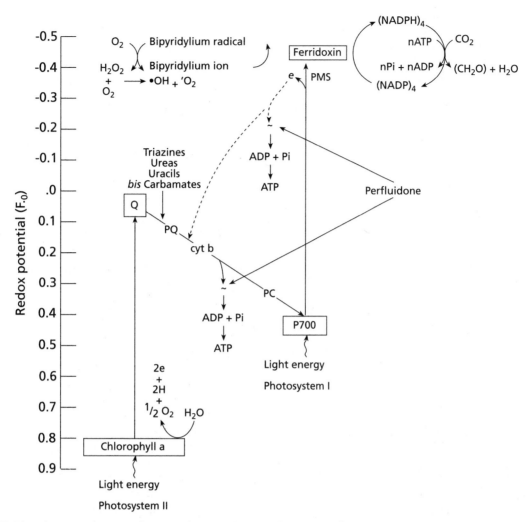

FIGURE 7.1 ■ The electron transport chain in photosynthesis and the sites of action of herbicides that interfere with electron transfer in this chain (Q = electron acceptor; PG = plastoquinone).

SOURCE: F. M. Ashton and A. S. Crafts. 1981. *Mode of action of herbicides,* 2nd ed., p. 55. New York: John Wiley & Sons, Inc. Reprinted with permission of John Wiley & Sons, Inc.

All currently registered herbicides that inhibit photosynthesis do so by blocking the flow of electrons through PS II, with the exception of diquat and paraquat (bipyridinium herbicides). Diquat and paraquat interfere with electron transport through PS I, probably by accepting electrons from the iron/sulfur protein F_aF_b (Fuerst and Norman, 1991).

Although inhibitors of electron transport through PS II and PS I electron acceptors have very different primary sites of action, both types of herbicides are phytotoxic due to peroxidation of membrane lipids. Plants treated with herbicides that inhibit electron transport in PS II, or act as electron transport acceptors in PS I, produce quantities of membrane-damaging triplet chlorophyll, singlet oxygen

($^1O_2$), superoxide (O_2^-), and hydrogen peroxide (H_2O_2). These products interact with fatty acids (linolenic and linoleic) of the lamellae through the process of lipid peroxidation. The lipid peroxidation process destroys membrane integrity, leading to the loss of cellular compartmentation and phytotoxicity (Fuerst and Norman, 1991).

Inhibition of ATP Formation

Herbicides that inhibit ATP formation deprive the plant of biological energy derived from PS II and PS I, and the plant dies because it no longer has the available biological energy needed to drive metabolic processes. Herbicides that inhibit photosynthesis also prevent ATP formation, but prevention of ATP formation, in this case, is considered a secondary mode of action. Inhibition of ATP formation *per se* is a *primary mode of action*. The *site of action of inhibition of ATP formation is the thylakoid membranes of the stroma*. The steps in photosynthesis and the formation and herbicide inhibition of ATP are shown in Figure 7.1. Of the herbicides included in this text, only *difenzoquat* inhibits ATP formation *per se*.

Of the herbicides included in this text, the following block electron transport through PS II:

Triazines	Ureas	Benzonitriles
Ametryn	Diuron	Bromoxynil
Atrazine	Fluometuron	**Nonfamily**
Cyanazine	Linuron	
Metribuzin	Tebuthiuron	Bentazon
Prometon		Difenzoquat
Prometryn	**Phenylcarbamates**	Pyridate
Simazine	Desmedipham	
Uracils	Phenmedipham	
	Phenyl Pyridazinones	
Bromacil		
Terbacil	Pyrazon	

Inhibition of Lipid Formation

Lipids (waxes, oils) are composed of fatty acids. Fatty acids are ubiquitous, requisite components of cellular membranes. Alteration of membrane fatty acid content results in changes in the function of chloroplasts and mitochondria.

There are three groups of herbicides that inhibit the biosynthesis of lipids in plants as a primary mode of action; they are *thiocarbamates, cyclohexanediones,* and *aryloxyphenoxypropionates*. Acid amide herbicides have been reported to inhibit lipid formation, but investigative evidence has been too contradictory to denote lipid inhibition as their primary mode of action. To varying degrees, substituted pyridazinone compounds alter the fatty acid composition of lipids, but none of these compounds are currently marketed as herbicides in the United States. Two phenyl pyridazinone herbicides (pyrazon and norflurazon) are used, but their primary modes of action are inhibition of photosynthetic electron transport and inhibition of carotenoid formation, respectively. Fatty acid biosynthesis and the sites of action of herbicides that inhibit lipid formation are diagrammed in Figure 7.2.

The thiocarbamate herbicides are applied preplant-incorporated, and they provide selective control of annual grass and some annual broadleaf weeds during germination and early seedling growth. The *mode of action* of the thiocarbamate herbicides is inhibition of lipid formation.

The thiocarbamates, through their inhibition of lipid synthesis, interfere with the biosynthesis of surface lipids (waxes, cutin, suberin), apparently due to their ability to inhibit acyl-coenzyme A (CoA) elongases. These enzymes are integral membrane proteins associated with endoplasmic reticulum and catalyze the condensation of malonyl-CoA with fatty acid acyl-CoA substrates to form very long fatty acids used in the synthesis of surface lipids. They are found in epidermal cells and cells at wound surfaces. The *site of action* of the thiocarbamate herbicides is generally assumed to be in the developing shoot, but its exact site of action has not been identified.

The cuticle, a waxlike layer of cutin coating the external surfaces of leaves, serves to greatly retard water loss from the leaves. Leaf surfaces with heavy waxy coatings are more difficult to wet with aqueous herbicide sprays than are leaves devoid of such coatings, and a thick waxy coating may inhibit herbicide penetration more effectively than a thin waxy coating. Herbicides that interfere with cuticular wax formation may contribute to the phytotoxicity of subsequently applied herbicides by enhancing their adsorption on, and absorption by, the pretreated foliage.

The cyclohexanedione and aryloxyphenoxypropionate herbicides are foliar-applied for selective control of grasses, with little or no effect on broadleaved (dicot) plants. Both groups of herbicides exhibit a similar spectrum of selectivity (phytotoxic to grasses but not dicots) and similar injury symptoms. Treated plants developing leaves are chlorotic and growth ceases and, within a few days, necrosis of the shoot apex and meristematic regions of leaves and shoots becomes apparent. *The primary mode of action* of the cyclohexanedione and aryloxyphenoxypropionate herbicides is inhibition of the biosynthesis of de novo fatty acids in grasses. Their *site of action* is the enzyme acetyl-CoA (ACCase), found in the stroma of plastids.

Of the herbicides included in this text, the following inhibit lipid formation as their primary mode of action:

Thiocarbamates	Aryloxyphenoxypropionates
Butylate	Diclofop-methyl
Cycloate	Fenoxaprop-ethyl
EPTC	Fluazifop-butyl
Molinate	Fluazifop-P-butyl
Pebulate	Haloxyfop-methyl
Thiobencarb	Quizalofop-ethyl
Triallate	Quizalofop-P-ethyl
Vernolate	

Cyclohexanediones

Clethodim
Sethoxydim

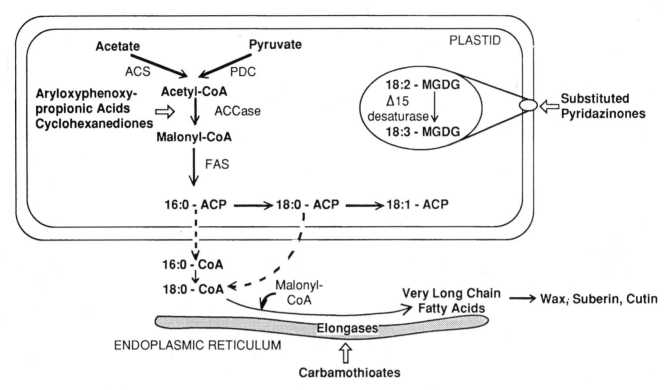

FIGURE 7.2 ■ Simplified schematic of fatty acid biosynthesis in higher plants illustrating proposed sites of action for the carbamothioate, substituted pyridazinone, aryloxyphenoxypropionic acid, and cyclohexanedione herbicides. (Abbreviations: ACCase, acetyl-CoA carboxylase; ACP, acyl carrier protein; ACS, acetyl-CoA synthetase; CoA, coenzyme A; FAS, fatty acid synthase; MGDG, monogalactosyldiacylglycerol; PDC, pyruvate dehydrogenase complex.)

SOURCE: J. W. Gronwald, 1991. Lipid biosynthesis inhibitors. *Weed* Sci. 39:435–449.

Inhibition of Acetolactate Synthase

The enzyme acetolactate synthase (ALS), also named aceto-hydroxyacid synthase (AHAS), catalyzes the first step in the biosynthesis of the branched-chain amino acids *valine*, *leucine*, and *isoleucine*. Amino acids are synthesized in plants by a common pathway located in the chloroplasts. They serve as the "building blocks" for proteins, which are made up of long chains of interconnected amino acids.

There are two groups of herbicides that inhibit acetolactate synthase as their *primary mode of action*; they are *imidazolinones* and *sulfonylureas*. Inhibition of ALS results in preventing the biosynthesis of the three previously mentioned branched-chain amino acids. This herbicidal effect can be reversed by an exogenous supply of these same three amino acids. However, it appears that the ultimate herbicidal effect (i.e., death of plant cells) is not due to death by amino acid starvation but, rather, to toxicity caused by the metabolite α-amino butyric acid. α-Amino butyric acid is produced from a precursor available from diversion in cells with inhibited acetolactate synthase. The

site of action of these herbicides resides with the enzyme ALS (AHAS) in the chloroplasts (Stidham, 1991).

Of the herbicides included in this text, the following inhibit ALS (AHAS) as their primary mode of action:

Imidazolinones	**Sulfonylureas**
Imazamethabenz	Bensulfuron-methyl
Imazapyr	Chlorimuron-ethyl
Imazaquin	Chlorsulfuron
Imazethapyr	Metsulfuron-methyl
	Nicosulfuron
	Primisulfuron-methyl
	Sulfometuron-methyl
	Thifensulfuron-methyl
	Tribenuron-methyl

Inhibition of the Shikimic Acid Pathway and Inhibition of δ-Aminolevulinic Acid

The shikimic acid pathway is unique to plants; it is not found in animals. End products of this pathway include the amino acids *phenylalanine, tyrosine,* and *tryptophane,* which are required for plant and animal growth and development. Animals obtain these amino acids from plants.

Although the exact pathway of δ-aminolevulinic acid synthesis in higher plants is a source of controversy, ALA is a unique precursor to the porphyrin ring in chlorophyll. The only known function of ALA is porphyrin ring synthesis. The formation of ALA is considered the rate-limiting step in chlorophyll synthesis.

Inhibition of Shikimic Acid Pathway

The herbicide glyphosate inhibits 5-enolpyruvyl-shikimate-3-phosphate (EPSP) synthase. The enzyme EPSP normally completes the second step in the "shikimic acid pathway" between shikimate and chlorismate. Inhibition of EPSP ultimately inhibits the syntheses of the amino acids phenylalanine, tryptophane, and tyrosine. Thus inhibition of the enzyme EPSP appears to be the *primary mode of action* of glyphosate, and the *site of action* is the shikimic acid pathway (Grossbard and Atkinson, 1985; Höllander and Amrheim, 1980).

Inhibition of δ-Aminolevulinic Acid

Glyphosate inhibits synthesis of δ-aminolevulinic acid (ALA). Since the only known function of ALA is porphyrin ring synthesis, inhibition of ALA synthesis by glyphosate blocks the syntheses of all compounds containing porphyrin rings in the plant such as chlorophyll, cytochromes, peroxidases, catalases, and phycobilins. Thus inhibition of ALA may be an integral component of the herbicidal action of glyphosate, possibly a *second mode of action* for this herbicide. The *site of action* of ALA inhibition by glyphosate may involve two enzyme pathways: (1) controlling the conversion of α-ketoglutarate to ALA and (2) controlling the condensation of glycine with succinyl-CoA to form ALA and CO_2 (Kitchen et al., 1981).

Of the herbicides included in this text, glyphosate is the only one that inhibits the shikimic acid pathway as the primary mode of action, and the only one that inhibits synthesis of ALA as a possible second mode of action.

Inhibition of the Biosynthesis of Cellulose from Glucose

A recently reported mode of action is the inhibition of the biosynthesis of cellulose from glucose by the herbicide isoxaben (Heim et al., 1990; Schneegurt et al., 1994). The herbicide dichlobenil also inhibits the biosynthesis of cellulose from glucose, but isoxaben is 40 times more active than is dichlobenil. The mechanisms of plant cell walls are presented in Delmar and Stone, 1987.

■ SELECTED REFERENCES

Akhavein, A. A., and D. N. Linscott. 1968. The dipyridinium herbicides, paraquat and diquat. *Residue Rev.* 23:97–145.

Ashton, F. M., and A. S. Crafts. 1981. *Mode of action of herbicides,* 2nd ed. New York: John Wiley & Sons Inc., 272–302.

Barrett, M. 1989. Protection of grass crops from sulfonylurea and imidazolinone toxicity. In *Crop safeners for herbicides,* K. K. Hatzios and R. E. Hoagland, eds., pp. 195–220. New York: Academic Press.

Cho, H-Y., J. M. Widholm, and F. W. Slife. 1986. Effects of haloxyfop on corn (*Zea mays*) and soybean (*Glycine max*) cell suspension culture. *Weed Sci.* 34:496–501.

Christianson, M. L. 1991. Fun with mutants: Applying genetic methods to problems of weed physiology. *Weed Sci.* 39:489–496.

Delmar, D. P., and B. A. Stone. 1987. Biosynthesis of plant cell walls. In *Plant biochemistry,* J. Preiss, ed., pp. 337–420. New York: Academic Press.

Duke, O. D., et al. 1991. Protoporphyrinogen oxidase–inhibiting herbicides. *Weed Sci.* 39:465–473.

Ebert, E., and K. Ramsteiner. 1984. Influence of metolachlor and the metolachlor protectant CGA 43089 on the biosynthesis of epicuticular waxes on the primary leaves of *Sorghum bicolor* Moench. *Weed Res.* 24:383–389.

Fuerst, E. P. 1987. Understanding the mode of action of the chloroacetamide and thiocarbamate herbicides. *Weed Technol.* 1:270–277.

Fuerst, E. P., and M. A. Norman. 1991. Interactions of herbicides with photosynthetic electron transport. *Weed Sci.* 39:458–464.

Gronwald, J. W. 1991. Lipid biosynthesis inhibitors. *Weed Sci.* 39:435–449.

Grossbard, E., and D. Atkinson, eds. 1985. *The herbicide glyphosate.* Boston: Butterworths.

Guttieri, M. J. 1992. DNA sequence variation in Domain A of the acetolactate synthase genes of herbicide-resistant and -susceptible weed biotypes. *Weed Sci.* 40:670–676.

Haehnel, W. 1984. Photosynthetic electron transport in higher plants. *Annu. Rev. Plant Physiol.* 35:659–693.

Harwood, J. Lipid metabolism in plants. In *Crit. Rev. Plant Sci.,* B. V. Conger, ed. Vol 8:1–43. Boca Raton, Fla.: CRC Press.

Harwood, J. L., S. M. Ridley, and K. A. Walker. 1989. Herbicides inhibiting lipid synthesis. In *Herbicides and plant metabolism,* A. D. Dodge, ed., pp. 73–96. New York: Cambridge University Press.

Heim, D. R., et al. 1990. Isoxaben specifically inhibits incorporation of ^{14}C-glucose into acid insoluble material. *Plant Physiol.* 93:695–700.

Höllander, H., and N. Amrheim. 1980. The site of the inhibition of the shikimic pathway by glyphosate. I. Inhibition by glyphosate of phenylpropanoid synthesis in buckwheat (*Fagopyrumm esculentum* Moench). *Plant Physiol.* 66:823–829.

Hoppe, H. H. 1989. Fatty acid synthesis—a target site of herbicide action. In *Target sites of herbicide action,* P. Boger, and G. Sandmann, eds. Boca Raton, Fla.: CRC Press.

Hosaka, H., and M. Takagi. 1987. Physiological responses to sethoxydim in tissues of corn (*Zea mays*) and pea (*Pisum sativum*). *Weed Sci.* 35:604–611.

Hosaka, H. and M. Takagi. 1987. Biochemical effects of sethoxydim in excised root tips of corn (*Zea mays*). *Weed Sci.* 35:612–618.

Kitchen, L. M., W. W. Witt, and C. E. Rieck. 1981. Inhibition of δ-aminolevulinic acid synthesis by glyphosate. *Weed Sci.* 29:571–577.

Kunert, K-J., and P. Boger. 1981. The bleaching effect of the diphenyl ether herbicide oxyflurofen. *Weed Sci.* 29:169–173.

Kunert, K-J., G. Sandmann, and P. Boger. 1987. Modes of action of diphenyl ethers. *Rev. Weed Sci.* 3:35–55.

McFarland, J. E., and F. D. Hess. 1985. Herbicide activity of acetanalides parallels alkylation potential. *WSSA Abstr.* 25:72.

Moreland, D. I. 1980. Mechanisms of action of herbicides. *Annu. Rev. Plant Physiol.* 31:597–638.

Penner, D., and F. M. Ashton. 1966. Biochemical and metabolic changes in plants induced by chlorophenoxy herbicides. *Residue Rev.* 14:39–113.

Ray, T. B. 1984. Site of action of chlorsulfuron. *Plant Physiol.* 75:827–831.

Robertson, M. M., and K. C. Kirkwood. 1970. The mode of action of foliage-applied translocated herbicides with particular reference to phenoxy acid compounds. II. The mechanism and factors influencing translocation, metabolism, and biochemical inhibition. *Weed Res.* 10:94–120.

Ruizzo, M. A., and S. F. Gorski. 1988. Inhibition of chloroplast-mediated reactions by quizalofop herbicide. *Weed Sci.* 36:713–718.

Sandmann, G., and P. Boger. 1989. Inhibition of carotenoid biosynthesis by herbicides. In *Target sites for herbicide action,* P. Boger and G. Sandmann, eds., pp. 25–44. Boca Raton, Fla.: CRC Press.

Schneegurt, M. A., D. R. Heim, and I. M. Larrinua. 1994. Investigation into the mechanism of isoxaben tolerance in dicot weeds. *Weed Sci.* 42:163–167.

Shaner, D. L., P. C. Anderson, and M. A. Stidham. 1984. Imidazolinone: Potent inhibitors of acetohydroxyacid synthase. *Plant Physiol.* 76:545–546.

Stidham, M. A. 1991. Herbicides that inhibit acetohydroxyacid synthase. *Weed Sci.* 39:428–434.

Wilkinson, R. E., and A. E. Smith. 1975. Thiocarbamate inhibition of fatty acid biosynthesis in isolated spinach chloroplasts. *Weed Sci.* 23:100–104.

Wesley, M. T., and D. R. Shaw. 1992. Interactions of diphenylether herbicides with chlorimuron and imazaquin. *Weed Technol.* 6:345–351.

Vaughn, K. C., and L. P. Lehnen, Jr. 1991. Mitotic disrupter herbicides. *Weed Sci.* 39:450–457.

8 HERBICIDE–PLANT SELECTIVITY

■ INTRODUCTION

Herbicides are chemicals that kill plants. Were they to kill all plants, however, they would be of very limited value to agriculture. Thus it is the ability of herbicides to kill certain plants without appreciable harm to others that makes them useful to agriculture. This characteristic is called *selectivity*.

Selectivity is basic to the use of herbicides in crop production. Through the judicious selection and application of herbicides, it is possible to selectively control:

1. *Grass weeds in grass crops*—for example, control of barnyardgrass in rice with propanil or wild oats in wheat or barley with diclofop-methyl

2. *Grass weeds in broadleaved crops*—for example, control of large crabgrass in cotton with trifluralin or rescuegrass in alfalfa with EPTC

3. *Broadleaved weeds in broadleaved crops*—for example, control of common cocklebur in cotton with DSMA or MSMA or common purslane in bell peppers with DCPA

4. *Broadleaved weeds in grass crops*—for example, control of coast fiddleneck in wheat with 2,4-D or common cocklebur in sorghum with dicamba

5. *Both grass and broadleaved weeds in grass and broadleaved crops*—for example, control of many annual grass and broadleaved weed species in field corn with atrazine or with diuron in cotton

Selectivity is dependent on many interrelated factors. It is a relative, rather than absolute, characteristic. It is influenced by environmental factors and by the kind and amount of herbicide applied, as well as when and how applied. Even the most tolerant plant species can succumb if the applied herbicide rate is great enough. Varieties and ecotypes of a plant species may respond differently to applications of a given herbicide. Selectivity may be lost through mistakes made by the user such as improper application or applying the wrong herbicide.

Many factors contribute to selectivity, and they may be grouped as follows:

1. Physical factors
2. Biological factors
3. Inherent properties of herbicides
4. Chemical plant-protectants
5. Chemical herbicide-protectants

6. Herbicide–agrochemical interactions
7. Herbicide-resistant weed biotypes

As each of these factors is discussed, bear in mind that there is considerable interdependence among groups as well as among the factors within groups.

■ PHYSICAL FACTORS OF SELECTIVITY

The physical factors of selectivity influence both the contact and retention of the applied herbicides on plant surfaces. To be effective, an applied herbicide must contact the targeted plant and be retained on its surfaces long enough to be absorbed in amounts great enough to induce the desired effect. The physical factors influencing this process vary depending on whether the herbicide is applied to the soil or to the plant foliage.

Soil-Applied Herbicides

The physical factors influencing selectivity of *soil-applied* herbicides are as follows:

1. Herbicide application rate
2. Herbicide formulation
3. Herbicide placement
4. Stage of plant growth

Rate of Application

A herbicide's application rate must be such that the problem weeds are effectively controlled but with little or no injury to the crop plants. For example, the phenylurea and triazine herbicides were first introduced as nonselective soil sterilants and used at rates of 20–40 lb/A (22–44 kg/ha), or more. It was later found that they could be used selectively in certain crops such as mint, alfalfa, and cotton, when applied at 1–2 lb/A (1–2 kg/ha).

The herbicide hexazinone is used for selective weed control in dormant or semidormant alfalfa at 0.45–1.35 lb/A (0.5–1.5 kg/ha). However, when used for brush control in noncrop areas, its recommended rate is 3.6–7.2 lb/A (4–8 kg/ha).

The recommended rate for trifluralin for selective control of annual grasses and certain broadleaf weeds in cotton is 0.5–0.75 lb/A (0.5–0.84 kg/ha). However, control of rhizome johnsongrass in cotton requires that the rate of trifluralin be doubled. In this case, manipulation of the application rate affects selective control of a weed (johnsongrass) rather than the crop, as in the previous examples.

In general, the rate of soil-applied herbicides in croplands varies from about 0.4 oz ai/A (28 g ai/ha) (e.g., triasulfuron in spring and winter wheat) to 10 lb ai/A (11 kg ai/ha) (e.g., DCPA in vegetable crops).

Herbicide Formulation

Selectivity with preemergence herbicides is obtained by using granular or pelleted formulations to control non-emerged weeds among emerged crop plants. Foliar retention of the applied herbicide is avoided by the crop plants, while susceptible weed seedlings absorb the herbicide through the soil. The granules or pellets fall through the foliar canopy of the crop onto the soil below. Granules or pellets remaining on the foliage or in leaf axils may be shaken free by using canvas drags mounted behind the applicator.

Herbicide Placement

Herbicide placement achieves selectivity by preventing the applied herbicide from contacting the underground absorptive sites of the crop plants. This is accomplished by applying herbicides to:

1. The soil surface (not incorporated)
2. The soil surface, shallowly incorporated
3. The area between seed rows

To achieve selectivity by soil-surface placement (not incorporated), the herbicide must remain on or near the surface so as not to gain entry into the crop seedlings yet be absorbed by weeds germinating or emerging through the layer of treated soil. However, such selectivity may be lost if the surface-applied herbicide leaches into the soil and is absorbed by the crop seedling. For example, diuron is a very effective surface-applied herbicide for controlling annual weeds in cotton, but on coarse soils it may be leached into the soil by rainfall or irrigation, resulting in death of the cotton seedlings.

Selectivity can be obtained by shallow soil incorporation (1–2 in. or 2.5–5 cm deep) of the herbicide above the more deeply seeded crop. For example, trifluralin may be applied in spring wheat, seeded 3–4 in. deep (7.5–10 cm), and soil-incorporated 1.5 in. deep. Selectivity is dependent on the fact that herbicide-absorptive sites of the wheat seedlings are located below the zone of treated soil and that trifluralin resists leaching.

Selectivity may be obtained by applying the herbicide to the area between crop rows. For example, in nonirrigated cotton, EPTC may be soil-applied and incorporated 2–3 in. deep (5–7.5 cm) as a band treatment between the cotton rows after the cotton has 2–4 leaves, taking care to apply the herbicide no closer than 4 in. (10 cm) on either side of the cotton drill.

Stage of Plant Growth

Selectivity may be obtained with soil-applied herbicides by taking advantage of differences in the development and location of underground absorptive sites through which herbicides gain entry into weed and crop plants. Selectivity achieved in this manner is dependent on herbicide placement with respect to the sites of herbicide entry into the weed and crop plants. Refer to the section on morphology in this chapter for additional information on this topic.

Foliar-Applied Herbicides

The following physical factors influence the selectivity of *foliar-applied* herbicides:

1. Herbicide application rate
2. Herbicide retention
3. Herbicide placement
4. Differences in stages of growth between weed and crop plants at time of herbicide application
5. Specialty herbicide-application equipment

Herbicide Application Rate

When foliar-applied at excessively high rates, herbicides that are normally foliar-absorbed become toxic to most plants, both weeds and crops. However, plant species do vary in their susceptibility to a given herbicide. By first determining the susceptibility of a given plant species to a given herbicide over a broad range of rates, selectivity between species may be obtained by applying the herbicide at a suitably low rate—one that will kill the weed species without significant injury to the crop plants. For example, coast fiddleneck was found to be about 30 times more susceptible to bromoxynil than was wheat.

Selective foliar-applied herbicides are used at rates as low as 0.06 oz ai/A (1.7 g ai/A) (e.g., metsulfuron-methyl in wheat) to as high as 3.6 lb ai/A (4 kg ai/ha) (e.g., DSMA in cotton).

Herbicide Retention

Foliar-applied herbicides must be retained on leaf surfaces long enough to be absorbed in amounts large enough to

bring about the desired effect. Differences in retention between plant species are responsible for the selective action of some herbicides. In general, herbicide retention on leaf surfaces is largely influenced by:

1. Leaf-surface characteristics such as waxiness and hairiness
2. Surface tension (cohesiveness) of the spray droplets

Leaf surfaces with heavy, waxy coatings are more difficult to wet with aqueous herbicide sprays than are those devoid of such coatings, and they retain less herbicide than do surfaces that are wetted comparatively easily. Some herbicides (e.g., EPTC and triallate) interfere with cuticular wax formation. Pretreatment with such herbicides will adversely affect normal differential wetting and retention with aqueous herbicide sprays, and it may result in loss of the desired selectivity.

Leaf surfaces with few, if any, hairs and those that are densely hairy are more difficult to wet with an aqueous herbicide spray than those that are moderately hairy.

Aqueous spray droplets that do not contain a spray adjuvant, such as a wetting agent or crop oil concentrate, tend to bounce or roll off of some leaf surfaces more than others. The differential wetting obtained with herbicidal aqueous sprays accounts for the selectivity of some herbicides. For example, the selective control of mustard with bromoxynil in stands of barley or peas is due largely to differential wetting and retention. Selectivity is lost when a wetting agent is added to the spray mixture.

Oil sprays have low surface tension and they readily wet leaf surfaces. When oil is used as the carrier for herbicides, selectivity based on differential wetting and herbicide retention is usually lost.

Herbicide retention on leaf surfaces may be influenced by climatic conditions such as temperature and rainfall. High temperatures (in the 90s Fahrenheit, 30s Celsius, or above) tend to increase the loss of some herbicides by volatilization from leaf surfaces. Rainfall occurring during about the first hour of herbicide application may wash some of the herbicide from the leaf surfaces.

Herbicide Placement

With respect to foliar-applied herbicides, herbicide placement achieves selectivity by preventing the herbicides from contacting the crop plants at sites of herbicide absorption such as the leaves or buds, while making contact with the emerged weeds. This is commonly achieved by use of (1) directed basal sprays, and (2) shielded sprays.

Directed basal sprays are usually applied after the crop plants are tall enough to permit a directed spray to be applied under the foliar canopy, avoiding contact with the leaves and axillary buds. At this time, the weeds should be small (less than 2 in. or 5 cm tall) so that they can be thoroughly covered with the applied spray—for example, the application of MSMA in cotton.

Directed sprays are applied with spray nozzles located low enough on the application equipment to be below the crop foliar canopy and directed toward the base of the crop plants. Drop nozzles (spray nozzles located on adjustable arms) are often used for this purpose. For example, drop nozzles are used in the application of 2,4-D in field corn to avoid reaching the apical bud in the leaf whorl with the spray mixture. Thus the spray pattern passes below the foliar canopy and toward, and slightly beyond, the base of the crop plants; yet it is high enough to adequately treat the weeds growing under the canopy. The use of such basal sprays will control weeds growing within and between rows. Basal sprays are generally used only in the taller erect crops such as corn, cotton, soybeans, grapes, and fruit orchards. They are not used on low-growing crops devoid of elongated aboveground stems.

The tolerance of cotton plants to basal-sprayed herbicidal oils is due to a protective barrier on their stems that prevents oil absorption. In young cotton plants (about 21 days old), this protective barrier is in the form of the unbroken (intact) epidermis of the young hypocotyl, whereas in older plants (62 days old), resistance is provided by the presence of oil-impervious cork layers surrounding the hypocotyls. The hypocotyls of cotton plants 36–50 days old are susceptible to herbicidal oils because at this time small cracks appear in their bark, and these cracks serve as channels for the oil to penetrate the stem and destroy vital tissue within.

Shielded sprays are commonly used in crops not suited to basal sprays or when weeds are too tall. A physical barrier (metal, wood, plastic, rubber), commonly mounted on the spray equipment, is used to shield the crop plants from the herbicide spray. However, weeds growing in the crop rows are also protected. Wind occurring at the time of application may cause spray droplets to swirl around the shields and drift onto the crop plants, resulting in crop injury. Such drift is minimized by using low spray pressures and relatively large nozzle orifices, thereby increasing spray-droplet size.

Stage of Plant Growth

In general, annual weeds are most susceptible to foliar-applied herbicides when in the seedling stage, and susceptibility tends to decrease as the plants grow older. For example, common cocklebur, common morningglory, common

ragweeds, and pigweeds are most susceptible to 2,4-D when in the 2- to 4-leaf stage. In contrast, some plants are more susceptible at later stages of growth. For example, soybeans are more susceptible to 2,4-D when they are about 18 in. tall than when 3–9 in. tall. Seedlings of johnsongrass are unaffected by foliar sprays of MSMA, whereas established johnsongrass plants are killed by MSMA. Field bindweed is most susceptible to glyphosate when it is foliar-applied in the fall about 1 month before first frost.

Specialty Application Equipment

Various "wiper" application devices for selectively controlling weeds that have grown taller than the crop plants are in commercial use. Basically, these wiper devices are used to apply a herbicide directly to the foliage of the weeds while avoiding herbicide contact with the foliage of the crop plants.

A rope-wick wiper utilizes a herbicide-wetted nylon fabric, or other abrasion-resistant material or combination of materials, braided into a rope and mounted on a vehicle, with different forms that vary by the composition and arrangement of the rope wicks and the manner in which the aqueous herbicide solution wets the wicks. This type of herbicide applicator selectively controls tall weeds overtopping the crop plants without injury to the crop plants. Rope-wick applicators are in use in cotton and soybean crops for the selective control of johnsongrass, pigweed species, and other tall-growing weeds that over-top the crop plants.

There are hand-held versions of the rope-wick applicator for use in spot-treating weeds or where use of vehicular-mounted applicators is impractical. There are a number of variants to the rope-wick applicator such as ground-driven carpet rollers (for use in orchards and lawns) and power-driven carpet rollers (for control of tall weeds in croplands). The use of the fabric-type applicators is suitable for selective weed control in many crops.

■ BIOLOGICAL FACTORS OF SELECTIVITY

The selective control of plant species with a given herbicide can be explained in many instances by differences between species in (1) *morphology,* and (2) *physiology.* These differences are referred to collectively as the *biological factors of selectivity.*

Morphology

Morphological factors influencing herbicide selectivity are those associated with the characteristic *external structure* of plants and, for purposes here, the development and location of the sites of herbicide entry into plants. Such morphological differences are concerned primarily with:

1. Leaves
2. Growing points (buds and regions of cell elongation)
3. Coleoptilar and crown nodes and those plant parts enclosed by the coleoptile (leaves and stem apex)
4. Roots
5. Underground reproductive organs

Leaves

Leaf characteristics that influence herbicide selectivity are primarily those that affect the *interception* and *retention* of herbicide sprays such as leaf surface, angle, shape, number, and arrangement. Leaf blades that form an angle of 45 degrees or more with the horizontal plane retain less spray than do those that are more parallel to this plane.

Leaf number and arrangement on plants affect the herbicide spray's penetration of the foliar canopy. Open canopies allow for greater spray penetration and therefore more thorough wetting of the entire plant. Dense foliar canopies tend to intercept the sprays, interfering with canopy penetration.

Growing Points

Exposure, or lack of exposure, of the growing points (buds and regions of cell elongation) of plants to contact-type herbicide sprays is often vital to herbicide–plant selectivity. Unless contact-type herbicides come in direct contact with these plant parts, the target plants may survive the herbicide treatment. The growing points of grass plants are more protected from contact-type herbicide sprays than are those of broadleaved plants.

Coleoptilar and Crown Nodes

The *coleoptilar node* is commonly the first node on the stem of grass seedlings, located at the upper end of the first internode. It is the node from which the coleoptile develops and to which the base of the coleoptile is attached. The coleoptilar node and the leaves and stem apex enclosed by the coleoptile serve as primary sites of entry for certain soil-applied herbicides (e.g., the carbamates and dinitroanilines).

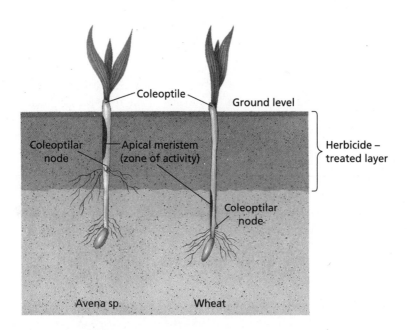

FIGURE 8.1 ■ Comparative location of the coleoptilar nodes of wild oat and wheat with respect to a soil layer of shallowly incorporated herbicide (e.g., trifluralin). The seed of both plants is buried to the same depth. The coleoptilar node is a primary site of entry for some soil-applied herbicides. Their sites of action may be in the apical meristems, depending on the herbicides involved.

SOURCE: J. D. Fryer and S. A. Evans, eds. 1968. *Weed control handbook,* 5th ed. 1:51. Oxford, England: Blackwell Scientific Publications.

The coleoptilar node always remains below the soil surface, but its exact location relative to the surface varies among grass species and even between varieties of some species. For example, the coleoptilar nodes of wild oat, barnyardgrass, and green foxtail seedlings are located (positioned) in the upper 0.5 in. (1.3 cm) of the soil, regardless of the depth (within practical limits) at which the seed is planted. In contrast, the coleoptilar nodes of barley and rice seedlings remain within about 0.5 in. of the seed, so that the positioning of their coleoptilar nodes below the soil surface is determined by the depth to which the seed is planted. The rice variety *Nato* is an exception, as its coleoptilar node develops just below the soil surface, even though the seed is planted relatively deep. The coleoptilar nodes of corn and sorghum seedlings are also located just below the soil surface, while that of wheat is intermediate between corn and barley. The relative location of the coleoptilar nodes of oat and wheat to the soil surface and their positioning in soil shallowly treated with herbicides (e.g., trifluralin) is shown in Figure 8.1. Thus selectivity due to location of the coleoptilar node, stem apex, and developing leaves enclosed by the coleoptile is closely associated with herbicide placement in the soil. *Broadleaved (dicot) plants have neither coleoptilar nodes nor coleoptiles.*

The *crown node* is a compressed structure of a variable number of nodes and internodes located on the main axis of grass seedlings just above the second internode. Due to elongation of the second internode, the ultimate position of the crown node is always at or just below the soil surface, whether the seed is buried deeply or shallowly. The crown node and coleoptilar node may be indistinguishable when the coleoptilar node is also located at or near the soil surface. Adventitious roots and, in some grass species, sprouts (tillers) arise from the crown node, and the first whorl of such roots appear at about the same time as the coleoptile breaks through the soil surface. These adventitious roots eventually form the fibrous root system characteristic of grass plants and, in others, the "brace roots" that are so apparent on corn plants. The positioning of the crown node of corn from seed planted at different depths in the soil is illustrated in Figure 8.2. Herbicide selectivity relative to the crown node is dependent on herbicide placement and susceptibility of the grass species to the applied herbicide.

Roots

Seeds having well-developed embryos with large numbers of unexpanded cells in their radicles can, upon germination, make considerable root growth without further cell division by the elongation of these cells in the radicle (Figure 8.3). Such cell elongation can, in some cases, push the root tip (meristem) out of the zone of herbicide-treated soil. However, to be of practical value, the root tip must be moved

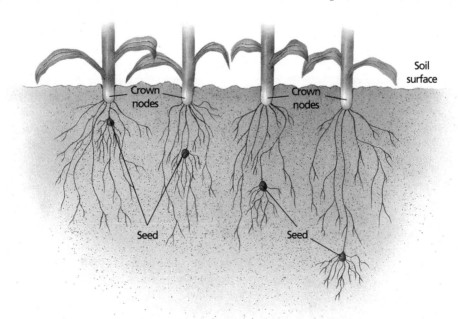

FIGURE 8.2 ■ Positioning of the crown node of corn from seed planted at depths of 2, 4, 6, and 10 in. (5, 10, 15, and 25 cm).

SOURCE: R. B. Mitchell, 1970. *Crop growth and culture*, p. 187. Ames, Iowa: The Iowa State University Press. Also reprinted with permission of Macmillan Publishing Co., Inc. from *Principles of field crop production* by J. H. Martin and W. H. Leonard. Copyright 1949 by Macmillan Publishing Co., Inc.

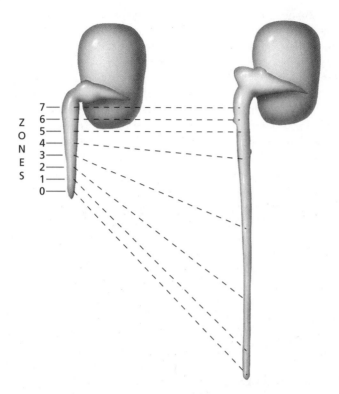

Z O N E S

7 —
6 —
5 —
4 —
3 —
2 —
1 —
0 —

FIGURE 8.3 ■ Root growth by cell elongation. Elongation of the radicle of a bean seedling (left), marked with India ink, is indicated by the widely spaced markings on the same radicle a day later (right).

SOURCE: A. J. Brook. 1964. *The living plant.* Edinburgh, Scotland: Edinburgh University Press.

out of the treated soil before the tip absorbs sufficient herbicide to impair its later growth. Certain large-seeded plant species, such as cotton, snapbeans, and soybeans, have this capability. The growth and development of roots vary among plant species; for example, the roots of tumblegrass grew 11 in. (27.5 cm) in 15 days, while those of cocklebur grew 33 in. (82 cm).

Underground Reproductive Organs

Perennial weeds, especially herbaceous weeds and certain woody species, commonly have underground vegetative reproductive organs such as rhizomes, roots, tubers, and basal buds. These reproductive organs are often difficult to control with herbicides. It is not unusual to kill the aerial parts of perennial weeds with herbicides without significantly affecting their underground reproductive parts. Examples of perennial weeds capable of rejuvenating from underground reproductive organs are johnsongrass, quackgrass, and bermudagrass from buds at nodes on rhizomes; field bindweed and Texas blueweed from adventitious buds

on an extensive root system; yellow and purple nutsedges from adventitious buds on tubers; and the woody plant mesquite from basal buds. Species selectivity is achieved, intentionally or unintentionally, with herbicides that fail to reach and kill these underground reproductive organs, even though their aerial parts may be killed.

Physiology

Selectivity to herbicides may be achieved by physiological differences between plant species. Such differences have to do with a herbicide's entry into plants and its subsequent effect on the plants once it has gained entry. Physiological factors that influence herbicide selectivity are categorized as follows:

1. Translocation
2. Conjugation
3. Herbicide accumulation and secretion
4. Metabolism (herbicide detoxification)

Translocation

Following absorption, herbicides are translocated in plants from their sites of entry to their sites of activity. Some herbicides are translocated more readily in certain plant species than in others. Selectivity may be achieved by taking advantage of differential movement in plants, as the following examples show.

Prometryn is absorbed equally by the roots of soybean, a susceptible species, and by cotton, a tolerant species. However, it is translocated more readily in soybean plants than in cotton plants, with four times more prometryn moved into the shoots of soybean than into the shoots of cotton. Differential translocation between the two species appears to be the principal factor responsible for this species selectivity.

Pyrazon is translocated primarily in the xylem of plants, and it is phytotoxic if a sufficient concentration of the herbicide reaches its site of action in the shoot. When root-absorbed, pyrazon is translocated most readily in tomato, a susceptible species, and much less so in red beet, a tolerant species.

Glyphosate effectively controls the free-floating aquatic form of alligatorweed, but not the terrestrial and rooted forms. Imazapyr controls the terrestrial and rooted forms much more effectively than does glyphosate. The greater control of the terrestrial and rooted forms of alligatorweed by imazapyr is due to the greater absorption and translocation of imazapyr from treated leaves. Eight days after treatment, more than twice as much imazapyr as glyphosate was absorbed by treated leaves, and about 20 times more imazapyr than glyphosate was translocated to the roots and underground storage tissues. In addition, imazapyr is phytotoxic to alligatorweed at lower rates than is glyphosate.

Terbacil is not translocated from treated leaves of peppermint, a tolerant species, whereas it is readily translocated from treated leaves of ivyleaf morningglory, a susceptible species (Figure 8.4). Terbacil is readily translocated upward to all plant parts following root absorption by peppermint and ivyleaf morningglory (Figure 8.5).

Conjugation

Upon entering a plant cell, a herbicide may interact with chemicals present in the cell to form conjugates, or complexes, which are insoluble or nontranslocatable or both. Such complexes immobilize the herbicide within the cell, either in the cytoplasm or vacuole. Similarly, a nontranslocatable herbicide may form a conjugate within the cell that enhances, rather than prevents, translocation. In either case, differential transport due to conjugation occurs, and such differences contribute to herbicide selectivity between plant species.

For example, only the "free" (nonconjugated) form of chloramben (a discontinued herbicide) is translocated in plants. Chloramben is immobilized in the roots of wheat plants, a tolerant species, by forming a nontranslocatable conjugate in the wheat root cells. The conjugate has been identified as N-(carboxyl-2,5-dichlorophenyl) glycosylamine, or glycosylamiben. In barnyardgrass, a susceptible species, chloramben does not form a nonmobile conjugate, and it is readily translocated from the roots into the shoots.

Herbicide Accumulation and Secretion

Some plant species tend to channel herbicides toward "sinks" that are metabolically insensitive to them and that serve as reservoirs for their accumulation. For example, prometryn is accumulated in the lysigenous glands (cavities formed by the disintegration and dissolution of cells) in cotton, a tolerant species, whereas it is uniformly distributed throughout soybean plants, a susceptible species. In jimsonweed, a 2,4-D tolerant plant, foliar-applied 2,4-D initially moves upward into the young leaves and stem apex, as well as downward into the lower portion of the stem. A little later, however, the herbicide moves downward from the leaves and stem apex to the basal stem tissues and roots; some of the downward-moving 2,4-D accumulates in the basal stem tissue, while the remainder moves into the roots.

Some plant species are able to prevent the accumulation of a herbicide within their system by ridding themselves of the herbicide. Some species channel the herbicide into their leaves, which subsequently drop from the plant prematurely. Herbicides may be secreted from plant roots into the surrounding medium. For example, 2,4-D is secreted from roots of jimsonweed and dicamba from roots of snapbeans. Herbicides lost from the plant by root secretions may be absorbed by roots of adjacent plants or reabsorbed by the plant from which they were secreted, or both.

Metabolism (Herbicide Detoxification)

Selectivity among plant species can be achieved by plants of one species rapidly detoxifying a particular herbicide, while other species are unable to do so and are killed.

Differential metabolism of diclofop-methyl accounts for the selectivity of this herbicide between wheat (tolerant) and wild oat (susceptible). Following plant absorption, wheat metabolizes diclofop-methyl to nontoxic metabolites by irreversible aryl hydroxylation, while wild oat is unable to detoxify the herbicide. Similarly, selectivity between

(a) Peppermint.

(b) No translocation.

(c) Ivyleaf morningglory.

(d) Translocation above treated leaves.

FIGURE 8.4 ■ Translocation of leaf-applied terbacil. The autoradiographs in parts (a) and (c) were taken 9 days after ^{14}C-terbacil was applied to two leaves on each plant, as indicated by triangular arrows pointing to the treated leaves in parts (b) and (d). The herbicide applied to the peppermint leaves did not translocate from the treated leaves. Applied to the ivyleaf morningglory leaves, the herbicide moved upward to all plant parts above the treated leaves but did not translocate below the treated leaves.

SOURCE: J. L. Barrentine and G. F. Warren. 1970. Selective action of terbacil on peppermint and ivyleaf morningglory. *Weed Sci.* 18:373–377. Permission to copy granted by the authors and the Weed Science Society of America. Photos provided by the authors.

(a) Peppermint.

(b) Complete translocation.

(c) Ivyleaf morningglory.

(d) Complete translocation.

FIGURE 8.5 ■ **Translocation of root-applied terbacil.** The autoradiographs in parts (a) and (c) were taken 4 days after ^{14}C-terbacil was applied to the roots of each plant and show that the herbicide was absorbed by the roots of both plants and translocated upward to all plant parts. Parts (b) and (d) are photographs of the same plants autoradiographed.

SOURCE: J. L. Barrentine and G. F. Warren. 1970. *Weed Sci.* 18:373–377. Permission to copy granted by the authors and by the Weed Science Society of America. Photos provided by the authors.

longspine sandbur (tolerant) and barnyardgrass (susceptible) is due to differential metabolism and conjugation of diclofop-methyl.

The soybean cultivar Essex is tolerant to metribuzin, while the cultivar Coker 102 is susceptible. Essex is tolerant of metribuzin, primarily because of its capacity to rapidly detoxify metribuzin metabolically to nontoxic compounds; Coker 102 does not have this capability.

Wheat and barley are tolerant to postemergence applications of chlorsulfuron due to its rapid detoxification by these plants, with more than 90% of the foliar-absorbed chlorsulfuron metabolized to a nonphytotoxic metabolite within 24 hours by both species. Detoxification of chlorsulfuron was accomplished by hydroxylation of its phenyl ring, followed by conjugation with a plant carbohydrate. Susceptible plant species are unable to metabolize chlorsulfuron or do so at too slow a rate to prevent being killed.

Eastern black nightshade is resistant to foliar-applied chlorsulfuron, while velvetleaf is highly susceptible. The two species differ in their susceptibility because they detoxify (metabolize) the herbicide differently. Eastern black nightshade rapidly detoxifies chlorsulfuron, with 70% of the absorbed herbicide metabolized within 24 hours after application to a nontoxic metabolite. Velvetleaf detoxifies only 7% of the absorbed chlorsulfuron 72 hours after application. Similarly, leafy spurge, a tolerant species, metabolizes all of the absorbed chlorsulfuron within 72 hours after application, while Canada thistle, a susceptible species, metabolizes less than 2% during this same period.

Hot pepper (*Capsicum chinense*) is tolerant to foliar-applied bentazon, while sweet pepper (*Capsicum annuum*) is susceptible. A significantly faster rate of bentazon metabolism occurred in hot pepper, as compared to sweet pepper. Similarly, differential metabolism has been a factor in bentazon-tolerant barley, navy bean, soybean variety Clark 63, rice, and wheat. Susceptible species, such as black nightshade, common cocklebur, Canada thistle, and white mustard, metabolize bentazon at a slower rate. However, differential metabolism of bentazon may not entirely account for its selectivity.

Rice plants are tolerant to propanil, while barnyardgrass plants are susceptible. Selectivity occurs because of a difference in arylacylamidase enzyme levels between the two plant species. Arylacylamidase hydrolyzes the amide bond of propanil, resulting in the formation of two nonphytotoxic metabolites, 2,4-dichloroanilin and propionic acid. Rice plants have a high level of arylacylamidase and rapidly metabolize propanil to these nontoxic metabolites, whereas barnyardgrass plants have a low level of this enzyme and are unable to detoxify the herbicide before being killed.

The tolerance of certain plants such as corn, sorghum, and sugarcane to the triazine herbicides is due to their ability to detoxify the herbicides to nontoxic metabolites. Susceptible plant species are unable to detoxify these herbicides and are killed. Tolerant plants detoxify the triazines by (1) *hydroxylation at the 2-chloro position*, (2) N-*dealkylation of the N-alkyl side chains*, and (3) *conjugation to the peptide glutathione*. With few exceptions, conjugation with glutathione appears to be the predominant means by which tolerant plants detoxify the triazines.

Sometimes a nonphytotoxic chemical may be metabolized to a phytotoxic compound. For example, the nonphytotoxic 4-(2,4-dichlorophenoxy)butyric acid (2,4-DB), following foliar absorption, is changed by β-oxidation to the phytotoxic 2,4-dichlorophenoxyacetic acid (2,4-D) by some plant species. Plant species that make this conversion are killed by applications of 2,4-DB, while ones that cannot convert 2,4-DB to 2,4-D, such as alfalfa and clover plants, are not adversely affected by the applied 2,4-DB.

■ INHERENT PROPERTIES OF HERBICIDES

The inherent properties of herbicides (phytotoxicity, solubility, volatility, and absorption) are predetermined by the *composition* (kinds of atoms) and *configuration* (arrangement of atoms or groups of atoms) of the herbicide molecules. The inherent phytotoxicity of a herbicide may be altered by seemingly slight changes in its molecule. For example, the phytotoxicity of the herbicide 2,4-D is lost when the chlorine atom at the 4-position of the phenyl ring is shifted to the 6-position of the ring. The ester forms of 2,4-D are soluble in oil but only slightly soluble in water, while the dimethylamine salt is soluble in water but not in oil. The esters of 2,4-D are volatile, while the salts are nonvolatile (at practical temperatures). The esters of 2,4-D are more readily foliar-absorbed than are the salt forms.

The plant selectivity of 2,4-D may be altered by replacing the chlorine atom at the 2-position of the phenyl ring with a methyl group to form the herbicide MCPA. MCPA is less injurious to small grains and legumes than 2,4-D, and it is more effective on some broadleaf weed species. The herbicide mecoprop (similar in chemical structure to MCPA, except that a methyl group instead of a hydrogen atom is bonded to the α-carbon of the acetic acid side chain) resists enzymatic detoxification in plants, whereas MCPA is enzymatically detoxified, in time resulting in the achievement of differential selectivity.

The triazine herbicides prometon and prometryn have identical chemical structures, except for the atom forming the connecting "bridge" between the triazine ring and the attached methyl group. In the case of prometon, the atomic bridge is oxygen, while in prometryn it is a sulfur atom. This seemingly slight difference affects their use as herbicides. For example, prometon is a nonselective soil sterilant, while prometryn is used for selective weed control in cotton. If the chlorine atom at the 2-position of the triazine ring in the structures of atrazine, cyanazine, and simazine is replaced with a methyl group, the resulting compounds are nonphytotoxic.

■ CHEMICAL PLANT-PROTECTANTS

In this case, a chemical plant-protectant may be any chemical that serves to protect one or more species of plants from the toxic action of one or more herbicides. By the use of a chemical plant-protectant, an otherwise susceptible plant species may be made tolerant to a particular herbicide, without affecting the toxic action of the herbicide toward unprotected, susceptible species. Herbicidal plant selectivity may be achieved by protecting a desired plant species, such as corn or grain sorghum, from the toxic action of a particular herbicide through the use of a safener chemical that serves to protect the crop plants from the toxic action of the herbicide. Such safeners are referred to as *chemical plant-protectants, crop safeners, seed safeners, herbicide safeners, herbicide antidotes,* or *contratoxicants*—these terms are used interchangeably.

Herbicides are not made safe by herbicide "safeners," as the name might imply: herbicide safeners protect crop plants against herbicide injury. The term *herbicide safener* refers to a chemical compound that has limited phytotoxicity itself and selectively protects crop plants against injury from one or more herbicides. The term *herbicide antidote* persists in the scientific literature in spite of the fact that "antidote" describes a medication intended to reverse (counteract) the action of a toxicant already present in a human or animal.

Crop safeners prevent injury to crop plants from applied herbicides; they do not reverse such injury. So far, commercial success in crop safening against herbicides has been limited to three major grass crops: corn, sorghum, and rice. Research results with crop safeners for other grass crops, such as wheat, barley, oats, and ryegrass, is promising. Safeners intended to protect broadleaf crops against any herbicide or against herbicides that are photosynthetic inhibitors have not achieved commercial success. It is probable that in the foreseeable future, a candidate safener will only be developed and commercialized when the range of its usefulness is limited to proprietary herbicides of the parent company and avoids favoring the products of other companies.

The thiocarbamate and chloroacetamide (acid amide) herbicides are the only families of herbicides for which crop safeners are used commercially; these crop safeners are used to protect grass crops. An exception is the crop safener naphthalic anhydride, which has provided satisfactory protection to corn from chlorsulfuron, a sulfonylurea herbicide.

The application of crop safeners in the field does not involve an extra operation, as they are marketed as either formulated premixes with the respective herbicides or as pretreated crop seed (seed safeners). The timing of applications of safeners and herbicides is critical in order to avoid herbicide injury to the crop, and all currently available crop safeners are most effective when applied prior to, or simultaneously with, the herbicide. Thus formulated premixes or pretreated seed avoid misapplication as to timing and the particular safener used.

The exact mechanisms of action by which safeners protect crop plants from herbicides have not been determined, but the following two hypotheses seem most plausible:

1. Crop safeners induce rapid herbicide metabolism.
2. Crop safeners interact with the herbicides at their sites of action.

Crop safeners are used as *seed treatments* or included in premixed herbicide formulations and applied as *soil-incorporated treatments.* Common and chemical names of crop safeners are given in Table 8.1.

Seed Treatments

Some crop safeners provide protection to weed species that are botanically related to the crop. In such cases, selectivity is achieved by coating the crop seeds with the safener. In practice, seeds of corn and grain sorghum are coated with the safeners naphthalic anhydride, CGA-133205, cyometrinil, dichlormid, flurazole, or oxabetrinil.

At one time, naphthalic anhydride (NA) was used as a seed treatment to protect corn from the toxic action of soil-applied thiocarbamate herbicides such as butylate, EPTC, and vernolate. However, it also protects corn from a number of chemically unrelated herbicides such as phenylcarbamates, thiocarbamates, chloroacetanilides, sulfonylureas, imidazolinones, cyclohexanediones, and aryloxyphenoxypropionates. NA is the only crop safener capable of protecting grass crops such as corn from postemergence applications of certain herbicides (the herbicide families just noted). NA has effectively protected corn from postemergence applications of

TABLE 8.1 ■ Common and chemical names of crop safeners and one herbicide extender.

COMMON NAME	CHEMICAL NAME
CGA-133205	O-(1,3-dioxolan-2-yl-methyl)-2,2,2-trifluoro-4'-chloroacetophenone-oxime
CGA-154281	4-(dichloroacetyl)-3,4-dihydro-3-methyl-2H-1,4-benzoxazine
Cyometrinil (CGA-43089)	α-[(cyanomethoxy)imino]benzeneacetonitrile
Dichlormid (R-25788)	2,2-dichloro-N,N-di-2-propenylacetamide
Dietholate*	O,O-diethyl-O-phenyl phosphorthioate
Fenclorim (CGA-123407)	4,6-dichloro-2-phenyl-pyrimidine
Flurazole	phenylmethyl-2-chloro-4-(trifluoromethyl)benzene
MG-191	2-dichloromethyl-2-methyl-1,3-dioxolane
Naphthalic anhydride (NA)	naphthalene-1,8-dicarboxylic acid anhydride
Oxabetrinil (CGA-92194)	α-(1,3-dioxolan-2-yl-methoxy)-iminobenzeneacetonitrile
PPG-1292	2,2-dichloro-N-(1,3-dioxolan-2-yl-methyl)-N'-2-propenylacetamide
R-29148	3-(dichloroacetyl)-2,2,5-trimethyl-1,3-oxazolidine
*Herbicide extender.	

chlorsulfuron, a sulfonylurea herbicide, and its use may be commercially feasible for this purpose.

The seed safeners cyometrinil, oxabetrinil, and CGA-133205 are the active ingredients in Ciba's Concep I, Concep II, and Concep III, respectively. These safeners protect grain and forage sorghum from injury from the soil-applied herbicide metolachlor. Under field conditions, the performance of cyometrinil was inconsistent, and it frequently caused adverse effects on sorghum seed germination and stand establishment. Oxabetrinil was developed to replace cyometrinil, but it interacts unfavorably with the downy mildew of sorghum. CGA-133205 was developed to replace cyometrinil and oxabetrinil as the desired crop safener in Concep. Without Concep protection, metolachlor will generally cause severe sorghum injury.

Dichlormid is an effective seed safener, although it is more commonly used in a formulated premix with dithiocarbamate herbicides. Used as a wheat seed treatment, dichlormid provides considerable protection to wheat seedlings from triallate, a thiocarbamate herbicide, when triallate is applied preplant-incorporated (PPI).

Flurazole is a seed safener developed to protect grain sorghum from the chloroacetanilide herbicides alachlor and acetochlor. It is marketed by Monsanto as Screen, a 40% water-dispersible granular formulation.

Soil-Incorporated Treatments

The crop safener dichlormid appeared to be highly effective only as a safener of corn against the thiocarbamate and chloroacetanilide herbicides. However, it does safen corn to some degree against sethoxydim, indicating that it may be more diverse in its protective action than previously believed. Dichlormid is marketed as a formulated premix with the respective thiocarbamate herbicides, EPTC, butylate, and vernolate. These premix formulations are marketed under the trade names Eradicane (EPTC + dichlormid), Sutan+ (butylate + dichlormid), and Vernam+ (vernolate + dichlormid). Dichlormid is readily absorbed by the corn seedlings, and it serves to protect these seedlings against the phytotoxicity of the above thiocarbamate herbicides.

The crop safeners MG-191 and PPG-1292 safen corn against the dithiocarbamate herbicides EPTC and butylate. Fenclorim protects rice from the chloroacetanilide herbicide pretilachlor in Japan.

■ CHEMICAL HERBICIDE-PROTECTANTS

Chemicals may be used to protect certain soil-applied herbicides from too-rapid degradation by soil microbes. Such chemicals may be termed "herbicide extenders," as they prolong the effective life of the herbicide in the soil. This is in contrast to chemicals used as safeners to protect plants from the toxic action of certain herbicides.

Dietholate protects the herbicide EPTC from too-rapid degradation in the soil by microbe-secreted enzymes. Dietholate is one of three active ingredients in the product Eradicane Extra; the three active ingredients are EPTC (herbicide), R-29148 (crop safener), and dietholate (herbicide extender). The chemical and common names for this herbicide extender are given in Table 8.1.

■ HERBICIDE–AGROCHEMICAL INTERACTIONS

Simultaneous or sequential applications of herbicides with one or more agrochemicals (e.g., herbicides, insecticides, fungicides, plant growth regulators, fertilizers) are common in today's crop-protection practices. Herbicide interactions with agrochemicals are primarily expressed as increasing or decreasing (1) plant absorption, (2) translocation, (3) metabolism and conjugation, and (4) toxicity at site(s) of activity of one or more of the applied chemicals. These effects may be independent or interrelated. In addition, components of two or more product formulations may adversely interact with one another when mixed in the same carrier. Numerous herbicide interactions with agrochemicals have been reported. Symptoms of such interactions involving plants may range from a change in foliar pigmentation or carbon dioxide evolution to a change in internode length or death of the plants. The use of herbicide–herbicide mixtures broadens the spectrum of weed species controlled, but it may alter the selectivity of the respective herbicides involved.

The general effect on a plant or plant community resulting from the simultaneous or sequential application of two or more agrochemicals may be categorized as follows:

1. *Additive*—the total response induced by the applied pesticide mixture is *equal* to the sum total of the responses that would be induced if each pesticide were applied alone.

2. *Synergistic*—the total response induced by the applied pesticide mixture is *greater* than the expected sum total of the responses induced by each pesticide applied alone.

3. *Antagonistic*—the total response induced by the applied pesticide mixture is *less* than the expected sum total of the responses induced by each pesticide applied alone.

4. *Enhancement*—a pesticide applied in mixture with an adjuvant (e.g., nonionic surfactant or phytobland crop oil) induces a total response that is greater than that induced by the pesticide applied alone.

Additive

The herbicides desmedipham and phenmedipham are applied postemergence as a premix formulation in sugar beets for broadleaf weed control. Desmedipham is included in the mixture especially for redroot pigweed control. Their phytotoxicity is additive, each contributing to control of broadleaf weed species as if applied alone.

Synergistic

A tank mix of the herbicides clopyralid and triclopyr applied postemergence, each at 0.12 lb ai/A (16 g ai/ha), to honey mesquite killed 87% of the treated plants. Applied alone at this same dosage, neither herbicide killed more than 27% of the treated plants.

A mixture of the herbicides atrazine plus cyanazine, each at 0.75 lb ai/A, (0.8 kg ai/ha), applied early postemergence to wild oat plants in the 2- to 3-leaf stage controlled about 50% of the plants, but when tridiphane was added to the spray mixture, at 0.5 or 0.75 lb ai/A (0.6 or 0.8 kg ai/ha), wild oat control increased to about 92%. Applied alone, tridiphane had no effect against wild oat.

The herbicides sethoxydim and fluazifop-P-butyl applied postemergence in tank mix, each at 1.13 oz ai/A (79 g ai/ha), controlled green foxtail, wild oat, and volunteer wheat and barley by more than 90%. The mixture was synergistic, compared to the herbicides applied alone.

A premix formulation of the herbicides 2,4-D, mecoprop, and dicamba, containing 2.03, 1.08, and 0.21 lb ai/gal, respectively, applied postemergence, resulted in much greater control of broadleaf weeds than would be expected from the additive effects of each herbicide applied alone. The synergistic response from this mixture allows each herbicide to be applied at a lower dosage than would be required if applied alone. This herbicide mixture is recommended for selective weed control in grass lawns and turf. Although the degree of weed control is greatly increased, the selectivity of these herbicides in turf is not lost owing to their synergistic interaction.

The practice of applying herbicides and insecticides to the same crop is common. However, crop tolerance to a given herbicide may be reduced because of synergistic interaction with the insecticide, especially if the insecticide is an organophosphate. The herbicide metribuzin, applied alone to soybeans, had no detrimental effect on soybean grain yield, even at twice the recommended rate. Phorate and disulfoton are soil-applied, systemic, organophosphate insecticides registered for use in soybeans. Sequential soil-surface applications of metribuzin with either phorate or disulfoton to soybeans interacted synergistically, resulting in death of the soybean seedlings shortly after emergence.

Tank mixes of the herbicide thifensulfuron with chlorpyrifos or malathion (both organophosphate insecticides) or with carbaryl or methomyl (both carbamate insecticides) resulted in increased phytotoxicity to soybeans—a response that was synergistic. The organophosphate and carbamate insecticides are esterase inhibitors, suggesting that enhancement of thisulfuron phytotoxicity to soybeans is due to inhibition of thifensulfuron de-esterification. Soybean tolerance to thifensulfuron is due almost exclusively to its detoxification by a de-esterification reaction.

Primisulfuron is a postemergence herbicide for use in field corn. Applied to corn in the 3- to 5-leaf stage following the soil-applied organophosphate insecticides disulfoton, fonofos, isozophos, or turbufos, primisulfuron injured corn and significantly reduced grain yields. However, no adverse interaction occurred when primisulfuron application followed the soil-applied carbamate insecticides carbaryl or carbofuran.

Additions of fertilizers containing liquid ammonium [e.g., urea ammonium nitrate (28-0-0), ammonium polyphosphate (10-34-0)] to the aqueous carrier at a rate of 10% (v/v) can greatly increase the activity of certain foliar-applied herbicides (e.g., 2,4-D and glyphosate). However, the control of giant foxtail and volunteer corn by the graminicides sethoxydim or quizalofop was not enhanced by such additions.

Antagonistic

One would expect that a mixture of postemergence grass and broadleaf herbicides would be advantageous, controlling both kinds of weeds. However, some tank mixes of grass and broadleaf herbicides are antagonistic and result in poor weed control, especially with respect to grasses. Often, the antagonistic effect of broadleaf herbicides to graminicides can be overcome by increasing the dosage of the grass herbicide by 50% or more.

The control of annual grasses was reduced by 65% when sethoxydim, at 0.25 lb ai/A (0.28 kg ai/ha), was applied postemergence in tank mix with 2,4-DB (a broadleaf herbicide), as compared to sethoxydim applied alone. However, increasing the dosage of sethoxydim in the herbicide mixture to 0.37 lb ai/A (0.4 kg ai/ha) overcame the antagonistic effect of 2,4-DB (the amount of 2,4-DB was unchanged).

Johnsongrass control by haloxyfop-methyl was reduced when applied postemergence in tank mix with 2,4-D. The antagonistic interaction modified the action of haloxyfop-methyl by reducing absorption and translocation and increasing metabolism (detoxification). The phytotoxicity of haloxyfop-methyl to johnsongrass was unaffected when it was applied as a sequential treatment 24 hours after 2,4-D. When 2,4-D was applied to johnsongrass 72 hours after haloxyfop-methyl, control was equal to that of haloxyfop-methyl applied alone.

The phytotoxicity of the graminicide diclofop-methyl to wild oat was nullified when either 2,4-D, MCPA, or dicamba was applied in the tank mix with diclofop-methyl. The antagonistic effect occurred because of decreased absorption of diclofop-methyl, reduced hydrolysis of diclofop-methyl to diclofop (parent-acid form), increased conjugation of diclofop, and reduced translocation of diclofop. The phytotoxicity of the broadleaf herbicides was not impaired by the presence of diclofop-methyl.

Tank mixes of imazethapyr (a broadleaf herbicide) with the grass herbicides clethodim, fluazifop-P-butyl, quizalofop-ethyl, or sethoxydim applied postemergence resulted in antagonistic interactions, with reduced control of large crabgrass, fall panicum, and broadleaf signalgrass when compared to the control obtained with each graminicide applied alone. When imazethapyr was applied 5 days before or 1 day after application of the grass herbicides, there was no antagonistic effect and grass control was equal to that of each graminicide applied alone.

Tank mixes of chlorimuron or imazaquin with acifluorfen or fomesafen usually resulted in antagonistic interactions when applied to broadleaf weeds (e.g., common cocklebur), but response varied with the rate and plant species and the size of the plants at time of treatment.

The phytotoxicity of EPTC to sorghum and giant foxtail was reduced by 2,4-D when the two herbicides were applied in the same spray mixture. However, the phytotoxicity of the mixture to broadleaf weeds was additive.

The toxicity of glyphosate to johnsongrass was reduced by including 2,4-D or dicamba in the spray mixture. This antagonistic effect was due to reduced foliar absorption and translocation of glyphosate.

The control of perennial broadleaf weeds with glyphosate can be enhanced in certain situations by the addition of broadleaf herbicides (e.g., 2,4-D or dicamba) to the spray mixture. Applied to field bindweed, mixtures of glyphosate and 2,4-D or dicamba have additive, sometimes synergistic, interactions. Phytotoxicity is enhanced because of the greater absorption and translocation of both herbicides to the roots of field bindweed. Lower glyphosate rates combined with 2,4-D or dicamba were as effective as higher glyphosate rates applied alone. It is of interest to note that combinations of 2,4-D or dicamba with glyphosate *increased* control of field bindweed (a perennial broadleaf weed) while, as previously noted, similar combinations *reduced* control of johnsongrass (a perennial grass weed).

Enhancement

The enhanced phytotoxicity resulting from mixing herbicides with spray adjuvants (e.g., surfactants, phytobland oils) is well documented. Surfactants are a common ingredient of most herbicide formulations and, if not included in the formulation, the directions on the product label often advise the user to include a nonionic surfactant to the spray mixture. For example, in the application of bentazon, the addition of a phytobland oil to the spray solution is highly recommended for effective weed control. The use of spray adjuvants in herbicide spray mixtures enhances phytotoxicity largely because of the increased foliar absorption of the herbicide.

■ HERBICIDE-RESISTANT WEED BIOTYPES

It is known that not all plants of a given weed species are killed by a particular herbicide or by herbicides having the exact same mode of action. It is apparent that biotypes (genotypes) of a number of weed species exist, some of which are susceptible to an applied herbicide, while others are tolerant. Thus susceptible biotypes are selectively controlled by the applied herbicides, while tolerant biotypes are not killed and continue to propagate themselves. During the past 10 years or so, great strides have been made in understanding herbicide-resistant weed biotypes and in the genetic development of crop varieties that are in themselves tolerant of certain applied herbicides.

Although the resistance of plants to herbicides results in selective plant control, the topic may logically be included in a chapter devoted to selectivity. However, herbicide resistance and transgenic crop tolerance are worthy of a chapter of their own, and the next chapter is devoted to a discussion of these topics.

■ SELECTED REFERENCES

Ahrens, W. H. 1990. Enhancement of soybean (*Glycine max*) injury and weed control by thifensulfuron–insecticide mixtures. *Weed Technol.* 4:524–528.

Ahrens, W. H., and R. J. Ehr. 1991. Tridiphane enhances wild oat (*Avena fatua*) control by atrazine–cyanazine mixtures. *Weed Technol.* 5:799–804.

Baltazar, A. M., and T. J. Monaco. 1984. Uptake, translocation, and metabolism of bentazon by two pepper species (*Capsicum annuum* and *Capsicum chinense*). *Weed Sci.* 32:258–263.

Baltazar, A. M., T. J. Monaco, and D. M. Peele. 1984. Bentazon selectivity in hot pepper (*Capsicum chinense*) and sweet pepper (*Capsicum annuum*). *Weed Sci.* 32:243–246.

Barrentine, J. L., and G. F. Warren. 1970. Selective action of terbacil on peppermint and ivyleaf morningglory. *Weed Sci.* 18:373–377.

Beckett, T. H., E. W. Stoller, and L. E. Bode. 1992. Quizalofop and sethoxydim activity as affected by adjuvants and ammonium fertilizers. *Weed Sci.* 40:12–19.

Biediger, D. L., et al. 1992. Interactions between primisulfuron and selected soil-applied insecticides in corn (*Zea mays*). *Weed Technol.* 6:806–812.

Bolt, P. F., and A. R. Putnam. 1981. Selectivity mechanisms for foliar applications of diclofop-methyl. II. Metabolism. *Weed Sci.* 29:237–241.

Bovey, R. W., and S. G. Whisenant. 1992. Honey mesquite (*Prosopis glandulosa*) control by synergistic action of clopyralid–triclopyr mixtures. *Weed Sci.* 40:563–567.

Chandler, J. M., and F. E. Fulgham. 1978. Influence of placement of charcoal on protection of cotton (*Gossympium hirsutum*) from diuron. *Weed Sci.* 26:239–244.

Christianson, M. L. 1991. Fun with mutants: Applying genetic methods to problems of weed physiology. *Weed Sci.* 38:489–496.

Derting, C. W. 1987. Wiper application. In *Methods of applying herbicides*, C. G. McWhorter and M. R. Genhardt, eds. Monograph No. 4, pp. 207–229. Champaign, Ill.: Weed Science Society of America.

Flint, J. L., and M. Barrett. 1989. Effects of glyphosate combinations with 2,4-D or dicamba on field bindweed (*Convolvulus arvensis*). *Weed Sci.* 37:12–18.

Flint, J. L., and M. Barrett. 1989. Antagonism of glyphosate toxicity to johnsongrass (*Sorghum halepense*) by 2,4-D and dicamba. *Weed Sci.* 37:700–705.

Foley, M. E. 1985. Response differences of wheat (*Triticum aestivum*) and barley (*Hordeum vulgare*) to chlorsulfuron. *Weed Sci.* 34:17–21.

Fuerst, E. P. 1987. Understanding the mode of action of the chloroacetamide and thiocarbamate herbicides. *Weed Technol.* 1:270–277.

Green, J. M., and S. P. Bailey. 1987. Herbicide interactions with herbicides and other agricultural chemicals. In *Methods of applying herbicides*, C. G. McWhorter and M. R. Gebhardt, eds. Monograph No. 4, pp. 37–61. Champaign, Ill.: Weed Science Society of America.

Grichar, W. J. 1991. Sethoxydim and broadleaf herbicide interaction effects on annual grass control in peanuts (*Arachis hypogyaea*). *Weed Technol.* 5:321–324.

Hageman, L. H., and R. Behrens. 1984. Basis for response differences of two broadleaf weeds to chlorsulfuron. *Weed Sci.* 32:162–167.

Hall, J. C., and W. H. Vanden Horn. 1988. The absence of a role of absorption, translocation, and metabolism in the selectivity of picloram and clopyralid in two plant species. *Weed Sci.* 36:9–14.

Harker, K. N., and P. A. O'Sullivan. 1991. Synergistic mixtures of sethoxydim and fluazifop-butyl on annual grass weeds. *Weed Technol.* 5:310–316.

Hatzios, K. K. 1983. Herbicide antidotes: Development, chemistry, and mode of action. *Adv. Agron.* 36:265–316.

Hatzios, K. K., and R. E. Hoagland, eds. 1989. *Crop safeners for herbicides.* New York: Academic Press.

Hatzios, K. K., and D. Penner. 1985. Interactions of herbicides with other agrochemicals in higher plants. *Rev. Weed Sci.* 1:1–63.

Hayes, R. M., et al. 1979. Interaction of selected insecticide–herbicide combinations on soybeans (*Glycine max*). *Weed Sci.* 27:51–54.

Jensen, K. I. A., G. R. Stephenson, and L. A. Hunt. 1977. Detoxification of atrazine in three *Gramineae* subfamilies. *Weed Sci.* 25:212–220.

Mangeot, B. L., F. E. Slife, and C. E. Rieck. 1979. Differential metabolism of metribuzin by two soybean (*Glycine max*) cultivars. *Weed Sci.* 27:267–269.

McMullen, P. M., and J. D. Nalewaja. 1991. Triallate antidotes for wheat (*Triticum aestivum*). *Weed Sci.* 39:57–61.

Mueller, T. C., M. Barrett, and W. W. Witt. 1990. A basis for the antagonistic effect of 2,4-D on haloxyfop-methyl toxicity to johnsongrass (*Sorghum halepense*). *Weed Sci.* 38:103–107.

Myers, P. F., and H. D. Coble. 1992. Antagonism of graminicide activity on annual grass species by imazethapyr. *Weed Technol.* 6:333–338.

Pallos, F. M., and J. E. Casida, eds. 1978. *Chemistry and action of herbicide antidotes.* New York: Academic Press.

Rowe, L., J. J. Kells, and D. Penner. 1991. Efficacy and mode of action of CGA-154281, a protectant for corn (*Zea mays*) from metolachlor injury. *Weed Sci.* 39:78–82.

Shimabukuro, R. H., W. C. Walsh, and R. A. Hoerauf. 1979. Metabolism and selectivity of diclofop-methyl in wild oat and wheat. *J. Agri. Food Chem.* 27:615–623.

Sweetzer, P. B., G. S. Schow, and J. M. Hutchison. 1982. Metabolism of chlorsulfuron by plants: Biological basis for selectivity of a new herbicide for cereals. *Pestic. Biochem. Physiol.* 17:18–32.

Swisher, B. A., and M. R. Weimer. 1986. Comparative detoxification of chorsulfuron in leaf disks and cell cultures of two perennial weeds. *Weed Sci.* 34:507–512.

Tucker, T. A., K. A. Langeland, and F. T. Corbin. 1994. Absorption and translocation of ^{14}C-imazapyr and ^{14}C-glyphosate in alligatorweed *Alternanthera philoxeroides*. *Weed Technol.* 8:32–36.

Waldrop, D. D., and P. A. Banks. 1983. Interactions of herbicides with insecticides in soybean (*Glycine max*). *Weed Sci.* 31:730–734.

Wesley, M. T., and D. R. Shaw. 1992. Interactions of diphenylether herbicides with chlorimuron and imazaquin. *Weed Technol.* 6:345–351.

9

WEED RESISTANCE AND TRANSGENIC CROP TOLERANCE

■ INTRODUCTION

The repeated use of herbicides with similar modes of action on the same site over a period of years has resulted in weed biotypes that are resistant to such herbicides. Since 1970, herbicide-resistant weed biotypes have become well known in scientific and agricultural communities around the world, and the rapid increase in these biotypes is, of course, of great concern to agriculturalists and herbicide manufacturers. In some cases, manufacturers have made changes in product label recommendations.

Research into the mechanisms responsible for herbicide-resistant weed biotypes has identified the mechanisms as genetic. These mechanisms can apparently be attributed to specific (one or very few) genes, and different levels of herbicide resistance have also been identified. This research has enabled scientists to transfer such genetic traits to crop plants, thereby strengthening their resistance to certain herbicides.

The most serious weed problems for a crop can be solved by developing that crop's tolerance to the herbicides that best control the problem weeds. In the past, development of herbicide-tolerant crops has relied on conventional genetic selection and breeding practices. It is now feasible to isolate and insert the genes that carry the herbicide-resistant characteristic from one plant or microorganism to another—a process termed *genetic engineering*. Interest in developing herbicide-tolerant crop cultivars, clones, or hybrids has been spurred by (1) the rising cost of developing new herbicides, (2) fewer new herbicides being discovered, and (3) advances in genetic engineering technology (Harrison, 1992). As with any new technology, there is both concern and optimism about the use of transgenic, herbicide-resistant crop cultivars.

■ WEED RESISTANCE

Pesticides are major technological tools that are used successfully throughout the world. An adverse consequence of their use has been the emergence of pesticide resistance among target pest populations. Such resistance is worldwide, and it exists for all forms of pesticides such as bactericides, fungicides, herbicides, insecticides, nematicides, and rodentacides.

Some weed species have biotypes that are resistant to a particular herbicide or group(s) of herbicides. Since 1970, following the discovery of the resistance of a biotype of common groundsel to the s-triazine herbicides simazine and atrazine, the phenomenon of herbicide resistance in weeds has become well known in scientific and agricultural communities around the world. Prior to 1970, the only confirmed instance of herbicide resistance attributed to selection by repeated use of the same herbicide was the differential susceptibility of wild carrot to 2,4-D, which occurred in Canada and was reported in 1963. At the time, this report was deemed a curiosity rather than an omen by the scientific community.

The differential response of plants to herbicides has generally been attributed to differences in morphological and physiological factors between and within species that affect the accumulation of the herbicide in toxic amounts at their sites of action. The failure of a herbicide to provide the anticipated control is often first attributed to misapplication or unsuitable environmental conditions. When these factors are successfully eliminated, herbicide resistance is often the reason for poor weed control.

More than 57 weed species, including at least 40 dicots and 17 monocots, have been reported to have biotypes resistant to the *triazine* herbicides. Examples of weed species reported to have triazine-resistant biotypes are given in Table 9.1. These biotypes have been found in 33 states, 4 provinces of Canada, 18 countries in Europe, and also in Australia, Israel, Japan, Spain, and New Zealand. These are countries from which information is readily available and in which large amounts of herbicides, often the same or similar herbicides, are used repeatedly in monocultural and limited-rotation cropping. In addition, 47 weed species have been reported to have biotypes resistant to one or more of 14 herbicides or groups of herbicides other than the triazines (Table 9.2). The rapid increase in the number of cases of herbicide resistance reported over the past few years in locations distributed throughout the world verify that the phenomenon is global.

Almost all instances of herbicide-resistant weed biotypes have occurred in situations where the same herbicide or herbicides with the same mode of action were applied repeatedly on the same site for a period of years—for example, in the cases cited previously, 5–10 years for triazine herbicides and 3–5 years for the sulfonylureas.

There is no evidence that any herbicide-resistant weed biotype has occurred through mutations caused by a particular herbicide, nor is there evidence to show whether or not these resistant biotypes were already present prior to herbicide treatment at the sites where they were detected. Herbicide-resistant weed biotypes are presumed to arise, through naturally occurring mutations, from small, preex-

TABLE 9.1 ■ Dicot and monocot weed species with triazine-resistant biotypes.

DICOTS	MONOCOTS
Beggarticks, bur	Barnyardgrass
Burcucumber*	Bluegrass, annual
Chickweed, common	Brome, downy
Galinsoga, hairy	Brome, false
Groundcherry, longleaf*	Canarygrass, hood
Groundsel, common*	Crabgrass, large
Horseweed	Foxtail, rigid
Kochia*	Polypogon, rabbitfoot
Lambsquarters, goosefoot*	Ryegrass, rigid
Mustard, birdsrape	Witchgrass
Mustard, wild	
Nightshade, black*	
Orach, spreading	
Pigweed, amaranth*	
Pineappleweed	
Ragweed, common*	
Shepherd's-purse	
Sowthistle, spiny	
Starwort, water*	
Thymeleaf, sandwort	
Velvetleaf*	
Willowweed	
*Denotes resistant biotypes reported in the United States.	

SOURCE: Adapted from J. S. Holt and H. M. LeBaron. 1990. Significance and distribution of herbicide resistance. *Weed Technol.* 4:145.

isting populations of the species. Resistance becomes apparent when herbicide selection kills off the herbicide-susceptible plants, leaving the herbicide-resistant biotypes.

Although the widespread occurrence and multiplication of herbicide resistance in many troublesome weed species common to crop and noncrop areas has raised concern among weed scientists and others in the agricultural community, these weeds can be controlled. Herbicide resistance in weed croplands was brought about by relying heavily on one method (chemical control), rather than utilizing a combination of weed-control practices. Suggested recommendations to prevent or reduce the occurrence of herbicide-resistant weed biotypes include: (1) increased cultivation, (2) crop rotation, (3) rotation of herbicides with different modes of action, (4) use of herbicide mixtures with different modes of action, and (5) preventing seed-set of resistant biotypes. In spite of the global extent of herbicide resistance in weeds, there have

TABLE 9.2 ■ Weed species with other herbicide-resistant biotypes.*

HERBICIDE	WEED SPECIES
Amitrole	Bluegrass, annual
	Ryegrass, rigid
Bromoxynil	Lambsquarters, common
Carbamates	Pigweed, redroot
	Pigweed, smooth
Chlortoluron (substituted urea)	Foxtail, rigid
	Horseweed
	Lambsquarters, common
	Pigweed, redroot
	Pigweed, smooth
Diclofop-methyl	Foxtail, rigid
	Oat, wild
	Ryegrass, Italian**
	Ryegrass, rigid
DSMA and MSMA (arsenicals)	Cocklebur, common**
Paraquat and diquat	Barley, hare
	Barley, wall
	Bluegrass, annual
	Capeweed
	Fleabane, hairy
	Fleabane, Philadelphia
	Hawksbeard, Asiatic
	Horseweed
	Oharechinoqiku
	Ryegrass, perennial
	Willowherb, American
Phenoxys:	
Mecoprop	Chickweed, common
2,4-D	Carrot, wild
	Gooseweed
	Thistle, Canada
Propanil (acid amide)	Barnyardgrass
	Junglerice
Pyrazon	Lambsquarters, common
Sulfonylureas, imidazolines, and trizolopyrimidine	Chickweed, common
	Cocklebur**
	Hatico
	Kochia**
	Lettuce, prickly**
	Ryegrass, Italian**
	Ryegrass, rigid
	Thistle, Russian**
Trifluralin (dinitroanilines)	Foxtail, green
	Goosegrass**
Uracils	Pigweed, redroot
	Pigweed, smooth

*Herbicides other than the triazine family.
**Denotes resistant biotypes reported in the United States.

SOURCE: Adapted from J. S. Holt and H. M. LeBaron. Significance and distribution of herbicide resistance. 1990. *Weed Technol.* 4:146.

been no reported instances where the resistant biotypes could not be controlled by alternative means.

■ PLANT RESPONSES

There are three types of plant responses to applied herbicides, and they are typically characterized as: (1) *susceptibility,* (2) *tolerance,* and (3) *resistance.* Susceptibility results in the plant being damaged or killed by the applied herbicide. Tolerance and resistance both describe a condition whereby a plant withstands a herbicide, but, when unqualified, their use often results in confusion. Until the occurrence of what is now termed "herbicide resistance," the term *tolerance* was most often used to designate crop response to a herbicide, and a "tolerant crop" was one that was not significantly injured by a herbicide applied at a recommended dosage. The terms *tolerance* and *resistance* were used interchangeably.

It would appear desirable, in view of recent developments, to restrict the use of the term *tolerance* to crop plants. A workable definition of tolerance is *the ability of a crop plant to withstand a predetermined dosage of a herbicide, which may be overcome by higher dosages.* Thus tolerance is rate dependent, and it is largely attributed to differential factors—*absorption, translocation,* and/or *detoxification* (metabolism)—factors that are morphological or physiological.

Where herbicide resistance has been established in a crop species *via* genetic engineering, it seems appropriate to use the term *transgenic tolerance* to denote this phenomenon and to delineate it from herbicide resistance brought about by a naturally occurring mutant as in weed species.

Susceptibility

Plant susceptibility is a positive response to an applied herbicide. The degree of response is a measure of a plant's susceptibility to the applied herbicide, under the conditions involved. The degree of susceptibility may range from an imperceptible response, to a marked growth abnormality, or to death of the plant.

Seedlings and young plants in the rosette stage or less than 4 in. (10 cm) high are more susceptible than older plants, especially to contact-type herbicides. However, older, established plants may be very susceptible to certain herbicides, especially those that translocate downward in the phloem tissue. Plant species commonly differ in their susceptibility to herbicides in general, as well as to a given herbicide. Susceptibility is influenced by:

1. Stage of plant growth when treated
2. Herbicide concentration absorbed by plant
3. Inherent toxicity of applied herbicide
4. Environmental factors such as light, moisture, and temperature

Plants are most susceptible to applied herbicides during the following stages of growth: when rapid growth is taking place and when rapid growth has ended and food reserves are temporarily depleted or exhausted. To be effective, a herbicide must be absorbed and moved from the sites of absorption to the sites of phytotoxicity in sufficient concentration to bring about the desired effect. The inherent toxicity of a herbicide is determined by its molecular composition and configuration.

Plants growing in subdued light are generally more susceptible to foliar-applied herbicides than are those growing in high light intensity, apparently owing to the formation of a thinner, more permeable cuticle formed under subdued light. Water-stressed plants are less susceptible to foliar-applied herbicides than those not water-stressed, due to a less expanded, and therefore less permeable, cuticle. In general, plants are less susceptible to herbicides applied when temperatures are below or above those favoring growth. The specific temperatures influencing susceptibility vary with plant species; for example, winter annuals may respond to an applied herbicide at a temperature as low as 50°F (10°C), while summer annuals may not respond until temperatures are 60°F (16°C) or higher. Soil characteristics primarily influence the *availability*, rather than the susceptibility, of soil-applied herbicides.

Annual plants are most susceptible to herbicides in their seedling stage and during periods of vigorous growth while less than 4 in. (10 cm) high. Applied to annuals in the bud to early-bud stage, herbicides may prevent seed formation, even though the plants are not killed. In the event that seeds do form, the herbicides may accumulate in the seed, adversely affecting subsequent seed germination and seedling growth.

Perennial plants are most susceptible to herbicides in their seedling stage. However, older, established perennials are also susceptible to foliar-applied, phloem-translocated herbicides when vigorous growth is *not* taking place. With certain perennials (e.g., Canada thistle), such growth stages correspond to periods when rosettes are fully developed, but before bolting (stem elongation) occurs, or when plants are in bud to early bloom. In either case, a period of rapid growth has ended, depleting food reserves, and the plant has yet to replenish these reserves. In addition, large leaf areas are present to intercept and absorb the applied herbicides.

Resistance

The discovery that resistance to the triazine herbicides was due to a site-of-action mutation in resistant weed biotypes led to the widespread use of the term "resistance" to denote this phenomenon. In view of this, it seems appropriate to restrict the use of resistance to weed species, denoting herbicide resistance in a biotype of the species. A workable definition of resistance is *the ability of a biotype to survive treatment with a given herbicide to which the weed species is normally susceptible.* Resistance is dependent on the selection or evolution of a mutant mechanism within a biotype that allows it to withstand repeated exposure to a herbicide. Resistance is then passed on to the mutant biotype's progeny.

As with tolerance and resistance, there is a need for consistency in terms used to denote the various aspects of herbicide resistance in weed biotypes. There appear to be three types of resistance as it relates to herbicides and weeds, namely: (1) *herbicide resistance,* (2) *cross resistance,* and (3) *multiple resistance.*

Herbicide resistance refers to a weed biotype that is resistant to only one specified herbicide, as in the case of resistance to amitrole or glyphosate. However, the term may also be used in a general sense to denote the phenomenon of herbicide resistance in weed biotypes.

Cross resistance refers to a weed biotype that is resistant to two or more *chemically similar herbicides,* as are those grouped in the same herbicide family, and that have the *same mode of action.* For example, a Powell amaranth biotype is cross-resistant to the uracil herbicides bromacil and terbacil.

Multiple resistance refers to weed biotypes resistant to two or more individual or series of *chemically unrelated herbicides,* as are those in different herbicide families, which have *different modes of action.*

A common feature of herbicides with resistant biotypes is their specificity for a single target site, which is often determined by a single gene or by very few genes. These randomly occurring genetic mutations may be easily selected in situations where a herbicide is used repeatedly. The mechanisms of resistance have become the focus of bioengineering, and many such transfers to crop plants have already taken place.

Mechanisms of Resistance

Weed-biotype resistance to herbicides is attributed to the following mechanisms: (1) modified photosystem II protein-binding site (triazines and uracils) (2) modified ALS-binding site (sulfonylureas and imidazolinones), (3) enhanced detoxification (aryloxyphenoxypropionates and substituted ureas), (4) enhanced detoxification and/or sequestration (bipyridiniums), and (5) hyperstabilized microtubules (dinitroanilines).

Resistance to s-Triazines

The mode of action of the s-triazine herbicides is inhibition of electron transport in photosystem II (PS II) of photosynthesis. Herbicides that inhibit PS II appear to compete for a common binding site on the chloroplast thylakoid membrane. *A triazine-binding-site alteration (mutation) in the chloroplast membrane of triazine-resistant biotypes is responsible for triazine resistance.*

In the fall of 1968, it was observed that simazine and atrazine failed to control common groundsel in a nursery in western Washington after having been used once or twice annually since 1958. Simazine had been satisfactorily used preemergence for the first 8 years, and atrazine had been used postemergence in the last 2 years. Subsequently, seedlings from seed collected from the nursery site were not controlled by preemergence applications of either simazine or atrazine at rates up to 16 lb/A (18 kg/ha). Seedlings from seed collected at a site where triazines had not been used were killed by only 1 or 2 lb/A (1 or 2.25 kg/ha) of either herbicide. Postemergence application of atrazine at 1, 2, or 4 lb/A had no effect on seedlings from the nursery source, but killed seedlings from the other source. The resistant biotype was controlled by alternative herbicides and other means, and the resistant biotype did not spread out of western Washington. It was not then considered of practical importance.

Similar triazine-resistant biotypes were reported within species of Powell amaranth in 1968 in a Washington sweet corn field and near Montrose, Ontario, in 1979. By 1989, triazine resistance had been confirmed in eight *Amaranthus* species. The reader is referred to Ahrens et al. (1981) for corrected identification of triazine-resistant *Amaranthus* species reported as *A. retroflexus*. As stated earlier in the chapter, more than 57 weed species are now known to have biotypes resistant to the s-triazine herbicides.

Resistance in triazine-resistant weed biotypes occurs at the herbicides' site of action (i.e., the site of electron-transport inhibition on the reducing side of PS II). The molecular basis for s-triazine resistance is a mutation in the D1 thylakoid protein (also called Q^b protein). The D1 protein is coded by the chloroplast *psb*A gene. At least five sites have been identified in the D1 protein where amino acid substitution results in reduced herbicide binding and resistance, but only one naturally occurring mutation has been identified in triazine-resistant weed biotypes. Thus triazine resistance in weed biotypes is due to a point mutation on the chloroplast *psb*A gene that encodes for an altered protein in the thylakoid membrane in the chloroplast. A polypeptide of about 32 kilodaltons[1] (kd) in the thylakoid membrane has been shown to contain the atrazine-binding site, and a change in a single amino acid (from serine to glycine) in the polypeptide is the alteration conferring triazine resistance in both smooth pigweed and black nightshade. Photosynthetic electron transport is roughly 1000 times *less sensitive* to the triazine herbicides in chloroplasts of resistant biotypes, compared to chloroplasts from nonresistant plants of the species.

The pattern of cross resistance is similar for the triazine-resistant biotypes of smooth pigweed, common lambsquarters, common groundsel, and canola (a crop plant), each with cross resistance to bromacil, pyrazon, and buthidazole at the whole-plant level. They exhibit extreme cross resistance to the triazine, triazinone, phenyl pyridazinone, uracil, and quinazoline herbicides (all having the same mode of action) evaluated on chloroplast thylakoid membranes. Triazine-resistant biotypes of smooth pigweed do not exhibit cross resistance to herbicides that do not inhibit photosynthesis.

Corn contains the same 32 kd polypeptide in its chloroplast membrane as in triazine-susceptible weed species, but it exhibits a natural tolerance to atrazine. However, the basis for corn tolerance is herbicide detoxification, rather than to a modified binding site. Detoxification has not been shown to be a primary basis for resistance in triazine-resistant weeds.

Resistance to Sulfonylurea and Imidazolinone Herbicides

The mode of action of the sulfonylurea and imidazolinone herbicides is the inhibition of the enzyme acetolactate synthase (ALS). *Resistance to ALS-inhibitor herbicides is due to an altered form of ALS that no longer binds the herbicide.* Resistance is controlled by a single nuclear gene

---

1. Recall from Chapter 7 that the terms *dalton* and *molecular weight* are used interchangeably; for example, a 32 kd polypeptide has a molecular weight of 32,000.

with incomplete dominance. The specific alteration responsible for resistance has been further identified as an amino acid substitution at position 197, which results from a point mutation in the gene encoding ALS. ALS-inhibitor–resistant biotypes exhibit cross resistance to the sulfonylurea and imidazolinone herbicides.

The DNA sequence of a 196-base-pair region of the acetolactate synthase genes encompasses the coding sequence for Domain A, a region of the amino acid sequence that plays a pivotal role in conferring ALS-inhibitor resistance to herbicides. The Domain A DNA sequence may vary among resistant weed biotypes, depending on species. For example, a chlorsulfuron-resistant biotype of kochia differed from a susceptible biotype by a single-point mutation, with substitution of the amino acid threonine for proline, by histidine for proline in prickly lettuce, by serine for proline in mouseearcress, and glutamine for proline in tobacco, a crop plant. Thus differences in ALS sensitivity of various herbicide families may be due to slightly different binding domains within a common binding site on the protein.

Weed species resistant to ALS inhibitors are an increasing problem in the western United States and Canada, and they include kochia, prickly lettuce, rigid ryegrass, Russian thistle, common chickweed, and perennial ryegrass. Examples of biotype resistance to ALS-inhibitor herbicides in the first two of these species are presented here.

Kochia In the United States in 1987, chlorsulfuron was applied to over 6.2 million A (2.5 ha) of wheat, most of which was in monoculture production. The first report of a sulfonylurea-resistant biotype of kochia (a major broadleaf weed in this acreage) was made in 1990. It was isolated in a Kansas wheat field that had been used for monoculture wheat production and treated with chlorsulfuron for 5 consecutive years, from 1982 to 1987, with a total amount of 1.5 oz/A (105 g/ha) of chlorsulfuron applied over the 5-year period. By late 1991, kochia-resistant biotypes were identified in 411 agricultural and 24 vegetation management sites. In Montana alone, kochia-resistant biotypes were confirmed in 259 agricultural sites. Field observations indicate that sulfonylurea-resistant kochia may germinate more rapidly and/or at lower soil temperatures than susceptible kochia, apparently because of the elevated levels of branched-chain amino acids available for cell division and growth during early germination. Due to concern over the chlorsulfuron-resistant biotypes of kochia, chlorsulfuron is no longer available for weed control in dryland wheat production for much of the Great Plains, and in Utah and southern Idaho, a 48-month application interval is required between chlorsul-

furon applications. Herbicides with different modes of action (e.g., atrazine, bromoxynil, MCPA, and diuron) effectively controlled the resistant biotypes of kochia.

Prickly lettuce In 1987, the first sulfonylurea-resistant prickly lettuce biotype was discovered in a field of monoculture winter wheat near Lewiston, Idaho. The winter wheat had been grown under no-till cultivation. Since 1982, chlorsulfuron had been used for control of prickly lettuce and other broadleaf weeds and applied at 6- to 14-month intervals to the crop and for chemical fallow; no other broadleaf herbicides were used during this period. Since spring 1987, other herbicides were used that effectively controlled prickly lettuce. There was a decrease in the proportion of sulfonylurea-resistant plants from 1988 to 1990, a period when sulfonylurea herbicides were no longer in use.

The resistant-biotype of prickly lettuce exhibited cross resistance to eight sulfonylurea herbicides and to the imidazolinone herbicides imazapyr and imazethypyr, but not to imazaquin. It showed no resistance to atrazine, bromoxynil, diuron, MCPA, picloram, or clopyralid plus 2,4-D—herbicides with modes of action different from the sulfonylurea and imidazolinone herbicides.

Resistance to Aryloxyphenoxypropionate and Cyclohexanedione Herbicides

The mode of action of the aryloxyphenoxypropionate and cyclohexanedione herbicides is inhibition of the biosynthesis of *de novo* fatty acids in grass species. *The mechanism in resistant biotypes of rigid ryegrass appears to be the enhanced ability to metabolize (detoxify) herbicides by elevated activity of the polysaturate mono-oxygenase suite of enzymes.*

Rigid ryegrass Rigid ryegrass, an annual grass weed, is possibly the most important weed of Australian crops. Since 1978, diclofop-methyl had been widely and successfully used for postemergent control of rigid ryegrass in cereal and legume crops. Since 1982, the sulfonylurea herbicide chlorsulfuron had become widely used for selective control of rigid ryegrass and other weeds in cereal crops.

Resistance to the aryloxyphenoxypropionate herbicide diclofop-methyl was first discovered in a biotype of rigid ryegrass in the Bordertown region of South Australia in 1982. Subsequently, it was found that the resistant biotype exhibited cross resistance to other aryloxyphenoxypropionate and cyclohexanedione herbicides and multiple resistance to the sulfonylurea and dinitroaniline herbicides and to the triazinone herbicide metribuzin. The multiple resistance exhibited by rigid ryegrass was the first report of this type of herbicide resistance.

The primary site of action of the aryloxyphenoxypropionate and cyclohexanedione herbicides is the plastid enzyme acetyl-CoA carboxylase, and the primary site of action of the sulfonylurea herbicides is the plastid enzyme acetolactate synthase. Multiple resistance in rigid ryegrass is not related to changes in either of these enzymes.

Resistant biotypes of rigid ryegrass can be controlled in croplands and pastures with simazine, oxyfluorfen, isoproturon, carbetamide, glyphosate, or paraquat. To avoid evolution of additional herbicide-resistant weed biotypes, a judicious, integrated weed-control program should be followed.

Interestingly, a mixture of amitrole plus atrazine was applied once annually for more than 10 years (to 1988) for the control of rigid ryegrass along 3000 mi (5000 km) of railway track in Western Australia. By 1988, a biotype of rigid ryegrass had developed resistance to both amitrole and atrazine applied pre- or postemergence. In addition, the biotype exhibited cross resistance to other triazine, triazinone, and phenylurea herbicides to which the weed population had never been exposed. Studies with isolated chloroplast thylakoids from the rigid ryegrass biotype resistant to the triazole, triazine, and phenylurea herbicides show that photosystem II activity is normally inhibited by exposure to triazine and phenylurea herbicides. Therefore, a modified PS II is not the mechanism of resistance in this cross-resistant biotype. The mechanism of cross resistance may be an elevated capacity to metabolize these herbicides.

An alternate herbicide combination of glyphosate and sulfometuron is now in use to control rigid ryegrass and other weeds along the railway tracks in Western Australia. These herbicides have different modes of action, and both control the amitrole- and atrazine-resistant biotype. However, it is not generally appreciated that if glyphosate and sulfometuron are used repeatedly in the same manner as amitrole and atrazine, then resistance to these herbicides will also develop.

Wild oat Resistance to aryloxyphenoxypropionate and cyclohexanedione herbicides have been reported in biotypes of wild oat in the United States, Canada, and Australia. In Australia, wild oat biotypes exhibited cross resistance to eight aryloxyphenoxypropionate herbicides and to three cyclohexanedione herbicides. These cross-resistant biotypes have not exhibited multiple resistance to herbicides with different modes of action. The resistant biotypes of wild oat were 2.7 times more resistant to diclofop-methyl than were nonresistant plants. However, resistant wild oat biotypes tend to vary in their levels of resistance and in cross resistance to the aryloxyphenoxypropionate and cyclohexanedione herbicides.

Resistance to Substituted Urea Herbicides

The mode of action of the substituted urea herbicides is inhibition of PS II. *Resistance to the substituted urea herbicides is due to the enhanced ability of plants to degrade the herbicide by the oxidative processes of N-dealkylation and ring alkyloxidation.* There is no evidence that resistance occurs at the site of herbicide action.

In Great Britain in 1985, resistance to substituted urea herbicides was reported in biotypes of slender foxtail (called blackgrass in Great Britain). These biotypes exhibited multiple resistance (resistance to chlorsulfuron, diclofop-methyl, pendimethalin, and several triazines), even though they had been exposed for many years to only the substituted urea herbicides chlortoluron or isoproturon. *The suspected mechanism of multiple resistance in blackgrass is enhanced degradation by cytochrome P450 monooxygenase enzymes.*

Resistance to Bipyridinium Herbicides

The mode of action of the bipyridinium herbicides is contact kill. *Resistance to the bipyridinium herbicides paraquat and diquat may be due to rapid sequestration (compartmentalization) of the herbicides in cell walls of treated foliage, which reduces herbicide levels at their site of action in the chloroplasts; alternatively, resistance may be due to the rapid enzymatic detoxification of superoxide and other toxic forms of oxygen.* The presence of both mechanisms acting concurrently in bipyridinium-tolerant biotypes has not been discounted.

Two mechanisms for the sequestration of paraquat have been proposed: (1) that paraquat is adsorbed to a cellular component by ionic interactions, and (2) that paraquat is actively transported into a membrane-enclosed organelle, possibly the vacuole, since many other cations are sequestered in the vacuole.

Resistance to paraquat was discovered in the 1970s in several weed species. By 1992, paraquat resistance had been reported in 12 weed species from seven locations around the world, though none in North America. Resistance in weed biotypes has typically evolved where paraquat was applied 5–10 times per year for 5 or more years, as to control weeds in vineyards and orchards.

Biotypes of capeweed, hare barley, and wall barley evolved resistance to paraquat and diquat following an annual application of one or both herbicides to control these and other weeds in an alfalfa field in Victoria, Australia, from 1963 to 1986. By 1983, control of these three weed species by paraquat and diquat was no longer satisfactory, and this one field hosted three weed species with biotypes resistant to paraquat and diquat. However, other weed species in the field remained susceptible to the herbicides. The repeated use of paraquat or diquat for 24 years without cultivation or other herbicides provided selection pressure, which allowed the resistant biotypes to dominate.

In 1982, a biotype of wall barley was found to be resistant to paraquat in an alfalfa field near Ararat in the Western District of Victoria. The resistant biotype was dominant in the alfalfa field, which had been sprayed once annually with paraquat for the past 24 years. This resistant biotype was found to be 250 times more resistant to paraquat than nonresistant plants. A paraquat-resistant biotype of hairy fleabane discovered in Egypt was found to be 100 times more resistant to paraquat and 10 times more resistant to diquat than nonresistant plants. Biotypes resistant to paraquat and diquat have not exhibited resistance to other herbicides with different modes of action.

Resistance to Dinitroaniline Herbicides

The dinitroaniline herbicides are mitotic disrupters that bind to tubulin and prevent the formation of microtubules. *Resistance to the dinitroaniline herbicides is due to a site-of-action mutation in the form of hyperstabilized microtubules in resistant biotypes, a modification that renders these plants resistant to the dinitroaniline herbicides.* Dinitroaniline-resistant biotypes have been found among populations of goosegrass and green foxtail.

Goosegrass   A dinitroaniline-resistant goosegrass biotype was suspected of being present in goosegrass populations in the northeastern coastal plain region of South Carolina in the early 1970s. However, its presence was not confirmed until 1984. The resistant biotype was found in a monoculture cotton field where trifluralin had been used on an annual basis for about 10 years. Tests with goosegrass confirmed biotype resistance to the dinitroaniline herbicides, and cross resistance to seven dinitroaniline herbicides was also exhibited. There was no multiple resistance exhibited by the resistant biotype to herbicides from six other chemical families. The dinitroaniline-resistant biotype is now found throughout the southeastern United States. In 1990, two kinds of dinitroaniline-resistant biotypes of goosegrass had been found in counties of South Carolina. One biotype resists trifluralin about 1,000–10,000

times that of susceptible plants. The other biotype is much less resistant, with resistance to trifluralin about 50 times greater than susceptible plants.

Green foxtail   Trifluralin had been applied in fields of wheat, barley, rapeseed, and flax for control of green foxtail in western Canada since the late 1960s. In 1988, it was confirmed that a biotype of green foxtail was resistant to trifluralin. Under field tests, the resistant biotype was seven times more resistant to trifluralin than were normal plants. The trifluralin-resistant biotype exhibited cross resistance to eight dinitroaniline herbicides tested, with a sevenfold difference in response between resistant and nonresistant plants, but the resistant biotype did not exhibit multiple resistance to a selection of nine herbicides belonging to seven chemical families, each with different modes of action.

■ TRANSGENIC CROP TOLERANCE

Technology that can develop crops with tolerance to selected herbicides is now available. The techniques are many and involve *plant breeding, tissue culture selection,* and *genetic engineering.* Herbicide-tolerant crop genotypes have arisen from (1) selection from naturally varying populations within a crop species, (2) selection of herbicide-tolerant mutants within a crop cultivar at the cell to whole-plant level, and (3) insertion of genes conferring tolerance. To be most useful, herbicide tolerance should be transferrable between crop cultivars through conventional methods of plant breeding.

For the past 50 years, herbicide candidates have been screened in field and greenhouse tests for crop tolerance and weed susceptibility, and the most suitable have been commercially marketed. Thus naturally occurring herbicide-tolerant crop plants have been a key part of the successful use of herbicides in croplands. The molecular basis for natural herbicide tolerance in crop plants has been (1) lack of absorption and/or translocation, (2) herbicide detoxification, and (3) enzyme insensitivity at the herbicide site of action. These mechanisms of herbicide tolerance are operative during germplasm selection, either by screening or mutational breeding. In many respects, they offer an alternative to the transfer of genes in conferring herbicide tolerance in crop plants.

Great strides were made in the 1980s relative to understanding herbicide modes of action, resistance in weeds, and safeners to protect crop plants from herbicides. Recent progress in gene transfer technology permits the transfer of genes from related and unrelated organisms into crop plants, opening the way for developing herbicide-tolerant

TABLE 9.3 ■ Some crop species that have been transformed by gene transfer.

AGRONOMIC CROPS	HORTICULTURAL CROPS
Alfalfa	Apple
Canola	Asparagus
Corn	Cabbage
Cotton	Carnation
Flax	Carrot
Orchardgrass	Cauliflower
Potato	Celery
Rice	Chrysanthemum
Rye	Cucumber
Soybean	Eggplant
Sugarbeet	Grape
Tobacco	Licorice
	Muskmelon
	Pea
	Petunia
	Strawberry
	Tomato

crop plants. With a few exceptions, most such gene transfers confer narrow spectrum tolerance to a single or select class of herbicides.

By 1991, genes conferring herbicide resistance had been introduced into most crops. Nearly 50 crop species, including both monocots and dicots, have been transformed by gene transfer in laboratories around the world. Some of these crops are listed in Table 9.3. There is no difference in morphology between crops with a natural tolerance to herbicides and transgenic herbicide-tolerant crops.

Two major methods used for plant gene transfer are (1) those based on *Agrobacterium tumefaciens,* and (2) physical methods of DNA transfer (Kishore et al., 1992). The *Agrobacterium* method is routinely used in transforming crop plants such as canola, cotton, sugarbeet, and tomato. For crops unresponsive to *Agrobacterium* technology, physical methods of DNA delivery have been used extensively. Of several physical methods, particle gun technology has provided the best DNA delivery, and this method has been used for the production of transgenic corn and soybean plants. Electroporation and protoplast transformation are other physical methods.

Controversy

As in other areas of genetic engineering, controversy has surrounded the development of transgenic herbicide-

tolerant crops. It has been argued that the possibility of gene transfer from transgenic herbicide-tolerant crop plants to unrelated weeds may create "superweeds" and even "new" weeds. What the critics fail to appreciate is that the genes conferring herbicide tolerance are already present in the weed or crop species, but not in the desired crop species, and that the projected transfer of genes has not hitherto occurred without the help of science—crop plants cannot by themselves transfer genes to weed plants. The use of transgenic technology in developing herbicide-resistant crops is not appreciably different from using traditional breeding methods to achieve the same end.

Critics of transgenic technology argue that it promotes greater use of herbicides. Actually, the introduction of a transgenic herbicide-tolerant crop plant generally favors the use of a specific herbicide for use on that crop, avoiding the necessity of using multiple herbicides and/or multiple herbicide applications. However, the exclusive and repeated use of a specific herbicide may well allow resistant weeds to develop and/or proliferate and, in some cases, enhance herbicide carryover problems.

It has been said that, through transgenic herbicide-tolerant crops, science is now attempting to solve problems that it created. In truth, the development of transgenic herbicide-tolerant crops is really only a refinement of the existing technology that relies almost exclusively on herbicides. Instead of introducing a new herbicide for use in a crop that has a natural tolerance to the herbicide, an otherwise susceptible crop is transformed into a herbicide-tolerant crop by insertion of a gene or genes that confer tolerance to a specific herbicide. Thus it would appear that concern should not be directed at the source or means by which herbicide crop tolerance is achieved, but instead at how the products of this technology are being used.

Concerns

Concerns expressed relative to the development of herbicide-tolerant crops (HTCs) include the following: (1) HTCs will accelerate the selection of resistant weed biotypes through the widespread use of herbicides with the same mode of action; (2) HTCs may themselves become weed problems; (3) HTCs may transfer resistance to related weed species; (4) HTCs may result in increased use of herbicides; (5) HTCs may lead to abandonment of alternative weed-control practices; (6) HTCs may increase groundwater contamination and other environmental problems through misuse of herbicides; (7) the use of HTCs may be adversely affected by the general public sentiment against the release of genetically engineered organisms derived from gene insertion techniques (R. Goldberg et al. as cited in Harrison, 1992). In view of these concerns, the introduction of transgenic HTCs does present problems in registration, regulation, and use.

Benefits

Transgenic herbicide-tolerant crops are creating new opportunities for decreasing the risk of crop injury, for decreasing herbicide-carryover problems, for broadening the spectrum of weeds controlled, and for using herbicides that present less risk to the environment.

Apparently, the economic interests of chemical companies in expanding the markets for their herbicides is spurring advances in this technology. Thus the companies most certain to accrue the benefits from transgenic herbicide-tolerant crops are those companies that develop superior marketing strategies and who, by doing so, displace other companies from that market.

Seeds of herbicide-tolerant crop varieties and herbicides are being marketed as one package; this can increase a company's market share and volume. At the grower level, reduced herbicide costs may be offset by higher costs for the herbicide-tolerant crop seed. The cost of transgenic herbicide-tolerant crop seed may be greater because of increased production and development costs.

Acceptance

The introduction of herbicide-tolerant crop cultivars is having the same impact on cropping systems as has the introduction of a new herbicide with the same mode of action and use pattern of herbicides already in use. Confusion at the crop-producer level may well be the major deterrent to the use of herbicide-tolerant crops. The confusion results from the fact that not all cultivars of a crop will be tolerant to the same herbicide and the tolerant and nontolerant cultivars cannot be distinquished from one another. A genetic morphological marker on the transgenic herbicide-tolerant crop cultivar would be desirable so that the tolerant cultivar can be easily identified. To avoid costly mistakes, the growers must be provided with adequate information to ensure that transgenic herbicide-tolerant crops are used intelligently. They are, after all, but another tool in the industry's overall endeavor to control weeds and produce crops efficiently and economically.

"Volunteer" crop plants arising from seed of the preceding crop can be a major problem, as in volunteer corn in soybeans. If a transgenic herbicide-tolerant crop cultivar appears in a succeeding crop, will there be a problem in its control? Will it be necessary to engineer herbicide suscep-

tibility into a transgenic herbicide-tolerant crop cultivar so that the cultivar can be controlled with a specific herbicide? These and other questions need to be addressed by the industry before use of this technology becomes widespread.

■ EXPANDING COMMERCIAL USES

Commercial Transgenic Herbicide-Tolerant Crops

In 1993, Calgene/Rhone Poulenc was field-testing a picker variety of cotton (Stoneville BXN 48) that exhibited a high degree of tolerance to bromoxynil. Monsanto was field-testing both picker and stripper types of cotton (e.g., Coker 312) that were glyphosate-tolerant. DuPont was field-testing cotton varieties for tolerance to sulfonylurea herbicides. Ciba, Land O'Lakes, Pioneer, and Zeneca were marketing corn hybrids approved for use with the imidazolinone herbicide imazethapyr (Pursuit). The use of Ciba's corn hybrid as a rotation crop following soybeans is focused on the soil residual of the imidazolinone herbicide imazaquin following its use in soybeans the previous year.

Transgenic Crop Tolerance Research

It is now feasible to introduce herbicide tolerance into agronomically important crops because of successful isolation and transfer of genes. Once it has been demonstrated that a gene can perform this service—that is, confer tolerance—in one crop, it can be introduced into a number of crops.

Corn

Inbred lines of corn differ in their sensitivity to some sulfonylurea herbicides, with one line completely tolerant to primisulfuron, while other lines are killed. DuPont and Ciba have worked cooperatively with 45 seed companies to field-evaluate more than 1000 inbred lines of corn for sensitivity to sulfonylurea herbicides. In a study of 94 inbred lines of corn for sensitivity to the sulfonylurea herbicides nicosulfuron, primisulfuron, and thifensulfuron, most inbreds were highly tolerant. Of these 94 inbred lines, 37 represented all major inbred families now used in hybrid seed production; 21 were defined as hybrids; and 36 were commercially coded hybrids. However, some inbreds used to produce hybrids were injured, especially in sweet corn and short-season hybrids. Nicosulfuron had the widest corn safety margin and the fewest sensitive reactions. The difference in sensitivity to primisulfuron of some lines were

4000-fold with others as much as 40,000-fold. Most corn hybrids planted today have more than enough tolerance to nicosulfuron, primisulfuron, and thifensulfuron for these herbicides to be used safely in this crop.

Sensitivity to sulfonylurea herbicides is controlled by a single recessive gene. Sensitive inbreds produce tolerant hybrids when crossed with tolerant inbreds. Sensitive hybrids occur when both parents are sensitive. Breeders can eliminate sensitivity by back-crossing sensitive inbreds with tolerant inbreds or by always using at least one tolerant hydrid parent. Backcrossing can transfer a single, simply inherited trait such as sulfonylurea tolerance from one inbred to another and thus create a nearly isogenic version of the original inbred. The single-gene characteristic is similar to studies with modified acetolactate synthase (ALS) genes.

ALS-modified genes, such as XA-17 in Pioneer transgenic varieties or a similar gene in Ciba's Funks Seed Company transgenic varieties, offer the opportunity to use sulfonylurea and imidazolinone herbicides that are not normally tolerant to corn. XA-17 confers a 200- to 1000-fold increase in corn tolerance at the ALS enzyme level and very high tolerance at the whole-plant level to sulfonylurea herbicides not normally tolerant to corn. At 10 or more times the rate needed for weed control, corn with the ALS-modified XA-17 gene was not inhibited by the ethyl ester of chlorimuron, chlorsulfuron, and the methyl esters of metsulfuron, sulfometuron, and tribenuron, while corn without XA-17 was strongly inhibited. The ALS-modified XA-17 gene overcame sensitivity to sulfonylurea herbicides, even when the organophosphate insecticide terbufos was present. Other ALS-modified genes, such as X112 and QJ22, are more specific to imidazolinone herbicides and did not impart much tolerance for sulfonylurea herbicides.

Thus corn breeders have three options to eliminate sensitivity to sulfonylurea herbicides: (1) back-cross sensitive inbreds with tolerant inbreds, (2) always use at least one tolerant hybrid parent, or (3) use the XA-17 gene.

Soybean

Screening of soybean germplasm has resulted in the identification of metribuzin-tolerant lines, which were used to cross into other cultivars. These metribuzin-tolerant lines appear to be tolerant because of a detoxification mechanism that results in the inactivation of metribuzin.

The soybean cultivar "Tracy" was selected for tolerance to 2,4-DB and a high level of tolerance to phytophthora rot, a soil-borne disease. Subsequently, Tracy was found to be susceptible to the herbicide metribuzin. Tracy has two major genes controlling tolerance to phytophthora rot, and

one of these genes is closely linked with a gene controlling tolerance to metribuzin. Genes at a single locus control tolerance to metribuzin. Metribuzin tolerance was subsequently incorporated in this cultivar, which is now marketed as "Tracy-M." The cultivar Tracy-M retains the high level of tolerance to phytophthora rot.

Where simple germplasm selection has not led to the identification of herbicide-tolerant lines, tolerance has been induced by mutagens followed by selection in the presence of herbicides. Thus *in vitro* selection of soybean has resulted in plant lines that are tolerant to the sulfonylurea and imidazolinone herbicides. In the tolerant lines, the molecular basis for resistance appears to be an altered ALS enzyme at the site of action of these herbicides.

Sugarbeets

Sugarbeets are extremely sensitive to both the sulfonylurea and imidazolinone herbicides, with severe stand and root reductions in fields treated 2 or 3 years earlier with either chlorimuron, imazaquin, or imazethapyr. Generation and development of sugarbeet cultivars that are tolerant to the herbicides is a potential solution to this carryover problem and may increase the herbicide options for weed control in this crop.

A chlorsulfuron-tolerant sugarbeet clone (CR1-B) has been successfully generated via somatic cell selection in tissue culture using a successful self-fertile clone (REL-1). CR1-B is cross-tolerant to postemergence applications at, or exceeding, field-use rates of primisulfuron and thifensulfuron. The physiological basis of tolerance to the sulfonylurea herbicides is an altered ALS enzyme. Genetic inheritance of this altered enzyme is via a single dominant nuclear gene, heterozygous in CR1-B.

Studies with sulfonylurea herbicide tolerance in soybeans and imidazolinone tolerance in corn indicate that heterozygous-tolerant plants may have a lower level of tolerance than homozygous-tolerant plants. Thus development of crop cultivars with the highest level of tolerance may, in some cases, require that all parents be homozygous-tolerant. Most modern sugarbeet hybrid seed production consists of a three-way cross [A × B] × C. The inheritance of the sulfonylurea herbicide tolerance trait and the differential tolerance between homozygotes and heterozygotes is of crucial importance to sugarbeet breeders. *Homozygous* sugarbeet lines tolerant to sulfonylurea herbicides and sprayed with increasing rates of primisulfuron, thifensulfuron, and chlorimuron were respectively 377, 269, and 144 times more tolerant to these herbicides than were susceptible homozygous lines, while *heterozygous*-tolerant lines were respectively 107, 76, and 57 times more tolerant to these herbicides than were susceptible heterozygous lines. Thus, compared to the heterozygous-tolerant lines, the homozygous-tolerant sugarbeet lines appear to be at least twice as tolerant of these sulfonylurea herbicides, and the magnitude of tolerance varies with the specific herbicide involved. However, at four times the field rate of primisulfuron applied in corn and of thifensulfuron applied in soybeans, there was no difference in the tolerance between heterozygous and homozygous sugarbeet lines. Thus, at practical levels of application, heterozygous- or homozygous-tolerant sugarbeet lines would have a safe margin of tolerance that would prevent them being injured by carryover from these herbicides applied the previous year. It also would appear that they have a high enough level of tolerance to primisulfuron and thifensulfuron to allow postemergence use of these herbicides.

Tomato

In May 1994, a tomato developed by Calgene gained the FDA's Center For Food Safety approval for consumption and marketing. This tomato was genetically engineered and is called FLAVR-SAVR. Its genetic changes will enable growers to allow it to ripen longer on the vine, let it get redder and more flavorful, and still have time to ship it long distances.

Brassica *spp.*

Tolerance to atrazine has been introduced into *Brassica* cultivars by breeding with less agriculturally important *Brassica* species. Atrazine tolerance was introduced into the *Brassica* species canola, broccoli, and cauliflower by genetic crosses with an atrazine-tolerant mutant of rape, an oilseed crop. In this case, the tolerance mechanism was the result of a mutation in the 32 kd protein, the site of action of atrazine. *In vitro* selection has resulted in the identification of canola lines that are tolerant to the imidazolinone and sulfonylurea herbicides. In this instance, the tolerance mechanism is due to a modification at the plant's site of action for these herbicides.

Glyphosate Tolerance

Research with glyphosate has been directed toward production of glyphosate-tolerant canola, cotton, soybean, and corn. The objective of this research is to demonstrate glyphosate tolerance at twice the agronomic-use rate. Glyphosate inhibits the enzyme 5-enolpyruvylshikimate-3-phosphate synthase (EPSPS), and tolerance is due to the failure of glyphosate to inhibit EPSPS. Glyphosate binds to an EPSPS enzyme/shikimate-3-phosphate complex and not to the free enzyme. Inhibition of the enzyme EPSPS is the major mechanism by which glyphosate kills plants; other effects attributed to glyphosate are secondary.

A highly glyphosate-tolerant mutant was isolated from *E. coli* by Monsanto. The *aro*A gene encoding this mutant EPSPS was isolated, sequenced, and shown to contain a single amino acid substitution at position 96, which results in the replacement of a glycine with alanine. The mutant EPSPS enzyme has an 8000-fold increase in tolerance to glyphosate, compared to the glyphosate-susceptible *E. coli* enzyme. Mutation of the corresponding residue in plant EPSPS to alanine has been shown to result in highly glyphosate-tolerant plant EPSPS enzymes. Although plant EPSPS GA (glycine to alanine) mutants are capable of conferring glyphosate tolerance to plants, the levels of tolerance are not commercial.

Further research with transgenic glyphosate tolerance in canola has led to the introduction of useful levels of tolerance in this crop. The glyphosate-tolerant genes have also been introduced into cotton, soybean, and sugarbeet. Soybean plants that show complete glyphosate tolerance at commercial-use rates have been produced and, at this writing, are in advanced agronomic and yield tests.

■ SELECTED REFERENCES

Ahrens, W. H., L. M. Wax, and E. W. Stoller. 1981. Identification of triazine-resistant *Amaranthus* spp. *Weed Sci.* 29:345–348.

Alcocer-Ruthling, M., D. C. Thill, and C. Mallory-Smith. 1992. Monitoring the occurrence of sulfonylurea-resistant prickly lettuce. *Weed Technol.* 6:437–440.

Alcocer-Ruthling, M., D. C. Thill, and B. Shafii. 1992. Differential competitiveness of sulfonylurea-resistant and -susceptible prickly lettuce (*Lactuca serriola*). *Weed Technol.* 6:303–309.

Alcocer-Ruthling, M., D. C. Thill, and B. Shafii. 1992. Seed biology of sulfonylurea-resistant and -susceptible biotypes of prickly lettuce. (*Lactuca serriola*). *Weed Technol.* 6:858–864.

Bandeen, J. D., G. R. Stephenson, and E. R. Cowett. 1982. Discovery and distribution of herbicide-resistant weeds in North America. In *Herbicide resistance in plants*, H. M. LeBaron and J. Gressel, eds., pp. 9–30. New York: John Wiley & Sons.

Beckie, H. J., and I. N. Morrison. 1993. Effect of ethalfluralin and other herbicides on trifluralin-resistant green foxtail (*Setaria viridis*). *Weed Technol.* 7:6–14.

Beckie, H. J., and I. N. Morrison. 1993. Effective kill of trifluralin-susceptible and -resistant green foxtail. *Weed Technol.* 7:15–22.

Beversdorf, W. D., and L. S. Kott. 1987. Development of triazine resistance in crops by classical plant breeding. *Weed Sci.* 35(Suppl. 1):9–11.

Birschbach, E. D., M. G. Myers, and R. G. Harvey. 1993. Triazine-resistant smooth pigweed (*Amaranthus hybridus*) control in field corn (*Zea mays*). *Weed Technol.* 7:431–436.

Boydston, R. A., and K. Al-Khatib. 1992. Terbacil and bromacil cross-resistance to Powell amaranth (*Amaranthus powellii*). *Weed Sci.* 40:513–516.

Burnet, M. W. M., et al. 1991. Amitrole, triazine, substituted urea, and metribuzin resistance in a biotype of rigid ryegrass (*Lolium rigidum*). *Weed Sci.* 39:317–323.

Burnside, O. C. 1992. Rationale for developing herbicide-resistant crops. *Weed Technol.* 6:621–625.

Casely, J. C., G. W. Cussans, and R. K. Atkins, eds. 1991. *Herbicide resistance in weeds and crops.* Oxford, England: Butterworth-Heinemann, Ltd.

Christianson, M. L. 1991. Fun with mutants: applying genetic methods to problems of weed physiology. *Weed Sci.* 39:489–496.

Comai, L., L. C. Sen, and D. M. Stalker. 1983. An altered *aro*A gene product confers resistance to the herbicide glyphosate. *Science.* 221:370–371.

Darmency, H., and J. Gasquez, 1990. Appearance and spread of triazine resistance in common lambsquarters (*Chenopodium album*). *Weed Technol.* 4:173–177.

De Prado, R., C. Dominguez, and M. Tena. 1989. Characterization of triazine-resistant biotypes of common lambsquarters (*Chenopodium album*), hairy fleabane (*Conyza bonariensis*). and yellow foxtail (*Setaria glauca*) found in Spain. *Weed Sci.* 37:1–4.

Duvick, D. N. 1992. Concerns of seed company officials with herbicide-tolerant cultivars. *Weed Technol.* 6:640–646.

Dyer, W. E., P. W. Chee, and P. K. Fay. 1993. Rapid germination of sulfonylurea-resistant *Kochia scoparia* L. accessions is associated with elevated seed levels of branched-chain amino acids. *Weed Sci.* 41:18–22.

Friesen, L. F., et al. 1993. Response of chlorsulfuron-resistant biotype of *Kochia scoparia* to sulfonylurea and alternative herbicides. *Weed Sci.* 41:100–106.

Fuerst, E. P., et al. 1986. Herbicide cross-resistance in triazine-resistant biotypes of four species. *Weed Sci.* 34:344–353.

Fuerst, E. P., and K. C. Vaughn. 1990. Mechanisms of paraquat resistance. *Weed Technol.* 4:150–156.

Gerwick, B. C., L. C. Mireles, and R. J. Eilers. 1993. Rapid diagnosis of ALS/AHAS-resistant weeds. *Weed Technol.* 7:519–524.

Giaquinta, R. T. 1992. An industry perspective on herbicide-tolerant crops. *Weed Technol.* 6:653–656.

Goldburg, R. J. 1992. Environmental concerns with the development of herbicide-tolerant plants. *Weed Technol.* 6:647–652.

Goodman, R. M. 1987. Future potential, problems, and practicalities of herbicide-tolerant crops from genetic engineering. *Weed Sci.* 35(Suppl. 1):28–31.

Green, J. M., and J. F. Ulrich. 1993. Response of corn (*Zea mays* L.) inbreds and hybrids to sulfonylurea herbicides. *Weed Sci.* 41:508–516.

Green, M. B., H. M. LeBaron, and W. K. Moberg, eds. 1990. *Managing resistance to agrochemicals.* ACS Symposium Series 421. Washington. D.C.: American Chemical Society.

Gressel, J., and L. A. Segel. 1982. Interrelating factors controlling the rate of appearance of resistance: The outlook for the future. In *Herbicide resistance in plants,* H. M. LeBaron and J. Gressel, eds., pp. 325–347. New York: John Wiley & Sons.

Guttieri, M. J., et al. 1992. DNA sequence variation in Domain A of the acetolactate synthase genes of herbicide-resistant and -susceptible weed biotypes. *Weed Sci.* 40:670–676.

Harrison, H. F. 1992. Developing herbicide-tolerant crop cultivars: Introduction. *Weed Technol.* 6:613–614.

Hart, S. E., S. Glenn, and W. W. Kenworthy. 1991. Tolerance and the basis for selectivity to 2,4-D in perennial *Glycine* species. *Weed Sci.* 39:535–539.

Hart, S. E., J. W. Saunders, and D. Penner. 1992. Chlorsulfuron resistant sugarbeet: Cross-resistance and physiological basis of resistance. *Weed Sci.* 40:378–383.

Hart, S. E., J. W. Saunders, and D. Penner. 1993. Semidominant nature of monogenic sulfonylurea herbicide resistance in sugar beet (*Beta vulgaris*). *Weed Sci.* 41:317–324.

Hartwig, E. E. 1987. Identification and utilization of variation in herbicide tolerance in soybean (*Glycine max*) breeding. *Weed Sci.* 35(Suppl. 1):4–8.

Heap, I. M., et al. 1993. Resistance to aryloxyphenoxypropionate and cyclohexanedione herbicides in wild oat (*Avena fatua*). *Weed Sci.* 41:232–238.

Hensley, J. R. 1981. A method for identification of triazine resistant and susceptible biotypes of several weeds. *Weed Sci.* 29:70–73.

Holt, J. S. 1992. History of identification of herbicide-resistant weeds. *Weed Technol.* 6:615–620.

Holt, J. S., and H. M. LeBaron. 1990. Significance and distribution of herbicide resistance. *Weed Technol.* 4:141–149.

Holt, J. S., S. B. Powles, and J. A. M. Hottum. 1993. Mechanisms and agronomic aspects of herbicide resistance. *Ann. Rev. Plant Physiol., Plant Mol. Biol.* 44:203–229.

Horne, D. M. 1992. EPA's response to resistance management and herbicide-tolerant crop issues. *Weed Technol.* 6:657–661.

Horstmeier, G. D. 1993. Herbicide-tolerant hybrids make a hit. *Farm J.* Mid-Jan. Issue, pp. 18–19.

Joseph, O. O., S. L. A. Hobbs, and S. Jana. 1990. Diclofop resistance to wild oat (*Avena fatua*). *Weed Sci.* 38:475–479.

Kishore, G. M., et al. 1992. History of herbicide-tolerant crops, methods of development and current state of art—Emphasis on glyphosate tolerance. *Weed Technol.* 6:626–634.

Knake, E. L. 1992. Technology transfer for herbicide-tolerant weeds and herbicide tolerant crops. *Weed Technol.* 6:662–664.

LeBaron, H. M. 1987. Genetic engineering for herbicide resistance. *Weed Sci.* 35(Suppl. 1): 2–3.

LeBaron, H. M., and J. Gressel, eds. 1982. *Herbicide resistance in plants.* New York: John Wiley & Sons.

LeBaron, H. M., et al. 1992. International organization for resistant pest management (IOPRM)—A step toward rational resistance management recommendations. *Weed Technol.* 6:765–770.

Leidner, J. 1993. Defiant weeds rebel against herbicides. *Progressive Farmer.* March, pp. 38, 40.

Mallory-Smith, C. A., D. C. Thill, and M. J. Dial. 1990. Identification of sulfonylurea herbicide-resistant prickly lettuce (*Lactuca seriola*). *Weed Technol.* 4:163–168.

Mallory-Smith, C. A., et al. 1990. Inheritance of sulfonylurea herbicide resistance in *Lactuca* spp. *Weed Technol.* 4:787–790.

Mansooji, A. M., et al. 1992. Resistance of aryloxyphenoxypropionate herbicides in two wild oat species (*Avena fatua* and *Avena sterilis* spp. *ludoviciana*). *Weed Sci.* 40:599–605.

Maxwell, B. D., M. L. Roush, and S. R. Radosevich. 1990. Predicting the evolution and dynamics of herbicide resistance in weed populations. *Weed Technol.* 4:2–13.

Morrison, I. A., B. G. Todd, and K. M. Nawolsky. 1989. Confirmation of trifluralin-resistant green foxtail (*Setaria viridis*) in Manitoba. *Weed Technol.* 3:544–551.

Moss, S. R. 1990. Herbicide cross-resistance to slender foxtail (*Alopecurus myosuroides*). *Weed Sci.* 38:492–496.

Mudge, L. C., B. J. Gosset, and T. R. Murphy. 1984. Resistance of goosegrass (*Eleucine indica*) to dinitroaniline herbicides. *Weed Sci.* 32:591–594.

Mueller, T. C., M. Barrett, and W. W. Witt. 1990. A basis for the antagonistic effect of 2,4-D on haloxyfop-methyl toxicity to johnsongrass (*Sorghum halepense*). *Weed Sci.* 38:103–107.

Newhouse, K. E., T. Wang, and P. C. Anderson. 1991. The imidazolinone-resistant crops. In *The imidazolinone herbicides,* D. L. Shaner and S. L. Conner, eds., pp. 139–150. Boca Raton, Fla.: CRC Press.

Perkins, E. J., C. M. Stiff, and P. F. Lurquin. 1987. Use of *Alcalagenes eutrophus* as a source of genes for 2,4-D resistance in plants. *Weed Sci.* 35(Suppl. 1):12–18.

Powles, S. B., and P. D. Howat. 1990. Herbicide-resistant weeds in Australia. *Weed Technol.* 4:178–185.

Primiani, M. M., J. C. Cotterman, and L. L. Saari. 1990. Resistance of kochia (*Kochia scoparia*) to sulfonylurea herbicides. *Weed Technol.* 4:169–172.

Radosevich, S. R., C. M. Ghersa, and G. Comstock. 1992. Concerns a weed scientist might have about herbicide-tolerant crops. *Weed Technol.* 6:635–639.

Ryan, G. F. 1970. Resistance of common groundsel to simazine and atrazine. *Weed Sci.* 18:614–616.

Saari, L., J. C. Cotterman, M. M. Primiani. 1992. Mechanism of sulfonylurea herbicide resistance in the broadleaf weed *Kochia scoparia. Plant Physiol.* 93:55–61.

Stephenson, G. R., and G. Ezra. 1987. Chemical approaches for improving herbicide selectivity and crop tolerance. *Weed Sci.* 35(Suppl. 1):24–27.

Thompson, G. A., et al. 1987. Expression in plants of a bacterial gene coding for glyphosate resistance. *Weed Sci.* 35(Suppl. 1):19–23.

Tucker, E. S., and S. B. Powles. 1991. A biotype of hare barley (*Hordeum leporinum*) resistant to paraquat and diquat. *Weed Sci.* 39:159–162.

Vaughn, K. C., and L. P. Lehnen, Jr. 1991. Mitotic disrupter herbicides. *Weed Sci.* 39:450–457.

Vaughn, K. C., M. A. Vaughn, and P. Camilleri. 1989. Lack of cross-resistance of paraquat-resistant hairy fleabane (*Conyza bonariensis*) to other toxic oxygen generators indicates enzymatic protection is not the resistant mechanism. *Weed Sci.* 37:5–11.

Vaughn, K. C., M. A. Vaughn, and B. J. Cossett. 1990. A biotype of goosegrass (*Eleusine indica*) with an intermediate level of dinitroaniline herbicide resistance. *Weed Technol.* 4:157–162.

Whitehead, C. W., and C. M. Switzer. 1963. The differential response of strains of wild carrot to 2,4-D and related herbicides. *Can. J. Plant Sci.* 43:255–262.

Wyse, D. L. 1992. Future impact of crops with modified herbicide resistance. *Weed Technol.* 6:665–668.

10 FORMULATIONS AND SURFACTANTS

■ INTRODUCTION

Pure forms of herbicides are practically useless to the general consumer, and they are rarely available as pure chemicals to the consumer. To make them of practical value, herbicides are *formulated*—that is, placed in a usable form composed of the herbicide and appropriate solvents, diluents, and surfactants. The end result is called a *formulation*. The herbicide is the "active ingredient" in the formulation. Formulations are mixtures of one or more herbicides carefully designed to provide maximum effectiveness for the proscribed use. The formulations are packaged and sold as trade-named products, and the term *product* is synonymous with the term *formulation*—purchasers are really buying a formulation when they buy a trade-named product. The number and variety of trade names can be bewildering.

The primary purpose of formulating a herbicide is to enable the user to disperse it in a convenient carrier, such as water, so that a small amount of herbicide may be dispersed uniformly over a relatively large area. Other purposes include: (1) enhancing the phytotoxicity of the herbicide; (2) supplying the herbicide in a handy, economical form; (3) improving the shelf life (storage) of the herbicide; and (4) protecting the herbicide from environmental extremes while in transit or storage.

■ CARRIER

As the formulation serves as the diluent for the herbicide, so does the *carrier* serve as the diluent for the formulation. The carrier enables a small amount of formulation to be uniformly distributed over a relatively large area. The carrier may be liquid, solid, or foam. *Water* is the most commonly used carrier for herbicide formulations. *Diesel fuel* may be the carrier with certain formulations for woody plant control.

With a few exceptions, the formulation is not considered the "carrier." But, in the case of certain ultra-low-volume (ULV) herbicide applications, a liquid formulation may be applied directly, so the formulation is technically considered the carrier in this situation. Also, in certain dry formulations (granules, pellets) specifically designed to be applied dry, the formulation again doubles as the carrier.

Liquid fertilizers, such as 28-0-0 and 32-0-0 nitrogen solutions, may be used as a carrier for many herbicides (as noted on their respective product labels), and their use involves few mixing or application problems. However, the use of the more viscous, clear liquid fertilizers, such as 7-21-7 or 10-34-0, may cause problems. The compatibility of the liquid fertilizer/herbicide mixture should be tested prior to tank mixing. Directions for a simple compatibility test are usually included on the product label. The use of a compatibility agent such as Compex or Unite may enhance uniform mixing. In some cases, dry bulk fertilizers impregnated with the herbicide serve as the carrier.

■ FORMULATIONS

Formulations vary according to (1) the solubility of the active ingredient (the herbicide) in water, oil, and organic solvents; and (2) the manner in which the formulation is to be applied (sprayed in a carrier or applied as the dry formulation). The kinds of herbicide formulations and their identifying abbreviations are as follows:

1. Water-soluble (S, WS)[1]
2. Oil-soluble (OS)
3. Emusifiable concentrate (E, EC)
4. Ultra-low-volume concentrate (ULV)
5. Wettable powder (W, WP)
6. Liquid-flowable (L, LF)
7. Dry-flowable (DF)
8. Water-soluble granules (SG)
9. Water-dispersible granules (DG, WDG)
10. Water-soluble pellets (SP)
11. Herbicide-coated or -impregnated particles:
    a. Granules (G)
    b. Pellets (P)
12. Ultra-low-weight granules (ULW)
13. Controlled release (CR)

It is not unusual for the same herbicide to be available in two or more kinds of formulations. For example, the herbicide 2,4-D is available in water-soluble, oil-soluble, and emusifiable-concentrate formulations; atrazine is available as a liquid and as water-dispersible granules; and hexazinone is formulated as a water-soluble powder, water-dispersible liquid, and water-soluble granules.

---

1. In some cases, formulations are available in "water-soluble packets," but this is a variant in packaging, not in formulation. Due to drift hazards, herbicides should not be applied in dust or aerosol formulations.

Product labels for certain herbicides include a proprietary designation to denote a particular formulation; for example, Ciba designates the water-dispersible granular formulation of atrazine and simazine as AAtrex NINE-O and Princep CALIBER 90, respectively. DowElanco denotes their emusifiable-concentrate formulation of trifluralin as Treflan E.C.; their multiple-temperature formulation of trifluralin as Treflan M.T.F.; and their granular formulation of trifluralin as Treflan TR-10. (Note: Treflan TR-10 contains 5% w/w trifluralin, rather than 10% w/w as one would assume from the name.) DuPont offers three formulations of hexazinone: water-soluble powder, water-dispersible liquid, and a special granular formulation, respectively named Velpar, Velpar L, and Velpar ULW.

The solubility of an acid herbicide can be varied by making salt and ester forms of the parent acid. For example, 2,4-D acid is essentially insoluble in water and toluene, but its salts are usually highly water-soluble (and insoluble in oil) and its esters are highly oil-soluble (and insoluble in water). However, as an exception to the rule, a few amine salts of 2,4-D (dodecylamine, tetradecylamine, and N-oleyl-1,2-propylenediamine—all no longer marketed) are oil-soluble (and insoluble in water).

Water-Soluble Herbicides

Water-soluble herbicides dissolve in water to form true solutions. A typical formulation of a water-soluble herbicide consists of the herbicide; water as the solvent for the herbicide and the bulk liquid for the formulation; selected surfactants to improve wetting and penetration properties of the formulation/carrier solution; and antifreeze to protect the formulation from freezing during shipment and storage. *Water is the carrier for water-soluble formulations.*

Oil-Soluble Herbicides

Oil-soluble herbicides dissolve in oil to form true solutions. A typical formulation of an oil-soluble herbicide consists of the herbicide and an oil (such as kerosene) or another organic solvent (such as xylene) as the solvent for the herbicide and the bulk liquid of the formulation. Surfactants may be included in the formulation to impart one or more desired properties. *Oil, such as diesel or burner oil, is the carrier for oil-soluble formulations.*

Emulsifiable Concentrates

Water-insoluble herbicides are often formulated as emulsifiable concentrates to enable them to be uniformly dispersed in water. A typical emulsifiable-concentrate formulation consists of the herbicide; solvent for the herbicide; emulsifier (usually a nonionic surfactant); selected surfactants to improve wetting, sticking, antifoaming, and other desired properties; and oil (such as kerosene) or an organic solvent (such as xylene) as the bulk liquid of the formulation. *Water is the most usual carrier for emulsifiable-concentrate formulations,* but oil or oil/water mixtures may also be used.

Emulsifiable-concentrate formulations mixed with water form an *emulsion,* rather than a true solution. In an emulsion, the water is the continuous or external phase of the mixture, while the herbicide/solvent droplets are in the dispersed or discontinuous phase. Emulsions possess low viscosity and present no problems in mixing, pumping, or spraying. Mild agitation of the mixture is necessary to keep the formulation uniformly dispersed in the carrier.

Invert Emulsions

To reduce spray drift, highly viscous emulsions may be used in which oil is the external or continuous phase and water droplets are the discontinuous phase. Such formulations are known as *water-in-oil* or *invert emulsions.* They are generally too viscous to be applied with conventional sprayers, and special equipment is available in which the two phases are actually mixed as they exit from separate orifices in the spray nozzle. With water-in-oil (invert) emulsions, the herbicide may be located in either the oil or water phase, whereas in oil-in-water (normal) emulsions, the herbicide is always contained in the oil phase. Mayonnaise and butter are examples of invert (water-in-oil) emulsions in which water is dispersed in vegetable oil or animal fat, respectively.

Wettable Powder

Herbicides of minimal solubility in water, oil, and common solvents are formulated commonly as *wettable powders.* A typical wettable-powder formulation consists of the herbicide (finely ground if crystalline), a diluent (finely ground hydrophilic clay, such as bentonite or attapulgite) for the herbicide, and various surfactants. Commonly, the herbicide makes up 50–80% (by weight) of a wettable-powder formulation, with the remainder made up of the clay diluent and the surfactants. Surfactants used in wettable-powder formulations are solids (nonliquid) that are noncaking and nonhygroscopic, and they enhance uniform dispersal and stabilization of the formulation when mixed with its carrier and in the wetting, spreading, sticking, and penetration of the herbicide on plant foliage when so applied. *Water is the carrier for wettable-powder formulations.*

Mixed with water, wettable-powder formulations form *unstable suspensions* rather than true solutions or emulsions. Unstable suspensions require continuous, vigorous agitation to prevent the suspended particles from settling.

The most important physical properties of a wettable-powder formulation are its suspension characteristics. These formulations must be conditioned to prevent lumpiness or fusion (flocculation) of the finely ground ingredients. The suspending agent (surfactant) in the formulation functions to (1) deflocculate the individual particles so that they form a smooth dispersion, and (2) stabilize the dispersed system and reduce sedimentation in the spray tank. Materials that are used as suspending agents include sodium lignin sulfate (sulfite lye), methylcellulose, aluminum silicate, polyvinyl acetate, and products of naphthalene sulfonic-acid formaldehyde condensation. These agents have very little wetting ability and must be supplemented in the formulation with surfactants having this property. Wetting agents used in wettable-powder formulations are generally anionic solids of low hygroscopicity; such wetting agents include sodium lauryl sulfate, sodium dialkylsulfosuccinates, alkylated naphthalene, and benzene sulfonates.

Care must be used in the selection of the hydrophilic diluent included in wettable-powder formulations to avoid incompatibility problems. Certain diluents may inactivate a herbicide by strongly adsorbing it or by bringing about its catalytic decomposition. Most wettable-powder formulations are ground fine enough to pass through a 50-mesh screen but not a 100-mesh screen.

Although the term *wettable powder* is widely used to designate the formulation just described, another term that may be used is *water-dispersible powder.* This term is more correct in that the important feature of wettable-powder formulations is not that they can be "wetted" but that they are converted into a suspension in water with a certain degree of stability.

Liquid-Flowable (Pourable or Liquid Suspension) Formulations

Liquid-flowable formulations are highly viscous (not easily poured) liquids, similar to a slurry. A slurry is a liquidy paste formed by mixing a wettable powder in a relatively small volume of water. This suggests a wettable-powder formulation to which the formulator has already added sufficient water to make a slurry. Although the suspending liquid is usually water, any liquid capable of maintaining a stable, suspended concentrate that can go into solution or suspension with water may be used. Liquid-flowable formulations tend to separate in storage or during transport and require careful remixing before use. When a liquid-flowable formulation separates, the heavier particles move downward, and the lighter particles and liquids (such as oil) collect near the top. *Water is the carrier for liquid-flowable formulations.*

Dry-Flowable Formulations

Dry-flowable formulations are herbicide-impregnated granules that, when added to water as the carrier, are designed to break up and disperse in a manner similar to that of wettable powders. The advantage of the dry-flowable formulation over a wettable powder is greater ease in handling—the granules are easier to measure and pour, and wind is less likely to blow the granules as it does wettable powders when they are added to the carrier under field conditions. *Water is the carrier for dry-flowable formulations.* In some cases, as with DuPont's Glean FC Herbicide, the dry-flowable granular formulation is designed to mix with, and be applied in, a carrier of either water or liquid nitrogen fertilizer.

Water-Soluble Granules

Water-soluble granules are a dry formulation that, when mixed with water, dissolve readily and form a true solution. The solution does not require further agitation once it is mixed.

Water-Dispersible Granules

Water-dispersible granular formulations serve the same purpose as dry-flowable formulations. The ingredients are in the form of granules that, when mixed with water as the carrier, break down and disperse in the carrier in a manner similar to that of a wettable powder. Their advantages over wettable powders are convenience and ease of handling.

Water-Soluble Pellets

Water-soluble pellet formulations are herbicide-impregnated pellets, usually relatively large in size, that are applied directly to the soil surface. Rainfall breaks down the pellets, and the water-soluble herbicide is leached into the soil. *The formulation itself is the carrier.*

Herbicide-Coated Granules/Pellets

Certain herbicides are formulated as granules or pellets to obtain greater crop selectivity or convenience in application. The principal difference between the two kinds of formulations is that the size of the pellets is usually larger than the granules. Granular and pelleted formulations consist of the herbicide coating or adhering to the surface of a relatively large particle such as vermiculite, clay, or sand. The herbicide commonly makes up less than 10% of the weight of the formulation. The formulations are applied as the dry granules or pellets. Again, *the formulation is itself the carrier.*

Controlled Release

Herbicide *controlled release* is defined as the alteration of herbicide entry into either soil or water in a manner that enhances weed control. Enhancement may be improved initial control, protracted control, or a combination of the two. The potential benefits of controlled release include enhanced weed control, greater crop safety, a reduction in the amount of herbicide available to move into the groundwater by run-off or leaching, and the ability to meter herbicide availability to prolong its effective residual life, among others. Disadvantages of controlled release include more costly formulations, modified toxicological characteristics that hamper herbicide registration, and having to establish a need for such formulations. As of this writing, the concept of controlled release is still in its infancy. (The interested reader is referred to Riggle and Penner, 1990.)

■ MIXTURES OF PESTICIDE FORMULATIONS

Herbicide formulations are at times mixed with other pesticide formulations (such as herbicide, fungicide, insecticide) or fluid fertilizer to save time, conserve fuel, and reduce costs in pest control and fertilizer application. It is important that the ingredients of such mixtures be compatible. Incompatibility may result in poor pest control, clogged sprayers, and problems associated with cleaning the spray equipment and disposal of the unusable conglomerate, accompanied by financial loss.

Incompatibility of formulation mixtures results from either *chemical reactions* or *physical reactions* between components. Examples of chemical reactions include the formation of a gel (a thickened colloid) following the mixing of two or more formulations and the preferential adsorption of one toxicant onto the diluent (such as the clays of wettable powders) of another. The flocculation of suspensions is an example of a physical reaction. Reactions between surfactants used to stabilize emulsions and suspensions can destroy the stability of formulation mixtures as when the addition of a cationic surfactant to an emulsion stabilized by an anionic emulsifier results in precipitation of the cation–anion complex and a breakdown of the emulsion.

Compatibility charts are available for various pesticide mixtures. However, compatibility of formulation additives (solvents, surfactants, diluents) are not usually provided by these charts, and it is often these very components that are incompatible—that is, they are the ones that interact. Such an omission from compatibility charts is understandable when one considers the multitude of available pesticide formulations.

Products that aid in achieving compatibility and stability of mixtures of liquid formulations or formulations mixed with liquid fertilizer are readily available today. Such products are composed of selected surfactants. When used, they should be added to the carrier (water or fluid fertilizer) before the herbicide formulation. The effectiveness of such compatibility adjuvants can be determined by utilizing the jar test described in the following section; in this instance, one jar would contain appropriate ingredients of the desired mixture but without the compatibility adjuvant and a second jar would contain the desired mixture plus the compatibility adjuvant. Comparing the mixtures in the two jars for uniformity and stability should readily indicate the value of compatibility adjuvants. The compatibility adjuvant should be added to the carrier (water or liquid fertilizer) in the jar prior to the pesticide formulations, with 0.25 teaspoon or 1.2 mL of compatibility adjuvant added per pint of carrier.

When mixing two or more kinds of herbicide formulations, the order in which they are mixed is important to uniform mixing. Unless the label states otherwise, add wettable powders first to the carrier or carrier plus compatibility adjuvant, followed by flowables, water-solubles, additional surfactants, and last, emulsifiable concentrates. This order should be followed in the jar test and when mixing pesticide formulations in the tank mix.

Compatibility Jar Test

Prior to mixing two or more herbicide formulations in a sprayer tank, a simple jar test of the intended mixture should be made to determine compatibility. The steps involved in such a test are as follows:

1. Add 1 pint (473 ml) of the carrier (usually water or fluid fertilizer) to a 1 qt (946 ml) jar.
2. Add each herbicide formulation to the water in the jar in the same proportion as it will be mixed with water in the sprayer tank; cap and shake vigorously after each addition.
3. Shake or invert the capped jar 10 times; immediately inspect for uniform mixing and inspect again after the mixture has stood for 30 minutes.

The mixture is compatible and usable if it remains uniformly mixed, without coagulation. The mixture is incompatible and should not be used if it is nonuniform, with clumps, aggregates, or sludge apparent. The mixture is acceptable for use if partial separation has occurred after standing 30 minutes, without formation of clumps, aggregates, or sludge, and the separation readily remixes with 10 jar inversions.

■ SURFACTANTS (SPRAY ADJUVANTS)

Spray adjuvants are substances that modify or enhance the performance of herbicides. Each spray adjuvant is, in itself, a surfactant. Surfactants facilitate and enhance the emulsifying, dispersing, wetting, spreading, sticking, penetrating, and/or other surface-modifying properties of liquids. They are chemicals that produce physical changes at the surface of liquids, and such changes occur at the *interface* between two liquids or between a liquid and a gas or solid. The practical applications of surfactants are due to their tendency to be adsorbed to the interface between the solution and the adjacent liquid, solid, or gaseous phase. Because of the changes they produce at surfaces, the term *surfactant* was coined as an acronym for *surface-active agents*.

Surfactants are commonly used in herbicide formulations to impart or enhance desired properties of the formulation and the ultimate spray mixture. They are used in liquid (soluble, emulsifiable) and dry (wettable powders, others) formulations that are applied in aqueous sprays. They may also be added as an additional adjuvant to the aqueous spray mixture.

Types of Surfactants

All surfactants possess the common feature of a water-soluble (hydrophilic) group attached to a long, oil-soluble (lipophilic) hydrocarbon chain. These two groups may be linked directly together, indirectly by an intervening group. There are more than 3000 chemicals available as surfactants, and many of the patented novel features of these chemicals lie with these intervening, connecting groups. Small differences in the structure of surfactants often greatly affect their behavior. In anionic surfactants, the closer the hydrophilic group is to the end of the hydrocarbon chain, the better it is as a detergent, whereas similar chemicals with this group located near the middle of the hydrocarbon chain are excellent wetting agents. Surfactants themselves may be phytotoxic. The phytotoxicity of isomers of anionic alkylbenzenesulfonate surfactants increases (1) as the position of the benzene ring is moved to midchain (carbon atom 5 or 6) of the *n*-dodecyl alkyl group, (2) through the use of a highly branched dodecylalkyl, or (3) by substitution of an ester oxygen atom for the benzene ring. When the structural characteristics of surfactants and a herbicide were incorporated into the same molecule by forming long-chain alkylamine salts of 2,4-D, it was found that structural features associated with herbicidal activity in aqueous sprays were those that contributed to hydrophilic characteristics of surfactants; lipophilic structural characteristics contributed to inactivity. In aqueous solutions, all of these salts of 2,4-D exhibited surface activities characteristic of surfactants.

Any one surfactant possesses to some degree more than one, if not all, properties characteristic of surfactants in general. One property usually predominates, however, forming the basis for classifications such as cosolvents (coupling agents), stabilizing agents (emulsifiers, dispersants), wetting agents, spreaders, penetrants, stickers (deposit builders), hygroscopic agents, activators or synergists, and detergents. Surfactants are commonly separated into four major groups, based on their ionization in water:

1. Anionic
2. Cationic
3. Nonionic
4. Amphoteric

Anionic and cationic surfactants ionize when mixed with water, and they owe their surface-active properties to their anions and cations. Nonionic surfactants do not ionize in aqueous solutions. Amphoteric surfactants act as either anionic or cationic surfactants, depending on the acidity of the solution. Surfactants generally are categorized as *soap* or *synthetic surfactants*.

Soap

Possibly the oldest and best known of the surfactants are *soaps*. Soaps are usually sodium or potassium salts of weak fatty acids with straight hydrocarbon chains of 12–18 carbon atoms. Sodium stearate, a typical soap, has as its molecular structure

$$CH_3{-}(CH_2)_{16}{-}\overset{\overset{\displaystyle O}{\|}}{C}{-}O^-\,Na^+$$

The polar, hydrophilic end (—COO⁻Na⁺) of the molecule is water-soluble (Figure 10.1), whereas the nonpolar, lipophilic portion [CH$_3$(CH$_2$)$_{16}$] is oil-soluble. Soaps owe their

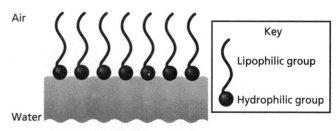

Air

Water

FIGURE 10.1 ■ Carboxylic acid film on water showing orientation of the polar end of each molecule toward the surface of the water.

SOURCE: Adapted from H. Hart and R. D. Scheutz. *Organic chemistry: A short course,* 4th ed. Boston: Houghton Mifflin. Copyright 1972 by Houghton Mifflin Co. Redrawn by permission of the publisher.

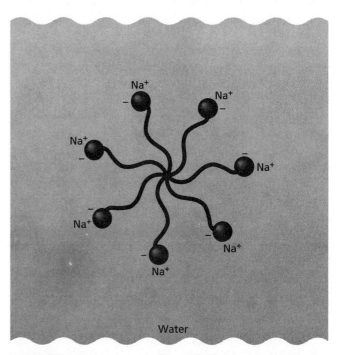

Water

FIGURE 10.2 ■ A soap micelle. When soap molecules dissolve in water, they form colloidal (rather than true) solutions. These soap solutions contain aggregates of soap molecules or *micelles.*

SOURCE: Adapted from H. Hart and R. D. Schuetz. *Organic chemistry: A short course,* 4th ed. Boston: Houghton Mifflin. Copyright 1972 by Houghton Mifflin Co. Redrawn by permission of the publisher.

surfactant properties to their anionic (negatively charged) portion; soaps are thus anionic surfactants.

When soap dissolves in water, a colloidal rather than a true solution is formed—it contains aggregates or micelles of soap molecules (Figure 10.2). When a small amount of oil is added to a soap solution and the mixture shaken, an emulsion of the oil in the soap solution is formed (Figure 10.3). In such a solution, the soap molecules surround the

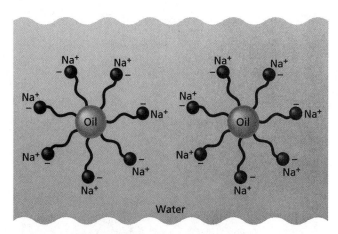

FIGURE 10.3 ■ **Soap molecules surrounding an oil droplet to form an oil-in-water emulsion.** The hydrocarbon "tails" (lipophilic) of the soap molecules are soluble in oil, whereas their hydrophilic portions stabilize the oil droplet in the water. The surface charge of the droplets prevents coalescence.

SOURCE: Adapted from H. Hart and R. D. Schuetz. *Organic chemistry: A short course,* 4th ed. Boston: Houghton Mifflin. Copyright 1972 by Houghton Mifflin Co. Redrawn by permission of the publisher.

fine oil droplets, with their lipophilic "tails" embedded in the oil and their hydrophilic ends stabilizing the droplets in the water solution. The surface charge of the droplets keeps them separated, preventing their merging or coalescing. Soap solutions have unusually low surface tension. It is a combination of surface action and emulsifying properties that enable soap solutions to serve as cleansing agents, surrounding and carrying away dirt, grease, and oil particles. The action of soap as a surfactant is illustrated in Figure 10.4.

Soaps have the disadvantage of readily forming insoluble salts with calcium, magnesium, and ferric ions in hard water. These salts precipitate from solution and form a scum on the water, wasting the soap. Soaps cannot be used in acidic solutions because they form insoluble salts with ions in the solutions.

Synthetic Surfactants

To overcome the disadvantages of soaps, chemists developed synthetic surfactants. Synthetic surfactants do not form precipitates with calcium, magnesium, and ferric ions, and they can be used with essentially the same efficiency in either hard or soft water. Synthetic surfactants are similar to soap in that they possess a nonpolar, lipophilic group composed of a long hydrocarbon chain of 12–18 carbon atoms, and a highly polar hydrophilic group. It is in the hydrophilic group that they differ from soap. The hydrophilic end of soap is composed of the polar carboxyl

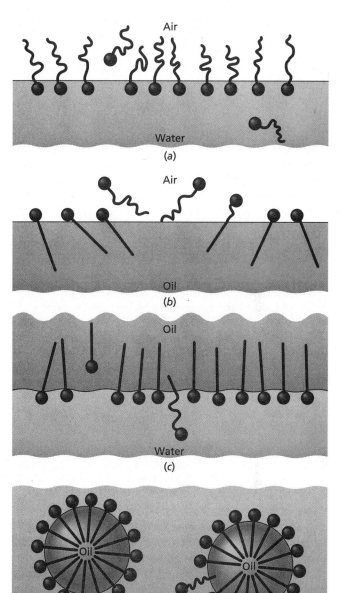

FIGURE 10.4 ■ **The surfactant action of soap.** (a) The "heads," or hydrophilic portion, of soap molecules are soluble in water, but their "tails," or hydrocarbon portions (lipophilic), are not. (b) In oil, the reverse is true. (c) At an oil–water interface, the hydrophilic and lipophilic parts of the soap molecules orient themselves with respect to the liquid in which they are soluble. (d) When oil droplets are mixed with a soap solution, the soap molecules surround the oil droplets to form an emulsion.

SOURCE: Adapted from R. W. Gerard. 1949. *Unresting cells.* New York: Harper & Row. Copyright 1940, 1949 by Harper & Row Publishers Inc. Redrawn by permission of the publishers.

(—COOH) group, whereas that of synthetic surfactants is composed of a highly polar sulfonic (—SO_3H) or sulfuric ester (—OSO_3H) group, or, in the case of the nonionic surfactants, to chains of two-carbon units linked by connecting oxygen atoms. These two-carbon units with one oxygen atom each (—CH_2—CH_2—O—) are called *oxyethylene* or, less correctly but more commonly, *ethylene oxide.*

Synthetic surfactants are the only surfactants used in herbicide formulations, and the term *surfactant* will be used synonymously with synthetic surfactant throughout this text, unless otherwise specified.

Most anionic synthetic surfactants fit into one of three basic types: alkylaromatic sulfonates (alkylarylsulfonates), sulfuric esters (fatty alcohol sulfates), and alkylsulfonates. The most important hydrophilic groups associated with the anionic surfactants are sulfonic acid (—SO_3H), sulfuric ester (—OSO_3H), and carboxyl (—COOH). Each hydrophilic group is attached directly or indirectly to the lipophilic portion of their respective molecules. Some anionic surfactants contain two different lipophilic groups; for example, some sulfated fatty acids contain both a carboxyl group and a sulfuric ester group on the same molecule. The molecular structure of sodium lauryl sulfate, a typical anionic surfactant of the sulfuric ester type, is

$$CH_3-(CH_2)_{11}-O-\overset{\displaystyle O}{\underset{\displaystyle O}{\overset{\|}{\underset{\|}{S}}}}-O^- \, Na^+$$

The typical structure of an alkylaromaticsulfonate is

where R is a hydrocarbon chain of about 12 carbon atoms.

Anionic surfactants are commonly used in herbicide formulations alone or in blends with nonionic surfactants. The anionic surfactants have the disadvantage of potentially reacting with other ions, possibly the herbicide itself, in the formulation or spray solution. As a group, their outstanding surface-active property is that they are *excellent wetting agents.* The anionic surfactants are also good detergents; until the development of the nonionic surfactants, they were the principal type of detergent. The action of an anionic detergent as a cleansing agent is illustrated in Figure 10.5.

Cationic surfactants are derived from ammonia (NH_3), and they are often called *quaternary ammonium salts.* In

FIGURE 10.5 ■ The cleansing action of an anionic detergent. Detergents remove grease because they contain a "double-headed" molecule, one example of which is shown here. To a long chain of —CH_2— units is added a sulphonic group —SO_3H, an atom of sulphur bonded to three oxygen atoms, one of which is linked by an ionic bond to a hydrogen atom, which is broken when the compound dissolves in water. This group has a strong affinity for water, whereas the chain has a strong affinity for grease and is like it chemically. The chains of a large number of molecules attach themselves to molecules of grease (as on the surface of a plate) and, pulled by the attraction of the water for the other ends of the molecules, break up the grease into small globules and float with them in the water.

SOURCE: Adapted from Sir Bernard Lovell and T. Margerison. 1967. *The explosion of science,* p. 191. New York: Meredith Press.

this class of surfactants, the most prevalent groups capable of ionizing are the primary, secondary, and tertiary amino groups and the quaternary ammonium groups; phosphonium and sulfonium groups are of far less importance.

An amine is formed when one or more of the hydrogen atoms of ammonia is replaced by hydrocarbon groups; when all three hydrogens are so replaced, the resulting compounds are called *tertiary amines.* Tertiary amines combine with certain alkyl halides to form quaternary ammonium salts. Quaternary ammonium cations also may be formed when all four of the hydrogen atoms of the ammonium ion (NH_4^+) are replaced by alkyl groups.

The quaternary ammonium salts are cationic surfactants in which one of the hydrocarbon chains is of the same order of length as those of the fatty acids, usually 12–20 carbon atoms. Cetyltrimethylammonium chloride is a typical cationic surfactant. Its molecular structure is

$$CH_3-(CH_2)_{15}-\overset{\overset{\displaystyle CH_3}{|}}{\underset{\underset{\displaystyle CH_3}{|}}{N^+}}-CH_3 \ \ Cl^-$$

The cationic surfactants are not usually used in herbicide formulations. They are generally phytotoxic and are highly effective bactericides. They precipitate readily in hard water and are poor detergents.

Nonionic surfactants are a relatively new group of surfactants, coming into general usage during the early 1960s. Although they were discovered and patented in 1930 by C. Schöller, a German chemist, it was not until 1957 that a pure nonionic surfactant, homogeneous with respect to alkyl chain length and oxyethylene chain length, was produced. The nonionic surfactants now constitute a sizable portion of the total production of surfactants.

The nonionic surfactants, do not ionize in aqueous solution. Consequently, they are unaffected by hard water; that is, they do not form insoluble salts with calcium, magnesium, or ferric ions. They can also be used in strong acid solutions. The nonionic surfactants have low mammalian toxicity and, in general, low phytotoxicity. *The outstanding property of the nonionic surfactants is as emulsifiers in the formation of stable emulsions.* They are good dispersing agents and excellent detergents. They foam less than anionic surfactants and are considered to be low-to-moderate foaming agents. *They are more soluble in cold water than in hot water*—an inverse temperature–solubility relationship. The nonionic surfactants are commonly used as emulsifiers in herbicide formulations of the emulsifiable-concentrate type. To improve the wetting properties of these formulations, anionic surfactants are often used in blends with the nonionics.

The term *nonionic surfactant* refers chiefly to derivatives of polyoxyethylene and polyoxypropylene. The polyoxyethylenealkylphenols comprise the largest production volume among nonionic surfactants, with a wide scope of applications in both household and industrial products. Other nonionic surfactants include anhydrohexitol derivatives, sugar esters, fatty alkanol amides, and fatty amine oxides.

TABLE 10.1 ■ Correlation between HLB* numbers and the predominant property of nonionic surfactants.

HLB RANGE	PROPERTY
3–6	Water/oil emulsifier†
7–9	Wetting agent
8–15	Oil/water emulsifier‡
13–15	Detergent
15–18	Solubilizer

*Hydrophilic–lipophilic balance.
†Invert emulsion.
‡Normal emulsion.

Nonionic surfactants consist of a lipophilic hydrocarbon chain [such as alkylphenols, aliphatic acids (especially fatty acids), and corresponding alcohols] attached to a second chain, branching from the first, that is hydrophilic and composed of oxyethylene units $(-CH_2-CH_2-O-)$. In aqueous solution, water molecules are affixed to the *oxygen*, which bonds adjacent oxyethylene units in the chain by hydrogen bonding. The more oxyethylene units forming the chain, the greater the total hydrogen bonding, which results in increased water solubility of the nonionic surfactant. Depending on the nature of the lipophilic group, at least 4–6 oxyethylene units per molecule are required to produce a water-soluble surfactant, with 10–12 such units being more common.

Nonionic surfactants may be identified by the relationship between the lengths of their hydrophilic and lipophilic chains, a relationship called the *hydrophile–lipophile balance* (HLB). In practice, the HLB is represented by an arbitrary scale in which the least hydrophilic surfactants have low HLB numbers, and increasingly larger numbers correspond to greater hydrophilic character. In general, the HLB numbers can be used in the selection of a nonionic surfactant for a particular purpose (Table 10.1).

The general structure of an alkylpolyoxyethylene glycol ether type of nonionic surfactant is

$$\begin{array}{c} CH_3 \\ | \\ (CH_2)_n \\ | \\ HC-O-(CH_2-CH_2-O)_xH \\ | \\ (CH_2)_{n'} \\ | \\ CH_3 \end{array}$$

where n and n' denote the number of (CH_2) groups of their respective units forming the alkyl chain and x denotes the

TABLE 10.2 ■ Examples of surfactants in each of the three classes of surfactants.

ANIONIC	CATIONIC	NONIONIC
sodium lauryl sulfate	lauryl trimethyl ammonium chloride	sec-butylphenoxy-polypropyleneoxy-poly-ethyleneoxy ethanol
amine salt of akylsulfonic acid	stearyl trimethyl ammonium chloride	alkylarylpolyoxyethylene glycols; free fatty acids; isopropanol
sodium dodecylbenzenesulfonate	dilauryl trimethyl ammonium chloride	
sulfonated alkyl ester	1-(2-hydroxyethyl-2-n-alkyl-2(or 3)	alkyl phenoxy polyethoxy
alkylarylsulfonate	benzyl-2-imadozolinium chloride	polyoxyethylene thioether
dioctyleser of sodium sulfosuccinic acid	heptadecyl imadazolinium chloride	dodecyletherpolyethylene glycol
		isooctylphenylpolyethoxy glycol
		octylphenylpolyethoxy ethanol
		polyoxyethylene sorbitan monolaurate
		nonylphenylpolyoxyethylene glycol ether

number of oxyethylene units forming the polyoxyethylene chain.

The general structure of an alkylphenol polyoxyethylene glycol ether nonionic surfactant is

$$C_nH_{2n+1} - \langle \text{ring} \rangle - O-(CH_2-CH_2-O)_xH$$

where n denotes the number of carbon atoms in the alkyl chain and x denotes the number of oxyethylene units in the polyethylene chain.

A list of some representative surfactants in each of the three principal classes is presented in Table 10.2.

Surfactants Often Found on Product Labels

Wetting agents enhance foliar wetting with aqueous spray mixtures.

Crop oil concentrates are phytobland petroleum or vegetable oils that increase absorption of the herbicide through the cuticle (waxy layer) of leaves. They contain an emulsifier (surfactant) that enables them to be miscible with water. Crop oil concentrates contain 83–85% oil and 15–17% emulsifier (surfactant). Directions on some product labels specify the use of a petroleum-based, rather than vegetable-based, crop oil concentrate.

Compatibility agents help maintain emulsion stability when two or more herbicides with incompatible formulations are added to the same spray mixture. They also aid solution stability when a herbicide is applied in liquid nitrogen fertilizer solutions.

Defoamers are used to minimize air entrapment in the spray mixture during agitation, which may cause problems in application of the mixture.

Drift reduction agents (thickeners) modify spray characteristics by minimizing formation of small spray droplets, thus reducing spray drift.

Stickers (adhesives) enhance the retention of the spray mixture on plant foliage.

Dash HC Spray Adjuvant (BASF Corporation) is used in water-based sprays to enhance the effectiveness of the herbicide sethoxydim (Poast; Poast Plus). Dash is used in place of crop oil cencentrate in spray mixtures of Poast or Poast Plus, whether used alone or in tank mix with Basagran in soybeans. Dash enhances the performance of Poast/Poast Plus by improving control of grasses such as volunteer corn, johnsongrass, quackgrass, and wild oats. Dash plus UAN or ammonium sulfate, added to the spray mixture, increases the performance of Poast in tank mix with Basagran. Directions for some other herbicides may specify the use of Dash in their spray mixtures.

Frigate (ISK Biotech Corporation) is used in place of nonionic surfactants in spray mixtures with glyphosate (Roundup) to improve foliar wetting and leaf absorption, resulting in improved control of quackgrass, johnsongrass, and annual weeds.

Lo-Dose (ISK Biotech Corporation) is a spray adjuvant to be used in formulations of glyphosate in annual cropping systems for increased control of quackgrass, johnsongrass, Canada thistle, and common bermudagrass. Lo-Dose should not be used with glyphosate (e.g., Rodeo) for aquatic weed control.

Liquid nitrogen fertilizers, such as urea–ammonium nitrate (UAN solution, commonly referred to as 28, 30, or 32% nitrogen solution) or ammonium sulfate increase phytotoxicity of foliar-applied herbicides. However, crop selectivity may be lost if the mixture contacts the crop foliage. Liquid nitrogen fertilizers are not surfactants, but they are included here for the reader's convenience, as they are commonly referred to on many herbicide product labels as an additive to the spray mixture or as the carrier for the herbicidal product (formulation).

Surfactants and Phytotoxicity

The addition of surfactants to herbicide formulations and spray solutions may have no effect on the herbicide's phytotoxicity, or it may enhance or retard its phytotoxicity. There may be interactions between the herbicide and the surfactants in the formulation or in the spray solution, and such reactions either have no effect on the subsequent phytotoxicity of the herbicide or they reduce its effectiveness.

In aqueous solutions, surfactants alone may be phytotoxic; they also may induce physiological, biochemical, and morphological effects that are either stimulatory or inhibitory to plants. The phytotoxicity of surfactants may adversely affect the activity of the primary herbicide by limiting its penetration and translocation in the plant.

The principal site at which surfactants enhance herbicide activity is the point of application and in the immediately underlying tissue of the plant. The enhancement of the phytotoxicity of a herbicide is due primarily to the surfactant aiding the herbicide in its penetration of the plant. Surfactants, apparently, do not facilitate herbicide transport *per se,* except possibly in the apoplast.

Herbicides are absorbed by the foliage of plants via two pathways, one of which is hydrophilic and the other lipophilic. Smith and Foy (1966) proposed the following hypothesis for the mode of action of surfactants in aiding the entry of herbicides into plants via the hydrophilic pathway: "Molecules of a surfactant diffuse from the spray droplet into the cuticle of the leaves, perhaps via imperfections and cracks, and then align themselves in monolayers with their nonpolar ends oriented in the cutin and wax. The polar ends will thus form a layer whose size depends on the length of the hydrophilic chain of the surfactant molecules. The layers, or 'hydrophilic channels,' will presumably attract water, causing swelling of the cuticle. Thus, channels or pores are formed along which herbicide molecules can diffuse according to their various chemical properties, such as solubility, residual chemical charge, polar properties, etc."

Hard Water

Hard water used as the carrier in the application of herbicides may adversely affect the dispersion of the herbicide in the carrier and, subsequently, its phytotoxicity.

Natural waters usually contain ions of calcium (Ca^{2+}), magnesium (Mg^{2+}), iron as ferric (Fe^{3+}), and possibly others. Water is considered "hard" when the total concentration of these ions is above a certain level. The *degree of hardness* is a measure of the total concentration, by weight, of the above ions in water, usually expressed as parts per million (ppm) or as grains per U.S. gallon (1 grain per gallon being equal to about 17.1 ppm). Water with a hardness of less than 50 ppm is considered soft water, 50–100 ppm as medium hard, and 100–200 ppm as hard water. Hard water is suitable for irrigation purposes; water with a hardness of less than 150 ppm is preferable for domestic purposes. Very soft water may be corrosive to water pipes.

The troublesome feature of hard water with respect to herbicides and their application and effectiveness is that the ions, especially Ca^{2+}, Mg^{2+}, and Fe^{3+}, react with salts of the herbicides and with some surfactants to form insoluble salts that precipitate, removing the herbicide or surfactant from the spray solution. The greatest benefit, over soap, derived from the synthetic anionic detergents is the relatively high solubility of the salts that they form with the ions of hard water. Soap reacts with these ions to form insoluble salts that precipitate.

■ SELECTED REFERENCES

Ahrens, J. F. 1972. Granular herbicides. *Weeds Today* 3(2):12–13.

Anonymous. 1958. Pesticide emulsifiers. *J. Agr. Food Chem.* 6:174–175.

Anonymous. 1961. How surfactants affect herbicides. *Agr. Res.* 9(12):3–4.

Anonymous. 1966. *Ucane alkylate—the biologically soft detergent alkylate.* New York: Union Carbide Corp.

Annonymous. 1994. Formulations. *Weed Control Manual.* Willoughby, Ohio: Meister Publishing. pp. 20–22.

Bayer, D. E. 1967. Effect of surfactants on leaching of subtituted urea herbicides in soil. *Weeds* 15:249–252.

Bayer, D. E. 1971. Choosing and using a surfactant. *Proc. Calif. Weed Conf.* 23:24–25.

Bayer, D. E., and H. R. Drever. 1965. The effects of surfactants on efficiency of foliar-applied diuron. *Weeds* 13:222–226.

Becher, P. 1967. Emulsification. In *Nonionic surfactants*, M. J. Schick, ed., pp. 604–630. New York: Marcel Dekker.

Behrens, R. W. 1964. The physical and chemical properties of surfactants and their effects on formulated herbicides. *Weeds* 12:255–258.

Crafts, A. S. 1957. The chemistry and mode of action of herbicides. *Adv. Pest Control Res.* 1:39–80.

Day, B. E., and L. S. Jordan. 1961. Spray retention by bermudagrass. *Weeds* 9:351–355.

Ebeling, W., and R. J. Pence. 1953. Pesticide formulation: Influence of formulation on effectiveness. *J. Agr. Food Chem.* 1:386–397.

Foy, C. L., and L. W. Smith. 1965. Surface tension lowering, wettability of paraffin and corn leaf surfaces, and herbicidal enhancement of dalapon by seven surfactants. *Weeds* 13:15–18.

Foy, C. L., and L. W. Smith. 1969. The role of surfactants in modifying the activity of herbicidal sprays. In *Pesticidal formulations research,* R. F. Gould, ed., pp. 55–69. Washington, D.C.: Adv. Chem. Series, American Chemical Society.

Fraser, R. P. 1958. The fluid kinetics of application of pesticidal chemicals. *Adv. Pest Control Res.* 2:1–106.

Freed, V. H., and M. Montgomery. 1958. The effect of surfactants on foliar absorption of 3-amino-1,2,4-triazole. *Weeds* 6:386–389.

Furmidge, C. G. L. 1962. Physical-chemical studies on agricultural sprays. IV. The retention of spray liquids on leaf surfaces. *J. Sci. Food Agr.* 13:127–140.

Gardner, L. R. 1953. Pesticide formulations: Basic principles of formulation. *J. Agr. Food Chem.* 1:521–522.

Garver, H. L. 1955. Water supplies for homes in the country. In *Water, yearbook of agriculture,* pp. 655–663. Washington, D.C.: USDA, Supt. of Doc.

Gerard, R. W. 1949. *Unresting cells,* chap. 4. New York: Harper & Row.

Gill, R. E. 1960. Formulations: Specifications–how they are used. *Span* 8:81–84.

Hart, H. and R. D. Schuetz. 1972. *Organic chemistry: A short course,* 4th ed. Boston: Houghton Mifflin.

Hayes, W. J., and G. W. Pearce. 1953. Pesticide formulations: Relations to safety in use. *J. Agr. Food Chem.* 1:466–469.

Jansen, L. L. 1964. Surfactant enhancement of herbicide entry. *Weeds* 12:251–255.

Jansen, L. L. 1965. Effects of structural variations in ionic surfactants on phytotoxicity and physical-chemical properties of aqueous sprays of several herbicides. *Weeds* 13:117–123.

Jansen, L. L. 1965. Herbicidal and surfactant properties of long-chain alkylamine salts of 2,4-D in water and oil sprays. *Weeds* 13:123–130.

Jansen, L. L., W. A. Gentner, and W. C. Shaw. 1961. Effects of surfactants on the herbicidal activity of several herbicides in aqueous spray systems. *Weeds* 9:381–405.

Kaupke, C. R. 1965. Studies on drift reduction by means of application techniques and low drift formulations. *Proc. Calif. Weed Conf.* 17:103–108.

Kelly, J. A. 1953. Commercial herbicides; Present methods of formulation. *J. Agr. Food Chem.* 1:254–257.

Lovell, Sir Bernard, and T. Margerison. 1967. *The explosion of science.* New York: Meredith Press.

McWhorter, C. G. 1963. Effects of surfactant concentration on johnson-grass control with dalapon. *Weeds* 11:83–86.

McWhorter, C. G. 1963. Effects of surfactants on the herbicidal activity of foliar sprays of dalapon. *Weeds* 11:265–269.

Parr, J. F., and A. G. Norman. 1965. Considerations in the use of surfactants in plant systems: A review. *Bot. Gaz.* 126:86–96.

Price, D. 1952. *Detergents.* New York: Chemical Publishing Co.

Rawlins, W. A. 1963. Granular pesticides. *Farm Technology.* 19(5):12–13.

Riggle, B. D., and D. Penner. 1990. The use of controlled-release technology for herbicides. *Rev. Weed Sci.,* Vol. 5:1–14 Champaign, Ill.: Weed Science Society of America.

Schick, M. J., ed. 1967. *Nonionic surfactants.* New York: Marcel Dekker.

Schwartz, A. M., and J. W. Perry. 1949. *Surface active agents,* Vol. I. New York: Interscience Publishers.

Schwartz, A. M., J. W. Perry, and J. Berch. 1958. *Surface active agents and detergents,* Vol. II. New York: Interscience Publishers.

Selz, E. 1953. Pesticide formulation: Liquid concentrate problems. *J. Agr. Food Chem.* 1:381–386.

Selz, E., and P. Lindner. 1959. Progress in liquid pesticide formulations. *J. Agr. Food Chem.* 7:540–543.

Shepard, H. H. 1953. Trends in production and consumption of pesticidal chemicals. *J. Agr. Food Chem.* 1:757.

Simpson, G. G., and W. S. Beck. 1969. *Life: An introduction to biology,* chap. 3. New York: Harcourt, Brace, & World.

Slabaugh, W. H., and T. D. Parsons, 1966. *General chemistry,* chaps. 10, 27. New York: John Wiley & Sons.

Smith, L. W., and C. L. Foy. 1966. The possible mode of action of nonionic surfactants in herbicide solutions. *Res. Prog. Rept., Western Weed Control Conf.,* pp. 139–140.

Smith, L. W., and C. L. Foy. 1966. Penetration and distribution studies in bean, cotton, and barley from foliar and root applications of Tween-20-C^{14}, fatty acid and oxyethylene labelled. *J. Agr. Food Chem.* 14:117–122.

Smith, L. W., C. L. Foy, and D. E. Bayer. 1966. Structure-activity relationships of alkylphenol ethylene oxide ether non-ionic surfactants and three water-soluble herbicides. *Weed Res.* 6:233–242.

Smith, L. W., C. L. Foy, and D. E. Bayer. 1967. Herbicidal enhancement of certain new biodegradable surfactants. *Weeds* 15:87–89.

Somers, E. 1967. Formulation. In *Fungicides, Vol. I: agricultural and industrial applications,* D. C. Torgeson, ed., pp. 153–193. New York: Academic Press.

Steward, F. C. 1966. *About plants: Topics in plant biology.* Reading, Mass.: Addison-Wesley.

Temple, R. E., and H. W. Hilton. 1963. The effect of surfactants on the water solubility of herbicides, and the foliar phytotoxicity of surfactants. *Weeds* 11:297–300.

Turk, A., H. Meislich, F. Brescia, and J. Arents. 1968. *Introduction to chemistry,* chaps. 10, 13. New York: Academic Press.

Van Valkenburg, J. W. 1969. The physical and colloidal chemical aspects of pesticidal formulations research: A challenge. In *Pesticidal formulations research,* G. F. Gould, ed. Washington, D.C.: Adv. Chem. Series, American Chemical Society.

White, E. H. 1970. *Chemical background for the biological sciences,* 2nd ed. Englewood Cliffs, N.J.: Prentice-Hall.

Wilcox, L. V. 1948. *Explanation and interpretation of analyses of irrigation waters. Circ. No. 784,* Washington, D.C.: USDA, U.S. Government Printing Office.

11 SPRAYER CALIBRATION AND HERBICIDE CALCULATIONS

■ SPRAYER CALIBRATION

The objective of calibrating any sprayer is to determine and adjust the amount of carrier that the sprayer will apply to a known area under a given set of conditions. The carrier is a liquid, usually water, but an oil or liquid nitrogen fertilizer may be used. Proper sprayer calibration is essential to applying the correct amount of herbicide.

In general, herbicides are applied in 5–40 gal water/A; lesser amounts may be applied by aircraft or greater amounts under conditions where thorough wetting of dense foliage is desired. The volume of carrier applied by a sprayer is governed by five factors:

1. Speed of the sprayer
2. Spray pressure at nozzles
3. Number of nozzles used
4. Size of nozzle orifice (opening)
5. Viscosity of the carrier

Increasing the speed of the sprayer results in less carrier being applied. Conversely, a slower speed results in a greater volume being applied. The more commonly used speeds of ground sprayers are between 2 and 5 mph (3 and 8 km/h). Speeds that are too fast can cause the spray boom to flex or bounce, resulting in an uneven spray pattern.

Increasing the spray pressure results in a greater volume of carrier being applied to a given area. Conversely, a lower spray pressure results in less carrier applied. In general, spray pressures of 15–40 pounds per square inch (psi) are used when applying herbicides. The spray pressure must be adequate to provide the recommended spray patterns from all nozzle tips on the spray boom.

Increasing the number of nozzles used with the sprayer results in increased output. However, additional nozzles do not necessarily increase the volume of carrier applied per acre; they may be added to a longer spray boom to increase the area sprayed during each pass of the sprayer. Similarly, a decrease in the number of nozzles usually results in a smaller area being sprayed, rather than a change in volume per acre. Flood-type nozzles cover a wider spray pattern than the more conventional fan-type nozzles.

The *use of larger-sized orifice nozzle tips* results in a greater volume of carrier applied to a given area. Conversely, smaller-sized orifice tips deliver a smaller volume of carrier.

The *viscosity of the carrier* may affect calibration when interchanging carriers such as water, oil, or liquid nitrogen fertilizer. Each of these carriers will have a different viscosity. The greater the viscosity, the less readily the liquid will flow. Oil will be the least viscous, water next, with liquid nitrogen fertilizer the most viscous. Therefore, the sprayer must be recalibrated when a carrier of different viscosity is used in place of another.

During sprayer calibration, sprayer output may be varied to meet the needs of the situation by adjusting any one or combination of the above factors. The most practical means is to adjust (1) *speed,* (2) *nozzle orifice size,* or (3) *both speed and orifice size.*

When adjusting sprayer output during calibration, remember that spray droplets become smaller as the spray pressure is increased, nozzle orifice size decreased, or both together. The smaller the spray droplets, the greater is their tendency to drift from the target area with air movement.

For uniform spray coverage, proper nozzle tips must be used, and the nozzle spacing and height must be properly adjusted for the conditions involved. For uniform output and coverage, nozzle tips of the same size and spray angle must be used in all nozzles on the spray boom. Care must be taken to discard nozzle tips that have been enlarged from the abrasive action of the spray material.

For convenience and safety, the sprayer is calibrated using only the carrier, usually water. If the herbicide is present in the carrier during sprayer calibration, care must be taken as to where the spray mixture is applied during the calibration procedure. It may be best to catch the output from each nozzle in a container, rather than allow the spray mixture to fall to the ground. The sprayer should be calibrated under the same conditions, including the softness of the ground over which the sprayer will pass. If calibrated on a firm surface such as a roadway or packed soil, the sprayer will cover a measured distance at a faster speed than it would if calibrated on soft or freshly tilled soil. In such a case, as the sprayer is moved from the firm surface onto soft soil in the field to be sprayed, the sprayer will be moving at a speed slower than that for which it was calibrated and an excess of spray mixture will be applied to the field.

Fan-type nozzles are commonly spaced and height-adjusted on the spray boom so that each, with proper overlapping, uniformly covers a width of 18–20 in. (45–50 cm).

When calibrating a sprayer with water as the carrier, it is convenient to know that sprayer output, expressed as gallons per acre (gal/A), is equal to the number of fluid ounces (fl oz) of water applied to an area of 340 square feet (sq ft or ft^2). One acre is composed of 128 units (plots) that are 340 ft^2 in size, and 1 gal of water is composed of 128 fl oz. Thus 32 fl oz (1 qt) of water applied per 340 ft^2 of area is equivalent to 32 gal/A (1 imperial gal/ha = 1 liter/ 10 m^2).

■ CALIBRATION PROCEDURES

Tractor-Mounted or -Pulled Sprayers

The following procedure may be used to calibrate a tractor-mounted or -pulled sprayer equipped with any type of spray nozzle designed to cover a given area uniformly with the spray mixture, *at constant speed and spray pressure,* with water as the carrier.

1. Fill spray tank completely full with water (or to an accurately measured depth).

2. Select a safe speed (usually between 2 and 5 mph) for the terrain to be sprayed (noting gear and throttle setting), and use this speed during calibration.

3. Mark off and measure any convenient distance. Generally, the greater the distance, the greater the accuracy in determining sprayer output.

4. Make one or more passes with the sprayer over the measured distance at the selected speed, operating the spray only over the measured distance.

5. To determine the volume of water applied, refill the spray tank to its original water level and carefully note the number of fluid ounces of water needed to refill the tank (1 pt = 16 fl oz).

6. Calculate the area sprayed by multiplying the width of the area sprayed by the distance traveled (sprayed), measured in feet.

7. Divide the number of square feet in 1 A (43,560 ft^2) by the number of square feet making up the area sprayed. The result is the number of such "plots" per acre.

8. Multiply the number of plots per acre by the number of fluid ounces of water applied to the sprayed area to obtain the equivalent number of fluid ounces of water applied per acre.

9. Divide the total number of fluid ounces of water applied per acre by the number of fluid ounces in 1 gal (128 fl oz) to obtain the sprayer output, expressed in gallons of water per acre.

10. For greater accuracy, repeat the procedure several times and average the sprayer output.

EXAMPLE:

A sprayer equipped with a single flood-type nozzle that covers an area 20 ft wide is calibrated over a distance of

400 ft. The volume of water applied during one pass of the sprayer was 950 fl oz. The sprayer output was 40 gal/A.

VERIFICATION:

Plot size sprayed = 20 ft × 400 ft = 8000 ft²

43,560 ft²/A ÷ 8000 ft²/plot = 5.4 plots/A

950 fl oz water/plot × 5.4 plots/A = 5130 fl oz/A

5130 fl oz/A ÷ 128 fl oz/gal = 40 gal/A

Once the sprayer output has been determined, the total number of acres that may be sprayed with any given sprayer may be calculated.

EXAMPLE:

A sprayer equipped with a tank capacity of 150 gal and calibrated to deliver 15 gal spray mixture/A will hold sufficient spray mixture to treat an area 10 A in size.

VERIFICATION:

Sprayer output = 15 gal/A

Tank capacity = 150 gal

150 gal ÷ 15 gal/A = 10 A

For sprayers equipped with more than one nozzle on the spray boom, the output, expressed in gallons per acre, of such a sprayer calibrated over a distance of 226.6 ft is equal to the number of fluid ounces of water applied to the total area *divided* by the number of nozzles used (each nozzle adjusted to cover an area 1.5 ft wide).

EXAMPLE:

At constant speed and spray pressure, a sprayer equipped with a spray boom with 16 fan-type nozzles, so spaced and adjusted to cover uniformly an area 24 ft wide, has an output of 512 fl oz (4 gal) when calibrated over a distance of 226.6 ft. The output is 32 gal/A (total fl oz applied, *divided* by number of nozzles).

VERIFICATION:

Plot area sprayed = 24 ft × 226.6 ft = 5438 ft²

43,560 ft²/A ÷ 5438 ft²/plot = 8 plots/A

512 fl oz water/plot × 8 plots/A = 4096 fl oz/A

4096 fl oz/A ÷ 128 fl oz/gal = 32 gal/A

A sprayer may also be calibrated by first determining the time it takes to cover a distance of 226.6 ft over the soil surface to be sprayed. The discharge from each nozzle on the spray boom must be collected and measured to ascertain that their output is the same. If not, worn or wrong nozzle tips must be replaced so that the spray pattern across the length of the spray boom is uniform. Then, with the sprayer stationary, catch the water discharge from one nozzle in a container over this same period of time, with the spray pump operating at the pressure that will be used during the spraying operation. The number of fluid ounces collected converts directly to gallons per acre for a nozzle covering a width of 18 in. Thus, if 30 fl oz are collected, the sprayer output is 30 gal/A.

When each nozzle on the spray boom covers a width of 20 in., time the sprayer over a distance of 203.6 ft and proceed as above. The volume of water collected from one nozzle, expressed as fluid ounces, converts directly to gallons per acre. Thus, if 32 fl oz are collected, the sprayer output is 32 gal/A.

Hand Sprayers

The following procedure may be used to calibrate a sprayer with a relatively small tank capacity, such as the 2- to 5-gal size commonly carried by hand, on the shoulder, or as a backpack. To properly calibrate such sprayers, they must be equipped with a pressure gauge and pressure-regulating valve located between the spray tank and the nozzle outlet so that a constant sprayer output is maintained.

1. Pour a known volume of water into the spray tank.
2. Close the spray tank and pump up pressure.
3. At constant pressure and uniform pace (speed), spray a known area.
4. Determine the sprayer output for the area by subtracting the amount of water remaining in the spray tank from the amount originally poured in the tank. (Drain the water remaining in the spray boom and hose back into the spray tank before measuring the amount of water applied.)
5. Sprayer output for a given area may be adjusted by changing the pace or the nozzle orifice size, or both. Generally, the spray pressure remains set at about 25 psi.
6. If necessary, the sprayer output can be expressed as gallons per acre by dividing the area of 1 A (43,560 ft²) by the area sprayed (ft²), multiplying this value by the volume (fl oz or ml) of water applied per area, and converting to gallons (1 gal = 128 fl oz = 3785 ml, 1 fl oz = 29.57 ml).

EXAMPLE:

A 3-gal hand sprayer is to be calibrated. Three quarts (96 fl oz, 2838 ml) of water are poured into the spray tank, the lid closed, and pressure pumped up. At constant pace (speed) and pressure, 68 fl oz of water remained in the spray tank after treating an area 1.5 ft wide by 80 ft long. Sprayer output was 28 fl oz (828 ml) of water per 120 ft^2 area, equivalent to 79.4 gal/A. Drain the water in the spray boom and hose back into the spray tank prior to measuring sprayer output.

VERIFICATION:

Plot area sprayed = 1.5 ft × 80 ft = 120 ft^2

43,560 ft^2/A ÷ 120 ft^2 = 363 plots/A

$$
\begin{array}{ccc}
\text{96 fl oz added to spray tank} & & \text{28 fl oz water applied} \\
minus\text{ 68 fl oz remaining in} = & & \text{per plot} \\
\text{tank after spraying} & & \text{(828 ml water/plot)}
\end{array}
$$

$$
\begin{array}{c}
\text{28 fl oz water/plot} \\
\text{× 363 plots/A}
\end{array} = \begin{array}{c}
\text{10,164 fl oz/A} \\
\text{(300,550 ml/A)}
\end{array}
$$

10,164 fl oz/A ÷ 128 fl oz/gal = 79.4 gal/A

300,550 ml/A ÷ 3785 ml/gal = 79.4 gal/A

Watering Can

Sometimes, when one doesn't have a sprayer, an ordinary watering can of about a 3-gal capacity equipped with a sprinkler spout may be used to apply herbicides mixed with water to areas of about 100 ft^2, as in the home garden or lawn. Calibrate the watering can by filling it with water to a marked level and then applying all of the water as uniformly as possible, using a smooth, sweeping motion with the arm holding the can; passing over the area twice for uniformity, the second time at right-angle to the first. The area covered, expressed as square feet, is determined and then divided into 43,560 ft^2 to obtain the number of areas (plots) per acre. Watering can output, expressed as gallons per acre, is calculated by multiplying the number of plots per acre by the volume of water applied per plot.

EXAMPLE:

A 2-gal watering can equipped with a sprinkler spout uniformly covered an area 100 ft^2 in size. The watering can is now calibrated. Calculate the correct amount of herbicide (product) and add to 2 gal water in the can and apply

as before. The gallons per acre is high. In this case, it is 871.2 gal/A (43,560 ft^2 divided by 435.6 plots/A × 2 gal/plot).

■ GROUND SPEED DETERMINATION

A general formula for determining ground speed for a sprayer is

$$
\text{mph} = 0.682 \times \frac{\text{distance (ft)}}{\text{time (sec)}}
$$

where 0.682 equals the time (sec) required for the sprayer to travel 1 ft traveling at 1 mph (0.682 sec/ft = 1 mph).

EXAMPLE:

A sprayer travels over a distance of 300 ft in 125 sec. Determine its ground speed.

$$
\text{mph} = 0.682 \times \frac{\text{distance (ft)}}{\text{time (sec)}} = 0.682 \times \frac{300}{125} = 1.6 \text{ mph}
$$

When determining the ground speed of a sprayer, it is convenient to know that a distance of 176 ft is equal to one-thirtieth of a mile; that is, there are 30 units 176 ft long per mile (1 mi = 5280 ft).

The following formula may be used to determine ground speed in miles per hour:

$$
\text{mph} = \frac{3600 \text{ sec/hr}}{30 \times \text{time (sec) to travel 176 ft}} = \frac{120}{\text{time (sec)}}
$$

EXAMPLE:

A sprayer travels over a distance of 176 ft in 40 sec. Determine its ground speed, expressed in miles per hour.

$$
\text{mph} = \frac{120}{\text{time (sec)}} = \frac{120}{40} = 3 \text{ mph}
$$

Similarly, a distance of 132 feet is one-fortieth of a mile; that is, there are 40 units 132 ft long per mile.

The following formula may be used to determine ground speed in miles per hour:

$$
\text{mph} = \frac{3600 \text{ sec/hr}}{40 \times \text{time (sec) to travel 132 ft}} = \frac{90}{\text{time (sec)}}
$$

EXAMPLE:

A sprayer travels a distance of 132 ft in 40 sec. Determine its ground speed, expressed in miles per hour.

$$\text{mph} = \frac{90}{\text{time (sec)}} = \frac{90}{40} = 2.25 \text{ mph}$$

■ HERBICIDE CALCULATIONS

When calculating the amount of herbicide to apply to a given area, remember that the herbicide is only one component, the *active ingredient,* of a formulation (the product). Before a calculation can begin, one must know the amount of herbicide in the formulation. This information is found on the product's label under "Active Ingredient," usually located at or near the top of the label (just below the product's trade name).

The amount of active ingredient present in *liquid formulations* (soluble, emulsifiable, liquid flowable) is usually given as pounds of active ingredient per gallon of product (lb ai/gal). In the event the active ingredient is present as a salt or ester of the parent acid of the herbicide, the *acid equivalent* of the herbicide in the formulation will be given as pounds of acid equivalent per gallon (lb ae/gal). Care must be taken not to confuse *lb ai/gal* with *lb ae/gal* when making a calculation, as the weight of the acid equivalent will always be less than its salt or ester. When making herbicide rate calculations involving liquid formulations, ignore the percent of active ingredient in the formulation: *use only lb ai/gal or lb ae/gal as part of your calculations.*

The amount of active ingredient present in *dry formulations* (wettable powders, dry flowables, granules, pellets) is usually given as a percent of the formulation by weight (% ai w/w). With dry formulations, the *% ai w/w* is used as part of your calculations.

The *rate* at which a herbicide is applied is, unless otherwise specified, expressed as pounds active ingredient per acre (lb ai/A). Some herbicides (e.g., sulfonylureas) are applied at rates of only a few ounces, or less, of active ingredient per acre, and these rates are abbreviated *oz ai/A. The expressed rate is based on a broadcast application per acre* (i.e., the entire surface area of an acre, or its equivalent, being treated). Sometimes, the rate may be based on a band treatment (over or between the crop row), with so much herbicide applied per acre but in a band of designated width, or for an area of so many square feet (e.g., 1000 ft^2). When this is the case, the variant from a broadcast treatment *must* be emphasized to avoid possible misapplication.

Weed-control publications are available from state experiment stations and other sources in the United States; these serve as guides for herbicide use in specific crop or noncrop areas and give respective application rates. The rates are usually expressed as lb ai/A, lb ae/A, or oz ai/A. Sometimes the equivalent amount of a specific product may be given, but this is usually avoided so as not to be construed as favoritism or bias for or against similar products not named.

The directions on a herbicide product label express the application rate as a given quantity of the product such as fluid ounces (fl oz), pints (pt) quarts (qt), or gallons (gal) per acre.

In general, the first steps in calculating the amount of herbicide to apply are as follows: (1) select the proper herbicide for best results under your particular conditions (crop, weed species, soil type, climatic conditions); (2) know the recommended rate for obtaining the best results from the selected herbicide applied under your conditions; (3) know the amount of herbicide (active ingredient) in the commercial product (formulation) that you have chosen to use; and (4) know the total area to be treated, expressed as acres. *Note:* This procedure is also applicable to regions where *metric* units are used. Usually, the labels on herbicide products offered in such regions express the active ingredient in metric terms, and the rate will be in kilograms per hectare (kg ai/ha).

Calculation Formulas

The following formulas may be used for solving most herbicide rate calculations:

For liquid formulations (products):

$$\frac{\text{pounds of active ingredient per acre}}{\text{pounds of active ingredient per gallon of formulation}} = \frac{\text{gallons of formulation required per acre}}{}$$

$$\frac{\text{grams of active ingredient per hectare}}{\text{grams of active ingredient per liter of formulation}} = \frac{\text{liters of formulation required per hectare}}{}$$

For dry formulations:

$$\frac{\text{pounds of active ingredient per acre}}{\text{percent active ingredient in formulation, expressed as its respective decimal}} = \frac{\text{pounds of formulation required per acre}}{}$$

$$\frac{\text{grams of active ingredient per hectare}}{\substack{\text{percent active ingredient in formulation,} \\ \text{expressed as its respective decimal}}} = \substack{\text{grams of} \\ \text{formulation} \\ \text{required} \\ \text{per hectare}}$$

Sample Problems for Areas 1 Acre or Larger

The following problems illustrate the procedure for calculating how much formulation (product) to use when applying a herbicide to an area 1 A or larger.

I. Problem: A grower plans to buy and apply Treflan preplant-incorporated at the recommended rate of 0.75 lb ai/A to 180 A of land to be planted to cotton. Treflan contains trifluralin as the active ingredient in the amount of 4 lb ai/gal. The sprayer has a tank capacity of 300 gal, and it has been calibrated to deliver 25 gal of spray mixture per acre.

A. How much Treflan must the grower purchase to treat all 180 A?

180 A × 0.75 lb ai/A = 135 lb ai

135 lb ai ÷ 4 lb ai/gal = 33.75 gal Treflan to be bought/180 A

B. How many acres can be treated with one tankful of spray mixture?

capacity ÷ output = acreage

300 gal ÷ 25 gal/A = 12 A/tankful

C. How much Treflan must be added to each full tank in order to apply the recommended lb ai/A of trifluralin?

12 A/tankful × 0.75 lb ai/A = 9 lb ai/tankful

9 lb ai/tankful ÷ 4 lb ai/gal Treflan = 2.25 gal Treflan/tankful

D. How many tankfuls are needed to treat all 180 A?

180 A ÷ 12 A/tankful = 15 tankfuls

II. Problem: A corn grower plans to apply AAtrex Nine-O to 250 A of cropland at the recommended rate of 2.4 lb ai/A. AAtrex Nine-O is a water-dispersible granular formulation containing atrazine as the active ingredient in the amount of 90% ai w/w.[1] The sprayer has a tank capacity of 250 gal, and it has been calibrated to deliver 15 gal spray mixture/A.

A. How much AAtrex Nine-O must the grower purchase to treat all 250 A?

250 A × 2.4 lb ai/A = 600 lb ai

600 lb ÷ 0.90 (% ai w/w) = 667 lb AAtrex Nine-O/250 A

B. How many acres can be treated with one tankful of spray mixture?

capacity ÷ output = acreage

200 gal capacity ÷ 15 gal/A output = 13.3 A/tankful

C. How much AAtrex Nine-O must be added to each full spray tank in order to apply the desired rate?

13.3 A/tankful × 2.4 lb ai/A = 31.9 lb ai/tankful

31.9 lb ai/tankful ÷ 0.90 (% ai w/w) = 35.5 lb AAtrex Nine-O/tankful

D. How many tankfuls of spray mixture are needed to treat all 250 A?

250 A ÷ 13.3 A/tankful = 18.8 tankfuls

III. Problem: A grower applied Classic early postemergence to 400 A of soybeans at the recommended rate of 0.5 oz ai/A. Classic is a water-dispersible granular formulation containing chlorimuron-ethyl as the active ingredient in the amount of 75% w/w.

A. How many pounds of chlorimuron-ethyl were applied to all 400 A?

400 A × 0.5 oz ai/A = 200 oz ai

200 oz ai ÷ 16 oz/lb = 12.5 lb chlorimuron-ethyl/400 A

B. How many pounds of Classic were applied to all 400 A?

12.5 lb ai/400 A ÷ 0.75 (% ai w/w) = 16.7 lb Classic/400 A

C. If the grower, by mistake, applied Classic at a rate of 0.5 lb ai/A, instead of the recommended rate of 0.50 oz ai/A, how many ounces of Classic were applied per acre?

0.5 lb ai/A ÷ 0.75 (% ai w/w) = 0.67 lb Classic/A

16 oz/lb × 0.67 lb Classic/A = 10.7 oz Classic/A

---

1. When dividing or multiplying with percent, be certain to convert the percent to its respective decimal—for example, 2.5% = 0.025, 80% = 0.80. Remember, *dividing* by a decimal number less than one yields a greater quantity; *multiplying* by a decimal number less than one yields a smaller quantity.

IV. Problem: A grower plans to apply Eptam 10-G to 70 A of alfalfa at the recommended rate of 3 lb ai/A. Eptam 10-G is a granular formulation that contains EPTC as the active ingredient in the amount of 10% ai w/w.

A. How much Eptam 10-G must be purchased to treat all 70 A?

$$70 \text{ A} \times 3 \text{ lb ai/A} = 210 \text{ lb ai}$$

$$210 \text{ lb ai} \div 0.10 \text{ (\% ai w/w)} = \frac{2100 \text{ lb}}{\text{Eptam/70 A}}$$

B. How many pounds of Eptam 10-G must be applied per acre to apply the recommended rate of EPTC?

$$3 \text{ lb ai/A} \div 0.10 \text{ (\% ai w/w)} = 30 \text{ lb Eptam 10-G/A}$$

Band Application Formulas

The following formulas are useful when calculating the amount of herbicide to apply in a band treatment:

1. $\dfrac{\text{band width (in.)}}{\text{row width (in.)}} \times 100 = \begin{array}{l}\text{percent of broadcast area} \\ \text{actually treated}\end{array}$

   $\dfrac{\text{band width (cm)}}{\text{row width (cm)}} \times 100 = \begin{array}{l}\text{percent of broadcast area} \\ \text{actually treated}\end{array}$

2. $\dfrac{\text{band width (in.)}}{\text{row width (in.)}} \times \text{lb/A broadcast} = \begin{array}{l}\text{lb/A of herbicide} \\ \text{applied} \\ \text{as band treatment}\end{array}$

   $\dfrac{\text{band width (cm)}}{\text{row width (cm)}} \times \text{kg/ha broadcast} = \begin{array}{l}\text{kg/ha of herbi-} \\ \text{cide applied as} \\ \text{band treatment}\end{array}$

Sample Problems for Areas Less Than 1 Acre

When calculating how much active ingredient or formulation to apply to an area less than 1 A, convert the area to (1) percentage of an acre, or (2) number of areas (plots) in 1 A.

If the area to be treated is nearly an acre in size, it is usually more convenient to convert the area to a percentage of the acre. For example, if an area to be treated is 41,160 ft², divide this value by 43,560 ft² (the area of 1 A) and multiply by 100 to get the percentage—94.5% in this case.

If the area to be treated is relatively small compared to an acre, it simplifies calculations to convert the area to number of plots per acre. For example, if an area to be treated is 400 ft², divide this area into 43,560 ft² to obtain number of plots per acre—108.9 plots/A in this case.

When working with areas smaller than 1 A, it is often more convenient and more precise to use metric units, rather than English units of measure, when calculating the amount of active ingredient or formulation to be applied.

I. Problem: A home owner plans to treat a lawn area with Banvel at the recommended rate of 0.33 lb ai/A. The area to be treated measures 70 ft by 300 ft (21,000 ft²). Banvel is a liquid formulation whose active ingredient is dicamba in the amount of 4 lb ai/gal.

A. How much dicamba must the home owner apply to the lawn in order to apply at the desired rate?

$$43,560 \text{ ft}^2 \div 21,000 \text{ ft}^2 = 2.07 \text{ plots/A}$$

$$0.33 \text{ lb ai/A} \div 2.07 \text{ plots/A} = 0.16 \text{ lb ai/plot}$$

or

$$21,000 \text{ ft}^2 \div 43,560 \text{ ft}^2/\text{A} = 0.482 \text{ or } 48.2\% \text{ of 1A}$$

$$0.33 \text{ lb ai/A} \times 0.482 = 0.16 \text{ lb ai/plot}$$

B. How much Banvel must be applied to treat the lawn with the desired amount of dicamba?

$$4 \text{ lb ai/gal} \div 128 \text{ fl oz/gal} = \frac{0.03 \text{ lb}}{\text{ai/fl oz Banvel}}$$

$$0.16 \text{ lb ai/A} \div \frac{0.03 \text{ lb}}{\text{ai/fl oz Banvel}} = \frac{5.3 \text{ fl oz}}{\text{Banvel/area}}$$

II. Problem: A home owner plans to treat a turfgrass lawn with Dacthal at the recommended rate of 10.5 lb ai/A. Dacthal is a wettable-powder formulation that contains DCPA as the active ingredient in the amount of 75% ai w/w. The lawn measures 30 ft by 80 ft, with an area of 2400 ft².

A. How much DCPA must be applied to the lawn area in order to apply the desired dosage?

$$43,560 \text{ ft}^2/\text{A} \div 2400 \text{ ft}^2 = 18.2 \text{ plots/A}$$

$$10.5 \text{ lb DCPA/A} \div 18.2 \text{ plots/A} = 0.58 \text{ lb DCPA/plot}$$

B. How much Dacthal must be applied to the lawn area to obtain the desired amount of DCPA?

$$0.58 \text{ lb DCPA/plot} \div \frac{0.75}{\text{(formulation \%)}} = \frac{0.77 \text{ lb}}{\text{Dacthal/plot}}$$

III. Problem: A home owner wants to apply Chipco Ronstar G to a lawn area 10 ft by 40 ft (400 ft²) at the recommended rate of 3 lb ai/A. Chipco Ronstar is a granular formulation that contains oxadiazon as the active ingredient in the amount of 2% ai w/w.

A. How much oxadiazon must be applied to the lawn area to obtain the desired amount?

$43,560 \text{ ft}^2/\text{A} \div 400 \text{ ft}^2 = 108.9 \text{ plots}/\text{A}$

$3 \text{ lb ai}/\text{A} \div 108.9 \text{ plots}/\text{A} = 0.0275 \text{ lb ai}/\text{plot}$

B. How much Chipco Ronstar G must be applied to treat the lawn area with the recommended amount of oxidiazon?

$0.0275 \text{ lb ai}/\text{plot} \div \dfrac{0.02}{(\% \text{ ai w/w})} = \dfrac{1.38 \text{ lb}}{\text{product}/\text{plot}}$

C. How much Chipco Ronstar G must be used on 1 A of lawn area in order to apply oxidiazon at 3 lb ai/A?

$3 \text{ lb ai}/\text{A} \div 0.02 \ (\% \text{ ai w/w}) = 150 \text{ lb product}/\text{A}$

IV. Problem: A cotton grower plans to treat both banks of an irrigation canal with Roundup at a rate of 3.0 pt/A. Roundup contains the herbicide glyphosate at a concentration of 3 lb ae/gal or 4 lb ai/gal. The active ingredient is the isopropylamine salt of glyphosate. The grower will apply Roundup in a band 6 ft wide on each embankment along 1000 ft of canal using a sprayer calibrated to deliver 25 gal/A of spray mixture. The grower intends to add a nonionic surfactant to the spray mixture at a concentration of 0.5% by volume (2 qt/100 gal). He is also adding dry ammonium sulfate at the equivalent rate of 12 lb/100 gal. (*Note:* 1 pint = 2 cups or 16 fl oz; 1 fl oz = 29.56 ml.)

A. What is the total area to be treated, expressed as acres?

$6 \text{ ft} \times 1000 \text{ ft} \times 2 \text{ banks} = 12,000 \text{ ft}^2$

$12,000 \text{ ft}^2 \div 43,560 \text{ ft}^2/\text{A} = 0.275 \text{ A}$

B. How much Roundup will be applied to the canal banks, expressed as fluid ounces, cups, and milliliters?

$3 \text{ pt}/\text{A} \times 16 \text{ fl oz}/\text{pt} = 48 \text{ fl oz}/\text{A}$

$48 \text{ fl oz}/\text{A} \times 0.275 \text{ A} = 13.22 \text{ fl oz}$

$13.22 \text{ fl oz} \div 8 \text{ fl oz}/\text{cup} = 1.65 \text{ cups}$

$13.22 \text{ fl oz} \times 29.56 \text{ ml}/\text{fl oz} = 390.78 \text{ ml}$

C. How many pounds of isopropylamine salt of glyphosate will be applied to the treated area?

$4 \text{ lb ai}/\text{gal} \div 128 \text{ fl oz}/\text{gal} = \dfrac{0.031 \text{ lb}/\text{fl oz}}{\text{of Roundup}}$

$0.031 \text{ lb}/\text{fl oz} \times 13.22 \text{ fl oz}/\text{area} = 0.41 \text{ lb ai}/\text{area}$

D. How many pounds of the acid equivalent of glyphosate will be applied to the treated area?

$3 \text{ lb ae}/\text{gal} \div 128 \text{ fl oz}/\text{gal} = 0.0234 \text{ lb ae}/\text{fl oz}$

$0.0234 \text{ lb ae}/\text{fl oz} \times 13.22 \text{ fl oz}/\text{area} = 0.31 \text{ lb ae}/\text{area}$

E. What is the percent difference in the weights of the active ingredient and the acid equivalent?

$4 \text{ lb ai}/\text{gal} - 3 \text{ lb ae}/\text{gal} = 1 \text{ lb}/\text{gal}$

$1 \text{ lb}/\text{gal} \div 4 \text{ lb ai}/\text{gal} = 0.25 = 25\%$

The active ingredient (isopropylamine salt of glyphosate) is 25% heavier than the acid equivalent.

F. How many gallons of spray mixture will be applied to the treated area?

$25 \text{ gal}/\text{A} \times 0.275 \text{ A} = 6.9 \text{ gal}$

G. How many cups of surfactant must be added to the spray mixture?

$6.9 \text{ gal} \times 0.005 \ (\text{or } 0.5\%) = 0.034 \text{ gal}$

$128 \text{ fl oz}/\text{gal} \times 0.034 \text{ gal} = 4.4 \text{ fl oz}$

$4.4 \text{ fl oz} \div 8 \text{ fl oz}/\text{cup} = 0.55 \text{ cup of surfactant}$

H. How many milliliters of surfactant must be added to the spray mixture?

$4.4 \text{ fl oz} \times 29.56 \text{ ml}/\text{fl oz} = 130 \text{ mL of surfactant}$

I. How many ounces of ammonium sulfate must be added to the spray mixture?

$6.9 \text{ gal} \times 1216/100 \text{ gal} = 0.83 \text{ lb}$

$0.83 \text{ lb} \times 16 \text{ oz}/\text{lb} = \dfrac{13.25 \text{ oz}/6.9 \text{ gal of}}{\text{ammonium sulfate}}$

Sample Problems Using Metric Units

I. Problem: A grower in Canada plans to apply 840 g ai/ha of the herbicide trifluralin to 70 ha of land to be planted to soybeans. The grower intends to use the product Treflan, which contains 545 g ai/L. The sprayer that will be used has a spray-tank capacity of 250 gal and is calibrated to deliver 60 gal/ha. (*Note:* 1 gal = 1 imperial gallon; 1 imperial gal = 4.546 L.)

A. How many gallons of Treflan must be used per hectare to obtain the desired application rate?

$840 \text{ g ai}/\text{ha} \div 545 \text{ ai g}/\text{L} = 1.54 \text{ L}/\text{ha of Treflan}$

$1.54 \text{ L}/\text{ha} \div 4.546 \text{ L}/\text{gal} = 0.34 \text{ gal}/\text{ha of Treflan}$

B. How much Treflan must the grower purchase to treat all 70 ha?

1.54 L/ha × 70 ha = 108 L of Treflan

or 108 L ÷ 4.55 L/gal = 23.7 gal

or 70 ha × 840 g ai/ha = 58,800 g ai

58,800 g ai/70 ha ÷ 545 g ai/L = 108 L

108 L ÷ 4.55 L/gal = 23.7 gal

C. How many hectares can be treated with one tankful of spray mixture?

$$250 \text{ gal/tankful} \div 60 \text{ gal/ha} = \frac{4.2 \text{ ha treated with}}{1 \text{ tankful}}$$

D. How much Treflan must be added to each tankful to obtain the desired application rate?

4.2 ha × 1.54 L/ha = 6.5 L of Treflan

or 6.5 L ÷ 4.55 L/gal = 1.4 gal

E. How many times must the spray tank be filled to treat all 70 ha?

70 ha ÷ 4.2 ha/tankful = 16.67 times

or 17 times (with 82.5 gal left over)

II. Problem: A grower in Canada plans to apply the product Avadex BW to 170 ha of winter wheat at a rate of 3.5 L/ha just prior to seeding. Avadex BW contains 400 g/L of the herbicide triallate. The herbicide will be applied in 60 gal spray mixture/ha. The grower's tractor-mounted sprayer has a spray-tank capacity of 500 gal. (*Note:* 1 gal = 1 imperial gal; 1 imperial gal = 4.546 L.)

A. How much Avadex BW must the grower purchase to treat all 170 ha at the desired rate?

170 ha × 3.5 L/ha = 595 L of Avadex BW

or 595 L ÷ 4.55 L/gal = 130.8 gal

B. How many grams of triallate will be applied per hectare?

3.5 L/ha × 400 g ai/L = 1400 g ai/ha

C. How many kilograms of triallate will be applied to all 170 ha?

595 L × 400 g ai/L = 238,000 g

238,000 g ÷ 1000 g/kg = 238 kg of triallate

or

1400 g ai/ha × 170 ha = 23,800 g or 238 kg ai

D. How many hectares of crop land can be treated with one tankful of spray mixture?

500 gal/tankful ÷ 60 gal/ha = 8.3 ha/tankful

E. How much Avadex BW must be added to each tankful to obtain the desired application rate of herbicide?

$$3.5 \text{ L/ha} \times 8.3 \text{ ha/tankful} = \frac{29.05 \text{ L of Avadex}}{\text{BW/tankful}}$$

F. How many times must the spray tank be filled to treat all 170 ha?

170 ha ÷ 8.3 ha/tankful = 20.5 times

III. Problem: A rancher in Canada plans to apply 1080 g/ha of the herbicide picloram to 45 ha of rangeland infested with Canada thistle. The product Tordon 22K, which contains 240 g ai/L, is going to be used. The herbicide will be aerially applied in 9 gal spray mixture/ha. The airplane that will be used to apply the herbicide has a spray-tank capacity of 50 gal. (*Note:* 1 gal = 1 imperial gallon; 1 imperial gal = 4.546 L.)

A. How much Tordon 22K must be used per hectare to obtain the desired application rate?

1080 g/ha ÷ 240 g ai/L = 4.5 L/ha of Tordon 22K

B. How much Tordon 22K must the rancher purchase to treat all 45 ha?

4.5 L/ha × 45 ha = 202.5 L of Tordon 22K

or 202.5 L/45 ha ÷ 4.55 L/gal = 44.5 gal

C. How many hectares of rangeland can be treated with one tankful of the spray mixture?

$$50 \text{ gal/tankful} \div 9 \text{ gal/ha} = \frac{5.6 \text{ ha treated with}}{1 \text{ tankful}}$$

D. How much Tordon 22K must be added to each tankful to obtain the desired application rate?

4.5 L/ha × 5.6 ha/tankful = 25.2 L/tankful

or 25.2 L/tankful ÷ 4.55 L/gal = 5.5 gal/tankful

E. How many times must the spray tank be filled to treat all 45 ha?

45 ha ÷ 9 ha/tankful = 5 times

IV. Problem: A grower in Canada plans to apply 841 g/ha of the herbicide ethalfluralin to 40 ha of dry beans. The grower is going to use the product Edge, a 5% ai w/w granular formulation.

A. How much Edge must be used per hectare to obtain the desired application rate?

841 g ai/ha ÷ 0.05 = 16,820 g/ha of Edge

or 16,820 g/ha ÷ 1000 g/kg = 16.82 kg/ha

B. How much Edge will the grower need to purchase?

40 ha × 16.82 kg/ha = 672.8 kg of Edge

C. Which one of the following carriers is the correct type for Edge?

(a) water

(b) oil

(c) liquid fertilizer

(d) the formulation itself

V. Problem: A landscape gardener in Canada plans to renovate a residential lawn that measures 30 m × 50 m. The gardener will use the product Roundup for weed control prior to reseeding with the desired turfgrass. Roundup contains the herbicide glyphosate at a concentration of 356 g ae/L, or 480 g ai/L. The active ingredient is the isopropylamine salt of glyphosate. Roundup will be applied at a rate of 638.4 g ae/ha, or 840 g ai/ha with a hand-held sprayer with a spray-mixture capacity of 3 gal, calibrated to deliver 7 gal/1000 m^2. (*Note:* 1 gal = 1 imperial gallon; 1 imperial gal = 4.546 L.)

A. What is the area to be treated?

30 m × 50 m = 1500 m^2

B. How many grams of glyphosate must be used to treat the area at the desired application rate, expressed as acid equivalent?

638.4 g ae/ha ÷ 10,000 m^2/ha = 0.06384 g ae/m^2

0.06384 g ae/m^2 × 1500 m^2/area = 95.8 g ae/area

C. How many grams of glyphosate must be used to treat the area at the desired application rate, expressed as active ingredient?

840 g ai/ha ÷ 10,000 m^2/ha = 0.0840 g ai/m^2

0.0840 g ai/m^2 × 1500 m^2/area = 126 g ai/area

D. What is the percent difference between the weights of the active ingredient of glyphosate and its acid equivalent?

480 g ai/L − 356 g ae/L = 124 g

124 g ÷ 480 g ai/L = 0.26 or 26%

The active ingredient (isopropyl salt of glysophate) is 26% heavier than the acid equivalent.

E. How many liters of Roundup must be applied to the area to obtain the desired rate, based on acid equivalent?

356 g ae/L ÷ 95.8 g ae/area = 3.7 L

F. How many liters of Roundup must be applied to the area to apply the desired rate, based on active ingredient?

480 g ai/L ÷ 126 g ai/area = 3.8 L

(*Note:* The difference between parts E and F is apparently due to the manufacturer's rounding of numbers.)

G. How many gallons of spray mixture will be used on the treated area?

7 gal/1000 m^2 × 1.5 = 10.5 gal/1500 m^2

H. How many times will the spray tank be filled?

10.5 gal ÷ 3 gal/tankful = 3.5 times

I. How many milliliters of Roundup must be added to each tankful of spray mixture to obtain the desired application rate of Roundup, based on active ingredient?

126 g ai ÷ 3.5 tankfuls = 36 g ai/tankful

36 g ai/tankful ÷ 480 g ai/L = 0.075 L/tankful

0.075 L/tankful × 1000 ml/L = 75 ml/tankful

■ ACID EQUIVALENTS OF SALTS AND ESTERS

When making herbicide rate recommendations for herbicides that are available as either salts or esters or both, it is a common practice to make the recommendations on the basis of *pounds of the acid equivalent of the active ingredient per acre* (lb ae/A). *The acid equivalent of a salt or ester form of a herbicide is that portion of the molecule that represents the parent acid form of the molecule.*

Herbicide rate recommendations involving salt or ester forms of a herbicide are based on acid equivalents to achieve standardization, since the respective salt and ester forms are variable as to weight. Thus recommended dosages pertaining to herbicides applied as salts or esters are usually based only on the herbicidal (parent acid) portion of the salt or ester molecule involved, and they exclude the weight of that portion of the molecule that is herbicidally inactive—that is, the salt or ester portion.

When calculating the amount of a salt or ester form of a herbicide to apply to a given area on the basis of acid equivalent, it is necessary to adjust the rate to take into account the heavier weight of the respective salt or ester, as compared to its acid form. Product labels of most commercial formulations containing salts or esters of the respective herbicide generally specify the amount of acid equivalent present in the formulation (usually in terms of pounds of acid equivalent per gallon or quart). When the acid equivalent content of the formulation is not known, however, it is necessary to make this determination before following a rate recommendation based on acid equivalent.

The acid equivalent of a herbicide's acid form is always 100%. The acid equivalent of a herbicide's salt or ester form is equal to the difference in weight, expressed as percent, between the parent acid molecule (minus a value of 1, representing the loss of the H^+) and that of its salt or ester molecule: it is always less than 100%. The acid equivalent of a salt or ester form of a herbicide molecule is found by dividing the molecular weight (mol wt) of the acid form (minus a value of 1) by the molecular weight of its salt or ester and multiplying by 100.

The *formula* for determining the acid equivalent of a salt or ester form of a herbicide is

$$\text{ae (\%)} = \frac{\text{mol wt of acid form} - 1}{\text{mol wt of salt or ester form}} \times 100$$

EXAMPLE:

The molecular weight of 2,4-dichlorophenoxyacetic acid (2,4-D) is 221, and the molecular weight of the triethanolamine salt of 2,4-D is 370. Determine the acid equivalent of this salt of 2,4-D as follows:

$$\text{ae (\%)} = \frac{221 - 1}{370} \times 100 = 59.5\% \text{ of the salt}$$

Figure 11.1 compares the acid and salt portions of the triethanolamine salt of 2,4-D.

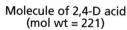

Molecule of 2,4-D acid (mol wt = 221)　　Molecule of triethanolamine salt of 2,4-D acid (mol wt = 370)

Parent- acid portion of "salt" molecule (59.5% of molecule)　　"Salt" portion of molecule (40.5% of molecule)

FIGURE 11.1 ■ Triethanolamine salt of 2,4-D. Comparable size, on a volume per molecular weight basis, of a molecule of 2,4-D acid with that of its triethanolamine salt and the respective portions of the triethanolamine salt of 2,4-D that are equivalent to its parent acid and salt parts.

■ PARTS PER MILLION

Herbicides are usually not applied on the basis of parts per million, but their concentration in soil or plant material is often expressed as parts per million (ppm). A concentration of 1 ppm is equal to 1 fl oz of a liquid uniformly dispersed in 7812.5 gal of an aqueous mixture, a rather awkward concept when dealing with most herbicide concentrations. A 1-ppm concentration is also equal to 1 mg of a chemical uniformly dispersed in 1 L (1000 ml) of an aqueous mixture or in 1000 g of dry matter such as soil or plant material.

A more manageable concept of parts per million is that it *is equal to the number of micrograms of a chemical uniformly dispersed in 1 g of a liquid or dry matter diluent*, or simply:

$$\text{ppm} = \mu\text{g of chemical/g of mixture}$$

■ SELECTED REFERENCE

Bode, L. E. 1987. Spray application technology. In *Methods of Applying Herbicides*. C. G. McWhorter and M. G. Gebhardt, eds. Monograph No. 4, pp. 85–109. Champaign, Ill.: Weed Science Society of America.

Continued

12 ACID AMIDE HERBICIDES

■ INTRODUCTION

The acid amide herbicides are derivatives of acid amides. As with other herbicide families, the chemical nomenclature is nonuniform, and members of this family are variously identified as amides, acetamides, or anilides. An acid amide is a derivative of an acid in which the hydroxyl (OH) portion of the carboxyl group (COOH) is replaced by an amino group (NH_2). The general structure of an acid amide is:

acid amide

R = an alkyl group

The acid amides are named for their corresponding acid. When R (as shown above) is a methyl, the corresponding acid is acetic acid, and the resulting acid amide is called *acetamide*. The chemical structures of acetic acid and acetamide are:

acetic acid   acetamide

When one hydrogen of the amino group of an acid amide is replaced by a phenyl group, the resulting compound is named variously as *phenylamide, acylaniline,* or *anilide.* When R, in the general structure for anilide, is methyl, the resulting compound is called *acetamide.* The general structure for anilide and the structure for acetanilide are:

general stucture for an anilide   acetanilide

TABLE 12.1 ■ Common and chemical names of the acid amide herbicides.

COMMON NAME	CHEMICAL NAME
Acetachlor	2-chloro-*N*-(ethoxymethyl)-*N*-(2-ethyl-6-methylphenyl)acetamide
Alachlor	2-chloro-*N*-(2,6-diethylphenyl)-*N*-(methoxymethyl)acetamide
Dimethenamid	2-chloro-*N*-[(1-methyl-2-methoxy)ethyl]-*N*-(2,4-dimethyl-thien-3-yl)acetamide
Metolachlor	2-chloro-*N*-(2-ethyl-6-methylphenyl)-*N*-(2-methoxy-1-methylethyl)acetamide
Napropamide	*N,N*-diethyl-2-(1-naphthalenyloxy)propanamide
Pronamide	3,5-dichloro (*N*-1,1-dimethyl-2-propynyl)benzamide
Propachlor	2-chloro-*N*-(1-methylethyl)-*N*-phenylacetamide
Propanil	*N*-(3,4-dichlorophenyl)propanamide

The common and chemical names of the acid amide herbicides are given in Table 12.1. Their chemical structures are shown in Figure 12.1.

■ CHARACTERISTICS OF ACID AMIDE HERBICIDES

Principal Use

The principal use of the acid amide herbicides is the selective preemergence control of annual grass and broadleaf weeds and control or suppression of yellow nutsedge. However, propanil is used only postemergence to the weeds; it has no preemergence activity.

Mode of Action

The primary mode of action of the acid amide herbicides, except for pronamide, is not known. They inhibit several metabolic processes, including lipid biosynthesis and the syntheses of proteins, gibberellins, and products of the phenylpropanoid pathway (lignin, anthocyanin).

The mode of action of pronamide is inhibition of mitosis, but its apparent activity is that of a foliar contact–type herbicide. Pronamide arrests cell division at the prometaphase stage, binding to tubulin and preventing formation of microtubules.

Translocation

In general, the acid amide herbicides are readily root-absorbed and translocated upward and distributed throughout the plant.

Selectivity

The primary mechanisms contributing to plant selectivity of the acid amide herbicides are not known. They are metabolized relatively rapidly in plants, especially in resistant plants. For example, propachlor is metabolized within 5 days and alachlor within 10 days. Pronamide is metabolized slowly in both susceptible and tolerant plants.

Except for propanil, the activity of the acid amide herbicides is through absorption by the roots and shoots of weed seedlings as these parts push through the treated soil. In the case of grass seedlings, the coleoptile is the primary site of entry; with broadleaf seedlings, the roots are the primary site of entry. In the case of propanil, the foliage is the site of entry.

Common Characteristics

1. The acid amide herbicides do not control established weeds.
2. Soil-applied, the acid amide herbicides must be moved downward into the weed seed germination zone by tillage or leaching (rainfall or irrigation).
3. The acid amide herbicides are adsorbed to soil colloids, especially the humic portion. They resist leaching once they are incorporated into the upper 1- to 2-in. soil layer.
4. In general, the soil persistence of the acid amide herbicides is relatively short, 1–3 months. The average persistence of propanil, which has no herbicidal activity in soils, in warm, moist soil is just 1–3 days. However, pronamide persists 6–9 months, depending on soil type and climatic conditions. Microbial action is the principal means by which these herbicides are lost from soils.

■ INFORMATION OF PRACTICAL IMPORTANCE

Acetochlor

1. Acetochlor is the active ingredient in an emulsifiable-concentrate formulation (Zeneca's Surpass EC) con-

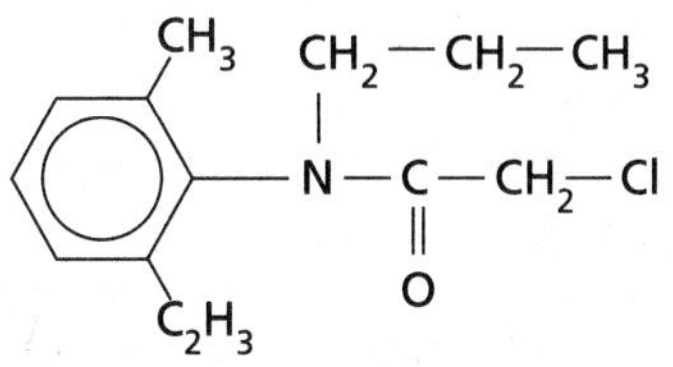

acetachlor (Surpass)

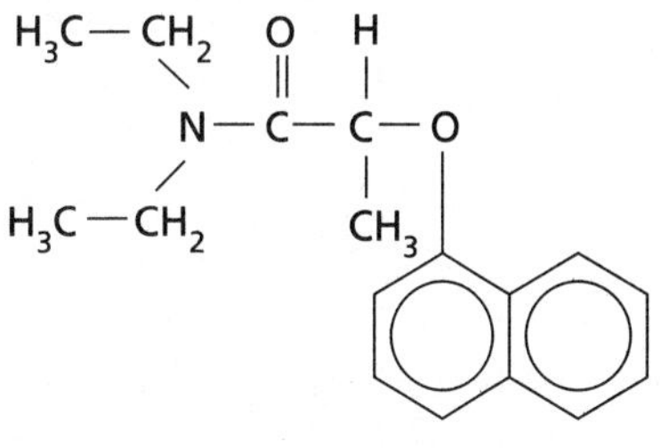

napropamide (Devrinol)

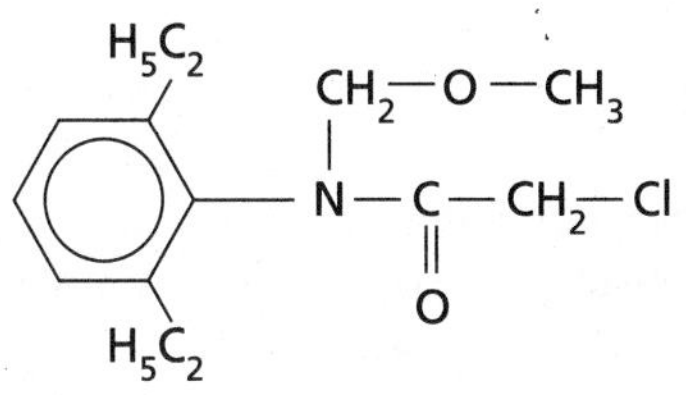

alachlor (Lasso)

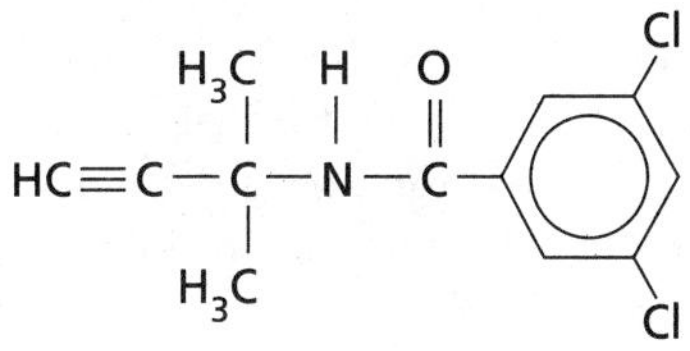

pronamide (Kerb)

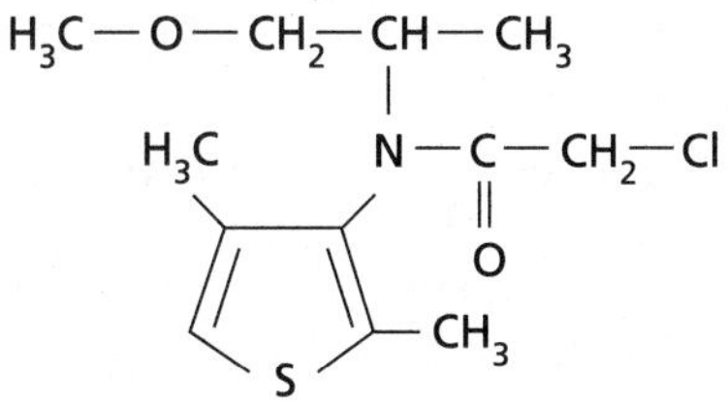

dimethenamid (Frontier)

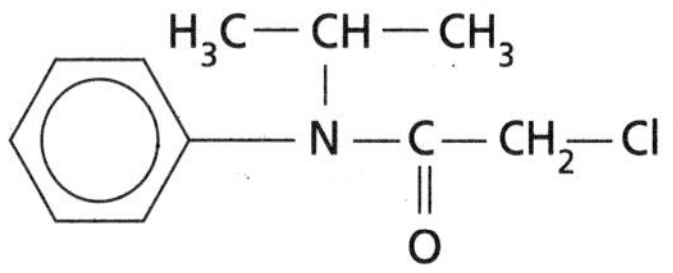

propachlor (Ramrod)

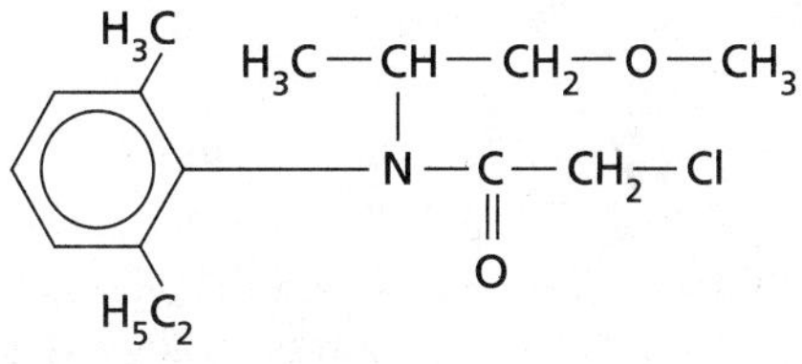

metolachlor (Dual)

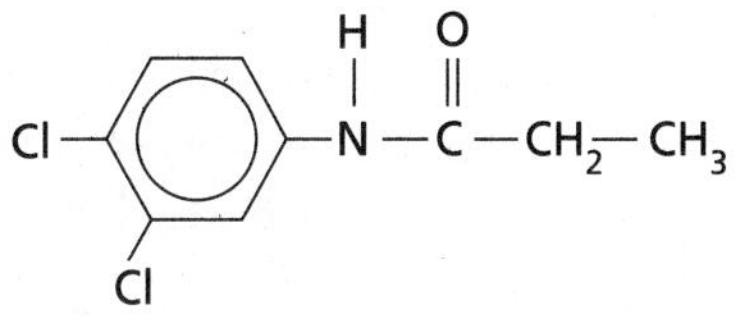

propanil (Stam)

FIGURE 12.1 ■ Chemical structures of the acid amide herbicides.

taining 6.4 lb ai/gal and the corn safener dichlormid (25788). Surpass EC is a Restricted Use Pesticide, which is toxic to fish.

2. In 1993, Surpass had EPA approval both for an Experimental Use Permit (EUP) and Temporary Tolerances for its use on corn.

3. Surpass is used preplant-incorporated or preemergence in corn. It may be used in conventional tillage corn or in reduced- or no-till corn. A burndown herbicide (such as Gramoxone Extra) and/or a residual herbicide (such as atrazine, Bladex, or Extrazine) as a tank mix with Surpass has proved to be beneficial when used in reduced- or no-till corn.

4. Water, liquid fertilizer, or dry bulk fertilizer may be used as the carrier for Surpass. It can be used alone or with broadleaf herbicides.

5. Surpass controls many annual grass and broadleaf weed species and yellow nutsedge and suppresses growth of some others.

6. Surpass is applied at 0.8–2.4 lb ai/A (1–3 pt product/A) in conventional tillage corn, and at 1.6–2.4 lb ai/A (2–3 pt product/A) in reduced- or no-till corn, depending on soil type and organic matter content.

7. *Precautionary Statement:* Surpass causes skin and eye irritation. It is harmful if swallowed, inhaled, or absorbed through the skin. Wear protective clothing and goggles or full-face shield when handling or mixing the formulation. Wash clothing before reuse.

   Do not apply spray mixture directly to water, areas where surface water is present, or to intertidal areas below the mean high tide mark. Do not contaminate water when disposing of equipment rinse water.

8. Surpass 100 is a premixed emulsifiable-concentrate formulation that contains acetachlor and atrazine as the active ingredients. It is used in corn grown under conventional, reduced-, or no-till systems.

Alachlor

9. Alachlor is a Restricted Use Pesticide due to its oncogenicity (causing tumors in laboratory animals).

10. Alachlor is the active ingredient in three kinds of formulations: (a) emulsifiable concentrate (Lasso and Micro-Tech) containing 4 lb ai/gal, (b) dry-flowable (Partner) containing 65% ai w/w, and (c) granular (Lasso II) containing 15% ai w/w.

11. Alachlor is used in a variety of crops such as corn (all types), beans (dry, lima, mung), grain sorghum, peanuts, soybeans, and woody ornamentals.

12. Alachlor formulations are applied in croplands preplant-incorporated, or preemergence before crop and weed emergence. In addition, Lasso can be applied early postemergence in corn before the weeds reach the 2-leaf stage and the crop is no more than 5 in. (12.7 cm) tall. Weeds past the two-leaf stage will not be controlled.

13. These formulations are applied in croplands at 2–3.25 lb ai/A, depending on soil type and organic matter content.

14. Alachlor may be tank-mixed with other appropriate herbicides to control a broader weed spectrum and also to control emerged weeds.

15. *Precautionary Statement:* Alachlor has been identified in groundwater sampling, and there is the possibility that it can leach through the soil to groundwater, especially where soils are coarse and groundwater is near the surface. Following application and during runoff following rainfall, alachlor may reach surface waters such as streams, rivers, and reservoirs.

Dimethenamid

16. Dimethenamid is the active ingredient in an emulsifiable-concentrate formulation (Frontier) containing 7.5 lb ai/gal.

17. Frontier is a selective preemergence (to the weeds) herbicide for control of most annual grasses, certain annual broadleaf weeds, and sedges in field corn, seed corn, and popcorn. Frontier does not control emerged weeds.

18. Frontier is used as a preplant surface (split application), preplant-incorporated, preemergence, or early postemergence (up to 8 in. or 20 cm tall) treatment to corn. Do not use Frontier on sweet corn.

19. For best results, Frontier should be soil-incorporated by rainfall, sprinkler irrigation, or mechanical tillage prior to weed seedling emergence from the soil. It may be applied in water or fluid fertilizer as the carrier, or it may be impregnated on, and applied with, dry bulk fertilizer. Do not use fluid fertilizer as the carrier after crop emergence.

20. Frontier controls annual grasses such as barnyardgrass, broadleaf signalgrass, crabgrasses (smooth and large), foxtails (giant, green, yellow), goosegrass, fall panicum, southwestern cupgrass, witchgrass and, at the highest rate, seedling johnsongrass and Texas panicum. Examples of broadleaved weeds controlled by Frontier include carpetweed, common purslane, Florida pusley, Palmer amaranth, pigweeds (prostrate, redroot, smooth, tumble), spurge (nodding, spotted) and, at the highest rate, nightshades (black, eastern black, hairy). Sedges controlled include rice flatsedge and yellow nutsedge (at highest rate).

21. The use rate of Frontier is 0.75–1.5 lb ai/A (13–25 fl oz product/A), depending on soil texture and organic matter content. The lower rate is for coarse-textured soils with less than 3% organic matter.

22. Frontier can be applied prior to, in tank mix with, or following many herbicides registered for use in corn.

Metolachlor

23. Metolachlor is the active ingredient in two types of formulations: (a) an emulsifiable concentrate (Dual 8E) containing 8 lb ai/gal, and (b) a granular formulation (Dual 25G) containing 25% ai w/w.

24. Dual 8E is used in beans (edible), corn, cotton, nonbearing citrus, nonbearing grapes, peanuts, potatoes, safflower, sorghum (grain and forage), and soybeans.

25. Dual 25G is used in corn, peanuts, potatoes, and soybeans.

26. These formulations are applied preplant surface, preplant-incorporated, or preemergence to the crop.

27. Metolachlor formulations are applied at rates of 1.5–3 lb ai/A, depending on crop and soil type involved.

28. When used for weed control in sorghum, the sorghum crop seed must be pretreated with Concep, a crop safener (herbicide antidote) specific for metolachlor, or severe injury will result.

29. Metolachlor may be tank-mixed with other appropriate herbicides to control a broader weed spectrum and to control emerged weeds.

30. Metolachlor is one of two active ingredients in Bicep, a premix of metolachlor and atrazine. Bicep is a Restricted Use Pesticide because of the presence of atrazine.

31. Metolachlor is one of two active ingredients in Cycle, a premix of metolachlor and cyanazine. Cycle is a Restricted Use Pesticide due to the presence of cyanazine. Bicep and Cycle are used in corn and sorghum. Do not use Cycle on forage sorghum.

Napropamide

32. Napropamide is the active ingredient in three kinds of formulations: (a) emusifiable concentrate (Devrinol 2E) containing 2 lb ai/gal, (b) dry-flowable (Devrinol 50-DF) containing 50% ai w/w, and (c) granular (Devrinol 10-G) containing 10% ai w/w.

33. Devrinol 2E is used in tomatoes, peppers, strawberries, and tobacco.

34. Devrinol 50-DF is used in citrus, nuts, fruits (pome, small, stone), vegetables, and tobacco.

35. Devrinol 10-G is used in citrus; nuts; fruits (pome, small, stone) and grapes.

36. Napropamide formulations are usually applied at 4 lb ai/A.

37. Napropamide is applied preplant. For best results, it should be soil-incorporated 1–2 in. (2–5 cm) deep.

38. Napropamide should be applied in the fall through early spring prior to weed emergence.

39. Devrinol 2E may be tank-mixed with Tillam 6E (pebulate) for use in tomatoes and tobacco.

Pronamide

40. Pronamide is a Restricted Use Pesticide because of its oncogenicity (causing tumors in laboratory animals).

41. Pronamide is the active ingredient in a wettable-powder formulation (Kerb 50-W) containing 50% ai w/w. The formulation is packaged in water-soluble pouches.

42. Kerb is used in a variety of crops such as blueberries, caneberries, lettuce, legumes, rhubarb, winter peas, and woody ornamentals.

43. Kerb is herbicidal when applied preemergence or postemergence to susceptible weeds, especially grass weeds.

44. Kerb is applied at broadcast rates of 0.25–3 lb ai/A (0.5–6 lb product/A), depending on the crop involved. Rainfall or irrigation following application is essential to soil-incorporate the herbicide for effective weed control.

45. Kerb is applied in the fall or early winter (mid-October to mid-December), when temperatures do not exceed 55°F (13°C). Do not apply to frozen ground.

46. Kerb selectively kills annual bluegrass in bermudagrass turf. It may be applied preemergence or postemergence to annual bluegrass. A postemergence application will kill annual bluegrass at any stage from early germination to maturity. It may be applied to actively growing or dormant bermudagrass (hybrid or improved strains), and it does not affect pegging. Kerb is applied at 0.5–1.5 lb ai/A (1.0–3.0 lb product/A), depending on stage of growth of the annual bluegrass.

Propachlor

47. Propachlor is the active ingredient in an emulsifiable-concentrate formulation (Ramrod) containing 4 lb ai/

gal, and in a granular formulation (Granular Ramrod 20) containing 20% ai w/w.

48. Both formulations are used in corn (field, hybrid seed, silage, and sweet) and grain sorghum.

49. Propachlor formulations are applied at a rate of 4–6 lb ai/A, depending on soil type and organic matter content.

50. These formulations are applied to the soil surface prior to crop and weed emergence. Rainfall or irrigation following application is necessary to move the herbicide into the upper 1- to 2-in. soil layer.

51. Propachlor is one of two active ingredients in a premix flowable formulation (Monsanto Ramrod/Atrazine) containing 3 lb propachlor and 1 lb atrazine/gal, and in a premix dry-flowable formulation (Monsanto's Ramrod + Atrazine DF) containing 48% propachlor and 15.5% atrazine w/w. Each of these products is a Restricted Use Pesticide due to the presence of atrazine.

Propanil

52. Propanil is the active ingredient in an emulsifiable concentrate (Stam M-4; Strel 4E) containing 4 lb ai/gal, and in an extruded dry-flowable formulation (Stam 80 EDF) containing 80% ai w/w, and in a dry-flowable (Wham DF80) containing 80% ai w/w.

53. Both formulations are used only in rice grown in the southern United States. All leading commercial varieties of rice are exceptionally tolerant to propanil.

54. Propanil formulations are applied only postemergence to the weeds. They have no soil activity. Annual grasses are controlled when treated in the one- to six-leaf stage. The formulations are applied at 3–4 lb ai/A.

55. Do not apply propanil to rice in mixture with carbamate insecticides, (such as carbaryl and methomyl) or with organophosphate insecticides (such as parathion, malthion, among others) as severe injury or death of the rice plants may result. Do not apply any of these insecticides to rice fields within 14 days before or after propanil application.

56. Propanil is one of two active ingredients in a premix emulsifiable-concentrate formulation (Stampede CM) containing 3 lb propanil and 1.4 lb MCPA/gal. This product is used only in barley and spring wheat.

13 ARYLOXYPHENOXY-PROPIONATE HERBICIDES

■ INTRODUCTION

The common nucleus of the chemical structures for the aryl-oxyphenoxypropionate herbicides is 4-oxyphenoxypropanoic acid, with each herbicide available as either a butyl, ethyl, or methyl ester. The herbicides differ in the various aryl-derived groups bonded to the oxygen at the 4-position of the phenoxy ring and, in some cases, in the ester group. The common and chemical names of the esters of these herbicides are given in Table 13.1, and their chemical structures are shown in Figure 13.1.

Diclofop-methyl, fenoxaprop-ethyl, fluazifop-ethyl, and quizalofop-ethyl are racemic mixtures, as denoted by (±) preceding their chemical names. A racemic mixture contains both *dextro* (+) and *levo* (–) enantiomers in equal proportions. The *levo* (–) form is nonherbicidal. Fluazifop-P-ethyl and quizalofop-P-ethyl are herbicidally active *dextro* enantiomers of the mixtures, and these enantiomers are denoted by either (+) or *R* preceding their chemical names.

■ CHARACTERISTICS OF ARYLOXYPHENOXYPROPIONATE HERBICIDES

Principal Use

The principal use of the aryloxyphenoxypropionate herbicides is the selective postemergence control of annual and perennial grasses in broadleaf crops.

Mode of Action

The primary mode of action of the aryloxyphenoxypropionate herbicides is inhibition of the biosynthesis of lipids in grasses. Their site of action is the enzyme acetyl-CoA (ACCase) in the stroma of plastids. They have little or no effect on dicots. The reader is referred to the text relative to modes and sites of action (Chapter 7) for more information.

TABLE 13.1 ■ Common and chemical names of the aryloxyphenoxypropionate herbicides.

Common name	Chemical name
Diclofop-methyl	Methyl(±)-2-[4-(2,4-dichlorophenoxy)phenoxy]propionate
Fenoxaprop-ethyl	Ethyl(±)-2-[4-[(6-chloro-2-benzoxazolyl)oxy]phenoxy]propionate
Fluazifop-butyl	Butyl(±)-2-[4-[[5-(trifluoromethyl)-2-pyridinyl]oxy]phenoxy]propionate
Fluazifop-P-butyl	Butyl(+)-R-2-[4-[[5-(trifluoromethyl)-2-pyridinyl]oxy]phenoxy]propionate
Haloxyfop-methyl	Methyl(±)-2-[4-[[3-chloro-5-(trifluoromethyl)-2-pyridinyl]oxy]phenoxy]propionate
Quizalofop-ethyl	Ethyl(±)-2-[4-[(6-chloro-2-quinoxalinyl)oxy]phenoxyl]propionate
Quizalofop-P-ethyl	Ethyl(+)-2-[4-[(6-chloro-2-quinoxalinyl)oxy]phenoxy]propionate

Translocation

The aryloxyphenoxypropionate herbicides are rapidly absorbed by plant foliage and translocated throughout the plant, accumulating in the meristems.

Selectivity

Selectivity of the aryloxyphenoxypropionate herbicides is due to the presence of a graminicide-tolerant enzyme, acetyl-CoA (ACCase), in tolerant dicot (broadleaf) plants and the absence of this enzyme in susceptible grass plants.

A few grass species are tolerant of some of these graminicides. Red fescue and Italian ryegrass are tolerant because of the presence of a graminicide-tolerant form of ACCase. Inheritance studies have determined that a single, partially dominant allele controls whether this form of ACCase will be present in grasses. However, wheat is tolerant to diclofop-methyl because the plant metabolizes the herbicide to nonherbicidal metabolites.

Common Characteristics

1. Haloxyfop-methyl (DowElanco's Verdict) is not marketed in the United States or Canada at this time. Its characteristic activity is similar to the other aryloxyphenoxypropionate herbicides. It will not be discussed further in the text.

2. Dicot (broadleaf), sedge, and rush plants are tolerant of aryloxyphenoxypropionate herbicides.

3. Because the aryloxyphenoxypropionate herbicides control only grasses, they are sometimes referred to as *graminicides.*

4. They are applied postemergence to the grass weeds. However, they have preemergence activity when applied to the soil at rates higher than used postemergence.

5. The aryloxyphenoxypropionate herbicides are applied as esters of their acids, which are readily foliar-absorbed. The esters are rapidly converted in the leaves to their respective acid form by carboxylesterase activity. The acid is considered the phytotoxic form of these herbicides.

6. Following conversion from the ester, the active acid form is readily translocated to meristematic regions.

7. Their injury symptoms are similar, with chlorosis in developing leaves and cessation of growth occurring initially, followed in a few days by necrosis of the shoot apex and meristematic regions of leaves and roots.

■ INFORMATION OF PRACTICAL IMPORTANCE

Diclofop-methyl

1. Diclofop-methyl is the active ingredient in an emulsifiable-concentrate formulation (Hoelon 3EC) containing 3 lb ae/gal. Hoelon 3EC is a Restricted Use Pesticide because it is toxic to fish.

2. In contrast to the other aryloxyphenoxypropionate herbicides, Hoelon has phytotoxic activity when soil-applied (preplant-incorporated or preemergence), as well as postemergence activity.

3. Hoelon is used preplant-incorporated or preemergence (at planting) to control annual grasses in winter wheat. Winter annual bromes (such as downy, Japanese, and ripgut) are controlled preplant; Italian ryegrass is controlled preemergence. Hoelon does not control brome grasses when applied postemergence.

4. Hoelon is recommended for postemergence control of annual grass weeds such as wild oats, Italian ryegrass,

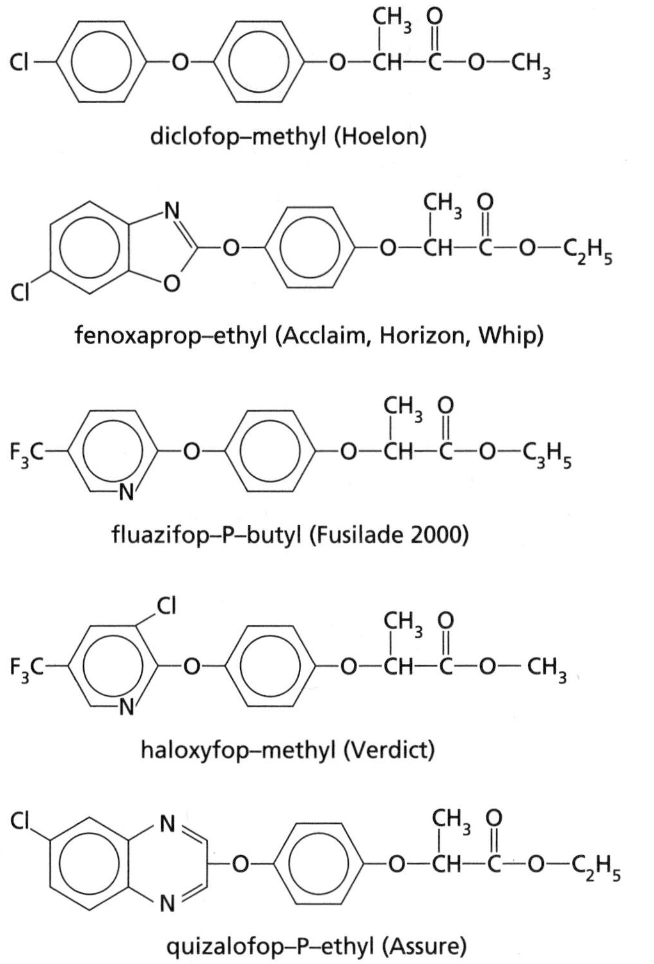

diclofop–methyl (Hoelon)

fenoxaprop–ethyl (Acclaim, Horizon, Whip)

fluazifop–P–butyl (Fusilade 2000)

haloxyfop–methyl (Verdict)

quizalofop–P–ethyl (Assure)

FIGURE 13.1 ■ Chemical structures of the aryloxyphenoxypropionate herbicides.

and foxtails, in the one- to four-leaf stage in barley, winter and spring wheats, flax, lentils, and peas.

5. Hoelon does not control established perennial grasses.

6. Hoelon is applied at 0.75–1.25 lb ae/A (1.0–1.67 qt product/A), with 1.0 lb ae/A (1.33 qt product/A) the optimal rate in most situations.

7. Hoelon can be tank-mixed with the broadleaf herbicides Buctril, MCPA, Harmony, and Glean, depending on the crop, to control a broader spectrum of weeds. The effectiveness of Hoelon is reduced by high rates of MCPA in tank mixes with Hoelon; very low rates of MCPA are sufficient to control mustards. Do not tank-mix Hoelon with any herbicide, other than those mentioned above, or reduced grass control may result. For example, tank mixes of Hoelon with Basagran,

2,4-D, or MCPA reduced grass control by as much as 60, 48, and 31%, respectively.

Fenoxaprop-ethyl

8. Fenoxaprop-ethyl is the active ingredient in emulsifiable-concentrate formulations (Whip, Acclaim, Horizon) containing 1.0 lb ae/gal.

9. Whip is recommended for the postemergence control of annual grasses and certain perennial grasses such as johnsongrass, in cotton, peanuts, rice, and soybeans. Whip does not control quackgrass.

10. Whip is applied at 0.1–0.2 lb ae/A in the aforementioned crops. It is applied before cotton or soybeans bloom, before pegging of peanuts, and when rice is in the 4-leaf to panicle-initiation stage. When used in rice, the field should be free of standing water. Certain rice varieties are susceptible to Whip.

11. Acclaim is recommended for certain established lawns and turfgrasses grown for sod. Tolerant lawn species include established annual bluegrass, fescues (fine, tall), Kentucky bluegrass, perennial ryegrass, and zoysia. Acclaim may be tank-mixed with other pesticides or fertilizer as noted on product label. Acclaim is applied at 0.08–0.35 lb ae/A on turfgrasses.

12. Horizon is recommended for annual grass control in perennial ryegrass, and fine and tall fescues in early spring (before April 1) after the turfgrass breaks winter dormancy and the annual grass weeds are in the 2-leaf to 2-tiller stage of growth. The recommended rate of Horizon is 0.15–0.25 lb ae/A.

13. Horizon is also recommended for highway rights-of-way to control annual grasses. It is applied at 0.15–0.25 lb ae/A.

Fluazifop-P-butyl

14. Fluazifop-butyl and fluazifop-P-butyl are, respectively, the active ingredients in the emulsifiable-concentrate formulation Fusilade, which contains 2 lb ae/gal, and Fusilade 2000, which contains 1 lb ae/gal. Fusilade 2000 is the more commonly used product.

15. Fusilade contains the racemic mixture of fluazifop-butyl, and its characteristics are the same as for Fusilade 2000, except that it is applied at twice the rate per acre. Fusilade will not be discussed further.

16. Fusilade 2000 contains the active (+) form, fluazifop-P-butyl, of the racemic mixture.

17. Fusilade 2000 is recommended for postemergence control of annual and established perennial weeds (bermudagrass, johnsongrass, and quackgrass) in many crops such as asparagus, cotton, grapes, lettuce, onions, peppers, rhubarb, soybeans, small fruits, tree fruits and nuts, and ornamental flowers.

18. Fusilade 2000 is applied at 0.094–0.375 lb ae/A depending on the targeted weed species and geographic area. For best results, always add a nonionic surfactant or crop oil concentrate to the spray mixture prior to use.

19. Fusilade 2000 may be tank-mixed with certain other herbicides, as noted on the product label. Tank-mixing herbicides other than those recommended may reduce the effectiveness of Fusilade 2000.

Quizalofop-P-ethyl

20. Quizalofop-ethyl and quizalofop-P-ethyl, respectively, are the active ingredients in the emulsifiable-concentrate formulation Assure, which contains 1.76 lb ae/gal and Assure II, which contains 0.88 lb ae/gal. Assure II is the more commonly used product.

21. Assure is a racemic mixture of quizalofop-ethyl, and its characteristics are similar to those of Assure II, except that Assure is applied at twice the rate as Assure II. Assure will not be discussed further.

22. Assure II contains the active (+) form, quizalofop-P-ethyl, of the racemic mixture.

23. Assure II is recommended for postemergence control of annual and perennial grasses (bermudagrass, johnsongrass, quackgrass) in soybeans and noncrop areas.

24. In soybeans, Assure II is applied at 0.03–0.05 lb ae/A for annual grass control and 0.05–0.0625 lb ae/A for perennial grass control, depending on targeted species.

25. In noncrop areas, Assure II is applied at 0.33 lb ae/A.

26. When Assure II is used in soybeans, always add a nonionic surfactant or crop oil concentrate to the spray mixture. When used in noncrop areas, add a nonphytotoxic petroleum oil concentrate (not a vegetable oil) to the spray mixture for best results.

14. BENZONITRILE HERBICIDES

■ INTRODUCTION

Benzonitrile herbicides have as their common molecular nucleus a phenyl ring with a nitrile (C≡N) group bonded to the ring at the 1-position. The herbicides vary from one another by the various substitutions at other positions on the phenyl ring. The herbicides bromoxynil and dichlobenil are benzonitrile herbicides, and their chemical names are:

Bromoxynil   3,5-dibromo-4-hydroxybenzonitrile

Dichlobenil   2,6-dichlorobenzonitrile

Bromoxynil is formulated as its octanoic acid ester. The chemical structures of bromoxynil, the octanoic ester of bromoxynil, and dichlobenil are shown in Figure 14.1.

Based on its chemical structure, bromoxynil can be considered a substituted phenol, as well as a benzonitrile. As a substituted phenol, its chemical name would be 4-nitrile-2,6-dibromophenol. Owing to the acidic nature of the hydroxy (OH) group bonded to the phenyl ring, salt and ester forms of bromoxynil can be made. The ester forms of bromoxynil are more herbicidal and less likely to be washed from foliage than are its salts. The octanoic acid ester of bromoxynil is the form used commercially.

Although bromoxynil and dichlobenil are benzonitrile herbicides, their modes of action and use as herbicides greatly differ.

■ CHARACTERISTICS OF BENZONITRILE HERBICIDES

Principal Use

Bromoxynil is used as a postemergence, contact-type herbicide for the selective control of small broadleaf weeds in seedling alfalfa, corn, flax, onions, mint, sorghum, small grains, seedling turfgrass, and grass grown for seed or sod production.

Dichlobenil is used as a soil treatment for control of annual grass and broadleaf weeds and certain perennials in established woody ornamentals, nut and fruit orchards, cranberries, blackberries, and raspberries. It can be used for nonselective weed control in industrial and noncrop areas and under asphalt pavements (applied prior to laying the asphalt) such as roadways, parking lots, and recreational areas. Dichlobenil can also be used to control rooted

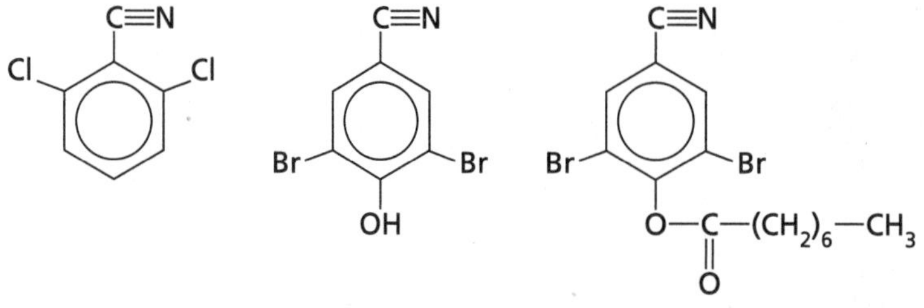

FIGURE 14.1 ■ **The chemical structures of the benzonitrile herbicides.**

aquatic weeds in nonflowing water such as ponds, reservoirs, and lakes. It kills rooted aquatic weeds by action through the soil rather than via the water, as is the case with most herbicides that control aquatic weeds.

Mode of Action

Bromoxynil inhibits photosynthesis by blocking electron transport through photosystem II (PS II).

Dichlobenil disrupts mitosis and inhibits the biosynthesis of cellulose from glucose in plants. Dichlobenil inhibits seed germination and cell division in meristems (regions of rapid cell division). As an inhibitor of cellulose formation, its mode of action is similar to that of isoxaben, but it is 40 times less effective than isoxaben in this regard.

Translocation

Bromoxynil does not translocate in plants, except for minute distances from site of entry.

Dichlobenil is slowly translocated upward from the roots via the transpiration stream. There is little or no downward movement via the photosynthate stream.

Selectivity

Bromoxynil does not control or injure grasses, seedling or established. Transitory leaf burn may occur, and recovery of the crop is generally rapid with no lasting effect. To reduce the potential for tempory leaf burn, applications should be made to dry foliage when weather conditions are not extreme.

Dichlobenil's selectivity is obtained by herbicide placement. In the treated soil zone, germination and growth of seeds and seedlings of both weeds and crops are adversely affected by dichlobenil. Marginal selectivity may be obtained by seeding large-seeded crops below the treated soil zone. Dichlobenil is absorbed by underground plant parts (roots and shoots) of seedlings and older plants. In general,

established crop plants tolerate dichlobenil if the herbicide is not in the root-absorption zone, but established plants with shallow roots may be injured or killed. Dichlobenil is not readily leached in soil. Its water solubility is 25 ppmw at 68°F (20°C).

Characteristics of Bromoxynil

1. Bromoxynil is a Restricted Use Herbicide because it causes birth defects in laboratory animals. Its retail sale and use is restricted to Certified Applicators or persons directly under their supervision and only for those uses covered by the Certified Applicator's certification.

2. Bromoxynil is applied only as a postemergence treatment to young broadleaf weeds. Bromoxynil does not control established annual or perennial broadleaved weeds.

3. For best results, apply bromoxynil before the broadleaved weeds are past the three- to four-leaf stage or when rosettes are less than 1.5 in. across.

4. Symptoms of bromoxynil activity are the greying and withering of treated foliage.

5. Bromoxynil has little or no soil activity and does not pose a soil residue problem.

Characteristics of Dichlobenil

1. Dichlobenil is applied only as a soil-incorporated treatment.

2. Under conditions of high temperatures and dry soils, dichlobenil may be rapidly lost from the soil surface by volatility. To avoid such loss, dichlobenil should be soil-incorporated by overhead irrigation or tillage immediately after application.

3. The soil persistence of dichlobenil is 2–6 months when applied at the recommended rates of 4–6 lb ai/A.

■ INFORMATION OF PRACTICAL IMPORTANCE

Bromoxynil

1. Bromoxynil is the active ingredient in an emulsifiable-concentrate formulation (Buctril), which contains 2.0 lb ai/gal.

2. Buctril is primarily a contact-type herbicide; therefore, thorough coverage of the weed seedlings and young plants is essential for optimal control.

3. As Buctril has little or no residual activity in the soil, subsequent flushes of weeds that appear after application of Buctril will not be controlled. Generally, crops that form a good foliar canopy will suppress subsequent weed emergence and growth by shading.

4. Buctril can be applied in tank mix with other broadleaf herbicides, such as 2,4-D, MCPA, and atrazine, to control residual growth and a broader spectrum of weeds. It can be applied in tank mix with liquid fertilizers and with many foliar-applied insecticides. Unless otherwise specified, it should not be applied in tank mix with foliar-applied grass herbicides as the activity of the grass herbicides may be reduced or nullified.

5. It is of interest to note that Buctril (a broadleaf herbicide) can be safely applied postemergence to seedling alfalfa (a broadleaf crop) for the selective control of seedling broadleaved weeds.

Dichlobenil

6. Dichlobenil is the active ingredient in a wettable-powder formulation (Casaron 50W) containing 50% ai w/w. It is also the active ingredient in two granular formulations: Casaron 4G (4% ai w/w), and Casaron 10G (10% ai w/w).

7. Casaron 50W is used to control grass and broadleaf weeds in woody ornamentals, shelterbelts, and under asphalt pavements. It is applied to soil that is free of emerged weeds or vegetative debris in early spring before weed seeds have germinated. Casaron 50W is applied at a rate of 4–6 lb ai/A (8–12 lb product/A).

8. Casaron 4G is used to control perennial weeds in established woody ornamentals, nut and fruit orchards and nurseries, and in noncrop areas. It is applied in late fall or early spring (before May 1) and immediately incorporated. Casaron 4G is applied at a rate of 4–8 lb ai/A (100–200 lb product/A), depending on crop and weeds involved.

9. Casaron 10G is used to control rooted aquatic weeds in nonflowing water (ponds, lakes, reservoirs), provided the user has control over the use of the water. It does not harm fish when used as directed. Casaron 10G is applied at rates varying from 4 to 15 lb ai/A (40–150 lb product/A), depending on the weeds present and whether it is applied to exposed bottoms and shorelines or through the water.

15

BIPYRIDINIUM HERBICIDES

■ INTRODUCTION

There are two herbicides in the bipyridinium herbicide family; they are paraquat and diquat. Paraquat and diquat are also referred to as "quaternary ammonium herbicides." These herbicides are active as their respective cations, but they are formulated as salts—paraquat dichloride and diquat dibromide. Their common and chemical names are given in Table 15.1. The chemical structures of the salts of paraquat and diquat are shown in Figure 15.1.

■ CHARACTERISTICS OF BIPYRIDINIUM HERBICIDES

Principal Use

The principal use of paraquat is the nonselective, early postemergence control of seedling grass and broadleaf weeds in croplands prior to crop planting or emergence and in certain dormant crops. Paraquat is also a crop dessicant.

Diquat is principally used to control aquatic weeds in ponds, lakes, and drainage ditches. It is also a crop dessicant.

Mode of Action

Paraquat and diquat are postemergence, contact-type herbicides, killing green tissue very rapidly. Their herbicidal activity is dependent on the presence of light, oxygen, and photosynthesis.

Paraquat and diquat are rapidly absorbed by green plant tissue, and they interact with the photosynthetic process

TABLE 15.1 ■ Chemical and common names of the bipyridinium herbicides.

COMMON NAME	CHEMICAL NAME
Paraquat	1,1'-dimethyl-4,4'-bipyridinium ion
Diquat	6,7-dihydrodipyrido[1,2-α:2',1'-c]-pyrazinediium ion

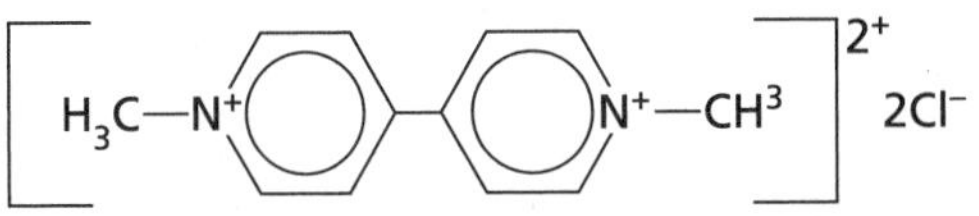

paraquat dichloride
(Gramoxone Extra; Cyclone)

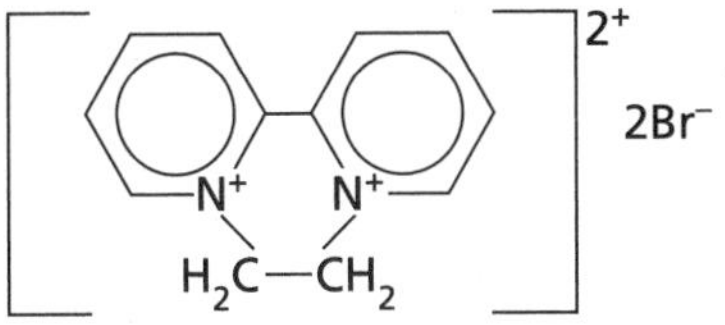

diquat dibromide
(Valent Diquat, Zeneca Diquat, Weedtrine –D)

FIGURE 15.1 ■ Chemical structures of the salts of paraquat and diquat.

to produce superoxides, which destroy the plant cells. Paraquat and diquat are capable of being reduced in plants to relatively stable, water-soluble free radicals. The energy for this reduction comes from photosynthesis and, to a lesser extent, respiration. The free radicals formed are readily reoxidized by molecular oxygen back to the original quaternary ions, generating hydrogen peroxide or intermediate radicals, which destroy plant cells. The herbicidal activity and organic chemical reactions of paraquat and diquat are dependent solely upon their respective cations.

Translocation

Paraquat and diquat are not translocated in plants, other than for local, short-distance transport in treated tissue.

Selectivity

Paraquat and diquat are nonselective herbicides.

Common Characteristics

1. Paraquat and diquat are photochemically degraded on plant surfaces and in aqueous solution upon exposure to ultraviolet light (strong sunlight).

2. Paraquat and diquat have no soil residual activity, as they are inactivated upon contact with the soil due to adsorption to organic and clay colloids.

3. Weed biotypes resistant to paraquat and diquat have occurred where these herbicides were applied repeat-

edly for 5 or more years. The reader is referred to Chapter 8, which addresses herbicide-resistant weed biotypes.

■ INFORMATION OF PRACTICAL IMPORTANCE

Paraquat

1. The paraquat cation is the active ingredient in a soluble-concentrate formulation (Gramoxone Extra), which contains 2.5 lb ai/gal. It is also the active ingredient in a liquid formulation (Cyclone) that contains 2 lb ai/gal.

2. Paraquat is poisonous to humans; it is fatal if swallowed, inhaled, or absorbed through skin. The Gramoxone Extra and Cyclone labels display the skull and crossbones symbol denoting a poison and the words "Danger—Poison." As safety precautions, both formulations emit a stench and an emetic (an agent that causes vomiting if swallowed).

3. Gramoxone Extra and Cyclone are used for early postemergence control of weeds 1–6 in. in height in croplands and fallow lands. They are applied preplant or preemergence to the crop.

4. Gramoxone Extra is applied at 1–3 pt/A (0.3–0.9 lb ai/A). Cyclone is applied at 1–4 pt/A (0.25–1 lb ai/A). Apply the low rate to weeds about 1 in. in height and the highest rate to weeds 6 in. tall. Add a nonionic surfactant or oil concentrate to the spray solution (1 pt/100 gal water) to enhance foliar wetting.

5. Gramoxone Extra and Cyclone are rapidly absorbed by plant foliage, and rain occurring 30 minutes after application has no effect on their activity.

Diquat

6. The diquat cation is the active ingredient in water-soluble concentrate formulations (Valent Diquat; Zeneca Diquat; Weedtrine-D), applied at a rate of 2 lb ai/gal. These products do not display the poison cautions as do the paraquat products (Gramoxone Extra, Cyclone), but they can still be fatal if swallowed, inhaled, or absorbed through the skin.

7. Products containing diquat are used primarily for the control of aquatic weeds in ponds, lakes, and drainage ditches where there is little or no outflow of water and

that are totally under the control of the product user. They may be used on submerged weeds, floating weeds, and marginal weeds (weeds along the water edges).

8. To control submerged weeds, the respective formulation (product) is poured directly from the container into the pond or lake water at 1–2 gal product/A of water surface (2–4 lb diquat cation/A) while moving slowly over the water surface in a boat. Do not add a surfactant to the formulation when using it on submerged weeds.

9. To control floating weeds, apply the formulation at a rate of 0.5–1.0 gal product/A of water surface mixed with 150–200 gal water plus 1 pt nonionic surfactant.

10. To control marginal weeds, use at a rate of 1 gal product/A in 100 gal water plus nonionic surfactant.

16 CYCLOHEXANEDIONE HERBICIDES

■ INTRODUCTION

The common nucleus of the cyclohexanedione herbicides is cyclohexene with an oxygen double-bonded to the ring at the 1-position (1-one), a hydroxy group bonded at the 3-position, and other substitutions at the 2- and 5-positions of the hexene ring. The common and chemical names of the cyclohexanedione herbicides are given in Table 16.1, and their chemical structures in Figure 16.1.

■ CHARACTERISTICS OF CYCLOHEXANEDIONE HERBICIDES

Principal Use

The principal use of the cyclohexanedione herbicides is the postemergence control of annual and perennial grasses in broadleaf crops.

Mode of Action

The primary mode of action of these herbicides is the inhibition of lipid biosynthesis in plants. Their site of action is the enzyme acetyl-CoA in the stroma of plastids. The cyclohexanedione, aryloxyphenoxypropionate, and thiocarbamate herbicides have the same mode of action. The reader is referred to Chapter 7, which is devoted to modes and sites of action, and Chapters 13 and 26 for more information.

Translocation

The cyclohexanedione herbicides are rapidly absorbed by plant foliage and translocated throughout the plant, where they accumulate in the meristems.

TABLE 16.1 ■ The common and scientific names of the cyclohexanedione herbicides.

COMMON NAME	CHEMICAL NAME
Clethodim	(±)-2-[1-[[(3-chloro-2-propenyl)-oxy]imino]propyl]-5-[2-(ethylthio)-propyl]3-hydroxy-2-cyclohexen-1-one
Sethoxydim	2-[1-(ethoxyimino)butyl]-5-[2-(ethylthio)propyl]-3-hydroxy-2-cyclo-hexen-1-one

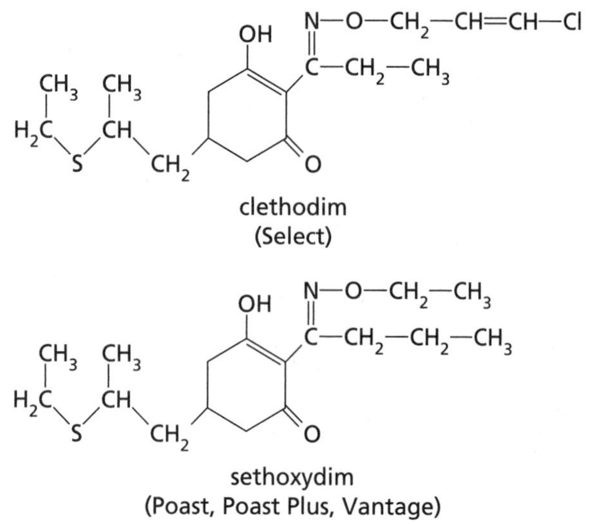

clethodim
(Select)

sethoxydim
(Poast, Poast Plus, Vantage)

FIGURE 16.1 ■ Chemical structures of the cyclohexanedione herbicides.

Selectivity

The selectivity of the cyclohexanedione herbicides is due to the presence of a graminicide-tolerant enzyme (ACCase) in tolerant broadleaf (dicot) plants and the absence of this tolerant form of ACCase in susceptible (grass) plants.

A few grass species, such as centipedegrass and fine fescues, are tolerant to these graminicide herbicides. This is due to the presence of an inherited tolerant form of ACCase. Inheritance is controlled by a single, partially dominant allele.

Common Characteristics

1. Because the cyclohexanedione herbicides control only grasses, they are referred to as graminicides.

2. Broadleaf plants (dicots), sedges, and rushes are tolerant of these herbicides; in contrast, grasses and grass crops, such as corn, sorghum, small grains, and turfgrasses (except as noted above), are highly susceptible.

3. Symptoms of activity are reduction in grass vigor and growth, with early chlorosis/necrosis of younger plant tissue followed by progressive collapse of the remaining foliage. Symptoms may require 7–14 days to be visible, depending on the grass species and environmental conditions.

4. A suitable spray adjuvant (oil concentrate, Dash, or nonionic surfactant) must be added to spray mixtures of the cyclohexanedione herbicides to enhance their foliar absorption and phytotoxicity.

■ INFORMATION OF PRACTICAL IMPORTANCE

Clethodim

1. Clethodim is the active ingredient in an emulsifiable-concentrate formulation (Select) containing 2.0 lb ai/gal.

2. Select is a selective postemergence herbicide that controls annual and perennial grasses in cotton and soybeans.

3. Select is applied at 0.09–0.125 lb ai/A (6–8 fl oz product/A) for control of annual grasses. It is applied at 0.125–0.25 lb ai/A (8–16 fl oz product/A) for perennial grass control.

4. In arid regions, a second application of Select will generally enhance control of perennial grasses.

5. Select is rainfast 1 hour after application.

6. Select may be applied as a spot treatment using hand sprayers. For example, add 2 fl oz of Select plus 4 fl oz crop oil to 3 gal water and mix. Apply mixture to thoroughly wet foliage to just before run-off.

Sethoxydim

7. Sethoxydim is the active ingredient in emulsifiable-concentrate formulations that contain 1.0 lb ai/gal (Vantage, Poast Plus) and 1.5 lb ai/gal (Poast). Poast Plus is formulated with components of Dash that provide increased foliar absorption and protection from ultaviolet light degradation of sethoxydim.

8. In contradiction to its nonselective toxicity to grasses, Vantage is used in seedling (first year) and established centipedegrass and in fine fescues (creeping red, chewings, sheep, hard fescue) for postemergence control of grass weeds. Do not apply Vantage on tall fescue or any other desired turfgrass than noted here, as they will be seriously injured.

9. Vantage is used in nonbearing food crops, nursery liners, trees, shrubs, bedding plants, cut flowers, and ground covers.

10. Poast Plus is used in alfalfa, cotton, peanuts, and soybeans. Poast Plus can be tank-mixed with Basagran for selective grass and broadleaf control in soybeans.

11. Poast is used in field crops (cotton, flax, soybean, sugar beets, sunflower), forage crops (alfalfa, birdsfoot trefoil, and sainfoil), vegetable crops (beans, lettuce, peas,

peppers, potato, tomato, others), tree fruits, strawberries, and noncrop areas.

12. Poast can be tank-mixed with Basagran or Blazer for selective grass and broadleaf weed control in soybean.

13. Poast is applied at 0.1–1.0 lb ai/A (0.5–5 pt product/A) for annual grass control and 0.5 to 1.0 lb ai/A for perennial grass control. The exact rate depends on the targeted species.

17 DINITROANILINE HERBICIDES

■ INTRODUCTION

The dinitroaniline herbicides have as their common nucleus 2,6-dinitroaniline, which is composed of an aniline ring structure with a nitro (NO_2) group bonded to the ring at the 2- and 6-positions (Figure 17.1). These herbicides can be subgrouped as *methylanilines* or *sulfonylanilines,* based on whether a *methyl* or *sulfonyl* group is bonded at the 4-position of the ring structure (Figure 17.2). The dimethylaniline herbicides vary in chemical structure through various substitutions on the 2,6-dinitroaniline nucleus. *Oryzalin* is the only available sulfonylaniline herbicide.

The common and chemical names of the 2,6-dinitroaniline herbicides are given in Table 17.1. Their chemical structures are shown in Figure 17.3.

■ CHARACTERISTICS OF DINITROANILINE HERBICIDES

Principal Use

The principal use of the dinitroaniline herbicides is the preemergence control of seedling grass and broadleaved weeds in many broadleaved crops and in corn (field, pop, sweet), spring wheat, sugarcane, and established turfgrass.

Mode of Action

The primary mode of action of the dinitroaniline herbicides is inhibition of mitosis. They inhibit development of roots and shoots by interfering with cell division in meristems, and they adversely affect development of cell walls and membranes during mitosis.

Translocation

The dinitroaniline herbicides are not translocated in plants following root or shoot absorption.

Selectivity

Herbicide placement that avoids contact with the roots of desired plants is the primary factor in the plant selectivity of the dinitroanilines.

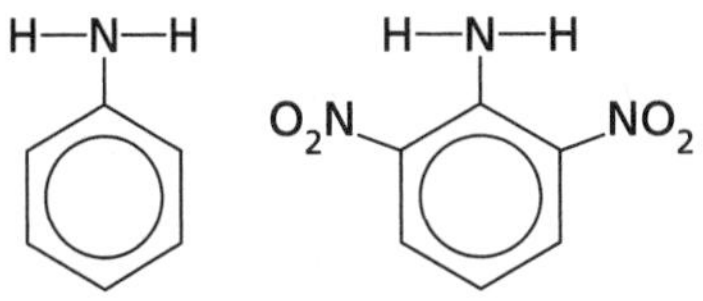

FIGURE 17.1 ■ Chemical structure for aniline and 2,6-dinitroaniline.

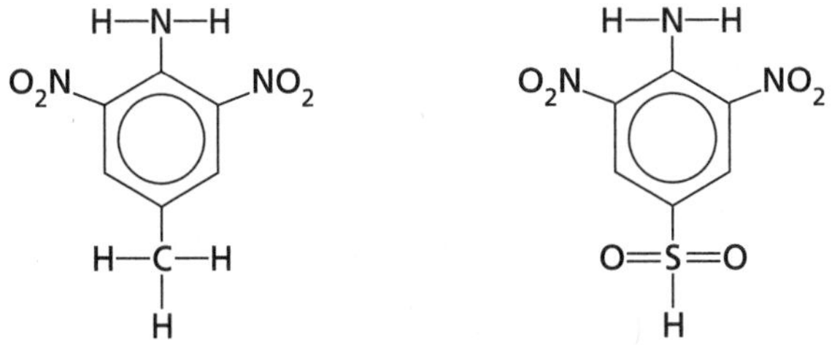

2,6–dinitro–4 methylaniline   2,6–dinitro–4–sulfonylaniline

FIGURE 17.2 ■ Chemical structures for the two subgroups of the 2,6-dinitroaniline herbicides.

Exceptional Characteristics

1. The methylaniline herbicides are highly volatile, whereas oryzalin is much less volatile.

2. The methylaniline herbicides undergo some local leaching in soil; oryzalin is more readily leached.

3. Oryzalin can be applied to the water surface of flooded rice for control of seedling barnyardgrass in the one- to two-leaf stage.

Common Characteristics

1. The dinitroaniline herbicides are applied prior to weed emergence, preferably prior to weed seed germination.

2. These herbicides have little or no postemergence activity.

3. They are absorbed by the roots and by underground portions of emerging shoots, especially shoots of grass seedlings.

4. They do not control established plants at rates used for selective weed control in croplands.

5. Their soil persistence is about 6 months or less, depending on soil type, temperature, and moisture (except for prodiamine, which persists longer).

TABLE 17.1. ■ **Common and chemical names of the 2,6-dinitroaniline herbicides.**

COMMON NAME	CHEMICAL NAME
Benefin	N-butyl-N-ethyl-2,6-dinitro-4-(trifluoromethyl)benzenamine
Ethalfluralin	N-ethyl-N-(2-methyl-2-propenyl)-2,6-dinitro-4-(trifluoromethyl)-benzenamine
Oryzalin	4-dipropylamino-3,5-dinitrobenzenesulfonamide
Pendimethalin	N-(1-ethylpropyl)-3,4-dimethyl-2,6-dinitrobenzenamine
Prodiamine	N,N-dipropyl-4-(trifluoromethyl)-5-amino-2,6-dinitroaniline
Trifluralin	2,6-dinitro-N,N-dipropyl-4-(trifluoromethyl)benzenamine

6. They are decomposed in soils by dealkylation and reduction reactions under both aerobic and anaerobic conditions.

7. The dinitroaniline herbicides are yellow dyes, due to the nitro groups on the phenyl ring of the aniline structure. They impart this color to their formulations and aqueous spray mixtures and to clothing and skin that they contact.

8. To be effective, the dinitroaniline herbicides must be mechanically soil-incorporated into the region of weed seed germination, with the exception of oryzalin, which can be leached into the soil by rainfall or irrigation.

9. Lateral root development of crop plants, such as cotton and soybean, is inhibited in soil treated with dinitroaniline herbicides. However, the growth of taproots in broadleaf crops is much less affected than the growth of lateral roots, and the taproots may elongate sufficiently to push the apical meristems below the zone of treated soil where normal lateral root development will occur. Thus herbicide placement is vital to weed/crop selectivity.

10. Lateral root growth of crop plants is prevented in soil treated with dinitroaniline herbicides soil-incorporated 4–6 in. deep at rates as low as 0.25 lb ai/A.

11. The dinitroaniline herbicides may be applied to emerged crops as an over-the-top or basal-directed spray to clean-tilled soil and incorporated with a rolling cultivator or other suitable tool as they have little or no postemergence activity.

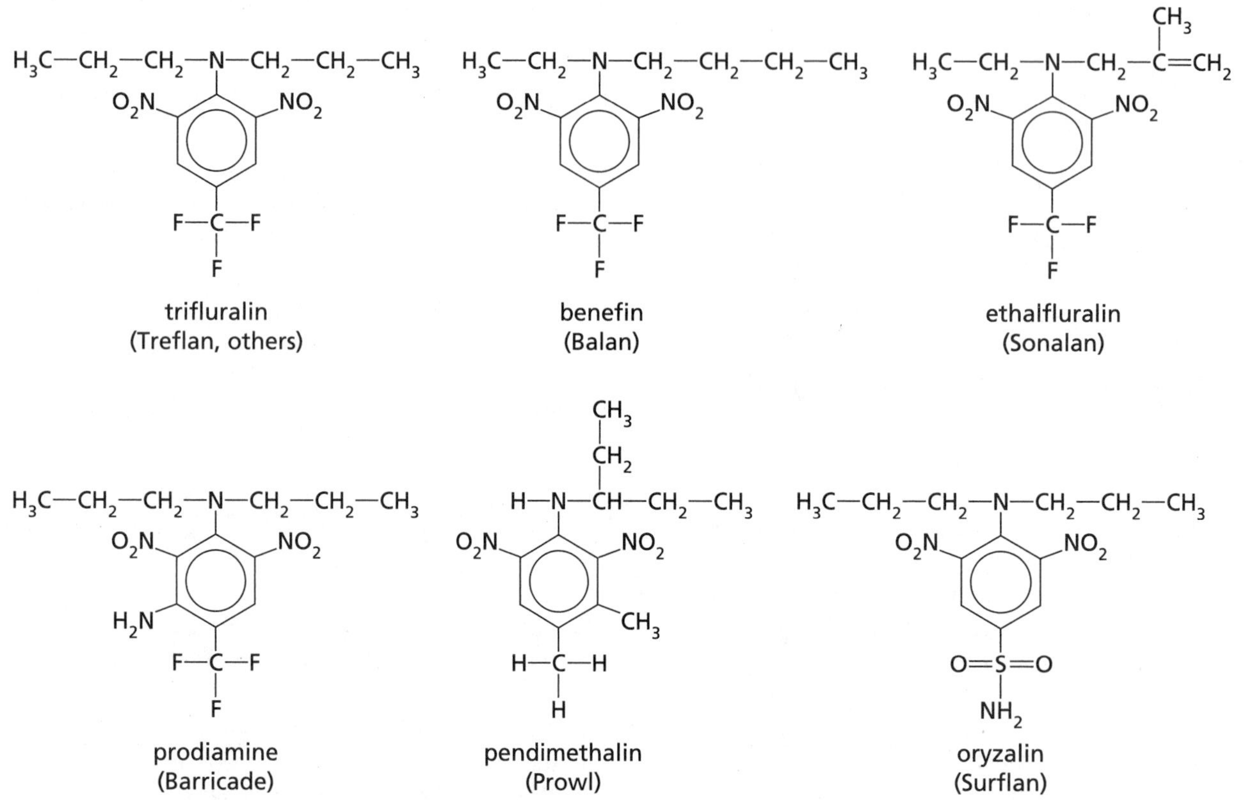

FIGURE 17 .3 ■ Chemical structures of the 2,6-dinitroaniline herbicides.

12. These herbicides are applied at rates of 0.3–3 lb ai/A, depending on herbicide, soil type, and weed species involved.

13. In arid, irrigated regions, rotational crops may be injured when planted in soil previously treated with dinitroaniline herbicides, due to herbicide persistence. To avoid such injury, planting of the following crops should be delayed 10–12 months in soil previously treated with dinitroaniline herbicides: grass crops (wheat, barley, oats, rye, corn, sorghum); onion; red beet; sugar beet; spinach; and various root crops.

14. The dinitroaniline herbicides can be tank-mixed with other herbicides to control a broader spectrum of weeds.

■ INFORMATION OF PRACTICAL IMPORTANCE

1. *Benefin* is the active ingredient in three types of formulations; they are (1) an emulsifiable concentrate (Balan EC) containing 1.5 lb ai/gal, (2) water-dispersible granules (Balan Dry Flowable) that contain 60% ai w/w, and (3) a granular (Balan 2.5G) containing 2.5% ai w/w.

Balan EC and Balan Dry Flowable are used preplant-incorporated in alfalfa, birdsfoot trefoil, clover (alsike, Ladino, and red), lettuce, and peanut at a rate of 1.1–1.5 lb ai/A. Balan 2.5G is used preemergence in established turfgrass at a rate of 1.5–3 lb ai/A.

2. *Ethalfluralin* is the active ingredient in two types of formulations; they are (1) an emulsifiable concentrate (Sonalan EC) containing 3 lb ai/gal and (2) a granular (Sonalan 10G) containing 10% ai w/w.

Sonalan EC is used preplant-incorporated in peanuts and soybeans at a rate of 0.38–1.2 lb ai/A (1.25–3 pt product/A), depending on soil texture. Sonalan 10G is used preplant-incorporated in dry bean and sunflower crops in North Dakota only, where it is marketed under a Special Local Needs registration.

3. *Oryzalin* is the active ingredient in an aqueous suspension (Surflan A.S.); it contains 4 lb ai/gal.

Surflan A.S. is used preemergence in many fruit and nut crops and in vineyards at a rate of 2–6 lb ai/A (2–

6 qt product/A). The lower rate will provide short-term (2–4 months) weed control, and the highest rate, long-term control (8–12 months).

4. *Pendimethalin* is the active ingredient in two emulsifiable-concentrate formulations; they are (1) Prowl 3.4 EC, containing 3.3 lb ai/gal, and (2) Prowl Herbicide, containing 4 lb ai/gal.

   Prowl 3.3 EC and Prowl Herbicide can be used preplant-incorporated in many crops such as beans (edible), corn (field, sweet), cotton, grain sorghum, forage legumes, peanuts, potatoes, rice, and sunflowers. The formulations are applied at a rate of 0.4–1.2 lb ai/A, depending on crop, soil type, and weed species involved.

5. *Prodiamine* is the active ingredient in a water-dispersible granular formulation (Barricade 65WG), which contains 65% ai w/w. Barricade 65WG is packaged in water-soluble packets containing 0.5 lb product/packet.

   Barricade 65WG is a selective preemergence herbicide that provides residual control of grass and broadleaf weeds in established turfgrass and landscape ornamentals. It is applied at a rate of 0.65–1.5 lb ai/A (1.0–2.3 lb product/A). The rate used in turfgrass is based on the particular turfgrass species involved. Barricade 65WG must be soil-incorporated by at least 0.5 in. of rainfall or irrigation within 14 days after application and prior to weed seed germination. Barricade 65WG will not control established weeds.

6. *Trifluralin* is the active ingredient in five formulations; they are (1) emulsifiable concentrates (Treflan EC, Tri-4, and Trilin EC) containing 4 lb ai/gal; (2) an emulsifiable concentrate (Treflan MTF) containing 4 lb ai/gal and described as a "multiple-temperature formulation" (MTF) that may be stored under winter conditions in unheated storage facilities; (3) emulsifiable concentrates (Treflan 5 and Trilin 5) containing 5 lb ai/gal; (4) water-dispersible granules (Tri-4 DF) containing 60% ai w/w; and (5) a granular formulation (Treflan TR-10 and Trilin 10G) containing 10% ai w/w.

   Treflan formulations are used as soil-incorporated treatments in many crops such as alfalfa (established), asparagus, beans (edible), carrot, pepper (transplant), cotton, soybean, fruit and nut crops, vineyards, and many others. The formulations are applied at a rate of 1–4 lb ai/A, depending on crop, soil type, and weeds involved.

7. *Trifluralin* is one of two active ingredients in premix formulations—for example: Freedom (0.33 lb trifluralin + 2.67 lb alachlor/gal); Salute (2.67 lb trifluralin + 1.33 lb metribuzin/gal); Team 2G (0.67% trifluralin + 1.33% benefin w/w); and Tri-Scept (2.57 lb trifluralin + 0.43 lb imazaquin/gal). Freedom is a Restricted Use Pesticide because of the alachlor present in the formulation.

18

DIPHENYL ETHER HERBICIDES

■ INTRODUCTION

Diphenyl ether herbicides have *4-nitro-diphenyl ether* as a common molecular nucleus. This nucleus is composed of two phenyl rings joined by an ether (—O—) bridge and a nitro (NO_2) group bonded to the 4-position (or *para-position*) of one of the phenyl rings. The herbicides differ in the substitutions on one or both phenyl rings. The chemical structure of 4-nitro-diphenyl ether is:

<!-- chemical structure -->

4-nitro-diphenyl ether

The common and chemical names of the diphenyl ether herbicides are given in Table 18.1, and their chemical structures are shown in Figure 18.1 on page 190.

■ CHARACTERISTICS OF THE DIPHENYL ETHER HERBICIDES

Principal Use

The principal use of the diphenyl ether herbicides is the selective control of broadleaf weeds in croplands.

Mode of Action

The primary mode of action of acifluorfen, fomesafen, and lactofen is inhibition of protoporphyrinogen oxidase. The primary mode of action of oxyfluorfen is inhibition of carotene biosynthesis. The diphenyl ether herbicides are light-dependent, fast-acting, contact, photobleaching herbicides. The reader is referred to Chapter 7 for a more detailed discussion of the mode of action of these herbicides.

TABLE 18.1 ■ The common and chemical names of the diphenyl ether herbicides.

COMMON NAME	CHEMICAL NAME
Acifluorfen	5-[2-chloro-4-(trifluoromethyl)phenoxy]-2-nitrobenzoic acid
Fomesafen	5-[2-chloro-4-(trifluoromethyl)phenoxy-N-(methylsulfonyl)-2-nitrobenzamide
Lactofen	(±)-2-ethoxy-1-methyl-2-oxoethyl 5-[2-chloro-4-(trifluoromethyl)-phenoxy]-2-nitrobenzoate
Oxyfluorfen	2-chloro-1-(3-ethoxy-4-nitrophenoxy)-4-(triflouromethyl)benzene

Translocation

The diphenyl ether herbicides undergo little or no translocation following root or foliar absorption, other than local, short-distance movement in treated tissue.

Selectivity

Tolerant plants rapidly metabolize the diphenyl ether herbicides to inactive metabolites, whereas susceptible plants are unable to this.

Common Characteristics

1. The diphenyl ether herbicides may be tank-mixed with appropriate herbicides to control a broader weed spectrum, especially grass weeds.
2. These herbicides are strongly adsorbed to soil colloids and undergo little or no leaching in soils.
3. The diphenyl ether herbicides have relatively low mammalian toxicity.

■ INFORMATION OF PRACTICAL IMPORTANCE

Acifluorfen-sodium

1. Acifluorfen-sodium is the active ingredient in a water-soluble formulation (Blazer) containing 2 lb ai/gal.
2. Blazer is used only in peanuts, rice, and soybeans.
3. In peanuts, Blazer is applied only as a preemergence treatment to the crop (at initiation of soil cracking) and postemergence to young, actively growing broadleaf weeds.

4. In rice, Blazer is applied as a postemergence over-the-top treatment, from late tillering up to early boot stage, and early postemergence to the weeds. Blazer may be tank-mixed with propanil (Stam M-4) and the mixture applied postemergence when the rice has at least three leaves.

5. Blazer is applied postemergence over-the-top to soybeans in the one- to two-trifoliate leaf stage and postemergence to the weeds. Adjust the height of the spray boom above the crop to give complete coverage of all weeds.

6. An appropriate nonionic surfactant must be included in all spray mixtures of Blazer applied postemergence when used alone or in tank mix with other herbicides. Add the surfactant at 1–2 pt/100 gal water. Urea ammonium nitrate (UAN) may be added in place of other spray adjuvants for improved weed control in soybeans.

7. Blazer may be tank-mixed with appropriate herbicides to obtain a broader spectrum of weed control, especially grasses.

8. Blazer is applied at 0.25–0.38 lb ai/A (1–1.5 pt product/A). Used in mixture with other recommended herbicides, the rate is 0.125–0.38 lb ai/A (0.5–1.5 pt product/A).

9. In case of crop failure, only peanuts, rice, or soybeans may be immediately replanted. Do not plant root crops (such as carrots, sweet potatoes, turnips, among others) in fields treated with Blazer for a period of 18 months following treatment.

10. Acifluorfen-sodium is one of two active ingredients in two premix water-soluble formulations (Galaxy, Storm). Galaxy contains 0.67 lb/gal acifluorfen-sodium and 3 lb/gal bentazon. Storm contains 1.33 lb/gal acifluorfen-sodium and 2.67 lb/gal bentazon.

11. Galaxy is used postemergence in soybeans to control emerged broadleaf weeds and yellow nutsedge with a maximum height of 3–6 in. (8–15 cm) generally corresponding to soybean growth stages of unifoliate to two expanded trifoliate leaves.

12. Storm is used in peanuts and soybeans. In peanuts, it is applied preemergence (soil cracking stage) or early postemergence (up to two expanded tetrafoliate leaves). In soybeans, it is applied postemergence, generally at the unifoliate to two expanded trifoliate leaf

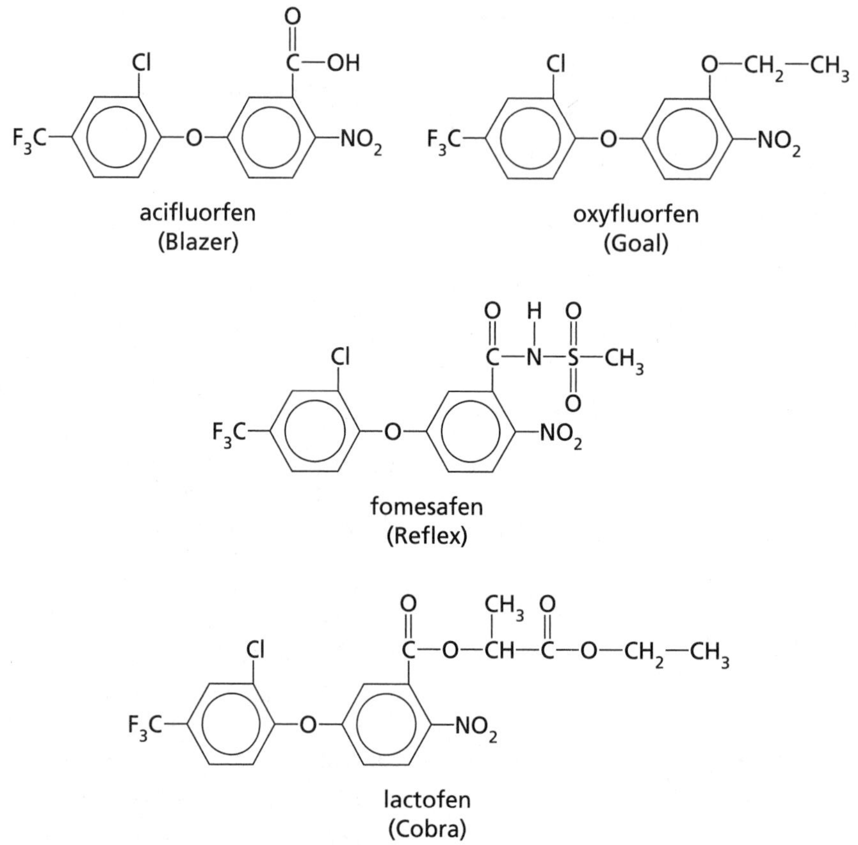

FIGURE 18.1 ■ Chemical structures of the diphenyl ether herbicides.

stage of the soybeans, for control of young broadleaf weeds and partial control of some grass species.

Fomesafen

13. Fomesafen is the active ingredient in an aqueous liquid concentrate formulation (Reflex 2LC) containing 2 lb ai/gal. It is formulated as its water-soluble sodium salt. Reflex 2LC does not contain a surfactant.

14. The principal use of Reflex is in the selective post-emergence control of broadleaf weeds in soybeans, applied prior to bloom. Soybeans are tolerant of Reflex.

15. Best results are obtained when Reflex is applied early postemergence to actively growing broadleaf weeds, usually 2–3 weeks after planting soybeans. Reflex may provide growth suppression of certain annual grass species.

16. Reflex is applied at 0.25–0.375 lb ai/A (1–1.5 pt product/A), depending on weed species present and their stage of growth.

17. Reflex is effective through contact activity; weed foliage must be thoroughly wetted by the herbicide spray solution. Always add a nonionic surfactant at 0.5–1 pt/25 gal spray solution or a crop oil concentrate at 1–2 pt/25 gal spray solution when Reflex is used alone or in tank mixes.

18. Reflex may be tank-mixed with other suitable post-emergence soybean herbicides for improved control of troublesome broadleaf weeds such as prickly sida and velvetleaf (Basagran 4E); cocklebur (Classic, Scepter, Butyrac 200); annual morningglory species and giant ragweed (Butyrac 200); and yellow nutsedge (Basagran 4E, Classic).

19. Reflex may be tank-mixed with Fusilade 2000 to control broadleaf and grass weeds in soybeans. Reflex may be applied as a sequential treatment with Fusilade 2000, with the Reflex treatment followed by Fusilade 2000.

20. The Reflex label shows four regions within the United States in which Reflex may be used. These regions coincide with the soybean-growing states, and each

region has its own specific restrictions and directions for use of Reflex. Refer to the product label for this information.

21. Fomesafen is one of two active ingredients in a premix liquid formulation (Tornado) containing 1.0 lb fomesafen and 0.75 lb fluazifop-P-butyl/gal.

22. Tornado is a selective, early postemergence herbicide used to control annual grass and broadleaf weeds in soybeans. Tornado has contact activity on broadleaf weeds (fomesafen) and systemic activity on grass weeds (fluazifop-P-butyl).

Lactofen

23. Lactofen is the active ingredient in an emulsifiable-concentrate formulation (Cobra) containing 2 lb ai/gal.

24. The principal use of Cobra is in the early postemergence control of annual broadleaf weeds in cotton, soybeans, and southern pine seedbeds. Cobra does not control grasses.

25. In general, broadleaf weeds with more than six leaves are not controlled by Cobra.

26. Always add a nonionic surfactant at 2 pt/100 gal spray mixture, or crop oil concentrate at 0.5–1 pt/A in 10–30 gal spray mixture, with Cobra applied alone or in tank mix with other herbicides.

27. Used in cotton, Cobra is applied as a sequential post-directed spray following the application of a suitable preplant-incorporated or preemergence herbicide for early control of grass and broadleaf weeds.

28. Cotton plants are not tolerant of Cobra. Do not make an over-the-top application to cotton. Apply Cobra as a directed spray when the cotton plants have reached a minimum height of 8 in. (20 cm) and the top of the broadleaf weeds are at least 5 in. (13 cm) below the lower leaves of the cotton plants. Cotton leaves contacted by the herbicide will appear burned or light brown to bronze in color. Leaves not contacted by the herbicide will not be affected, and all growth following herbicide application will be normal.

29. Cobra may be applied post-directed in cotton at lay-by after the cotton has reached a height of 12 in. (30 cm) or more. For enhanced control of late season weeds, Cobra may be tank-mixed with Bladex, Karmex, or MSMA.

30. Cobra is applied as an over-the-top application to soybeans, preferably in the first to second trifoliate leaf stage, when the broadleaf weeds are small and actively growing, growth stages usually corresponding to 2–3 weeks after planting.

31. Cobra is applied in cotton and soybeans at 0.2 lb ai/A (12.5 fl oz product/A).

32. Cobra may be tank-mixed with other postemergence herbicides recommended for soybeans.

33. Used in southern pine seedbeds, Cobra is applied over-the-top of pine seedlings to control emerged, actively growing broadleaf weeds. Pine seedlings will tolerate Cobra when applied after complete emergence. For maximum crop safety, wait until the seedlings have initiated primary shoot growth before applying Cobra. Do not apply Cobra with spray adjuvant until primary shoot growth has begun. Cobra is applied at a rate of 6.5 fl oz product/A at 1-week intervals (four applications maximum) or 12.5 fl oz/product/A at 2-week intervals (two applications maximum).

Oxyfluorfen

34. Oxyfluorfen is the active ingredient in an emulsifiable-concentrate formulation (Goal) containing 1.6 lb ai/gal.

35. Goal is used primarily for the selective preemergence or early postemergence control of broadleaf weeds in the two- to four-leaf stage in a variety of crops. Some grass weed species may be controlled with Goal applied preemergence or early postemergence before the grass seedlings are in the one- to two-leaf stage.

36. Goal is used post-directed in a variety of crops, among which are cotton, conifers, tree fruits and nuts, and certain vegetables. It may be applied broadcast on fallow beds for control of winter annual broadleaf weeds.

37. Goal is applied at 0.03–2 lb ai/A (2.4 fl oz to 10 pt product/A), depending on crop and weed species involved, regional location, time of year applied, and whether applied preemergence or postemergence to the weeds. In general, the most common rate is 0.25–1 lb ai/A.

38. Goal is used in cotton as a post-directed application for the control of broadleaf weeds. The cotton plants must have reached a minimum height of 6 in. and preferably higher at time of application. The post-directed spray must be kept from the cotton foliage to avoid severe injury to the cotton plants, and use of spray shields is advised. The recommended dosage rate is 0.25–0.5 lb ai/A (1.25–2.5 pt product/A), with the lower rate effective against susceptible weeds in the two- to three-leaf stage and the higher rate against weeds not exceeding four true leaves.

39. Goal is usually not applied as an over-the-top spray to actively growing crop plants, as severe injury may result. However, it is recommended as an over-the-top treatment to dormant mints and to onions that have at least two or three fully developed true leaves.

40. Goal is used in direct-seeded and transplanted onions to control broadleaf weeds in the two- to four-leaf stage and certain grass weeds. In seeded onions grown in the northeastern states, Goal is applied at 0.03–0.06 lb ai/A (2.4–4.8 fl oz product/A) when the onions have at least three true leaves. For seeded onions grown in the western states, the rate is 0.12–0.25 lb ai/A (0.6–1.25 pt product/A), applied when the onions have at least two true leaves. In transplanted onions grown in all states except the northeastern states, the rate is 0.5 lb ai/A (2.5 pt product/A), applied within 2 days after transplanting. Multiple applications of Goal may be made to onions at the above rates to control later germinating weeds, not to exceed a total of 0.5 lb ai/A in one season. Goal may cause necrotic lesions or other symptoms on the onions following over-the-top applications, but subsequent new leaves will be normal.

41. Goal may be tank-mixed with suitable herbicides.

19 GROWTH REGULATOR–TYPE HERBICIDES

■ INTRODUCTION

The growth regulator–type (hormone-type) herbicides are characterized by similarities in use, mode of action, translocation, and selectivity. Herbicides in this family have been grouped for ease of discussion by similarities in chemical structure and, in two cases, as individual herbicides. The groups are *phenoxy-carboxylic* and *pyridine,* and the individuals are *dicamba* and *quinclorac.* Their similarities are given in the following sections.

■ CHARACTERISTICS OF GROWTH REGULATOR–TYPE HERBICIDES

Principal Use

The principal use of the growth regulator–type herbicides is the postemergence control of annual and perennial broadleaf weeds and woody shrubs. Quinclorac is an exception in that it is used to control certain grass and broadleaf weeds, and only in rice.

Mode of Action

The primary mode of action of the growth regulator–type herbicides is not known. They have auxinlike properties, and their mode of action involves interference with nucleic acid metabolism and disruption of the normal transport systems in plants through induced massive cell proliferation.

Translocation

The growth regulator–type herbicides translocate systemically throughout the plant following absorption by underground and/or aboveground plant parts. They tend to accumulate in regions of high metabolic activity such as the meristems.

Selectivity

The growth regulator–type herbicides selectively control broadleaf weeds in grass crops (e.g., small grains, corn, turfgrass), rangeland and established pastures, and in non-

TABLE 19.1 ■ The common and chemical names of the phenoxy-carboxylic herbicides.

COMMON NAME	CHEMICAL NAME
2,4-D	(2,4-dichlorophenoxy)acetic acid
dichlorprop (2,4-DP)	(±)-2-(2,4-dichlorophenoxy)pro-panoic acid
dichlorprop-P	(+)-2-(2,4-dichlorophenoxy)pro-panoic acid
2,4-DB	4-(2,4-dichlorophenoxy)butanoic acid
MCPA	(4-chloro-2-methylphenoxy)acetic acid
mecoprop (MCPP)	(±)-2-(4-chloro-2-methylphenoxy)-propanoic acid
MCPB	4-(4-chloro-2-methylphenoxy)-butanoic acid

croplands. Quinclorac is an exception, as noted under principal use. As would be expected, susceptibility to these herbicides varies among species of broadleaf weeds and woody plants. If this were not so, there would only be need for one growth regulator–type herbicide.

■ INFORMATION OF PRACTICAL IMPORTANCE

Phenoxy-Carboxylic Herbicides

The phenoxy-carboxylic herbicides have molecular structures composed of (1) an aromatic (benzene) ring; (2) an oxygen atom substituted for one hydrogen bonded to the ring; (3) a carboxyl group bonded indirectly to the oxygen atom, separated from the oxygen atom by an aliphatic chain of one or more carbon atoms; and (4) various substituents on the ring.

Members of the phenoxy-carboxylic herbicide family are distinguished from one another by the length of the aliphatic chain of carbon atoms separating the carboxyl radical from the oxygen atom bonded to the ring, by the particular substituents and their location on the ring, or both.

The common and chemical names of the phenoxy-carboxylic herbicides are given in Table 19.1, and their chemical structures are shown in Figure 19.1.

Dichlorprop (2,4-DP) and mecoprop (MCPP) are racemic mixtures of their respective optical isomers. The herbicidally active enantiomer of dichlorprop is the *dextro* (+) form, and it has been named dichlorprop-P. Its herbicidal

characteristics are the same as for dichlorprop, but half as much dichloprop-P is required to produce the same effect. Mecoprop is only available as a racemic mixture.

The chemical structures of the acid forms of these herbicides are shown in Figure 19.1. The phenoxy-carboxylic herbicides are usually formulated as salts and esters of their respective parent acids. Chemical structures of water-soluble salts of 2,4-D are shown in Figure 19.2, and those of its emulsifiable esters are shown in Figure 19.3. The chemical structures of the salts and esters of other members of this herbicide family are similar to those of 2,4-D. In practice, the acid form of the phenoxy-carboxylic herbicides is rarely used. An exception is Weedone 638 in which the active ingredients are a mixture of the 2,4-D acid and the butoxyethyl ester of 2,4-D. Weedone 638 is more effective than 2,4-D amines in controlling hard-to-kill perennial weeds such as field bindweed, Russian knapweed, Canada thistle, leafy spurge, cattails, and tules.

Salts

The common salt forms of the phenoxy-carboxylic herbicides are the dimethylamine and sodium salts. Of these, the dimethylamine form is the more common.

Amine salts are ammonia–alcohol derivatives of the parent phenoxy-carboxylic herbicides. The ammonia–alcohol group is cationic, and it replaces the hydrogen ion on the acidic carboxyl group of the parent acid molecule. On the amine ion, one or more of the hydrogens of ammonia (NH_3) are replaced by suitable alcohols. For example, a substitution of one of the hydrogens with ethanol results in ethanolamine; substitution with two ethanol groups results in diethanolamine; substitution of all three hydrogens of ammonia with three ethanol groups results in trietholamine. Methyl and ethyl radicals also may be substituted for hydrogens of ammonia.

The amine salts of the phenoxy-carboxylic herbicides dissolve readily in water, forming true solutions. They, like other salt forms, dissociate in water; in hard water, the calcium and magnesium ions present may form insoluble salts with the anionic portion of these herbicides. Such insoluble salts will precipitate, resulting in a reduction of herbicide in the spray solution and the plugging of screens and nozzle orifices of the spray equipment. The amine salts are less affected by hard water than are the sodium and ammonium salts.

In aqueous mixtures with liquid fertilizers, cations of the fertilizer may replace the cationic portion of the salt forms of herbicides, resulting in the formation of insoluble precipitates.

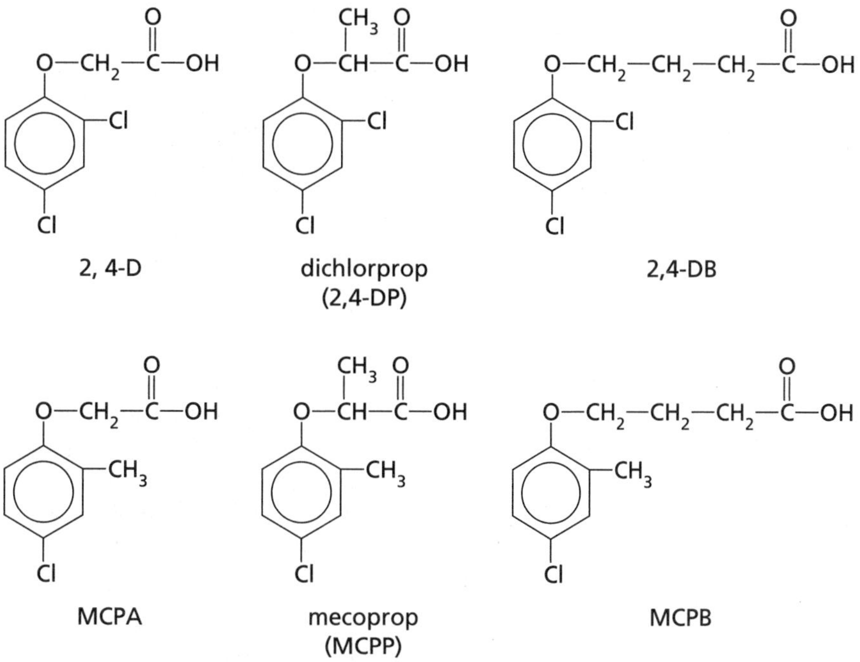

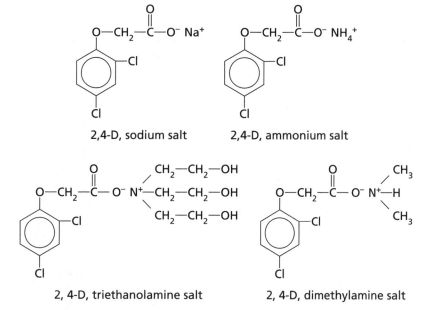

FIGURE 19.1 ■ Chemical structures for the parent acid form of the principal phenoxy-carboxylic herbicides.

FIGURE 19.2 ■ Chemical structures for four water-soluble salts of 2,4-D.

The salts of the phenoxy-carboxylic herbicides may be readily washed from the plant foliage by water. There is no serious loss of phytotoxicity, however, if the salt form of the herbicides remains in contact with the foliage for at least 6 hours. After being washed from the leaves by rainfall, the herbicide may still enter the plant via root absorption.

Under normal conditions of plant growth, the salt forms of the phenoxy-carboxylic herbicides are nonvolatile.

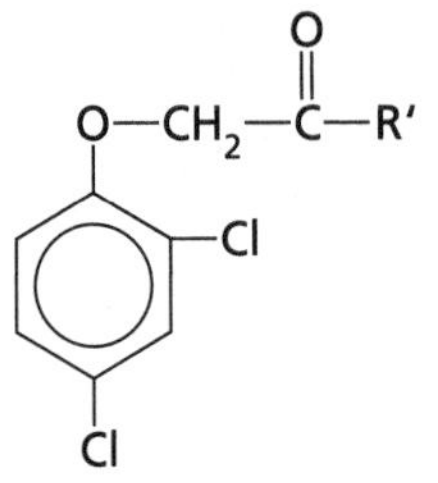

2, 4-D, ester

R′ denotes the position on the 2,4-D molecule of the following long-chain, low-volatile ester groups:

1. $-O-CH_2-CH_2-O-CH_2-CH_2-CH_2-CH_3$

butoxyethyl ester (Sometimes incorrectly written as "butoxyethanol ester")

2. $-O-CH_2-CH_2-CH_2-CH_2-CH_2-CH-CH_3$
                                              $|$
                                           CH_3

isooctyl ester

3. $-O-CH_2-CH-CH_2-CH_2-CH_2-CH_3$
                       $|$
                   CH_2-CH_3

isooctyl(2-ethylhexyl) ester

FIGURE 19.3 ■ Three low-volatile esters of 2,4-D.

However, disposal of their containers by burning has resulted in the vaporization of chemical residues, with subsequent plant injury caused by vapor drift.

Esters

Esters of the phenoxy-carboxylic herbicides are soluble in oils and insoluble in water. For practical use, they commonly are formulated as emulsifiable concentrates for application in either oil or water carriers.

Esters of the phenoxy-carboxylic herbicides are formed by the reaction of their acid form with an alcohol. The reaction replaces the hydroxyl (OH) of the carboxyl group with the respective alcohol. The hydroxyl group of the carboxyl combines with the hydrogen of the hydroxyl group on the alcohol to form water, and the alcohol is bonded to the carbon of the carboxyl group by its oxygen atom—the oxygen linkage characteristic of an ester. The resulting ester molecule is nonionic, does not dissociate in water, and thus does not react with Ca^{2+} and Mg^{2+} in hard water to form insoluble precipitates.

A particular ester of the phenoxy-carboxylic herbicides is identified by the alcohol reacting to replace the hydroxyl of the carboxyl group of the parent acid herbicide. For example, the methyl ester is formed when the alcohol is methanol, and the isopropyl ester is formed from isopropanol. In some cases, two or more alcohols, connected by an ether (—C—O—C—) linkage react with the parent acid herbicide to form long-chain esters such as butoxyethyl ester. The long-chain structure of butoxyethyl ester is $-O-CH_2-CH_2-O-CH_2-CH_2-CH_2-CH_3$.

The short-chain esters (methyl, ethyl, isopropyl, butyl) of the phenoxy-carboxylic herbicides are highly volatile; under ordinary temperatures of plant growth, their vapors present a real hazard to susceptible plants. The long-chain esters are much less volatile than the short-chain esters. It is well to emphasize, however, that the long-chain esters are *less* volatile, not *non*volatile, and that their use does present a potential hazard from subsequent vaporization and vapor drift. In general, the short-chain esters are commonly referred to as *high-volatile esters*, whereas long-chain esters are referred to as *low-volatile esters*.

1. Applied in their salt or ester form, the phenoxy-carboxylic herbicides are converted within the plant to their respective acid forms, and it is in this form that they are ultimately phytotoxic.

2. Most broadleaf crops are susceptible to the phenoxy-carboxylic herbicides.

3. The phenoxy-carboxylic herbicides are readily degraded in plants, except for mecoprop, which is degraded more slowly.

4. The butanoic acid herbicides 2,4-DB and MCPB are in themselves nonherbicidal. They are converted to their respective acetic acid form by β-oxidation within susceptible plants (but not in tolerant plants), and it is in the acetic acid form that they are phytotoxic. Plants incapable of rapid β-oxidation are not harmed by the butanoic acid herbicides, thereby providing a means of selectivity among broadleaf species.

5. Avoid spray drift of the phenoxy-carboxylic herbicides, as light to severe injury will occur to most broadleaf plants. Cotton, grapes, and tomatoes are extremely sensitive to these herbicides.

6. The esters of the phenoxy-carboxylic herbicides are more effective than their corresponding salts in the control of perennial weeds.

7. In croplands, the phenoxy-carboxylic herbicides are applied at a rate of 0.25–1.0 lb ae/A, depending on the crop and weeds involved. Used in noncroplands, rates of 3–4 lb ae/A can be used.

8. The salt forms of the phenoxy-carboxylic herbicides are readily leached in coarse-textured soils and less so in fine-textured soils.

9. In general, the phenoxy-carboxylic herbicides persist 1–4 weeks in warm, moist soil. Longer persistence usually occurs under cool, dry conditions. Their decomposition in soils is primarily by microbial attack.

10. Examples of products containing phenoxy-carboxylic herbicides formulated as salts are Weedar 64 (2,4-D, dimethylamine); Rhomene (MCPA, dimethylamine); Formula 40 (mixture of triisopropanolamine and diethylamine salts of 2,4-D); and Thistrol (MCPB, sodium salt).

11. Examples of products containing phenoxy-carboxylic herbicides formulated as esters are Weedone LV4 (2,4-D, butoxyethyl ester); Esteron 99C [2,4-D, isooctyl(2-ethylhexyl) ester]; Rhonox (MCPA, isooctyl ester); and Weedone (2,4-DP, butoxyethanol ester).

12. Combinations of 2,4-D or MCPA, mecoprop, and dicamba (a benzoic acid herbicide) in proper proportions are synergistic mixtures. TRIMEC products (PBI/Gordon) are examples of such combinations.

Pyridine Herbicides

Pyridine herbicides are of the growth regulator type. Their chemical structures have in common the pyridine ring. This ring structure is similar to that of benzene, except that a nitrogen atom replaces a carbon atom in the 1-position of the ring, resulting in a ring called *pyridine*. Individual pyridine herbicides differ, one from another, in the substituents bonded to the pyridine ring. The chemical structure of the pyridine ring is:

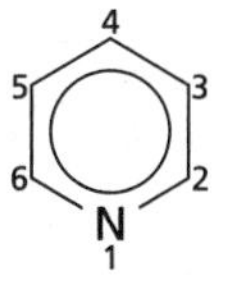

pyridine

The common and chemical names of the pyridine herbicides are given in Table 19.2, and their chemical structures are shown in Figure 19.4.

TABLE 19.2 ■ Common and chemical names of the pyridine herbicides.

COMMON NAME	CHEMICAL NAME
Clopyralid	3,6-dichloro-2-pyridinecarboxylic acid
Picloram	4-amino-3,5,6-trichloro-2-pyridine-carboxylic acid
Triclopyr	[(3,5,6-trichloro-2-pyridinyl)oxy]acetic acid

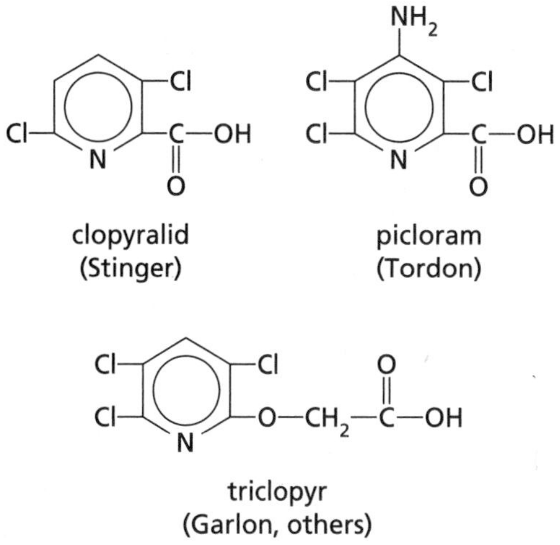

FIGURE 19.4 ■ Chemical structures of the pyridine herbicides.

Clopyralid

13. Clopyralid is the active ingredient in a liquid formulation (Stinger, Reclaim, Transline) containing 3 lbs ae/gal. It is formulated as its monoethanolamine salt.

14. Stinger is used for selective, postemergence control of broadleaf weeds in sugar beets, field corn, small grains (wheat, barley, and oats) not underseeded with a legume, and in fallow cropland, rangeland and permanent grass pastures, grass grown for seed, Christmas trees, and noncrop areas to control broadleaf weeds and woody brush.

15. Clopyralid is a highly effective herbicide for honey mesquite control either alone or in mixture with picloram. Clopyralid killed 80 percent or more of foliar-treated honey mesquite plants when applied at 0.25 lb ai/A (0.28 kg ai/ha).

16. Stinger is applied in crops at a rate of 0.09 to 0.25 lb ae/A (0.25 to 0.67 pt product/A), and in fallow and rangeland/pastures for perennial weed control at 0.25 to 0.5 lb ae/A (0.67 to 1.33 pt product/A). Stinger applications are rainfast within 6 to 8 hours.

17. Reclaim is a specialty product used to control mesquite, acacias, and other woody brush in rangelands and pastures only in New Mexico, Oklahoma, and Texas. Transline is a specialty product for broadleaf weed control in industrial sites, rights-of-way, and other noncrop areas.

18. Clopyralid is one of two active ingredients in the premixes Curtail (with 2,4-D), Curtail M (with MCPA), and Confront (with triclopyr).

Picloram

19. Picloram is the active ingredient in a liquid formulation (Tordon 22K) containing 2 lb ae/gal. It is formulated as the potassium salt of picloram. Tordon 22K is a Restricted Use Pesticide due to possible phytotoxicity to susceptible nontarget plants. See the precautionary statement, item 10.

20. Tordon is used in areas *west of the Mississippi River* to control broadleaf weeds and woody plants on rangeland and permanent grass pastures; fallow cropland; small grains (wheat, barley, oats) not underseeded with a legume; and flax on grainland (land not flooded or subirrigated and not rotated to broadleaf crops); noncropland; and on Conservation Reserve Program (CRP) areas.

21. Do not apply Tordon in residential areas or near desired ornamental trees or shrubs, as the root systems of desired plants may absorb the herbicide following its leaching in soil or as excretions from roots of nearby treated trees.

22. In spring-planted small grains, Tordon is applied when the crop is in the three- to five-leaf stage up to jointing. Do not treat durum wheat, as some varieties are susceptible to Tordon.

23. In winter-planted wheat and barley, Tordon is applied up to jointing after the crop has resumed active growth in the spring.

24. Tordon may be tank-mixed with 2,4-D or MCPA to control a broader weed spectrum.

25. In small grains, do not apply more than 0.023 lb ae/A (1.5 fl oz product/A) during the small-grain growing season.

26. In rangeland/pastures and noncroplands, Tordon is applied at a rate of 0.16–1.5 lb ai/A (0.5–6 pt product/A), depending on the weed species involved.

27. Picloram is one of two active ingredients in a liquid formulation (Tordon RTU) used undiluted as an injection, frill, girdle, or cut stump treatment. The other active ingredient is 2,4-D, and both herbicides are formulated as amine salts. Tordon RTU is *not* a Restricted Use Pesticide.

28. *Precautionary Statement:* Picloram can move through soil and, under certain conditions, it may contaminate groundwater used for irrigation and drinking purposes. Users are advised not to use picloram where soils have a rapid to very rapid permeability throughout the profile (such as loamy sand to sand); where the water table of an underlying aquifer is shallow; or where soils contain sinkholes over limestone bedrock, severely fractured surfaces, and substrates that would allow direct introduction into an aquifer.

Triclopyr

29. Triclopyr is the active ingredient in several liquid formulations; they are Garlon 3A (3 lb ae/gal), Garlon 4 (4 lb ae/gal), Redeem (3 lb ae/gal), Remedy (4 lb ae/gal), Grandstand (3 lb ae/gal), and Pathfinder (0.75 lb ae/gal).

30. Garlon 3A is used to control annual and perennial broadleaf weeds in noncropland. Garlon 4 is used to control broadleaf weeds, such as black medic, clover, goldenrod, ground ivy, mustard, oxalis, wild carrot, and wild violet, in noncroplands. The Garlon formulations are applied at a rate of 1–9 lb ae/A, depending on the weed species involved.

31. Redeem and Remedy are used for woody brush control, such as mesquite and acacia, in rangeland/pastures and noncroplands. In rangeland/pastures, Redeem is applied at a rate of 0.75–1.5 lb ae/A (1–2 qt product/A), and Remedy is applied at 1–3 lb ae/A (1–3 qt product/A).

32. Grandstand and Redeem are used in Christmas tree plantations to control annual and perennial broadleaf weeds and brush. Grandstand is applied at a rate of 0.75–1.9 lb ae/A (2–5 pt product/A). Redeem is applied at a rate of 0.5–1 lb ae/A (1.33–2.66 pt product/A).

33. Pathfinder is applied undiluted as a basal treatment to woody plants less than 6 in. (15 cm) in diameter and as a stump treatment. Spray to run-off around the entire circumference of the tree, thoroughly wetting the lower 15 in. (38 cm) of trunk. Wet the entire cambium and bark area and also the sides of freshly cut stumps.

Dicamba

Dicamba is a substituted benzoic acid herbicide. Its chemical name is *3,6-dichloro-2-methoxybenzoic acid*. The chemical structures of dicamba and its dimethylamine salt are:

dicamba

dicamba, dimethylamine salt

34. Dicamba is the active ingredient in a water-soluble formulation (Banvel) containing 4 lb ae/gal. It is formulated as the dimethylamine salt of dicamba. It is available in a premix with atrazine (Marksman) for corn and fallow land.

35. Banvel is used postemergence for selective control of annual and perennial broadleaved weeds in asparagus; corn (field corn, pop, silage); sorghum; small grains (wheat, barley, oats) not underseeded to legumes; turfgrass; pastures; and noncrop areas. It is not specifically used for grass control, but some seedling grass species may be killed or injured. It is used for brush control in pastures, rangeland, and noncrop areas, and as a dormant application for the control of multiflora rose.

36. Banvel selectivity appears to be caused by the capability of tolerant plant species to form nonphytotoxic conjugates more quickly than susceptible species and to tolerate higher concentrations of either herbicide within the plant.

37. Banvel is readily leached in soils.

38. Dicamba is excreted, or leaked, from the roots of plants following foliar or root absorption, and this exudate may be reabsorbed by roots of the same plant or by those of nearby plants in sufficient amounts to induce a plant growth regulator or herbicidal effect.

39. The soil persistence of Banvel is influenced by the rate of leaching and by conditions amenable to its rapid decomposition in soil. It may be leached below the zone of its activity in humid regions within a period of 3–12 weeks. Applied at 0.5 lb ae/A, dicamba may persist in a highly active state for more than 3 months.

40. Banvel is used in tank mix with other herbicides to control a broader spectrum of weeds.

41. Banvel is applied postemergence at a rate of 0.12–0.25 lb ae/A (0.25–0.5 pt product/A) for annual broadleaf weed control, and at a rate of 0.25–8 lb ae/A (0.5 pt to 2 gal product/A) for control of perennial broadleaf weeds and brush. The rate is dependent on the crop and weed species involved.

Quinclorac

Quinclorac is another growth regulator–type herbicide. Its chemical name is *3, 7- dichloro- 8-quinolinecarboxylic acid,* and its chemical structure is shown below.

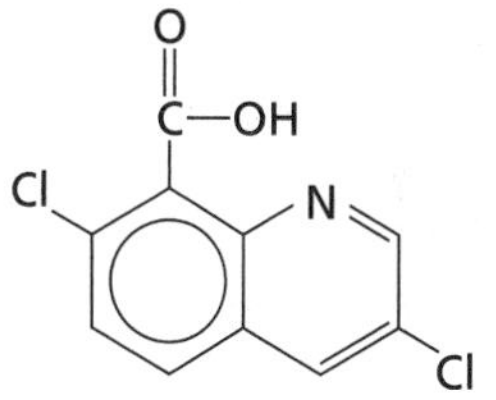

quinclorac
(Facet)

42. Quinclorac is the active ingredient in a wettable-powder formulation (Facet) containing 50% ai w/w.

43. Facet is used only in rice to control certain grass and broadleaf weeds. It can be applied preemergence, delayed preemergence, and early postemergence.

44. Facet is absorbed by the roots, coleoptiles, and leaves of grasses.

45. Facet controls barnyardgrass, broadleaf signalgrass, junglerice, large crabgrass, eclipta, and sesbania.

46. Facet is applied at a rate of 0.25–0.5 lb ai/A (0.5–1.0 lb product/A).

47. In case of crop failure, rice may be immediately replanted. However, do not plant any other crop for a period of 10 months after application. Do not plant carrots and tomatoes within 2 years following application.

20 IMIDAZOLINONE HERBICIDES

■ INTRODUCTION

The imidazolinone herbicides have as their common nucleus the imidazole ring with methyl and isopropyl radicals bonded to the 4-position of the ring and an oxygen double-bonded to the 5-position. The chemical structure of the imidazole ring and the nucleus common to the imidizolinone herbicides are shown in Figure 20.1.

The imidazolinone herbicides differ in the ring structure substitutions at the 2-position of the imidazole ring. These ring structures may be benzene, pyridine, or quinoline, with various radical substituents on their rings. The chemical structures of these substituent rings are shown in Figure 20.2.

The common and chemical names of the imidazolinone herbicides are given in Table 20.1, and their chemical structures are shown in Figure 20.3 on page 202.

■ CHARACTERISTICS OF IMIDAZOLINONE HERBICIDES

Principal Use

The principal use of the imidazolinone herbicides is the control of grass and broadleaf weeds in certain crops and warm-season turfgrass and for general weed and brush control in noncrop areas.

Mode of Action

The primary mode of action of the imidazolinone herbicides in susceptible plants is the inhibition of the enzyme acetolactate synthase (ALS), also named acetohydroxyacid synthase (AHAS). This enzyme catalyses the first step in the biosyntheses of the essential amino acids valine, leucine, and isoleucine, and its inhibition deprives the plant of these essential amino acids, interfering with DNA synthesis and cell growth. This enzymatic pathway is unique to plants. This particular mode of action is the same as that attributed to the sulfonylurea herbicides (see Chapter 25).

Translocation

The imidazolinone herbicides are absorbed by roots and foliage and translocated throughout the plant, accumulating in meristems such as apical buds at root and shoot tips and buds in the leaf axils.

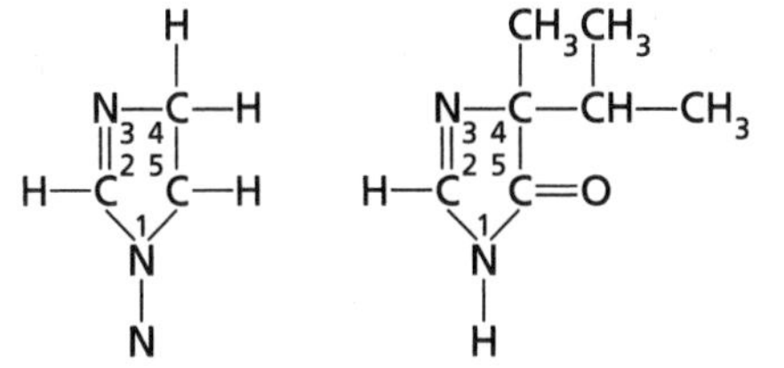

imidazole 4–methyl–4–isopropyl–
5–oxo–1H–imidazol

FIGURE 20.1 ■ Chemical structures of imidazole and the common nucleus of the imidazolinone herbicides.

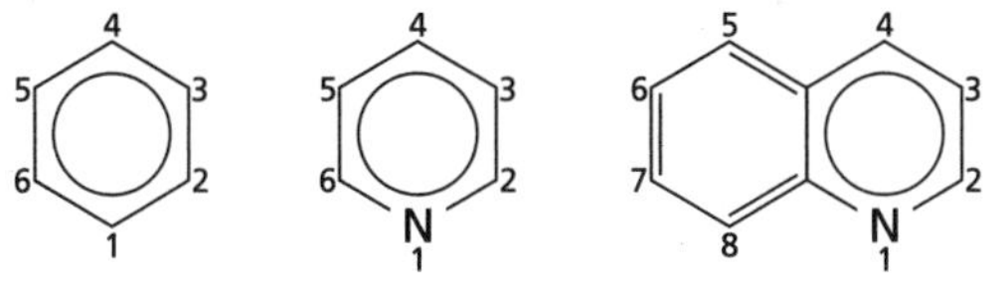

benzene   pyridine   quinoline

FIGURE 20.2 ■ Chemical structures of benzene, pyridine, and quinoline.   Hydrogen atoms bonded to the ring structures are not shown.

TABLE 20.1 ■ Common and chemical names of the imidazolinone herbicides.

COMMON NAME	CHEMICAL NAME
Imazamethabenz	(±)-2-[4,5-dihydro-4-methyl-4-(1-methylethyl)-5-oxo-1H-imidazol-2-yl]-4(and 5)-methylbenzoic acid(3:2)
Imazapyr	(±)2-[4,5-dihydro-4-methyl-4-(1-methylethyl)-5-oxo-1H-imidazol-2-yl]-3-pyridinecarboxylic acid
Imazaquin	2-[4,5-dihydro-4-methyl-4-(1-methylethyl)-5-oxo-1H-imidazol-2-yl]-3-quinolinecarboxylic acid
Imazethapyr	2-[4,5-dihydro-4-methyl-4-(1-methylethyl)-5-oxo-1H-imidazol-2-yl]-5-ethyl-3-pyridinecarboxylic acid

Selectivity

The plant selectivity of the imidazolinone herbicides derives from their rapid degradation to nontoxic metabolites in tolerant plants and the inability of susceptible species to do the same.

Plant Tolerance

Imazaquin is reported to have a half-life of 3, 12, and 30 days in tolerant soybeans, moderately tolerant velvetleaf,

and susceptible cocklebur, respectively. Increased tolerance of velvetleaf at the four-leaf stage, as compared to the cotyledonary stage, is attributed to a 72% decrease in imazaquin absorption and a 62% decrease in its half-life.

Susceptible grass species such as wild oat quickly metabolized the methyl ester of imazamethabenz to its phytotoxic acid form, while tolerant grass species such as corn and wheat did so much more slowly.

Plant tolerance to imidazolinone herbicides is, in some cases, due to the presence of an altered form of the enzyme ALS. In imidazolinone-susceptible corn, the ALS enzyme present was 3000 times more sensitive to inhibition by imidazolinone herbicides than was the ALS enzyme in tolerant corn.

Common Characteristics

1. The imidazolinone herbicides are translocated into, and kill, underground storage organs of susceptible perennial plants.

2. Susceptible plants stop growth soon after herbicide absorption and are no longer competitive with crop plants. The first injury symptom is chlorosis in the youngest leaves on the plant, with necrosis spreading from this point. These symptoms may not be apparent for 2 weeks or more after treatment, and complete kill may not occur for several more weeks.

3. Applied postemergence, it is essential to add a nonionic surfactant or crop oil concentrate, and, if needed, an anti-foaming agent and/or drift retardant to spray solutions containing the imidazolinone herbicides.

4. The imidazolinone herbicides may be mixed with liquid nitrogen fertilizers, with the fertilizer serving as the carrier for the herbicide.

5. These herbicides are adsorbed to soil colloids and not leached in soils.

6. The soil persistence of these herbicides vary from a few months to 2 years or more, depending on the herbicide and the application rate. Soil persistence presents a real hazard to rotational crops, and plans for rotating suitable crops must be made prior to the application of these herbicides. To avoid crop injury, crop-rotation restrictions noted on each product label must be followed.

7. Crops tolerant to one or more of the imidazolinone herbicides include small grains (wheat, barley); choen; edible beans; peanuts; safflower; soybean; sunflower; and certain warm-season turfgrasses.

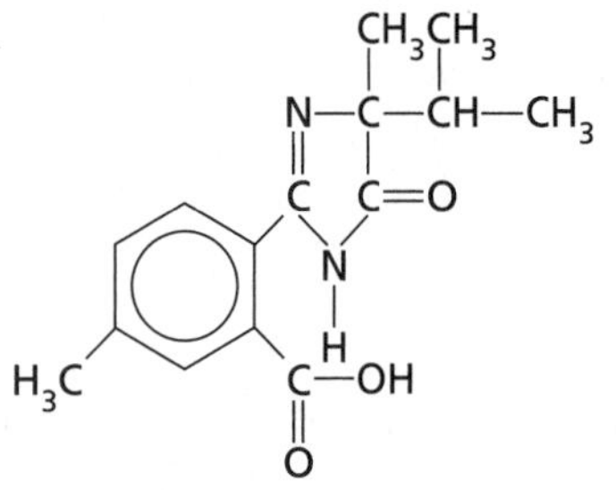

imazamethabenz[1]
(Assert)

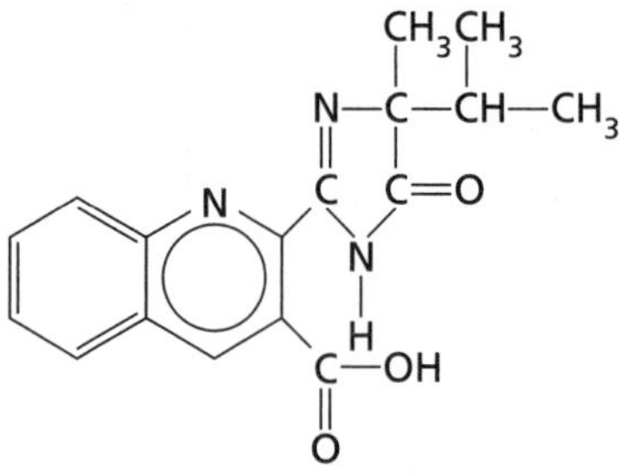

imazaquin
(Image, Scepter)

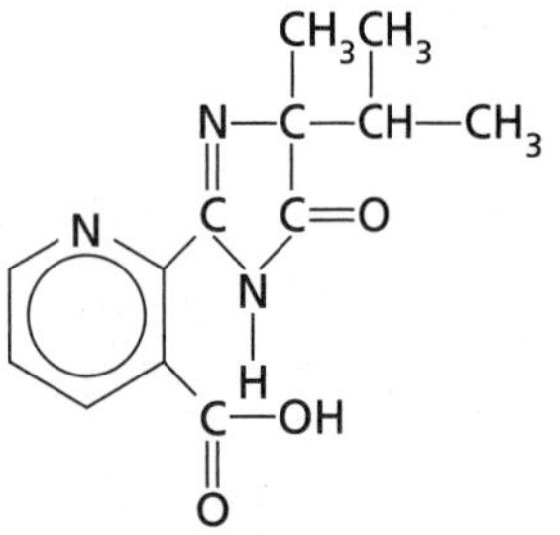

imazapyr
(Arsenal, Chopper)

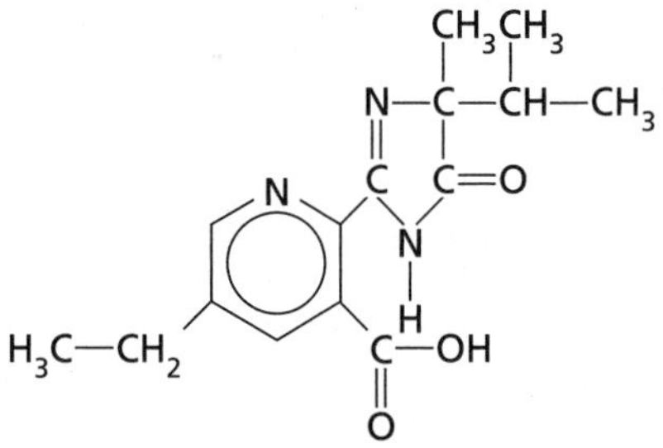

imazethapyr[2]
(Pursuit)

FIGURE 20.3 ■ Chemical structures of the acid forms of the imidazolinone herbicides.

8. Spray drift of the imidazolinone herbicides poses a real hazard to nearby susceptible crops and to the exposed soil of nearby croplands.

9. These herbicides may be used alone, as a tank mix with suitable herbicides, or applied sequentially with other suitable herbicides.

■ INFORMATION OF PRACTICAL IMPORTANCE

The following information is grouped by herbicide.

Imazamethabenz

1. Imazamethabenz is the active ingredient in an emulsifiable-concentrate formulation (Assert) containing 2.5 lb ae/gal. It is formulated as the methyl ester.

2. Assert is used postemergence in wheat (including durum), barley, and sunflower to selectively control wild oats, roughstalk bluegrass, interrupted windgrass, and certain annual broadleaved weeds.

3. Assert is applied to wheat and barley from the two-leaf stage to just before jointing (development of first internode). Do not apply Assert when freezing temperatures are forecast.

4. To control wild oats, Assert is applied to wild oats in the one- to four-leaf stage.

5. A nonionic surfactant must be added to the spray solution at 1 qt/100 gal water. Thorough, uniform wetting of emerged weeds is necessary to maximize weed control.

6. Assert is applied at a rate of 0.31–0.47 lb ae/A (1–1.5 pt product/A), depending on the weed species present. Do not make more than one application of Assert per growing season.

1. Structure shown is the *meta*-isomer; the *para*-isomer places the methyl group at the 4-position of the benzoic acid ring.

2. Imazethapyr is a racemic mixture of its *dextro* (+) and *levo* (−) enantiomorphs (optical isomers).

7. Assert may be applied in a liquid nitrogen fertilizer (28% N), with the fertilizer serving as the herbicide carrier. Assert alone or in combination with 2,4-D ester, MCPA ester, or bromoxynil + MCPA ester (Bronate) may be applied in the liquid fertilizer.

8. Assert may be tank-mixed with broadleaf herbicides to control a broader spectrum of weeds. However, do not tank-mix Assert with 2,4-D ester unless crop is fully tillered. Do not tank-mix Assert with dicamba (Banvel) or any product containing dicamba, or with MCPA amine or 2,4-D amine formulations.

9. Assert may be tank-mixed with the wild oat herbicide Avenge for improved control of wild oats. Assert should be tank-mixed with Avenge to apply 0.75 pt Assert and 2 pt Avenge/A.

10. The following rotational crops may be planted the crop season following the application of Assert: wheat, barley, corn, edible beans, soybean, safflower, and sunflower. Do not plant sugar beets for at least 20 months following Assert application. Do not plant other rotational crops, except those listed on product label, for 15 months after Assert application.

Imazapyr

11. Imazapyr is the active ingredient in an aqueous solution formulation (Arsenal, Chopper) containing 2.0 lb ae/gal. It is formulated as the isopropylamine salt. The formulation also contains a surfactant, and it is not necessary to add another adjuvant.

Arsenal

12. Arsenal is used for nonselective control of annual and perennial grass and broadleaf weeds in noncrop areas.

13. Arsenal is most effective applied postemergence to the weed foliage, particularly in the control of perennials. However, the herbicide does provide residual preemergence control of newly germinating weeds following a postemergence application.

14. Arsenal is translocated into, and kills, underground reproductive parts of susceptible perennial plants.

15. Arsenal is applied at 0.75–1.5 lb ai/A (3–6 pt product/A), depending on the weed species involved.

16. Arsenal can be used under asphalt, pond liners, and paved areas in industrial sites, or in areas where the roots of desired plants will not encroach. Apply Arsenal at a broadcast rate of 6 pt/A.

Chopper

17. Chopper is used for nonselective postemergence control of broadleaved weeds and woody brush in noncrop areas. Mix 8–12 fl oz of Chopper in 1 gal water, diesel oil, or penetrating oil and apply as a cut stump, tree injection, frill or girdle treatment, or as a low-volume basal treatment.

18. Do not use Chopper on lawns, driveways, tennis courts, or similar areas, especially where the roots of desired plants may encroach into the treated area. Do not use on ditch banks or in ditches used to transport irrigation water.

Imazaquin

19. Imazaquin is the active ingredient in a water-dispersible granular formulation (Scepter 70DG) containing 70% ai w/w. Scepter 70DG is packaged in small, water-soluble packets (ECO-PAK), which contain an amount that will treat 5 A of soybeans. Each packet is enclosed in a two-compartment plastic package to protect it from moisture prior to use.

   Imazaquin is also the active ingredient in a water-soluble liquid concentrate formulation (Image) that contains 1.5 lb ae/gal. It is formulated as the ammonium salt.

Scepter

20. Scepter is used only in soybeans, and it can be applied preplant-incorporated, preemergence, or early postemergence to control annual grass and broadleaf weeds. Crop-rotation restrictions must be followed.

21. When Scepter is soil-applied, susceptible weeds emerge; growth stops; and the weeds either die or are not competitive with the crop.

22. A timely cultivation about 7 days after application may aid in the control of certain weeds or may improve general weed control if adequate moisture does not occur to move the herbicide into the seed germination zone.

23. Scepter is applied at 2 oz ae/A (2.8 oz product/A) for preplant-incorporated or preemergence treatments. The postemergence treatment rate is 1–2 oz ae/A (1.4–2.8 oz product/A), depending on the geographic region and the height and species of weeds involved.

24. Applied postemergence, the soybean plants should be small so their foliage does not block the herbicide

spray from wetting the foliage of emerged weeds and reaching the exposed soil surface.

25. Add a nonionic surfactant to postemergence applications of Scepter at 1 qt/100 gal spray mixture. A suitable crop oil concentrate may be used in place of the surfactant.

26. Scepter may be tank-mixed with Prowl, Treflan, Lasso, or Dual.

27. Scepter can be applied as sequential treatments. Do not exceed 5.8 oz product/A per year.

Image

28. Image is used in established warm-season turfgrass for postemergence control of grass and broadleaf weeds, garlic, wild onion, purple nutsedge, and yellow nutsedge. Warm-season turfgrasses include bermudagrass, centipedegrass, St Augustinegrass, and zoysiagrass.

29. Image is applied postemergence at 0.5–1.0 fl oz/1000 ft^2 (14.8–29.6 mL/1000 ft^2). Add a nonionic surfactant to the spray solution at 1 qt/100 gal spray mixture.

30. Image can be tank-mixed with methanearsonate herbicides such as MSMA to control a broader weed spectrum. Image may also be mixed with other appropriate herbicides for preemergence or postemergence use in warm-season turfgrasses. When mixing Image with another herbicide, add the second herbicide to the carrier first and mix thoroughly and then add Image.

31. Do not apply Image to golf course greens, tees, aprons, or similar areas that are closely mowed.

32. Scepter O.T. is a premix formulation that contains 0.5 lb imazaquin and 2 lb acifluorfen/gal. It is used only in soybeans as a postemergence treatment.

33. Squadron is a premix formulation containing 0.33 lb imazaquin and 2 lb pendimethalin/gal for use preplant-incorporated or preemergence in soybeans.

34. Tri-Scept is a premix formulation that contains 0.43 lb imazaquin and 2.57 lb trifluralin/gal for use preplant-incorporated in soybeans.

Imazethapyr

35. Imazethapyr is the active ingredient in a water-soluble liquid concentrate formulation (Pursuit) containing 2.0 lb ae/gal. It is formulated as the ammonium salt.

Pursuit

36. Pursuit is used preplant, preplant-incorporated, at ground-cracking (peanuts), preemergence, or early postemergence in peanuts and soybeans to control annual grass and broadleaf weeds.

37. Pursuit is applied early postemergence when weeds are actively growing and less than 3 in. tall. Time the application on the stage of weed growth, not on stage of crop growth.

38. Always add a nonionic surfactant to postemergence applications of Pursuit at the rate of 1 qt/100 gal spray solution.

39. Pursuit is soil-applied or applied early postemergence in peanuts at 4 oz ae/A (0.25 pt product/A). It can be applied as a sequential treatment, with 2 fl oz product/A soil-applied followed by 2 fl oz product/A postemergence.

40. Pursuit can be tank-mixed with other soil-applied or postemergence peanut herbicides, or as a sequential treatment to such herbicides.

41. Pursuit is applied in soybeans at 4 oz ae/A (0.25 pt product/A). Make only one application of Pursuit in soybeans per year.

42. Pursuit can be soil-applied or applied early postemergence in tank mix with most soybean herbicides.

Pursuit Plus

43. Pursuit Plus is a premix, emulsifiable-concentrate formulation containing 0.2 lb imazethapyr and 2.7 lb pendimethalin/gal for use preplant-incorporated or preemergence in soybeans.

Passport

44. Passport is a premix formulation that contains 0.2 lb imazethapyr and 2.4 lb trifluralin/gal for use preplant-incorporated in soybeans.

21 METHANEARSONATE HERBICIDES

■ INTRODUCTION

Members of the methanearsonate herbicide family are salts of methylarsonic acid, varying in chemical structure only in their respective cations. Their common chemical nucleus is methylarsonic acid. The chemical structure of methylarsonic acid is:

$$H_3C - As \begin{matrix} O \\ \| \\ \end{matrix} \begin{matrix} OH \\ \\ OH \end{matrix}$$

methylarsonic acid

The herbicide cacodylic acid, or dimethylarsinic acid, is generally grouped with the methanearsonate herbicides and is thus included here. The common and chemical names of cacodylic acid and the methanearsonate herbicides are given in Table 21.1, on the following page. Their chemical structures are shown in Figure 21.1.

■ CHARACTERISTICS OF METHANEARSONATE HERBICIDES

Principal Use

Cacodylic acid is principally used as a nonselective post-emergence herbicide in noncrop areas. Methanearsonate herbicides are mostly used for selective postemergence control of weeds—especially grasses—in cotton, citrus (bearing and nonbearing), turf (lawn and ornamental), nonbearing vineyards and deciduous fruit and nut orchards, and noncrop areas.

Mode of Action

The apparent mode of action of cacodylic acid and the methanearsonate herbicides is contact-type kill. The exact mechanisms involved in their mode of action are not known.

TABLE 21.1 ■ Common and chemical names of cacodylic acid and the methanearsonate herbicides.

COMMON NAME	CHEMICAL NAME
Cacodylic acid	dimethylarsinic acid
CAMA	calcium acid methanearsonate
DSMA	disodium methanearsonate
MSMA	monosodium methanearsonate

cacodylic acid
(Phytar 560)

MSMA
(Ansar 6.6, Bueno 6, others)

CAMA
(Super Dal–E–Rad–Calar)

DSMA
(Drexel DSMA, others)

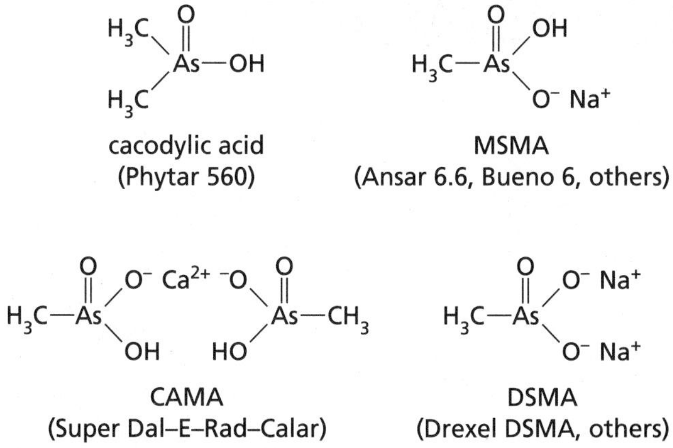

FIGURE 21.1 ■ Chemical structures of cacodylic acid and the methanearsonate herbicides.

Translocation

Cacodylic acid is translocated in the transpiration stream, but this is of little practical significance as it is foliar-applied.

The methanearsonate herbicides are translocated in the transpiration and photosynthate streams. As they are foliar-applied, transport in the photosynthate stream is of greater importance in moving phytotoxic amounts of the herbicides into underground reproductive propagules such as rhizomes and tubers.

Selectivity

Factors responsible for the selective action of these herbicides are not known.

Common Characteristics

1. The methanearsonate herbicides and cacodylic acid are pentavalent arsenicals (As^{5+}). They are of low toxicity to man and animals—about the same toxicity, or less, as aspirin.

2. The methanearsonate herbicides and cacodylic acid are highly soluble in water.

3. These herbicides are used only postemergence to the weeds to be controlled. They have no soil activity, as they are quickly adsorbed to soil colloids.

4. A nonionic surfactant added to the spray solution enhances the phytotoxicity of the methanearsonate herbicides and cacodylic acid. With some products, a surfactant is included in the formulation and an additional surfactant is not needed.

■ INFORMATION OF PRACTICAL IMPORTANCE

Cacodylic Acid

1. Cacodylic acid is the active ingredient in a water-soluble formulation with surfactant (Phytar 560); it contains 2.48 lb ai/gal.

2. Phytar 560 is a nonselective postemergence herbicide used in nonbearing citrus, lawn renovation, and non-crop areas such as rights-of-way; along sidewalks, driveways, and fence rows; and around buildings and ornamentals. Do not allow spray solution to contact desired plants.

3. In noncrop areas, Phytar 560 is applied to thoroughly wet the foliage of unwanted vegetation, using a mixture of 1–2 gal product/100 gal water.

4. Used in nonbearing citrus orchards, Phytar 560 is applied at 4.2–5 lb ai/A (1.5–2 gal/A). When used near young citrus trees or in nursery plantings, use a shield to avoid contacting young trees with spray solution.

5. Phytar 560 only provides top-kill of treated vegetation, and treatments must be repeated for season-long control of perennial weeds or weeds that emerge after application.

6. For lawn renovation, apply 10 fl oz of Phytar 560 in 4 gal water/1000 ft^2.

CAMA

7. CAMA is the active ingredient in a water-soluble formulation (Super Dal-E-Rad-Calar), which contains 1.0 lb ai/gal.

8. Super Dal-E-Rad-Calar is used in bermudagrass turf for selective postemergence control of weeds, espe-

cially crabgrass and dallisgrass. Do not use this product on carpetgrass, centipedegrass, or St. Augustinegrass.

9. Super Dal-E-Rad-Calar is applied at 2.5 lb ai/A (2.5 gal product/A).

DSMA

10. DSMA is the active ingredient in several formulations:

    ▪ As a water-soluble powder formulation without surfactant (Ansar 8100, Drexel DSMA 81 P) containing 81% DSMA w/w; add a suitable nonionic surfactant to this spray solution prior to use.

    ▪ As a liquid water-soluble formulation with surfactant (ISK-Biotech DSMA Liquid, Drexel DSMA Liquid) containing 3.6 lb ai/gal; do not add additional surfactant to this spray solution.

    ▪ As a slurry formulation with surfactant (Drexel DSMA Slurry) containing 7.2 lb ai/gal; do not add additional surfactant to this spray solution.

11. DSMA is a postemergence, foliar-translocated herbicide used to selectively control grass and broadleaf weeds and also to control or suppress purple and yellow nutsedges.

12. DSMA formulations are used in cotton, bearing and nonbearing citrus (except in Florida), grass lawns and ornamental turf, and noncrop areas.

13. To avoid crop injury, DSMA is applied only as a basal postdirected spray when cotton is at least 3 in. tall up to first bloom. Do not apply after first bloom. If necessary, a second application may be made 1–3 weeks after the first application, but it must be made prior to first bloom.

14. When used in cotton, DSMA is applied broadcast at a rate of 3.6 lb DSMA (using the desired formulation) in 40 gal water/A. If applied as a band treatment in cotton planted in rows spaced 40 in. apart, apply 1 gal/A of this mixture per inch of treated band. Thus, if a 4-in.-wide band is treated on each side of the cotton row, 8 gal/A of the mixture should be applied.

15. DSMA will often cause a slight discoloration or burning of the cotton foliage present at time of application, but the plants will continue to develop normally.

16. Using the desired formulation in citrus orchards, DSMA is applied at 3.6–7.2 lb ai/A as a directed spray around the base of the trees and in between trees. It is applied to wet the foliage of the weeds to just short of run-off. Up to two repeat applications may be made per year. Avoid contact of spray solution with foliage, stems, and green bark of the trees; use shields if necessary for young trees or in nursery plantings.

17. DSMA is applied postemergence in turfgrass at 3.6 lb/A (using the desired formulation) to control a variety of grass and broadleaf weeds, including bahiagrass, dallisgrass, johnsongrass, sandbur, chickweed, cocklebur, puncturevine, ragweed, and woodsorrel. Bluegrass, bermudagrass, and zoysiagrass are very tolerant of DSMA, although temporary discoloration may occur. Do not use on St. Augustinegrass or centipedegrass.

MSMA

18. MSMA is the active ingredient in several formulations:

    ▪ As a liquid water-soluble formulation without surfactant (Drexel MSMA 6.6, Ansar 6.6) containing 6.6 lb ai/gal. A suitable nonionic surfactant must be added to this spray solution prior to use.

    ▪ As a water-soluble liquid formulation with surfactant (Bueno 6, Daconate 6, Drexel MSMA 6 Plus) containing 6 lb ai/gal and Drexar 530 containing 4 lb ai/gal. Do not add additional surfactant to this spray solution.

19. MSMA is used in cotton, bearing and nonbearing citrus, nonbearing tree fruits and nuts, turfgrass, and noncrop areas, depending on the specific product selected.

20. It is a postemergence, foliar-translocated herbicide used to selectively control grass and broadleaf weeds and to control or suppress purple and yellow nutsedges.

21. Using the desired formulation, MSMA is applied in cotton at a dosage of 2 lb ai/A as follows: (a) preplant application to control emerged weeds, (b) one basal postdirected application to cotton treated previously with one application of DSMA, or (c) two postdirected applications to cotton not treated with DSMA. As a postdirected application, apply MSMA only when the cotton plants are at least 3 in. high up to first bloom. Do not apply after first bloom.

22. MSMA may be applied as an over-the-top broadcast spray to actively growing cotton as a salvage operation when the cotton is 3–6 in. tall or up to early first square stage, whichever occurs first. Do not apply MSMA to cotton beyond this stage, or injury will result. MSMA is applied at 0.75–1.0 lb ai in 40 gal water/A.

23. Cotton is more tolerant of MSMA than of DSMA. Do not apply DSMA over-the-top of cotton, or severe injury or death of crop plants will occur.

24. Used in turfgrass, MSMA is applied postemergence to actively growing turfgrass at 2–3 lb ai/A (using the desired formulation). The following turfgrasses are very tolerant of MSMA: bermudagrass, bluegrass, and zoysiagrass. Do not apply MSMA to St. Augustinegrass or centipedegrass.

25. Using the desired formulation in noncrop areas, MSMA is applied at a rate of 4.5–9 lb ai/A.

22 PHENYLCARBAMATE HERBICIDES

■ INTRODUCTION

The phenylcarbamate herbicides are esters of carbamic acid. The chemical structures of carbamic acid and the general structure of its esters are:

carbamic acid

ester of carbamic acid, general structure

The R_1 in the general ester structure above is an imino hydrogen (i.e., a hydrogen bonded to nitrogen), R_2 is a phenyl group, and R_3 is an alkyl group or a combination of alkyl and aryl groups.[1]

The common and chemical names of the phenylcarbamate herbicides are given in Table 22.1 on the following page, and their chemical structures are shown in Figure 22.1.

---

1. An alkyl group is an open-chain saturated hydrocarbon minus one hydrogen atom (e.g., methyl, ethyl, isopropyl). An aryl group is a closed-chain aromatic hydrocarbon minus one hydrogen (e.g., phenyl, a six-carbon ring structure).

■ CHARACTERISTICS OF THE PHENYLCARBAMATE HERBICIDES

Principal Use

The phenylcarbamate herbicides are principally used in the selective postemergence control of annual broadleaf weeds in sugar beets. They usually do not control grasses, but, applied together in a premix, they do control green foxtail and yellow foxtail.

Mode of Action

The phenylcarbamate herbicides principally inhibit photosynthesis by blocking electron transport through photosystem II (PS II), similar to the triazine and urea herbicides (Chapters 27 and 29, respectively).

Translocation

The phenylcarbamate herbicides undergo local, short-distance movement within the treated leaf.

TABLE 22.1 ■ Common and chemical names of phenylcarbamate herbicides.

COMMON NAME	CHEMICAL NAME
Desmedipham	ethyl[3-[[(phenylamino)carbonyl]oxy]phenyl]carbamate
Phenmedipham	3-[(methoxycarbonyl)amino]phenyl (3-methylphenyl)carbamate

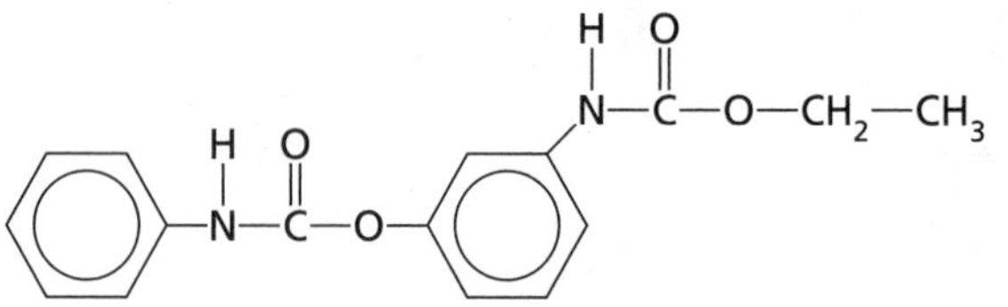

FIGURE 22.1 ■ Chemical structures of the phenylcarbamate herbicides.

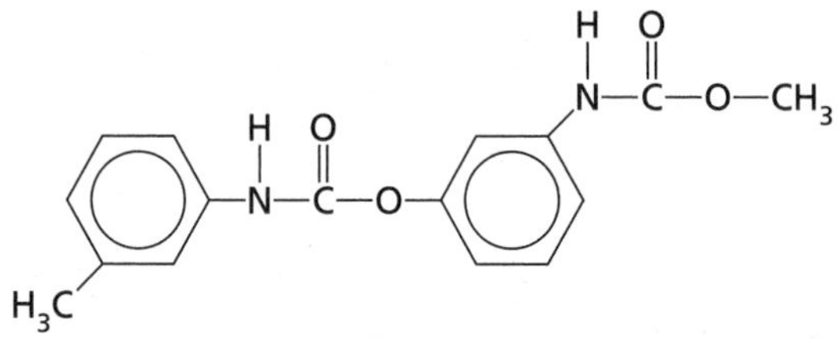

Selectivity

The selective action of the phenylcarbamate herbicides occurs because tolerant plants rapidly degrade the herbicides metabolically while susceptible plants fail to do so.

Common Characteristics

1. Desmedipham and phenmedipham are only applied postemergence to the weeds. They can also be applied as an over-the-top spray to the crops.

2. Desmedipham and phenmedipham control only seedling broadleaf weeds. They do not control grasses or established plants.

3. These herbicides are not leached deeply in soils; they tend to stay within the upper 2-in. (5-cm) soil layer. Their soil residual half-life is about 1 month.

■ INFORMATION OF PRACTICAL IMPORTANCE

Desmedipham

1. Desmedipham is the active ingredient in an emulsifiable-concentrate formulation (Betanex) that contains 1.3 lb ai/gal.

2. Betanex is used only in sugar beets to selectively control annual broadleaf weeds. Sugar beets are tolerant to Betanex when past two true leaves; they may be severely injured if treated before this stage. For best results, apply Betanex when the weeds are actively growing and are small, at the two-true-leaf stage. Weeds controlled by this herbicide include black nightshade, coast fiddleneck, common chickweed, common lambsquarters, common ragweed, common purslane, lanceleaf groundcherry, London rocket, pigweed species, shepherd's-purse, wild buckwheat, and wild mustard.

3. Betanex is applied at 0.73–1.2 lb ai/A (4.5–7.5 pt product/A); use the 7.5-pt rate only on well-established sugar beets. For later emerging weeds, a second application of Betanex can be applied at 4.5–6 pt product/A after the beets have at least four leaves.

Phenmedipham

4. Phenmedipham is the active ingredient in an emulsifiable-concentrate formulation (Spin-Aid) that contains 1.3 lb ai/gal.

5. Spin-Aid is used for selective weed control only in red (table) beets and spinach. Do not apply to spinach when temperatures exceed 75°F (24°C), or crop injury may result. Red beets and spinach are tolerant of Spin-Aid when they are past the six-leaf stage; they may be severely injured if treated before this stage. Best control is obtained when the weeds are actively growing and are in the two-true-leaf stage. Spin-Aid is applied at a rate of 0.5–1.0 lb ai/A (3–6 pt product/A), on a broadcast basis. Use the high rate only on well-established beets and spinach.

6. Weeds controlled by Spin-Aid include the same species as noted for Betanex.

Combined Formulation

7. Desmedipham and phenmedipham are the active ingredients in an emulsifiable-concentrate premix formulation (Betamix) that contains a total of 1.3 lb ai/gal.

8. Betamix is used for selective weed control in sugar beets; apply when the beets are past the two-true-leaf stage and the weeds are in the two-true-leaf stage. Betamix is applied at 0.73–1.2 lb ai/A (4.5–7.5 pt product/A). A second application of 4.5–6 pt of Betamix can be applied to control later emerging weeds after the sugar beets have at least four leaves. Use the high rate only on well-established sugar beets.

9. In addition to the weeds controlled by Betanex, Betamix controls kochia and two annual grasses (green foxtail and yellow foxtail).

PHENYL PYRIDAZINONE HERBICIDES

■ INTRODUCTION

Members of the phenyl pyridazinone herbicide family have as their common nucleus a phenyl ring bonded to the 2-position of a pyridazinone ring, a six-membered ring with two adjacent nitrogen atoms at the 1- and 2-positions of the ring, a carbon atom at each of the other four positions, and an oxygen bonded to the carbon at the 3-position of the ring. The chemical structure of phenyl pyridazinone is:

phenyl pyridazinone

The common and chemical names of the two members of the phenyl pyridazinone herbicide family are given in Table 23.1, and their chemical structures are shown in Figure 23.1. Although grouped in the same herbicide family, the characteristic activity of chloridazon and norflurazon differs.

■ CHARACTERISTICS OF THE PHENYL PYRIDAZINONE HERBICIDES

Principal Use

Chloridazon is principally used for selective preemergence control of broadleaf weeds in sugar beets and red beets. Chloridazon does not control grasses.

Norflurazon is mostly used for preemergence control of annual grass and broadleaf weeds in cotton; soybeans; tree fruits (citrus, deciduous); nuts; small fruits (caneberries, blueberries); and cranberry.

Mode of Action

Chloridazon inhibits photosynthesis by blocking electron transport through photosystem II (PS II).

Norflurazon inhibits the biosynthesis of carotenoids; it acts as a plant pigment inhibitor. Without carotenoid pigments to filter light, photodegradation of chlorophyll occurs. Susceptible plants become chlorotic (bleached).

TABLE 23.1 ■ Chemical and common names of the phenyl pyridazinone herbicides.

COMMON NAME	CHEMICAL NAME
Chloridazon*	5-amino-4-chloro-2-phenyl-3(2H)-pyridazinone
Norflurazon	4-chloro-5-(methylamino)-2-(3-(trifluoromethyl)phenyl)-3(2H)-pyridazinone

*Formerly pyrazon.

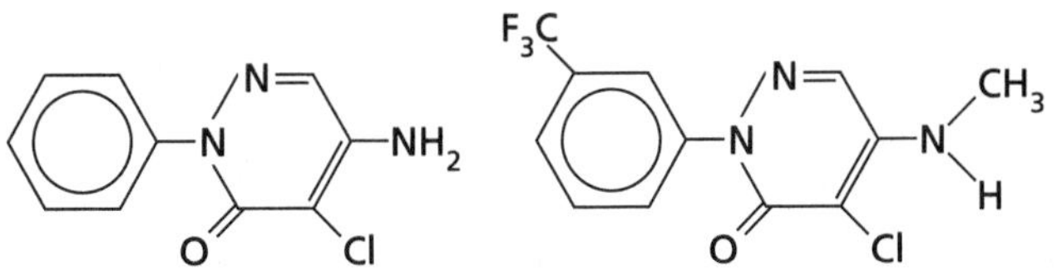

chloridazon
(Pyramin)

norflurazon
(Zorial, Solicam)

FIGURE 23.1 ■ Chemical structures of chloridazon and norflurazon.

Translocation

Chloridazon is rapidly root-absorbed and translocated to the leaves in the transpiration stream. Although foliar-absorbed, chloridazon is not translocated from the leaves to other plant parts.

Norflurazon is readily root-absorbed and translocated in the transpiration stream to the leaves.

Selectivity

Chloridazon forms a conjugate (*N*-glucosyl chloridazon) in the leaves of sugar beets and table beets (plants tolerant of chloridazon). In plants susceptible to chloridazon, this conjugate is not formed. The formation of this conjugate in tolerant plants but not in susceptible plants is thought to be the basis for chloridazon selectivity.

Mechanisms responsible for the selectivity of norflurazon are not known.

■ INFORMATION OF PRACTICAL IMPORTANCE

Chloridazon (formerly pyrazon)

1. Chloridazon is the active ingredient in a dry-flowable formulation (Pyramin DF) that contains 67.7% ai w/w.

2. Pyramin is used only in sugar beets and red table beets for preemergence control of seedling broadleaf weeds. Weeds controlled by this herbicide include common lambsquarters, common purslane, mustards, nettleleaf goosefoot, nightshades, pigweeds, ragweeds, shepherd's-purse, and velvetleaf.

3. Pyramin can be applied preplant-incorporated, preemergence, early postemergence, or sequentially to the beets. Apply postemergence after the beets have two fully expanded true leaves (do not count cotyledonary leaves, which are the first two leaves to appear) and before any weeds have emerged.

4. Applied to the soil or early postemergence, Pyramin is applied at a broadcast rate of 3.1–3.7 lb ai/A (4.6–5.4 lb product/A), depending on soil organic matter content and geographical region in which applied. If an over-the-row band treatment is used, the application rate is reduced proportionately.

5. Postemergence applications of Pyramin can follow preplant-incorporated or preemergence applications provided the total amount applied does not exceed 11.25 lb product/A (7.6 lb ai/A) per season.

6. To avoid crop injury, do not add a surfactant or other spray adjuvant to spray solutions containing Pyramin.

7. Pyramin can be tank-mixed with Antor 4ES (diethatyl) or Nortron EC (ethofumesate) to control a broader weed spectrum, including annual grasses.

8. In case of crop failure, do not plant other crops in the Pyramin-treated soil during the same crop season. Beets can be reseeded in the treated soil; do not apply a second preemergence treatment.

9. Do not apply Pyramin to soils classified as sands or sandy loams because of the potential for crop injury and leaching. Do not apply to peat or muck soils, as adsorption to the organic colloids precludes adequate weed control.

10. The soil persistence of Pyramin is 4–8 weeks, depending on soil moisture and temperature. Rapid microbial breakdown is the principal means by which Pyramin is degraded in soils.

Norflurazon

11. Norflurazon is the active ingredient in dry-flowable formulations (Zorial Rapid 80, Solicam) containing 80% ai w/w and a granular formulation (Evital 5G) containing 5% ai w/w.

12. Zorial is used preplant-incorporated or preemergence in cotton and soybean to control grass and broadleaf weeds. It can be applied as a split treatment, with one-half the rate applied preplant-incorporated and the other half applied preemergence. Zorial is applied prior to crop and weed emergence.

13. Zorial is applied at 1–2 lb ai/A (1.25–2.5 lb product/A).

14. Solicam is used to control weeds in tree fruits (citrus, deciduous); nut orchards; and small fruits (caneberries and blueberries).

15. Solicam is applied at 2–4 lb ai/A (2.5–5 lb product/A). For 1 year after its application, only cotton, peanuts, or soybeans can be planted in soil treated with Solicam.

16. Evital is used for weed control in cranberries only. It is applied at 4–8 lb ai/A (80–160 lb product/A).

17. Norflurazon is adsorbed to soil colloids and does not leach appreciably. The average half-life of norflurazon in soils is 45–180 days in the Mississippi Delta and Southeast regions of the United States.

18. Norflurazon dissipates from soils through volatilization and photodecomposition from the soil surface and also from microbial degradation.

24 PHTHALIC ACID HERBICIDES

■ INTRODUCTION

Phthalic acids may also be called *benzene dicarboxylic acids.* They exist as *ortho, meta,* and *para* isomers of benzene dicarboxylic acid; these isomers are named *phthalic acid, isophthalic acid,* and *terephthalic* acid, respectively. When the hydroxyl (OH) group of one of the carboxyl groups is replaced by an amino (NH_2) group, the resulting compound is called *phthalamic acid.* The structures of these four compounds are shown in Figure 24.1 on the following page.

Three herbicides make up this group, and their common and scientific names are as follows:

DCPA   dimethyl ester of tetrachloroterephthalic acid

Naptalam   *N*-1-naphthylphthalamic acid

Endothall   3,6-endoxohexahydrophthalic acid

DCPA is a *para* isomer of phthalic acid, also known as terephthalic acid, and it is used as its dimethyl ester. Naptalam is an *ortho* isomer of phthalamic acid and is used in its sodium salt form. Endothall is also an *ortho* isomer of its respective parent acid, and is used in its *disodium salt* form when used in turf and in its *dipotassium salt* form when used in sugar beets and for aquatic weed control. The chemical structures for these herbicides are shown in Figure 24.2 on page 217.

■ CHARACTERISTICS OF THE PHTHALIC ACID HERBICIDES

Principal Use

DCPA is principally used in the preemergence control of seedling grass and broadleaved weeds in vegetable crops, turf, and nursery ornamentals.

Naptalam is mostly used in the preemergence control of seedling broadleaved weeds in cucurbits (cantaloupe, cucumber, muskmelon, watermelon); peanuts; soybean; and woody nursery ornamentals.

Endothall is used in the preemergence and postemergence control of annual and perennial broadleaved weeds and certain grass species (e.g., green and yellow foxtails) in turfgrass and sugar beets. Endothall is also used as an aquatic herbicide and algicide and as a harvest aid in alfalfa, clover, cotton, and potatoes.

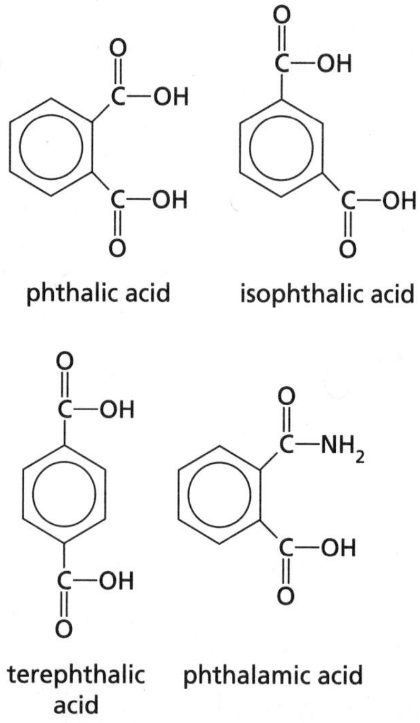

FIGURE 24.1 ■ Chemical structures of phthalic acid, iso-phthalic acid, terephthalic acid, and phthalamic acid.

Mode of Action

DCPA disrupts mitosis by disrupting phragmoplast microtubule organization and production. The process blocked most effectively is cell plate formation.

The exact mechanisms by which naptalam and endothall kill plants are not known, but both kill germinating seeds and interfere with root development. Naptalam blocks the action of indoleacetic acid (IAA) and induces unique anti-geotropic responses (e.g., roots grow upward rather than downward).

Endothall inhibits messenger RNA (mRNA) activity, thereby decreasing the rate of respiration and lipid metabolism, inhibiting protein synthesis, and interfering with normal cell division. Endothall also acts as a contact-type herbicide when used as an aquatic herbicide and algicide, with rapid penetration, desiccation, and browning of foliage.

Translocation

Following root absorption by grass and broadleaved plants, DCPA is not translocated upward in the xylem. However, DCPA is translocated upward in dicotyledous plants when absorbed by the hypocotyls; it is freely translocated in xylem and phloem conduits when absorbed by the hypocotyls of purslane and *Galinsoga*.

Naptalam has limited mobility in plants. Endothall is root-absorbed and translocated upward into the shoot *via* the transpiration stream. There is little or no translocation of endothall *via* the phloem following foliar absorption.

Selectivity

The exact mechanism contributing to the selective activity of these three herbicides is not known.

Common Characteristics

1. DCPA, naptalam, and endothall are applied preemergence to the weeds. They must be applied to the soil prior to weed seed germination; they do not control established plants.

2. DCPA and naptalam are not foliar-absorbed, and they may be safely applied directly over-the-top of established crop plants.

3. DCPA, naptalam, and endothall may be tank-mixed with other appropriate herbicides to control a broader spectrum of weeds.

Exceptional Characteristics

1. DCPA may be applied postemergence for control of creeping speedwell. *Note:* This is a marked variant from its usual use as a preemergence herbicide.

2. The salt forms of naptalam and endothall are water-soluble and are leached in soils. DCPA is insoluble in water and it is not leached in soils.

3. Applied preemergence, DCPA is absorbed by the roots and emerging shoots of both resistant and susceptible plants.

5. The absorption of DCPA *via* the developing shoot of grass seedlings is of greater importance than its absorption *via* the roots.

6. Naptalam is rapidly degraded in plants to form phthalic acid and α-naphthylamine.

■ INFORMATION OF PRACTICAL IMPORTANCE

DCPA

1. DCPA is the active ingredient in a wettable-powder formulation (Dacthal W-75; Dacthal W-75 Turf) and a

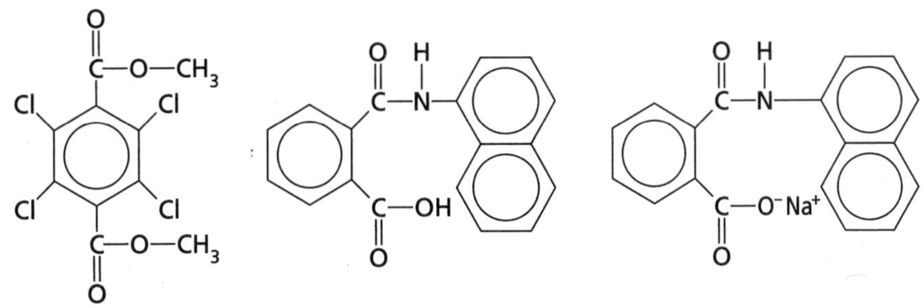

DCPA (Dacthal) | naptalam (Alanap) | sodium salt of naptalam

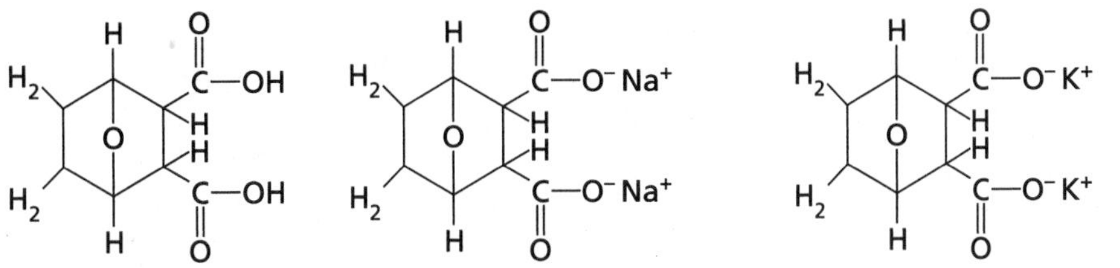

endothall (Endothal) | disodium salt of endothall | dipotassium salt of endothall

FIGURE 24.2 ■ Chemical structures of phthalic acid herbicides and their most common salts.

liquid-flowable formulation (Dacthal Flowable Herbicide Turf Care) that contains 6 lb ai/gal.

2. Dacthal Flowable Herbicide is the formulation to use for postemergence control of creeping speedwell. This product is a selective preemergence herbicide for professional lawn care only. It is compatible with liquid nitrogen fertilizers commonly used in the lawn care industry.
3. Dacthal W-75 is used on mineral soils for preemergence control of crabgrass and other annual grasses and certain broadleaf weeds in vegetables, strawberries, ornamentals, turf, and agronomic crops. It is applied at 4.5–10.5 lb ai/A (6–14 lb product/A).
4. Dacthal W-75 Turf is used on mineral soils for preemergence control of crabgrass and other annual grasses and certain broadleaf weeds in ornamental turf and nursery stock. It is applied at 10.5 lb ai/A (14 lb product/A).
5. Weeds that are particularly sensitive to DCPA include large and smooth crabgrasses, green and yellow foxtails, fall panicum, carpetweed, common chickweed, dodder, and purslane.
6. *Groundwater advisory:* Tetrachloroterephthalic acid, a breakdown product of Dacthal, is known to leach through the soil and has been found in groundwater that may be used as drinking water. Users are advised not to apply Dacthal to sand or loamy sand soils where the water table (groundwater) is close to the surface and where these soils are very permeable.

Naptalam

7. Naptalam is the active ingredient in a liquid formulation (Alanap-L) that contains 2.0 lb ai/gal.
8. Although Alanap is most effective in controlling broadleaf weeds, it also controls smooth and large crabgrass, foxtails, barnyardgrass, goosegrass, fall panicum, and red sprangletop.
9. Alanap-L should be applied immediately after seeding cucurbits. A second application may be made about a month after planting as an over-the-top postemergence treatment, just before the crop starts to vine, but before weeds have emerged. It may also be used as an over-the-top spray immediately after transplanting the cucurbits, but before weeds emerge.
10. Broadleaf weeds that are particularly sensitive to naptalam include common chickweed, cocklebur, lambsquarters, purslane, ragweed, redroot pigweed, and shepherd's-purse, among others.

11. Alanap is applied at 3–4 lb ai/A (6–8 qt product/A). On light sandy soils, the rate is reduced to 2–3 lb ai/A (4–6 qt product/A).

12. To be effective, DCPA and naptalam must be present in the germination zone of the weed seeds. If the herbicides are not moved from the soil surface into this zone within 5–7 days by rainfall or irrigation, they should be mechanically incorporated to a depth not deeper than 2 in. (5 cm).

Endothall

13. Endothall is the active ingredient in a water-soluble formulation (Endothal Turf Herbicide) that contains 1.46 lb ae/gal. It is formulated as the disodium salt of endothall. This formulation is recommended for selective weed control in lawns. It is applied at 0.37 lb ae/A (1.0 qt product/A). For best results, make two applications at 2- to 3-week intervals.

14. Note that the common name of endothall is spelled with two "l's," while the trade name is spelled with one "l."

15. Weeds controlled by Endothal Turf Herbicide include little barley, cheat and other annual bromes, rescuegrass, ryegrass, black medic, burclover, cranesbill and filaree, dichondra, puncturevine (goathead), henbit, knotweed, speedwell (veronica), and oxalis, among others.

16. Endothall is also available as the dipotassium salt for use in sugar beets (Herbicide 273, containing 3.0 lb ae/gal) and as an aquatic herbicide (Aquathol Granular, 10.1% ai w/w, and Aquathol K, containing 3.0 lb ae/gal).

17. Endothall is also available as the mono (*N,N*-dimethylalkylamine) salt (Hydrothol 191, containing 2 lb ae/gal) for use as an aquatic herbicide and algicide.

18. The product labels for Herbicide 273, Aquathol K, and Hydrothol 191 display the skull-and-crossbones symbol denoting their extreme toxicity.

19. Best results are obtained with endothall in sugar beets when applied during planting and soil-incorporated into the seed bed no deeper than 1.5 in.

20. In sugar beets, susceptible weeds include annual bluegrass, volunteer barley, barnyardgrass, bullgrass, cheatgrass, foxtails, burclover, kochia, henbit, purslane, ragweed, smartweed, wild buckwheat, wild carrot, volunteer sunflowers, shepherd's-purse, and Texas blueweed, among others.

21. Avoid spray drift; Herbicide 273 is known to be phytotoxic to a wide variety of plants, especially strawberries.

25 SULFONYLUREA HERBICIDES

■ INTRODUCTION

The structure of the chemical nucleus, *sulfonylurea*, common to the sulfonylurea herbicides is shown here, with R_1 and R_2 denoting various radical substitutions.

$$R_2 - \overset{\overset{O}{\|}}{\underset{\overset{\|}{O}}{S}} - \overset{\overset{H}{|}}{N} - \overset{\overset{O}{\|}}{C} - \overset{\overset{H}{|}}{N} - R_1$$

sulfonylurea

The sulfonylurea herbicides differ in one or more substituents bonded to the sulfonylurea nucleus. The common and chemical names of the sulfonylurea herbicides are given in Table 25.1 on the following page. Their chemical structures are shown in Figure 25.1 on page 221. As an aid to the reader, the chemical structures and names of the substituents are shown in Figure 25.2 on page 222.

■ CHARACTERISTICS OF SULFONYLUREA HERBICIDES

Principal Use

In general, the sulfonylurea herbicides are used to control broadleaved weeds and a few annual grasses in certain crops and in the nonselective control of annual and perennial grass and broadleaf weeds in noncrop and industrial areas. Weed control is by postemergence applications, but certain sulfonylurea herbicides must or may be applied preemergence.

Mode of Action

The sulfonylurea herbicides inhibit the enzyme acetolactate synthase (ALS), also called acetohydroxyacid synthase (AHAS). Since ALS is the catalyst in the first step of the biosynthesis of the essential amino acids valine, leucine,

TABLE 25.1 ■ Common and chemical names of the sulfonylurea herbicides.

COMMON NAME	CHEMICAL NAME
Bensulfuron	2-[[[[(4,6-dimethoxy-2-pyrimidinyl)amino]carbonyl]amino]sulfonyl]methyl]benzoic acid
Chlorimuron	2-[[[[(4-chloro-6-methoxy-2-pyrimidinyl)amino]carbonyl]amino]sulfonyl]benzoic acid
Chlorsulfuron	2-chloro-N-[[(4-methoxy-6-methyl-1,3,5-triazin-2-yl)amino]carbonyl]benzenesulfonamide
Metsulfuron	2-[[[[(4-methoxy-6-methyl-1,3,5-triazin-2-yl)amino]carbonyl]amino]sulfonyl]-2-benzoic acid
Nicosulfuron	2-[[[[(4,6,-dimethoxy-2-pyrimidinyl)amino]carbonyl]amino]sulfonyl]-N,N-dimethyl-3-pyridinecarboxamide
Primisulfuron	2-[[[[(4,6-bis(difluoromethoxy)-2-pyrimidinyl]amino]carbonyl]amino]sulfonyl]benzoic acid
Sulfometuron	2-[[[[(4,6-dimethyl-2-pyrimidinyl)amino]carbonyl]amino]sulfonyl]benzoic acid
Thifensulfuron	3-[[[[(4-methoxy-6-methyl-1,3,5-triazin-2-yl)amino]carbonyl]amino]sulfonyl]-2-thiophenecarboxylic acid
Triasulfuron	2-(2-chloroethoxy)-N-[[(4-methoxy-6-methyl-1,3,5-triazin-2-yl)amino]carbonyl]benzenesulfonamide
Tribenuron	2-[[[[(4-methoxy-6-methyl-1,3,5-triazine-2-yl)methylamino]carbonyl]amino]sulfonyl]benzoic acid

and isoleucine, the primary action of these herbicides thereby deprives the plant of these essential amino acids. The sulfonylurea and imidazolinone herbicides both have, as their primary mode of action, the inhibition of the enzyme ALS. However, the sulfonylureas are more active than the imidazolinones.

Translocation

The sulfonylureas are readily absorbed by both roots and shoots of plants and are translocated via xylem and phloem to meristematic regions of the plants.

Selectivity

Plant selectivity of the sulfonylurea herbicides occurs because the tolerant plants rapidly detoxify the herbicides metabolically to nontoxic metabolites; susceptible plants are unable to do this.

Chlorsulfuron-resistant grasses, such as annual bluegrass, barley, wheat, and wild oat, metabolized more than 90% of the chlorsulfuron in their leaves, while chlorsulfuron-sensitive broadleaf plants, such as cotton, mustard, soybean, and sugar beet, metabolized 20% or less of the chlorsulfuron in their leaves.

Weed resistance to sulfonylurea herbicides is associated with a genetically altered form of the enzyme ALS. A single nucleotide change in the gene coding for the ALS enzyme is responsible for the observed resistant plants; the altered ALS molecule is less sensitive to herbicidal inhibition.

Following repeated use of the sulfonylurea herbicides (or herbicides with a similar mode of action), resistant biotypes of certain broadleaf weeds have evolved. Examples are kochia, prickly lettuce, and Russian thistle. When such biotypes are present in areas about to be treated with sulfonylurea herbicides, the herbicides should be applied in tank mix with a suitable broadleaf herbicide that has a different mode of action (e.g., 2,4-D, MCPA, dicamba, bromoxynil, clopyralid, diuron, or metribuzin), or such herbicides should be applied as separate treatments.

Common Characteristics

1. Crop plant tolerance to sulfonylurea herbicides results from the rapid metabolic detoxification of the herbicides in these plants.

2. In general, the sulfonylureas are highly phytotoxic to most plants at dosages of less than 1 oz ai/A. In some cases, as little as 0.1 oz ai/A is the recommended dosage.

3. The sulfonylureas rapidly inhibit growth of susceptible weeds, but typical symptoms of dying (discoloration) may not be noticeable for 1–3 weeks after application. Death of leaf tissue may occur in some species, while others may remain green but stunted and noncompetitive.

4. In areas with a soil pH greater than 7.0 and with prolonged periods of low soil temperature and low annual rainfall, the sulfonylurea herbicides can remain active in the soil for 2–4 years or more and can thereby injure subsequent, nontolerant rotation crops. It is recommended that a bioassay of the soil be made of the upper 4-in. (10-cm) soil layer to determine the level of herbicide residue prior to planting susceptible rotation crops. In general, the sulfonylureas should not be applied to soils with a pH of 7.9 or greater, or crop injury may occur.

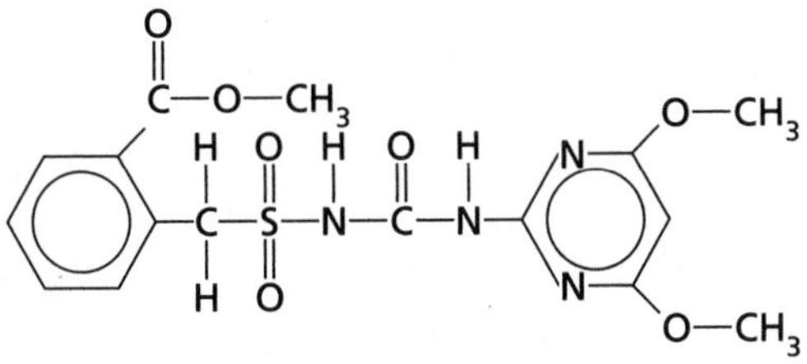

bensulfuron-methyl (Londax)

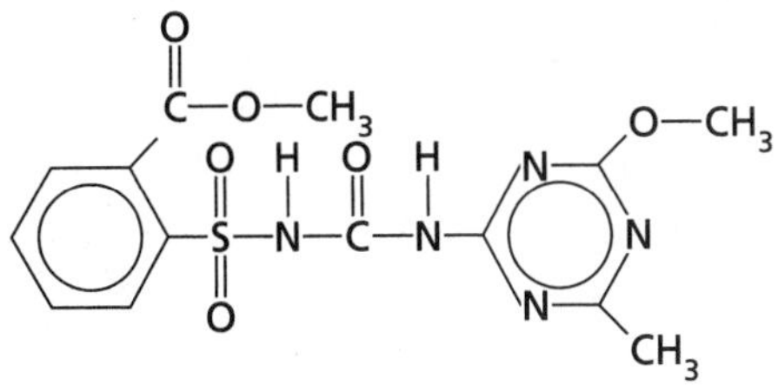

metsulfuron–methyl (Ally, Escort)

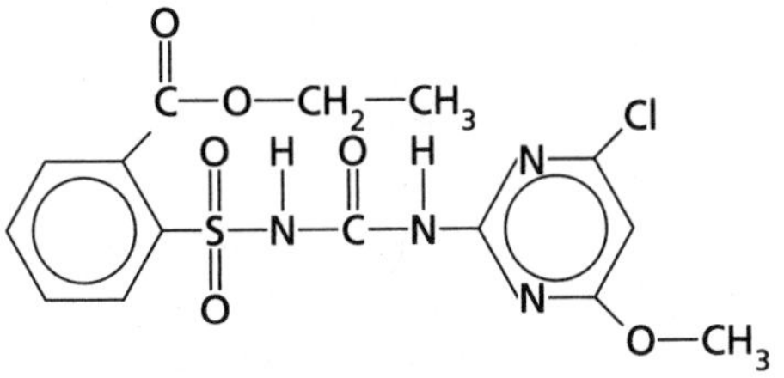

chlorimuron–ethyl (Classic)

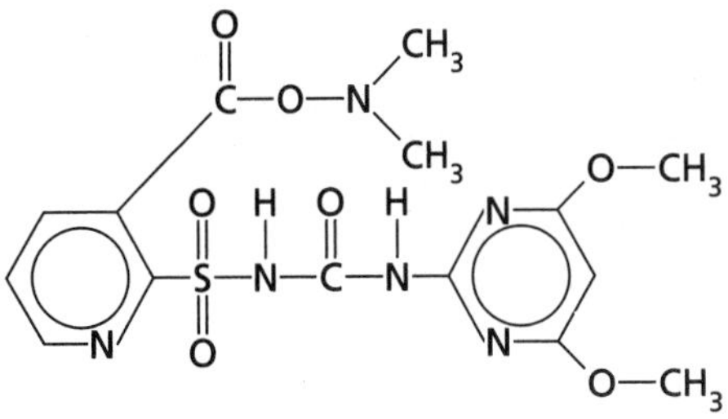

nicosulfuron (Accent)

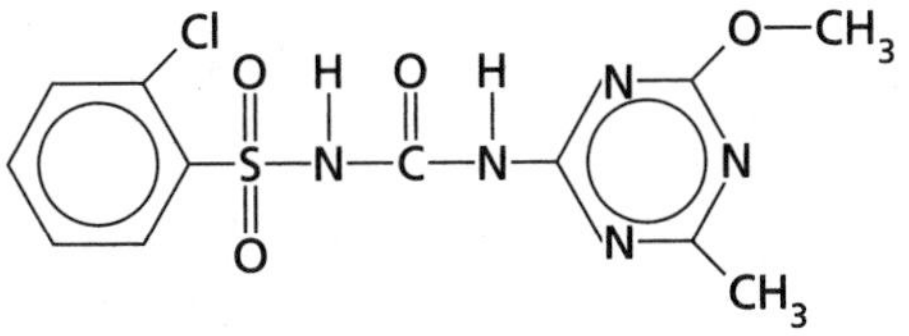

chlorsulfuron (Glean, Telar)

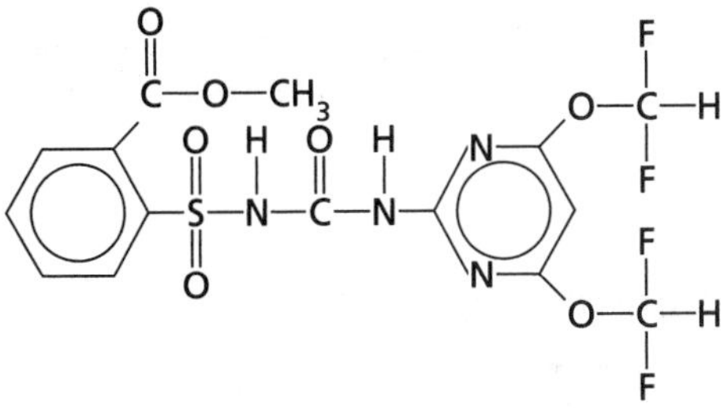

primisulfuron-methyl (Beacon)

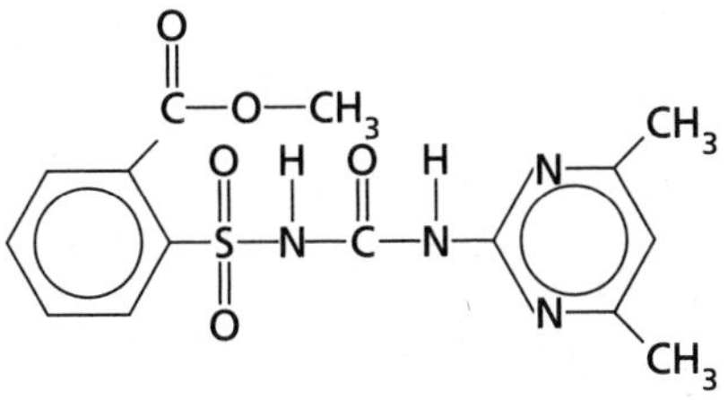

sulfmeturon–methyl (Oust)

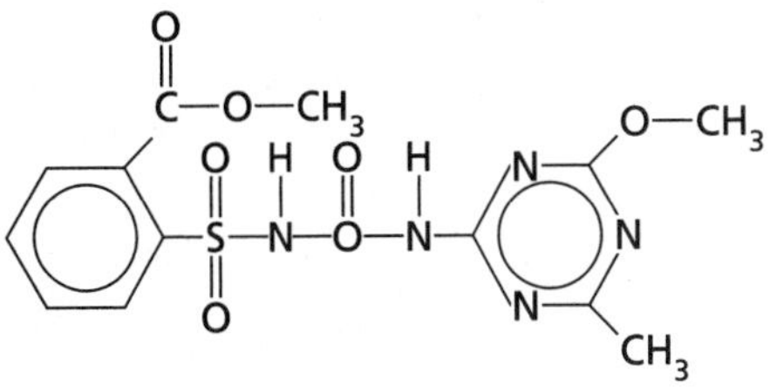

tribenuron–methyl (Express)

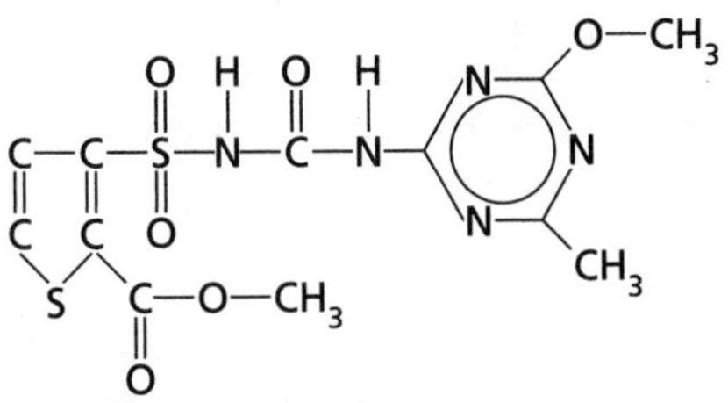

thifensulfuron–methyl (Pinnacle)

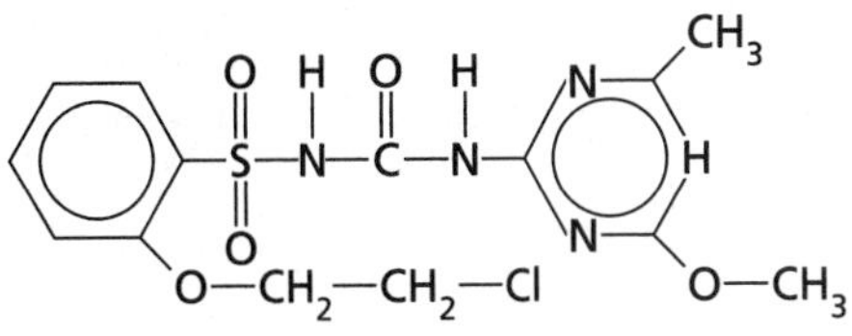

triasulfuron (Amber)

FIGURE 25.1 ■ Chemical structures of the sulfonylurea herbicides.

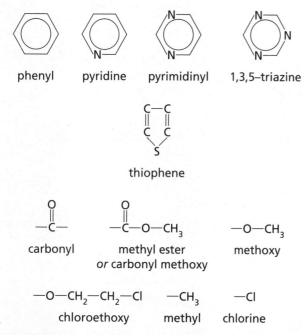

FIGURE 25.2 ■ Chemical structures of radicals substituted on the nucleus that is common to the sulfonylurea herbicides.

Exceptional Characteristics

1. Bensulfuron (Londax) is the only sulfonylurea herbicide labeled for use in rice, and, in fact, it is only used in rice.

2. Chlorimuron (Classic) is the only sulfonylurea herbicide labeled for use in peanuts, but it may also be used in soybeans and noncrop areas. Most other crops are highly sensitive to Classic.

3. Chlorsulfuron is the active ingredient in Glean and Telar, but Glean is applied at 0.13–0.25 oz ai/A and Telar at 0.19–2.25 oz ai/A.

4. Chlorsulfuron (Glean) is formulated to be mixed with water or with liquid nitrogen fertilizer solutions.

5. Metsulfuron (Ally), tribenuron (Express), and chlorsulfuron (Glean) are used in wheat and barley; Glean may also be used in oats.

6. Metsulfuron (Ally) is applied at the lowest dosage per acre (0.1 oz product/A) of any of the sulfonylurea herbicides.

7. Metsulfuron is the active ingredient in Ally and Escort, but Ally is applied at 0.06 oz ai/A and Escort at 0.18–1.2 oz ai/A.

8. Metsulfuron (Escort), sulfmeturon (Oust), and chlorsulfuron (Telar) are used in noncrop areas and industrial sites.

9. Escort and Telar control annual and perennial broad-leaved weeds, but not grasses. Oust controls annual and perennial grass and broadleaved weeds.

10. Nicosulfuron (Accent) and primisulfuron (Beacon) are used only in field corn to selectively control annual and perennial grasses.

11. Thifensulfuron (Pinnacle) poses the least hazard to rotational crops of any of the sulfonylurea herbicides; any crop may be planted 45 days after its application. Other sulfonylurea herbicides have crop-rotation restrictions measured in months or years.

12. The premixes Canopy (chlorimuron + metribuzin), Gemini (chlorimuron + linuron), and Preview (chlorimuron + metribuzin) are used only in soybeans.

13. The premixes Finesse (chlorsulfuron + metsulfuron) and Harmony Extra (thifensulfuron + tribenuron) are used only in wheat and barley.

Precautions

1. The prudent user of any of the sulfonylurea herbicides will do well to read the product label prior to its purchase or use. All of these herbicides are highly injurious to most plants; only a few crops have tolerance to some of these herbicides. Recommended rates in tolerant crops is usually less than 0.3 oz ai/A, and the recommended rate for Ally is 1.7 g ai/A (equivalent to 39 µg ai/ft^2).

2. Spray drift during the application of these herbicides poses a real hazard to most crops; even spray drift onto the surface of nearby fields that will be planted with a susceptible crop may be hazardous to that crop due to the long soil persistence of these herbicides and their high phytotoxicity at minute amounts per acre.

3. Because of possible chemical interactions that can result in crop injury or death, do not use sulfonylurea herbicides where the systemic organophosphate insecticide/nematicide terbufos (Counter) has been, or will be, applied any time during the growing season; use caution if other organophosphate insecticides have been applied.

4. Read the product label before purchase or use of any of the sulfonylurea herbicides, and follow the label directions with care!

■ INFORMATION OF PRACTICAL IMPORTANCE

Bensulfuron

1. Bensulfuron-methyl is the active ingredient in a dry-flowable formulation (Londax) that contains 60% ai w/w.

2. Londax is used for selective preemergence and early postemergence control of broadleaf and sedge weeds in rice. The targeted weeds and soil must be completely covered by about 3 in. of water at the time of treatment. Londax does not control grasses.

3. There are two Londax product labels: one written for use in California and the other for use in the five eastern rice-growing states.

4. Londax is applied at 1.0 oz ai/A (1.67 oz product/A).

5. For best results, Londax should be applied preemergence to early postemergence to submerged weeds prior to the three-leaf stage. The rice should be in the one- to three-leaf stage.

Chlorimuron

6. Chlorimuron-ethyl is the active ingredient in a water-dispersible granular formulation (Classic) that contains 25% ai w/w.

7. Classic is used for selective postemergence control of broadleaf weeds and yellow nutsedge in soybeans and peanuts, and for nonselective weed control in noncrop areas.

8. Classic is applied at 0.25–0.5 oz product/A (in tank mix with Pinnacle) in soybeans and 0.5 oz product/A applied alone in peanuts, depending on targeted species. In noncrop areas, apply 1–2 oz product/A.

9. In soybeans, Classic is usually applied in tank mix with Assure II or Pinnacle, and the tank mix can be applied any time after the first trifoliate leaf has opened, but no later than 60 days before soybean maturity.

10. In peanuts, Classic may be applied from 60 days after crop emergence to 45 days prior to harvest.

11. Classic provides excellent control of Florida beggarweed in peanuts when applied before the weed reaches 10 in. (25 cm) in height or begins to bloom.

Chlorsulfuron

12. Chlorsulfuron is the active ingredient in a dry-flowable formulation (Glean) that contains 75% ai w/w. It is also the active ingredient in a water-dispersible granular formulation (Telar) that contains 75% ai w/w.

Glean

13. Glean is a formulation compatible with liquid nitrogen fertilizers, and it may be mixed directly with these fertilizers, with the liquid fertilizer serving as the carrier for the herbicide.

14. Glean is used for selective preemergence and postemergence control of broadleaf weeds on land primarily dedicated to long-term production of wheat, barley, and oats.

15. It can be applied postemergence to the small grains after the crop is in the two-leaf stage but before boot stage, or, in the Pacific Northwest, Glean can be applied through the second joint stage.

16. For best preemergence results, at least 1–2 in. of rainfall or sprinkler irrigation is needed following Glean application to move the herbicide into the upper 3-in. soil profile before weed seeds germinate and develop an established root system. Weeds that germinate and form an established root system before the herbicide is moved into the seed germination zone may not be controlled.

17. For best postemergence results, apply Glean to actively growing broadleaf weeds less than 2 in. tall or with rosettes less than 2 in. across. Postemergence treatments control or suppress weeds through foliar and root absorption. Root absorption occurs when that portion of the postemergence treatment falling on the soil surface is carried into the upper soil profile and contacts weed seeds as they germinate.

18. The application rate of Glean is 0.17–0.5 oz product/A, depending on the geographical location (refer to product label). Only a single application of Glean may be made in one 18-, 36-, or 48-month period, depending on the geographical location (refer to product label).

Telar

19. Telar is used for preemergence and postemergence control of broadleaf weeds in noncrop areas.

20. For best results, apply Telar postemergence to actively growing weeds less than 2 in. tall or with rosettes less than 2 in. across. Telar is absorbed by both roots and foliage.

21. The application rate of Telar is 0.25–3 oz product/A, depending on the weed species targeted.

22. Telar may be tank-mixed with Karmex DF, Krovar I DF, Banvel, 2,4-D, or glyphosate to control a broader weed spectrum. Do not tank-mix Telar with DuPont Hyvar XL.

Metsulfuron

23. Metsulfuron-methyl is the active ingredient in several dry-flowable formulations (Ally, Escort); each contains 60% ai w/w.

Ally

24. Ally is used for postemergence control of broadleaf weeds in wheat (including durum wheat), barley, and on land in the Conservation Reserve Program.

25. The maximum use rate of Ally is 0.1 oz product/A in one 10- or 22-month period, depending on the geographic location (refer to product label).

26. In wheat and barley, Ally is applied postemergence after the crop is in the two-leaf stage but before the boot stage. For durum spring wheat and the Wampum variety of spring wheat, apply only in tank mix with 2,4-D ester and only after crop is tillering but before boot stage. To avoid crop injury, do not apply the mixture during boot stage or early heading.

27. Ally does not control wild oats or other grasses. It may be applied in tank mix with a suitable postemergence grass herbicide or as a sequential treatment.

Escort

28. Escort is used for preemergence and postemergence control of annual and perennial broadleaf weeds and woody brush in conifer plantations and noncrop areas.

29. Escort is applied at 0.5–1.5 oz product/A in conifer plantations, depending on the targeted species. The application rate in noncrop areas is 1–2 oz product/A.

Nicosulfuron

30. Nicosulfuron is the active ingredient in a water-dispersible granular formulation (Accent) that contains 75% ai w/w.

31. Accent may be applied as a broadcast or postdirected treatment to field corn up to 24 in. (61 cm) tall and to popcorn up to 20 in. tall. Do not apply Accent to sweet corn.

32. Accent is used in the selective postemergence control of annual and perennial grass and certain broadleaf weeds. The optimal stage at which the weeds are best controlled varies among species. For example, Texas panicum is treated at the one- to three-leaf stage, foxtail species at the two- to four-leaf stage, Italian ryegrass at the two- to six-leaf stage, quackgrass at the four- to ten-leaf stage, tall and pitted morningglories at the one- to four-leaf stage, and smooth amd redroot pigweeds at the one- to four-leaf stage.

33. The application rate of Accent is 0.67 oz product/A; higher rates may be required for weeds exceeding the optimal size.

Primisulfuron

34. Primisulfuron is the active ingredient in a water-dispersible granule formulation (Beacon) that contains 75% ai w/w. Beacon is packaged in water-soluble packets, each containing 1.52 oz of formulation (product). This is the correct dosage to treat 2 A of corn (0.76 oz product/A).

35. Beacon is used for postemergence control of johnson-grass, quackgrass, shattercane, sorghum-alum, and certain other grass and broadleaf weeds in field corn and popcorn.

36. Beacon may be applied postemergence over-the-top, directed, or semidirected when weeds are within the height range specified for optimal control (refer to product label). Do not apply Beacon to sweet corn.

37. Beacon may be applied as a sequential treatment following other suitable soil-applied or postemergence herbicides or as tank mixes with Banvel (dicamba), Buctril (bromxynil), or 2,4-D. Use only a nonionic surfactant with these tank mixes; do not use any other spray adjuvant.

38. Growth of susceptible weeds is inhibited following application of Beacon. The leaves turn yellow and/or red after several days, followed by death of the growing points. Complete plant death occurs 1–4 weeks after application, depending on the targeted species.

Sulfometuron

39. Sulfometuron-methyl is the active ingredient in a water-dispersible granular formulation (Oust) that contains 75% ai w/w.

40. Oust is used for preemergence and postemergence control of annual and perennial grass and broadleaf weeds in noncrop areas and industrial sites. It can be used in forest site preparation and release of several types of pines (e.g., loblolly, slash, longleaf, jack, and Virginia) and in dormant white pine, white spruce, and Douglas fir. To avoid injury to the trees, do not use a surfactant in the spray mixture.

41. In noncrop areas, the application rate of Oust is 1.33–2 oz product/A in arid areas and 3–5 oz product/A in areas of 20-in. rainfall or more. Greater rates are recommended for control of certain weeds. Rate selection is based on weed species, weed size, and soil texture.

42. Oust may be applied preemergence or postemergence to weeds, but for best results, apply Oust postemergence to young, actively growing weeds. It can be applied at any time of the year, except when the ground is frozen.

43. Oust is absorbed by both roots and foliage.

Thifensulfuron

44. Thifensulfuron-methyl is the active ingredient in a dry-flowable formulation (Pinnacle) that contains 25% ai w/w.

45. Pinnacle is used for selective postemergence control of annual broadleaf weeds in soybeans.

46. Pinnacle may be applied to soybeans anytime after the first trifoliate leaf has fully expanded, but no later than 60 days before harvest.

47. For best results, apply Pinnacle to actively growing weeds 2–6 in. (5–15 cm) tall, depending on targeted species.

48. The application rate of Pinnacle is 0.25 oz product/A.

49. Any rotational crop may be planted 45 days after an application of Pinnacle at the recommended dosage.

50. Pinnacle may be tank-mixed with Classic (chlorimuron) to control a broader spectrum of broadleaf weeds. The rate of application for this tank mix is 0.25 oz/A of each product. A nonionic surfactant must be included in the spray mixture, using 1–2 pt surfactant/100 gal water.

Triasulfuron

51. Triasulfuron is the active ingredient in a water-dispersible granular formulation (Amber) that contains 75% ai w/w.

52. Amber is used preemergence and postemergence in winter wheat and spring wheat. Applied preemergence, Amber controls Italian ryegrass and suppresses downy brome and cheat. Applied postemergence when weeds are young and actively growing, Amber controls many annual broadleaf weeds, especially mustards and kochia. Amber is applied at 0.42 oz ai/A (0.56 oz product/A).

53. Amber is used postemergence in barley to control young, actively growing broadleaf weeds, especially mustards and kochia. It is applied at 0.21–0.42 oz ai/A (0.28–0.56 oz product/A).

54. Adsorption of Amber to soil colloids is relatively low, and leaching can occur.

Tribenuron

55. Tribenuron-methyl is the active ingredient in a dry-flowable formulation (Express) that contains 75% ai w/w.

56. Express is used for selective postemergence control of annual broadleaf weeds in wheat (including durum) and barley. Make applications after the crop is in the two-leaf stage but before the flag leaf is visible.

57. The dosage rate for Express applied in wheat and barley is 0.17–0.25 oz product/A, depending on the stage of weed growth and the targeted species.

58. Sequential applications of Express may be made, provided the total amount of the product does not exceed 0.33 oz/A per season.

Premixes

In addition to the previously mentioned products, five premixes are available that contain, as their active ingredients, two sulfonylurea herbicides or one sulfonylurea herbicide plus metribuzin or linuron. These products and their recommended uses are given below, with their formulation percentages enclosed in parentheses.

59. *Canopy* is a water-dispersible granular formulation that contains chlorimuron and metribuzin (10.7 and 64.3% w/w, respectively) as the active ingredients. Canopy is used in soybeans as a preplant-incorporated or preemergence treatment to control broadleaf weeds and to suppress sedges. Canopy provides partial control of certain annual grasses and yellow and purple nutsedges.

60. *Finesse* is a dry-flowable formulation that contains chlorsulfuron and metsulfuron (62.5 and 12.5% w/w,

respectively) as the active ingredients. Finesse is used on land primarily devoted to the production of winter/spring wheat and winter/spring barley for the selective control of many broadleaf weeds. It may be tank-mixed with recommended herbicides, such as Karmex, Lexone, Buctril, or 2,4-D, to control specific weed problems.

61. *Gemini* is a water-dispersible granular formulation that contains chlorimuron and linuron (4.6 and 55.4% w/w, respectively) as the active ingredients. Gemini is used in soybeans for selective preplant-incorporated or preemergence control of broadleaf weeds and for suppression of certain grass weeds.

62. *Harmony Extra* is a dry-flowable formulation that contains thifensulfuron and tribenuron (50 and 25%

w/w, respectively) as the active ingredients. Harmony Extra is used in winter/spring wheat and spring barley for the selective postemergence control of wild garlic and annual broadleaf weeds and Canada thistle. Harmony Extra does not control grasses. To avoid severe crop injury, do not use Harmony Extra on any crop other than the ones noted here.

63. *Preview* is a water-dispersible granular formulation that contains chlorimuron and metribuzin (6.5 and 68.5% w/w, respectively) as the active ingredients. Preview is used in soybeans for selective preplant-incorporated or preemergence control of broadleaf weeds. Do not apply Preview if the soil pH is greater than 6.8.

26 THIOCARBAMATE HERBICIDES

■ INTRODUCTION

The thiocarbamate herbicides are compounds that contain sulfur in place of the hydroxyl oxygen of carbamic acid. The chemical nucleus common to the thiocarbamate herbicides is:

$$R_1 \quad O$$
$$\underset{R_2}{\overset{R_1}{N}} - \overset{\overset{O}{\parallel}}{C} - S - R_3$$

thiocarbamate

At present, all thiocarbamate herbicides have alkyl groups at positions R_1, R_2, and R_3 of the above structure. The common and chemical names of the thiocarbamate herbicides are given in Table 26.1 on the following page, and their structures in Figure 26.1 on page 229.

When sulfur atoms replace both the hydroxyl and carbonyl oxygens of thiocarbamate, the resulting compound is called a *dithiocarbamate*. Metham is a dithiocarbamate herbicide and, for convenience, is also discussed in this chapter (Table 26.1 and Figure 26.1).

■ CHARACTERISTICS OF THIOCARBAMATE HERBICIDES

Principal Use

The thiocarbamate herbicides are principally used in the preemergence control of annual grass weeds and, with certain of these herbicides, annual broadleaved weeds and purple nutsedge and yellow nutsedge. The thiocarbamate herbicides do not control established weeds.

Metham, a dithiocarbamate, is used as a nonselective soil fumigant to control weeds, insects, fungi, and nematodes. Metham is also used to kill roots in sewer pipes.

Mode of Action

The thiocarbamate herbicides inhibit lipid formation, interfering with the biosynthesis of surface lipids (waxes, cutin, suberin). This is apparently due to the inhibition of acyl-coenzyme A (acyl-CoA) elongases by these herbicides.

Metham is a nonselective, contact-type soil fumigant that kills germinating seeds, roots, rhizomes, tubers, and stems in the soil.

TABLE 26.1 ■ Common and chemical names of thio- and dithiocarbamate herbicides.

COMMON NAME	CHEMICAL NAME
Thiocarbamates	
Butylate	S-ethyl bis(2-methylpropyl)-carbamothioate
Cycloate	S-ethyl cyclohexylethylcarbamothioate
EPTC	S-ethyl dipropyl carbamothioate
Molinate	S-ethyl hexahydro-1H-azepine-1-carbothioate
Pebulate	S-propyl butylethylcarbamothioate
Thiobencarb	S-[(4-chlorophenyl)methyl]-diethylcarbamothioate
Triallate	S-(2,3,3-trichloro-2-propenyl) bis-(1-methylethyl)carbamothioate
Vernolate	S-propyl dipropylcarbamothioate
Dithiocarbamate	
Metham	methylcarbamodithioic acid

Translocation

The thiocarbamate herbicides are rapidly root-absorbed and translocated upward by crop plants. Applied prior to weed seed germination, translocation (other than short-distance, local movement) in seedling weeds is not a factor in the activity either of the thiocarbamate herbicides or of metham.

Selectivity

The selectivity of the carbamate herbicides is due to:

■ The rapid metobolic degradation of the herbicides by tolerant plants and the failure of susceptible plants to do so.

■ Herbicide placement with respect to seed germination and seedling development.

■ Differences in the stage of plant growth at time of application, with established plants most tolerant.

Common Characteristics

1. The thiocarbamate herbicides are applied only preemergence to the weeds. The site of action of the thiocarbamate herbicides is the developing shoot and coleoptile of grass seedlings.

2. In general, the thiocarbamate herbicides only control seedling grasses. However, butylate and cycloate also provide preemergence control or suppression of purple nutsedge and yellow nutsedge and certain broadleaf weeds.

3. Established crop plants (6–12 in. or taller when direct-seeded, or after new growth has started from transplants) are highly tolerant of the thiocarbamate herbicides.

4. The thiocarbamate herbicides are highly volatile and are readily lost from the surface of moist soils. To avoid such losses, they must be soil-incorporated by tillage or by leaching.

5. Applied to dry soils, the thiocarbamate herbicides are adsorbed to soil colloids and are not readily lost by vaporization. Following rainfall or irrigation, water molecules easily displace the herbicide molecules at adsorptive sites on mineral colloids, releasing the herbicide molecules into the soil water. The herbicide molecules displaced at the soil surface, or those that subsequently reach the soil surface, may be lost from the surface by volatilization.

6. Displaced from soil colloids, the thiocarbamate herbicides are readily leached in soils.

7. The persistence of the thiocarbamate herbicides in warm, moist soils is approximately 3–6 weeks. Their persistence is longer in dry or cold soils. In general, their half-life in soils at 70–80°F (21–26°C) is 1–4 weeks; that of cycloate is 4–8 weeks.

8. Microbial action is the primary means by which the thiocarbamate herbicides are lost from soils. Soil persistence can vary greatly in a given soil because of changes in microbial activity, temperature, and moisture level.

9. The ease with which the thiocarbamate herbicides are degraded in plants and soils is an advantage with respect to residues, but it is a disadvantage with respect to long-term weed control.

Crop Safeners

1. The crop safener dichlormid is included in one formulation of EPTC (Eradicane). Another formulation (Eradicane Extra) contains a mixture of R-29148 *plus* dietholate (crop safener and herbicide protectant, respectively). Both formulations are used for selective grass control in corn.

2. The crop safener R-29148 is used with butylate when it is formulated for use in corn (Sutan+).

3. The chemical names of the above crop safeners and herbicide protectant (extender) are:

Dichlormid 2,2-dichloro-N,N-di-2-propenylacetamide

R-29148 3-(dichloroacetyl)-2,2,5-trimethyloxazolidine

Dietholate O,O-diethyl O-phenyl phosphorthioate

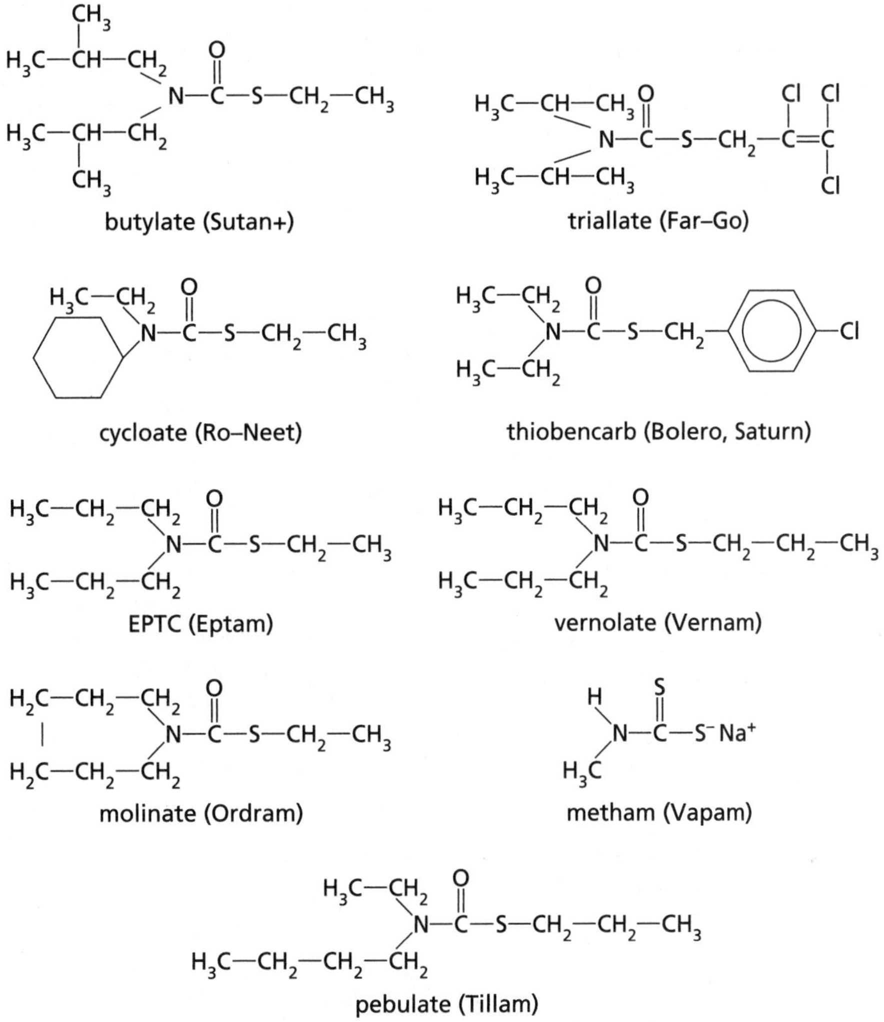

FIGURE 26.1 ■ **Chemical structures of the thiocarbamate herbicides and of metham, a dithiocarbamate herbicide.**

4. The reader is referred to the section "Chemical Plant-Protectants" in Chapter 8, Herbicide–Plant Selectivity, for more information on crop safeners.

■ INFORMATION OF PRACTICAL IMPORTANCE

Butylate

1. Butylate is the active ingredient in an emulsifiable-concentrate formulation (Sutan+ 6.7E) that contains 6.7 lb ai/gal *plus* R-29148.

2. Sutan+ is used only in corn (all types). R-29148 is included in the formulation to protect the corn seedlings from the phytotoxic effects of butylate.

3. Sutan+ is applied preemergence to the weeds. It controls grasses, purple nutsedge, yellow nutsedge, and volunteer sorghum, and suppresses growth of rhizome bermudagrass and johnsongrass.

4. Sutan+ is applied at 3–6 lb ai/A (3.75–7.33 pt product/A), with the low rate for annual grass and nutsedge control, and the high rate for suppression of perennials.

5. Sutan+ may be tank-mixed with atrazine, Bladex, or Princep to control a broader spectrum of weeds.

6. Sutazine+ is a premix formulation that contains butylate (4.8 lb/gal) and atrazine (1.2 lb/gal) as the active ingredients. It is used only in corn. Sutazine+ is a Restricted Use Pesticide because of the presence of atrazine.

Cycloate

7. Cycloate is the active ingredient in an emulsifiable-concentrate formulation (Ro-Neet 6E) that contains 6 lb ai/gal.

8. Ro-Neet 6E is used only in sugar beets, table beets, and spinach. It is applied preemergence-incorporated at 3–4 lb ai/A (4–5.3 pt product/A), depending on the crop treated. It may be tank-mixed with Eptam 7E to control a broader weed spectrum, including purple nutsedge and yellow nutsedge.

EPTC

9. EPTC is the active ingredient in several formulations:

   ■ As an emulsifiable-concentrate formulation (Eptam 7E) that contains 7 lb ai/gal.

   ■ As a granular formulation (Eptam 10-G) that contains 10 lb ai w/w.

   ■ As an emulsifiable-concentrate formulation (Eradicane 6.7E) that contains 6.7 lb ai/gal *plus* dichlormid.

   ■ As an emulsifiable-concentrate formulation (Eradicane Extra) that contains 6 lb ai/gal *plus* R-29148 *plus* dietholate.

10. Eptam 7E and Eptam 10-G are used in a variety of crops, depending on geographic location. Eptam 7E is applied at 1.5–6 lb ai/A (1.75–7 pt product/A), depending on crop and weeds involved.

11. Eradicane 6.7E and Eradicane Extra are used only in corn (field, sweet, silage, and popcorn). Eradicane 6.7E is applied at 3–6 lb ai/A (3.75–7.33 pt product/A). Eradicane Extra is applied at 3–6 lb ai/A (4–8 pt product/A).

Molinate

12. Molinate is the active ingredient in an emulsifiable-concentrate formulation (Ordram 8E) that contains 8 lb ai/gal. It is also available in two granular formulations (Ordram 10G and Ordram 15G), containing 10 and 15% ai w/w, respectively.

13. All three formulations of molinate are used only in rice to control barnyardgrass. Ordram 8E and Ordram 15G are used on rice grown only in the southern region of the United States; they are applied preemergence-incorporated or postflood. Ordram 10G is used in rice grown only in California; it is applied preplant or preflood.

14. When Ordram formulations are applied postflood, the rice must be in the seedling stage (emerged), barnyardgrass must be between 2–5 in. tall, and the soil flooded to a depth of not less than 2 in. Molinate moves from the flood water into the leaves of barnyardgrass, subsequently killing it; emerged rice seedlings are tolerant to the molinate.

Pebulate

15. Pebulate is the active ingredient in an emulsifiable-concentrate formulation (Tillam 6E) that contains 6 lb ai/gal.

16. Tillam 6E is used only in sugar beets, tobacco, and tomatoes for the preemergence control of annual grass and broadleaf weeds and purple and yellow nutsedges. It must be soil-incorporated immediately to a depth of 2–3 in. following application, using a power-driven tiller or a tandem disk followed by a spring-tooth harrow. For thorough mixing, work the soil in two directions. Directions for the use of Tillam varies with the crop involved.

Thiobencarb

17. Thiobencarb is the active ingredient in an emulsifiable-concentrate formulation (Bolero 8EC) that contains 8 lb ai/gal, and in a granular formulation (Bolero 10G) that contains 10% ai w/w.

18. Bolero 8EC and Bolero 10G are used only in rice. Bolero 8EC may be applied to weeds late preemergence or early postemergence to control annual grass and broadleaf weeds. For late preemergence, apply 4 lb ai/A (2 qt product/A) 1–5 days prior to rice emergence, about 5–9 days after seeding the rice. For early postemergence, apply 4 lb ai/A (2 qt Bolero 8EC/A) when barnyardgrass has developed no more than two leaves. Rice may be fully emerged at time of the postemergence application.

19. Bolero 10G is used only in rice as a postflood treatment to control seedling grass weeds in the two-leaf stage or less, depending on the species. The rice must be at least in or past the two-leaf stage. Weeds larger than specified will not be controlled and rice smaller than the two-leaf stage will be injured or killed. Water level should be maintained at a depth of 3–4 in., with

no exposed soil. Bolero 10G is applied at 3–4 lb ai/A (30–40 lb product/A).

Triallate

20. Triallate is the active ingredient in an emulsifiable-concentrate formulation (Far-Go) that contains 4 lb ai/gal, and in a granular formulation (Granular Far-Go) that contains 10% ai w/w.

21. Both formulations of triallate are used preemergence, shallowly incorporated immediately after application with a spiketooth or springtooth harrow (do not use disk implements for incorporation), to control annual grass weeds, such as wild oat, Italian ryegrass, and *Bromus* spp., in wheat, barley, peas, and lentils. Weeds that have emerged prior to treatment must be controlled by other means.

22. Far-Go is applied at 1–1.25 lb ai/A (1–1.25 qt product/A), depending on crop involved and time of year applied. Granular Far-Go is applied at 1–1.5 lb ai/A (10–15 lb product/A), depending on crop involved and time of year applied. Used in wheat or barley, Far-Go can be tank-mixed with Treflan for the added control of green foxtail and yellow foxtail.

23. Triallate is primarily absorbed by the shoots of wild oats located in the treated soil layer. The wild oat seedlings are usually killed before they emerge from the soil, but some may emerge and reach the three- to four-leaf stage before dying.

Vernolate

24. Vernolate is the active ingredient in an emulsifiable-concentrate formulation (Vernam 7E) that contains 7 lb ai/gal and in a granular formulation (Vernam 10G) that contains 10% ai w/w.

25. Vernam can be tank-mixed with other herbicides to control a broader spectrum of weeds.

26. Vernam 7E and Vernam 10G are used only in soybeans and peanuts to control annual grass and broadleaf weeds. Both products are soil-incorporated immediately after application at a rate of 2–3 lb ai/A (2.3–3.5 pt Vernam 7E/A or 20–30 lb Vernam 10G/A). The low rate is applied to sandy (light) soils and the high rate to clay (heavy) soils.

27. Both Vernam formulations kill the vegetative reproductive parts of bermudagrasss and purple and yellow nutsedges that have been thoroughly chopped up by tillage prior to treatment.

Dithiocarbamate Herbicides

Metham

28. Metham is the active ingredient in a water-soluble formulation (Vapam) that contains 3.18 lb ai/gal.

29. Metham is a water-soluble salt that, on contact with moist soil, breaks down to form the highly volatile and toxic chemical *methyl isothiocyanate* (H_3C—N═C═S). This then becomes the actual herbicide, which is rapidly dispersed (vapor form) by diffusion through the soil. Metham is changed to methyl isothiocyanate within 1–5 hours by chemical, rather than biochemical, means. Methyl isothiocyanate disappears almost completely from the soil in 2–3 weeks.

30. Vapam is a soil fumigant that is soil-applied and soil-incorporated; it provides nonselective control of germinating seeds, seedlings, rhizomes, tubers, roots, and stems in the soil and also acts as a fungicide, insecticide, and nematicide.

31. Vapam is applied to freshly worked soil by injection or leaching with water. The soil surface is water-sealed after application or covered with a tarp (plastic, paper, or fabric) to prevent loss of the herbicide by volatility. The soil surface should be kept covered for a minimum of 48 hours during treatment.

32. The rate of Vapam used varies according to specific usage. Usually, it is applied at 1 qt/100 ft^2 or 50–100 gal/A.

33. Vapam may be applied to small areas, such as in the garden, using a water sprinkling can by adding 1 pt Vapam to the can, filling with water, and applying uniformly over 50 ft^2 of loosely worked soil. Wet the soil immediately after application to seal the soil or cover with a tarp for 48 hours.

34. Cultivate the treated soil 7 days later to a depth of 2 in. to aerate the soil. Where water is used to seal the soil, crop seeding can be made 2–3 weeks after soil aeration; where a tarp was used, wait 3 weeks before seeding.

27 TRIAZINE HERBICIDES

■ INTRODUCTION

The chemical structures of the diamino-*s*-triazine herbicides are centered on a common six-membered ring structure composed of three nitrogen and three carbon atoms arranged symmetrically (in this case, alternately) about the ring, with an amino (NH_2) group bonded to the carbon atoms at the 4- and 6-positions of the ring, as shown here:

4,6–diamino–*s*–triazine

A six-membered ring structure with two or more nitrogen atoms in the ring is called "azine." When there are three nitrogen atoms arranged symmetrically in the ring structure, it is called "*s*-triazine."

Diamino-*s*-triazine herbicides may be placed in one of three subgroups, based on whether the substitution at the 2-position of the ring structure is a chlorine (Cl), a methoxy (—O—CH_3) group, or a methylthio (—S—CH_3) group. The terms *methylthio* and *methylmercapto* are synonymous, and the reader may encounter either term in the chemical nomenclature. The chemical structures of these three subgroups are shown in Figure 27.1.

The common and chemical names of the diamino-*s*-triazine herbicides are given in Table 27.1. Their respective chemical structures are shown in Figure 27.2 on page 234.

Hexazinone is a symmetrical triazine herbicide, but it is not a diamino-*s*-triazine. The chemical name of hexazinone is 3-cyclohexyl-6-(dimethylamino)-1-methyl-1,3,5-triazine-2,4(1*H*,3*H*)-dione.

Metribuzin is a triazine herbicide, but its chemical structure is asymmetrical. The chemical name of metribuzin is 3-methylthio-4-amino-6-*tert*-butyl-1,2,4-triazine-5(4*H*)-one.

The characteristics of these two herbicides are similar to those of the diamino-*s*-triazine herbicides. Their chemical structures are shown in Figure 27.2.

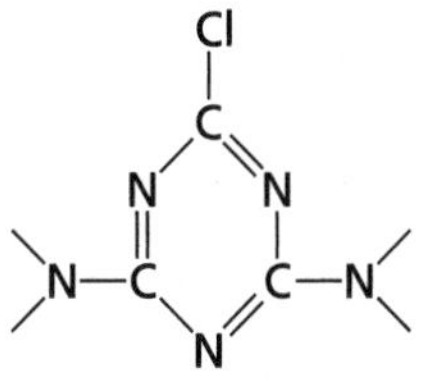

2–chloro–4,6–diamino–s–triazine 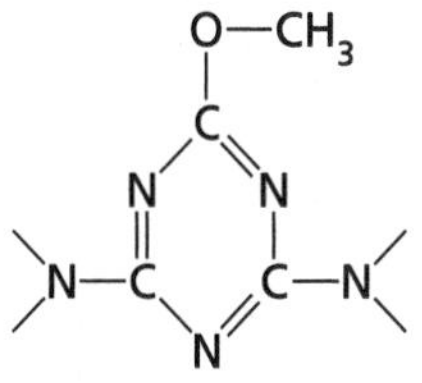   2–methoxy–4,6–diamino–s–triazine

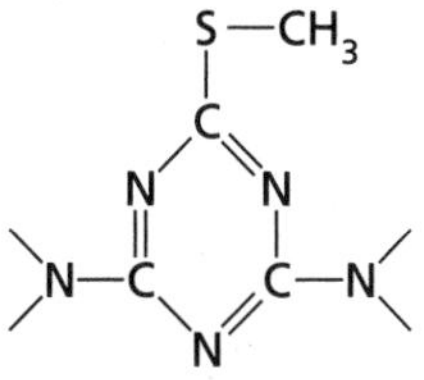

2–methylthio–4,6–diamino–s–triazine

FIGURE 27.1 ■ Chemical nucleus common to each of the subgroups of the diamino-s-triazine herbicides.

TABLE 27.1 ■ Common and chemical names of members of the diamino-s-triazine herbicides.

SUBGROUP	COMMON NAME	CHEMICAL NAME
2—Cl	atrazine	2-chloro-4-ethylamino-6-isopropylamino-s-triazine
2—Cl	cyanazine	2-chloro-4-(2-cyano-isopropylamino)-6-ethylamino-s-triazine
2—Cl	simazine	2-chloro-4,6-bis(ethylamino)-s-triazine
2—O—CH$_3$	prometon	2-methoxy-4,6-bis(isopropylamino)-s-triazine
2—S—CH$_3$	ametryn	2-methylthio-4-ethylamino-6-isopropylamino-s-triazine
2—S—CH$_3$	prometryn	2-methylthio-4,6,-bis(isopropylamino)-s-triazine

■ CHARACTERISTICS OF TRIAZINE HERBICIDES

Principal Use

The triazine herbicides are principally used for selective preemergence and postemergence control of seedling grass and broadleaved weeds in croplands and, with certain of these herbicides, for nonselective vegetation control in noncrop areas. Although the triazine herbicides control both grass and broadleaved weed seedlings, they are most effective in controlling the broadleaf weeds.

Mode of Action

The triazine herbicides inhibit photosynthesis; they achieve this by binding onto the pigment protein of the photosys-tem II complex in the thylakoid membrane of chloroplasts, thereby interfering with normal electron transport into the plastoquinone pool. The triazine and urea herbicides have the same modes of action.

In addition to preventing energy production *via* photosynthesis, secondary effects (e.g., production of free radi-cals) are most likely responsible for the rapid plant kill.

Translocation

The triazine herbicides are readily absorbed by plant roots and then translocated to the leaves *via* the transpira-tion stream. Applied postemergence, the triazine herbi-cides (except for simazine) are foliar-absorbed. However, they are not translocated from the leaves to other plant parts.

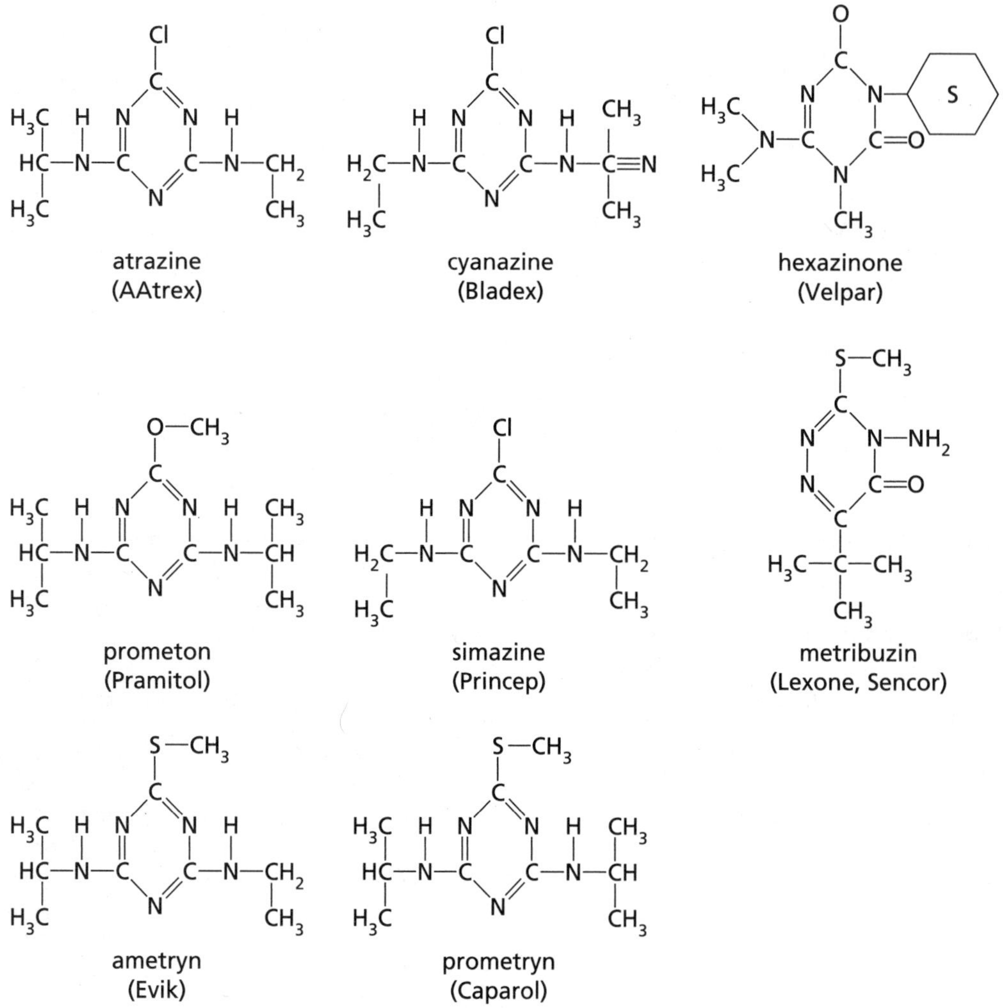

FIGURE 27.2 ■ Chemical structures of the diamino-s-triazine herbicides and of hexazinone and metribuzin.

Selectivity

Crop selectivity of these herbicides is achieved by their rapid detoxication by tolerant crop plants, the inability of susceptible weed species to detoxify them, and by the judicious placement of the herbicide relative to the root zone of the crop plants.

Common Characteristics

1. The triazine herbicides are readily absorbed by both the roots and foliage. The addition of a suitable surfactant to the spray mixture greatly enhances foliar wetting and herbicide absorption.

2. The triazine herbicides do not control established annual or perennial weeds or deep-rooted perennials when applied at the dosages recommended for selective weed control in croplands.

3. The triazine herbicides are readily adsorbed to soil colloids, and they resist leaching. However, leaching can occur, depending on soil texture, the volume of water percolating through the soil, and the herbicide involved.

4. The soil persistence of the triazine herbicides is about 1 month to more than 12 months, depending on the herbicide and application rate and environmental factors such as rainfall, soil organic matter, soil pH, and soil temperature. Soil persistence is longer under arid conditions than under humid ones.

5. Following many years of continuous use of the triazine herbicides (and other herbicides with the same mode of

action), resistant weed biotypes of some species that are not controlled by these herbicides have developed (see Chapter 9, Weed Resistance and Transgenic Crop Tolerance).

Exceptional Characteristics

1. Atrazine and cyanazine are designated as Restricted Use Pesticides. (Refer to items 12 and 16 under "Information of Practical Importance").

2. Simazine has little or no foliar activity, and it is used only as a preemergence herbicide.

3. With the exception of simazine, each of the triazine herbicides is foliar-absorbed.

4. Atrazine, applied at higher-than-recommended rates for selective weed control in croplands, is used for long-term nonselective weed control in noncrop areas.

5. Prometon is used *only* as a nonselective herbicide for general weed control in noncrop areas; it inhibits all plant growth for a year or more.

6. Atrazine, cyanazine, and simazine are detoxified in tolerant plants by a nonenzymatic reaction that replaces the chlorine atom at the 2-position of the triazine ring with a hydroxyl (—OH) group, resulting in a nonphytotoxic molecule. Eventually the molecules are degraded to simple compounds.

7. The differential rate of atrazine degradation in plants to nonphytotoxic metabolites is the primary factor accounting for its plant selectivity. The three metabolic pathways for atrazine degradation by corn are:

   - Hydrolysis of the chlorine at the 2-position of the triazine ring
   - N-dealkylation of the alkylamino side chains at the 4- and 6-positions of the triazine ring
   - Conjugation at the 2-position of the triazine ring with tripeptide glutathione

8. *Chemical hydrolysis* is the primary means by which atrazine, cyanazine, and simazine are detoxified in soils.

9. *Microbial degradation* is the primary mechanism by which ametryn, hexazinone, metribuzin, prometon, and prometryn are lost from the soil. Degradation occurs fastest under aerobic conditions and comparatively high temperatures.

10. The soil persistence of metribuzin is 2–4 weeks under optimal conditions.

11. The soil persistence of ametryn, cyanazine and prometryn is 1–3 months under warm, moist soil conditions

when applied at dosages recommended for selective weed control in croplands.

12. The soil persistence of hexazinone is about 12 months. Corn may be planted after alfalfa, 12 months after the last hexazinone treatment. In areas receiving 20 in. (51 cm) or less of rainfall, corn may be planted 12 months after the last treatment provided the applied rate did not exceed 0.75 lb ai/A.

13. The soil persistence of atrazine, applied at recommended rates for selective weed control for specific soil types, is such that most rotational crops can be planted 1 year after application, except under arid or semiarid conditions. Under certain conditions, atrazine carryover may injure alfalfa, small grains, and soybeans planted more than 12 months after the initial application.

■ INFORMATION OF PRACTICAL IMPORTANCE

1. Applied under dry soil conditions, the triazine herbicides will likely provide poor weed control. Moisture is essential in bringing about contact and absorption of the herbicides by the root systems of the weeds.

2. Applied preemergence, best results are obtained when sufficient rainfall or sprinkler irrigation (0.5–1 in. on moist soils and 1–2 in. on dry soils) occurs within 2 weeks after application, thereby enabling the herbicides to be carried into the root zone of the weed seedlings. If there is insufficient moisture, a shallow cultivation (preferably with a rotary hoe) should be made to soil-incorporate the herbicides and to control any weeds that have emerged.

3. To avoid injuring emerged crops, use directed sprays so that the herbicides do not contact the crop foliage.

4. Continuous agitation of the spray mixture is essential to keep the triazine herbicides in suspension in the water carrier.

5. To avoid crop injury, do not apply the triazine herbicides to highly permeable soils, such as sand or loamy sand (soils made up of more than 70% sand), with less than 1% organic matter.

6. To avoid injury or death to desired plants, do not apply the triazine herbicides near home plantings of trees, shrubs, or herbaceous plants, or on lawns, walks, driveways, tennis courts, or similar areas. Bear in mind that the triazine herbicides are root-absorbed and translocated to the foliage of ornamental plants just as easily as they are by weeds.

7. When applied to the same fields year after year at rates used for selective weed control, there is no buildup or accumulation of the triazine herbicides in the soil.

8. When triazine-resistant weed biotypes are known to be present in fields treated with triazine herbicides, switch to suitable herbicides with modes of action different from the triazine herbicides, or use tank mixes of triazines, or sequential treatments, with suitable herbicides that have different modes of action.

Ametryn

9. *Ametryn* is the active ingredient in a wettable-powder formulation (Evik 80W) that contains 80% ai w/w. Evik 80W is used to control annual grass and broadleaf weeds in corn, bananas, pineapple, sugarcane, and noncrop areas. It is applied preemergence or early postemergence to the weeds.

10. When used in corn, apply Evik 80W as a postemergence directed spray to weeds after the smallest corn plants are at least 12 in. (30 cm) tall (measured to the highest leaf surface on free-standing plants). Extreme care must be taken to keep the spray, spray drift, or spray "bounce-back" from contacting the corn leaves, especially the whorl of the corn plants. Use a minimum of 20 gal of water/A and a suitable surfactant, adjusting the spray pattern to provide uniform coverage and to completely wet the weed foliage. Used in corn, the dosage of Evik 80W ranges from 0.75 to 2.5 lb product/A (0.6–2.0 lb ai/A) depending on weed species present and the geographic region involved. Refer to the product label for recommended use in bananas, pineapple, sugarcane, and noncrop areas.

Atrazine

11. Atrazine is the active ingredient in two types of formulations: a liquid formulation (AAtrex 4L), which contains 4 lb ai/gal, and a water-dispersible granule formulation (AAtrex Nine-O), which contains 90% ai w/w.

12. Atrazine is a Restricted Use Herbicide because of its potential for leaching into and contaminating groundwater, since groundwater may subsequently be used as drinking water (see the following Precautionary Statement). The retail sale and use of atrazine is restricted to certified applicators, or persons under their direct supervision.

13. Atrazine is used to control many annual grass and broadleaf weeds in corn, sorghum, sugarcane, and certain other crops specified on the product label. It

may be used during the fallow period in certain crop rotations such as wheat–sorghum–fallow, wheat–corn–fallow, or wheat–fallow–wheat. It is also effective in noncrop areas and industrial sites to control most annual and many perennial grass and broadleaf weeds. Atrazine may be used preplant, preemergence, or early postemergence to the weeds. Formulations of atrazine may be tank-mixed with other appropriate herbicides to broaden the spectrum of weed control. Atrazine is applied at 1–4 lb ai/A, depending on the crop and soil type involved.

14. *Precautionary Statement:* AAtrex 4L and AAtrex Nine-O may not be mixed, loaded, or used within 50 ft of all wells, including abandoned wells, drainage wells, and sink holes. They may not be mixed or loaded within 50 ft of intermittent streams and rivers, natural or impounded lakes and reservoirs. They may not be applied aerially or by ground within 66 ft of the points where field surface water runoff enters perennial or intermittent streams and rivers or within 200 ft around natural or impounded lakes or reservoirs. If they are applied to highly erodible land, the 66-ft buffer or setback from runoff entry points must be planted to crop, seeded with grass or other suitable crop. Do not apply directly to water, to areas where surface water is present, or to intertidal areas below the mean high water mark. Do not apply when weather conditions favor drift from treated areas. Runoff and drift from treated areas may be hazardous to aquatic organisms in neighboring areas. Do not contaminate water when disposing of equipment wash water. These products are toxic to aquatic invertebrates. (These statements were taken from individual product labels.) The water solubility of atrazine is very low (33 ppmw at 80°F or 27°C).

Cyanazine

15. Cyanazine is the active ingredient in two types of formulations: a liquid formulation (Bladex 4L) containing 4 lb ai/gal, and a water-dispersible granule formulation (Bladex 90DF) containing 90% ai w/w.

16. Cyanazine is a Restricted Use Pesticide because birth defects occurred in laboratory animals following treatment at doses that also caused serious maternal illness (see Precautionary Statement). The retail sale and use of cyanazine is restricted to certified applicators, or persons under their direct supervision.

17. Cyanazine is used to control annual grass and broadleaf weeds in field corn, popcorn, sweet corn, cotton, and

sorghum. It may be applied preplant-incorporated, preemergence, and/or basally directed postemergence. It is applied at 1.25–4.75 lb ai/A, depending on soil type and organic matter content. The Bladex formulations may be tank-mixed with other suitable herbicides, such as atrazine, EPTC+ antidote, simazine, 2,4-D, paraquat, metolachlor, or butylate, depending on the crop and weeds involved. Refer to the product labels for specific recommendations for the appropriate crop.

18. Bladex can move through soil and contaminate groundwater that may be used as drinking water, and has been found in groundwater as a result of agricultural use. Users are advised not to apply Bladex where the water table (groundwater) is close to the surface and where soils are very permeable (such as loamy sands). (These precautions were taken from individual product labels.) The water solubility of cyanazine is 171 ppmw at 77°F (25°C).

19. *Precautionary Statement:* Wear long trousers when applying Bladex formulations. Wear chemical-resistant gloves extending above the wrist, a chemical-resistant apron, long trousers, and long-sleeved clothing when mixing or loading or when adjusting, repairing, or cleaning equipment. Protective gloves must be washed with soap and water after use and before removing from hands. Remove contaminated clothing and wash before reuse. Contaminated clothing should be laundered separately from household laundry to prevent contamination of other laundry. Heavily contaminated or drenched clothing and protective equipment must be discarded or destroyed in accordance with state and local regulations. Do not apply Bladex formulations in such a manner as to directly, or through drift, expose workers or other persons. The area being treated must be vacated by unprotected persons. Keep out of reach of domestic animals, especially cattle, to prevent serious illness or possible death.

Hexazinone

20. Hexazinone is the active ingredient in three types of formulations: a wettable-powder formulation (Velpar) containing 90% ai w/w; a water-dispersible liquid formulation (Velpar L) containing 2 lb ai/gal; and a water-soluble granular formulation (Velpar ULW) containing 75% ai w/w. Hexazinone is used for selective weed control in certain crops (dormant alfalfa, Christmas tree plantings, pineapple, sugarcane), weed and brush control in reforestation areas, and nonselective weed and brush control in noncrop areas.

21. Hexazinone is an effective general herbicide that provides both contact and residual control of many annual and biennial weeds, most perennial weeds (except johnsongrass), and woody plants. Applied as a soil treatment, rainfall or irrigation is required to move the herbicide downward to the roots in the soil. For best results, apply hexazinone just before, or soon after, weed emergence, when weeds are less than 2 in. in height or diameter. The soil should be moist at time of application, and a timely 0.25–0.5 in. of rainfall occurring within 2 weeks after application ensures good weed control.

22. Hexazinone does not control johnsongrass, and, when johnsongrass is present as a problem weed, a suitable herbicide must be used for its control. For enhanced brush control in areas of reforestation, hexazinone may be tank-mixed with picloram (Tordon). Bear in mind, however, that Tordon is a Restricted Use Pesticide.

23. On herbaceous plants, symptoms of hexazinone activity usually appear within 2 weeks after application under warm, humid conditions, while 4–6 weeks may be required when weather is cool. If rainfall or irrigation fails to move hexazinone into the root zone after foliar application, plants may recover from foliar-contact effects. Similarly, poor weed control will result if the herbicide is not moved downward in the soil by water such that it makes contact with roots. On woody plants, hexazinone symptoms usually appear within 3–6 weeks after the herbicide has been root-absorbed during a period of active growth. Defoliation and subsequent refoliation may occur, but susceptible plants will be killed. Although hexazinone has some foliar-contact activity, it must be root-absorbed and translocated upward to the foliage for complete kill.

24. Hexazinone is applied in croplands at 0.5–1.8 lb ai/A, depending on the weed species targeted and the specific crop involved. In noncroplands, hexazinone is applied at 1.8–4.5 lb ai/A for short-term (up to 3 months) weed control, 5.5–11 lb ai/A for season-long weed control, and 3.6–7.2 lb ai/A for brush control. Applied postemergence, a nonionic surfactant must be added to the spray solution (1 qt/100 gal) prior to use.

Metribuzin

25. Metribuzin is the active ingredient in water-dispersible granule formulations (Lexone DF, Sencor DF) that contain 75% ai w/w. Metribuzin is used for premergence or early postemergence control of annual grass

and broadleaf weeds in certain agronomic and horticultural crops and in bermudagrass turf. Formulations of metribuzin may be tank-mixed with certain other herbicides for more effective weed control (refer to product labels).

26. Metribuzin is *not* a Restricted Use Pesticide. However, it can leach in soils and has been found in groundwater as a result of agricultural use. Do not use metribuzin where the water table is close to the soil surface and where soils such as sand and loamy sands are very permeable. It can contaminate groundwater used as drinking water.

27. Metribuzin is used in alfalfa (dormant), field corn, potatoes, sainfoin (dormant), soybeans, sugarcane, and tomatoes to control susceptible emerged seedling weeds. Metribuzin may be applied preplant-incorporated, preemergence, or postemergence to soybeans. Applied postemergence, soybeans must be at least 12 in. tall and broadleaf weeds not over 3 in. in height and seedling grasses and common ragweed not over 1 in. in height. Apply as a directed spray to cover weeds with minimal or no contact to soybean plants.

28. Metribuzin is applied in croplands at 0.25–1.0 lb ai/A, depending on organic matter content in the soil, its texture, and the crop involved; a rate as high as 2.0 lb ai/A may be applied preemergence in early spring to asparagus.

29. To avoid possible crop injury, do not use metribuzin on sand, loamy sand, or soils with less than 0.5% organic matter content. Do not use on peat or muck soils, as poor weed control will result because this herbicide adsorbs to the organic colloids. The herbicidal activity of metribuzin increases as the soil pH increases (becomes more alkaline).

Prometon

30. Prometon is the active ingredient in a liquid formulation (Pramitol 25E) that contains 2 lb ai/gal (0.2 kg ai/L), and as one of four active ingredients in a pelleted formulation (Pramitol 5PS) that contains 5.0% prometon, 0.75% simazine, 40.0% sodium chlorate, and 50.0% sodium borate (percent is w/w). Prometon is used as a nonselective herbicide to provide total vegetation (annual and perennial grass and broadleaf weeds) control in noncrop areas such as industrial sites, around the farm, and under asphalt paving.

31. Used in noncrop areas, rates of Pramitol 25E for annual grass and broadleaf weed control vary from 4.0

to 6.125 gal product/A (8–12 lb ai/A) and for hard-to-kill perennial weeds, 7.5–10 gal product/A (15–20 lb ai/A). For annual grass and broadleaf weed control, rates of Pramitol 5PS range from 152 to 200 lb product/A and, to control most perennial grass and broadleaf weeds, the rate varies from 304 to 400 lb product/A, depending on the species involved.

32. Since prometon enters plants through their roots, its effectiveness depends on rainfall to move the herbicide into the root zone; very dry soil conditions and lack of rainfall may result in poor weed control.

33. Simazine, one of the active ingredients in Pramitol 5PS, may leach into and contaminate groundwater that may subsequently be used as drinking water. Thus users of Pramitol 5PS are advised not to apply this product where the water table is close to the soil surface and where soils such as sands and loamy sands are very permeable. However, the water solubility of simazine is very low (3.5 ppmw at 68°F or 20°C), which would negate leaching and subsequent groundwater contamination. Pramitol 5PS is *not* a Restricted Use Pesticide.

Prometryn

34. Prometryn is the active ingredient in a liquid formulation (Caparol 4L) that contains 4 lb ai/gal. Prometryn is used for selective annual grass or broadleaf weed control in cotton, celery, and pigeon peas (Puerto Rico only).

35. Caparol 4L is applied preemergence or early postemergence to the weeds. Used in cotton, Caparol 4L may be applied preplant-incorporated, preemergence, and/or directed postemergence to the crop. Postemergence applications may follow preplant-incorporated or preemergence treatments of Caparol 4L. Refer to product label for specific recommendations for Caparol 4L use in cotton, celery, and pigeon peas.

36. Caparol 4L may be tank-mixed or sequentially applied in cotton with pendimethalin, trifluralin, DSMA, or MSMA to control a broader spectrum of weed.

Simazine

37. Simazine is the active ingredient in two types of formulations: a liquid containing 4 lb ai/gal (Princep 4L), and a water-dispersible granule (Princep Caliber 90) containing 90% ai w/w.

38. Simazine is used for the selective preemergence control of annual grass and broadleaf weeds in fruit and nut crops, corn, sugarcane, turfgrass, nurseries, Christmas tree plants, shelterbelts, asparagus, and artichokes. It is soil-applied prior to weed emergence or after emerged weeds have been killed. Simazine is applied at 2–4 lb ai/A, with the lower rate applied to coarse-textured soils and the higher rate to organic soils (peat, muck, high-organic clay).

39. Simazine can be tank-mixed with glyphosate, paraquat, norflurazon, oryzalin, EPTC + antidote, or atrazine, depending on the crop and weed species involved.

40. Simazine may leach into and contaminate groundwater that may subsequently be used as drinking water. However, the water solubility of simazine is very low (3.5 ppmw at 68°C or 20°C). Do not use simazine where the water table is close to the soil surface or where the soils such as sand and loamy sands are very permeable. However, simazine is *not* a Restricted Use Pesticide.

28 URACIL HERBICIDES

■ INTRODUCTION

The chemical structures of the uracil herbicides are centered about a common six-membered ring composed of four carbon atoms and two nitrogen atoms separated by one of the four carbon atoms. One oxygen atom is attached to the ring structure by a double bond to the carbon separating the two nitrogen atoms, and a second oxygen atom is double-bonded to the carbon atom located at the 4-position of the ring. The chemical structure of this common nucleus is:

common nucleus of uracil herbicides

There are two members in the uracil herbicide family (Table 28.1), and they differ in the substituents at the 3-

and 5-positions on the ring structure. Each herbicide has a methyl group bonded at the 6-position. The chemical structures of these herbicides are shown in Figure 28.1.

■ CHARACTERISTICS OF URACIL HERBICIDES

Principal Use

Bromacil is used for selective weed control in citrus orchards (Florida and Texas only) and pineapple (Hawaii, Florida, and Puerto Rico) and also to control most annual and perennial weeds and woody plants in noncrop areas.

Terbacil is principally used in the selective control of annual and perennial weeds in crops such as alfalfa, asparagus, blueberries, caneberries, mint, sugarcane, and orchards of apples, peaches, and pecans.

Mode of Action

The uracil herbicides inhibit photosynthesis by blocking electron transport through photosystem II (PS II).

TABLE 28.1 ■ Common and chemical names of the uracil herbicides.

COMMON NAME	CHEMICAL NAME
Bromacil	5-bromo-3-*sec*-butyl-6-methyluracil
Terbacil	3-*tert*-butyl-5-chloro-methyluracil

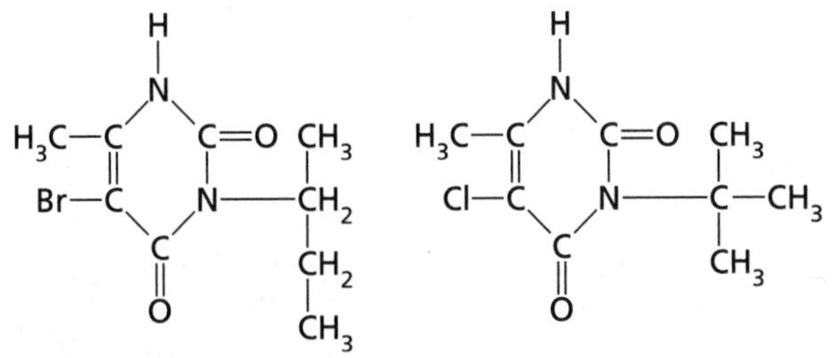

bromacil
(Hyvar)

terbacil
(Sinbar)

FIGURE 28.1 ■ Chemical structures of the uracil herbicides.

Translocation

Following root absorption, the uracil herbicides are translocated upward to the leaves of susceptible plants.

Selectivity

The selective action of the uracil herbicides appears to be the differential translocation between tolerant and susceptible plants and herbicide placement relative to plant roots and foliage.

Common Characteristics

1. The uracil herbicides are readily root-absorbed, less so through foliage and stems. The addition of a wetting agent to the spray mixture enhances foliar activity.

2. The uracil herbicides are not readily adsorbed to soil colloids, and they are leached in soils.

3. When using the uracil herbicides, avoid applying them in locations where the roots of desired plants (other than those specified) extend into the treated area, or where they may be washed onto, or moved into contact with, the roots of desired plants. Bear in mind that the roots of trees can extend far beyond the drip line of the foliar canopy.

Exceptional Characteristic

1. Bromacil is especially effective in the control of perennial grasses.

■ INFORMATION OF PRACTICAL IMPORTANCE

1. Best results are obtained when the uracil herbicides are applied shortly before or shortly after weed growth begins. They must be applied to the soil; if dense vegetation is present, remove it to expose the soil surface.

2. Moisture is necessary to carry the uracil herbicides into the weed root zone, and they are most effective when rainfall or irrigation follows within 2 weeks after application.

3. The uracil herbicides will not control hard-to-kill, deep-rooted perennial weeds in areas of limited rainfall (usually less than 4 in.).

4. In general, the effects of the uracil herbicides on perennial weeds are slow to appear, usually progressing over a period of several months.

5. Perennial weed control may be improved by cultivation *prior* to application of these herbicides. Avoid working the soil after application, which could reduce their effectiveness.

6. In general, the uracil herbicides are applied at 0.8–3.2 lb ai/A for annual weed control, 5.6–12 lb ai/A for perennial weed control, and 10–20 lb ai/A to control deep-rooted, hard-to-kill perennial weeds.

Bromacil

7. Bromacil is the active ingredient in two water-soluble formulations (Hyvar L and Hyvar X-L) that contain 3.2 and 2 lb ai/gal, respectively. In each of these formulations, bromacil is in the form of its lithium salt, but the active ingredient is based on the amount of bromacil in the salt. Bromacil is also the active ingredient in a wettable-powder formulation (Hyvar X) that contains 80% ai w/w.

8. Hyvar L is used in citrus orchards. It is particularly effective in controlling perennial grass weeds. For annual and perennial weed control, apply Hyvar L any time of the year, preferably shortly before or shortly after weed growth begins. Herbicidal activity may not

be apparent until the chemical has been carried into the root zone by rainfall or irrigation water. In most cases, Hyvar L is used in citrus crops at a rate of 1.6–2.4 lb ai/A (2–3 qt product/A).

9. Hyvar X-L is used in noncrop areas. Apply Hyvar X-L before or during the period of active weed growth when rainfall can be expected to carry the herbicide into contact with the weed roots. If the soil surface is covered by vegetation, remove it prior to application of the herbicide. For annual weed control, apply Hyvar X-L at 3–6 lb ai/A (1.5–3 gal product/A). To control easier-to-kill perennial weeds, apply Hyvar X-L at 6–12 lb ai/A (3–6 gal product/A). To control johnsongrass and other hard-to-kill perennial weeds, apply Hyvar X-L at 12–24 lb ai/A (6–12 gal product/A).

10. Hyvar X is used as a nonselective weed and brush herbicide in noncrop areas. For annual weed control, apply Hyvar X at 2.4–4.8 lb ai/A (3–6 lb product/A). For brush and perennial weed control in noncrop areas, apply Hyvar X at 5.6–12 lb ai/A (7–15 lb product/A). For small areas, apply Hyvar X at 0.25 cupful of product/250 ft^2, equivalent to about 15 lb product/A.

11. Hyvar X may also be used in citrus and pineapple. It is applied at 1.6–3.2 lb ai/A (2–4 lb product/A) in citrus, and 1.6–4.8 lb ai/A (2–6 lb product/A) in pineapple.

12. Hyvar X may be tank-mixed with dicamba, 2,4-D amine, MSMA, or paraquat for better control when emerged weeds are present. An appropriate surfactant should be added to the spray mixture to enhance foliar wetting.

13. Bromacil and diuron are available as premixes in dispersible granular formulations (Krovar I DF and Krovar II DF) that contain 40% bromacil *plus* 40% diuron w/w and 53% bromacil *plus* 27% diuron w/w, respectively. Krovar I and Krovar II are used for selective weed control in citrus crops and for nonselective weed control in noncrop areas.

Terbacil

14. Terbacil is available as the active ingredient in a wettable-powder formulation (Sinbar) that contains 80% ai w/w. Sinbar is used as a selective herbicide in alfalfa, asparagus, apples, peaches, pecans, blueberries, caneberries, mint, and sugarcane. The dosage rate of Sinbar varies from about 0.4 to 2.4 lb ai/A (0.5–3 lb product/A), depending on the crop involved and the amount of organic matter in the soil.

15. For best results, Sinbar should be soil-applied before or shortly after weed growth begins, such that rainfall or sprinkler irrigation will occur within 2 weeks after application (to move the herbicide into the weed root zone). In some cases, as in mint, Sinbar may be applied postemergence to broadleaf weeds less than 2 in. tall (or across) and to grasses less than 1 in. tall.

29 UREA HERBICIDES

■ INTRODUCTION

The chemical structures of the urea herbicides have as their common nucleus the structure of urea, a nitrogen fertilizer for plants and a protein source fed to cattle. The chemical structure of urea is:

$$H-N-C-N-H$$

urea

Members of the urea herbicide family vary one from another in the substitutions on one or both of the amino groups of the urea molecule. Their common and chemical names are given in Table 29.1 on page 244, and their chemical structures are shown in Figure 29.1.

■ CHARACTERISTICS OF UREA HERBICIDES

Principal Use

The substituted urea herbicides are used for selective preemergence and postemergence control of seedling grass and broadleaved weeds in croplands and the nonselective control of annual and perennial weeds in noncrop areas. Although the urea herbicides control both grass and broadleaved weed seedlings, they are most effective against broadleaved seedlings.

Mode of Action

The substituted urea herbicides inhibit photosynthesis by binding onto the pigment protein of the photosystem II complex in the thylakoid membrane of chloroplasts, thus

TABLE 29.1 ■ Common and chemical names of the urea herbicides.

COMMON NAME	CHEMICAL NAME
Diuron	N'-(3,4-dichlorophenyl)-N,N-dimethylurea
Fluometuron	N,N-dimethyl-N'-[3-(trifluoromethyl)-phenyl]urea
Linuron	N'-(3,4-dichlorophenyl)-N-methoxy-N-methylurea
Siduron	N-(2-methylcyclohexyl)-N'-phenylurea
Tebuthiuron	N-(2-methylcyclohexyl)-N'phenylurea

interfering with normal electron transport into the plasto-quinone pool. The urea and triazine herbicides have the same mode of action.

In addition to preventing energy production *via* photosynthesis, secondary effects (e.g., production of free radicals) are most likely responsible for the rapid plant kill.

Translocation

The urea herbicides are readily absorbed by plant roots and translocated to the leaves *via* the transpiration stream.

Applied postemergence, they are absorbed by plant leaves. However, they are not translocated from the leaves to other plant parts.

Selectivity

Crop selectivity with the urea herbicides is primarily due to herbicide placement rather than physiological tolerance of the crop plants.

Common Characteristics

1. Following many years of continuous use of the urea herbicides (and other herbicides having the same mode of action), resistant weed biotypes of some species have developed (see Chapter 9, Weed Resistance and Transgenic Crop Tolerance, for more information).

2. At dosages recommended for selective weed control in croplands, the urea herbicides do not control established perennial weeds.

3. In general, the urea herbicides are of low water solubility (18–90 ppmw). However, tebuthiuron has a water solubility of 2300 ppmw.

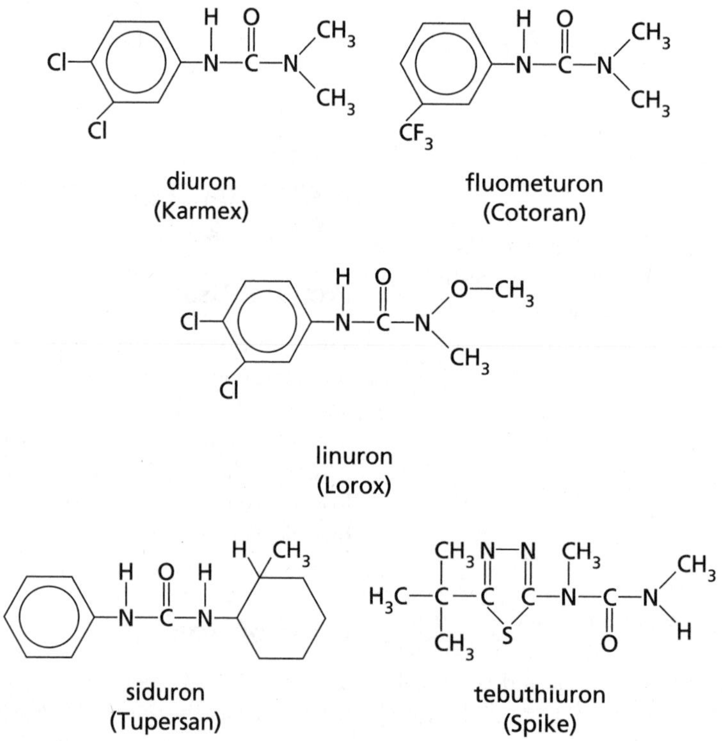

diuron
(Karmex)

fluometuron
(Cotoran)

linuron
(Lorox)

siduron
(Tupersan)

tebuthiuron
(Spike)

FIGURE 29.1 ■ Chemical structures of the substituted urea herbicides.

4. The urea herbicides are adsorbed to soil colloids, and they resist leaching. At dosages applied for selective weed control in croplands, they tend to remain in the upper 1-in. layer of soil.

5. In general, the urea herbicides are applied to the soil surface as a preemergence treatment to the weeds. Applied postemergence to young weeds, the addition of a suitable nonionic surfactant greatly enhances their postemergence activity.

6. There is little advantage to be gained from mechanically soil-incorporating the urea herbicides. In fact, mechanical incorporation can result in reduced weed control. However, since the herbicides must be present in the zone of root development of seedling weeds to be effective, best results are obtained when rainfall or irrigation occurs within 2 weeks after application.

7. Phytotoxic symptoms of the urea herbicides disappear within one crop season when applied at lower selective rates. At higher "soil sterilant" rates, phytotoxicity may occur for more than one crop year.

8. Microbial breakdown is the principal means by which the urea herbicides are degraded in soils. Photodecomposition may occur when the urea herbicides are exposed to sunlight for several days or weeks under hot, dry conditions.

9. Applied at rates greater than those recommended for selective weed control in croplands, the urea herbicides may be used for nonselective weed control in noncrop areas.

Exceptional Characteristics

1. At high rates, 20–60 lb ai/A, diuron is used as a soil sterilant in noncrop areas; however, it may not provide satisfactory control of deep-rooted perennials.

2. Fluometuron is registered for use in cotton only.

3. The mode of action of siduron is inhibition of root growth. It is not a potent inhibitor of photosynthesis, as are other urea herbicides.

4. Siduron selectively controls seedling grass weeds in turfgrass, both newly seeded and established plantings. Siduron does not control annual bluegrass or most broadleaf weeds. Do not use siduron on bermudagrass turf or as a preemergence treatment to dichondra.

5. The soil residual life of siduron is 4 months or less when applied at recommended rates.

6. Tebuthiuron is used to control brush and woody plants in rangelands and permanent pastures, highway and utility

rights-of-way, fence rows, industrial sites, and unimproved areas. It is not used for selective weed control in croplands.

■ INFORMATION OF PRACTICAL IMPORTANCE

1. Control of emerged weeds with the urea herbicides under drought conditions is impractical.

2. Adding a suitable nonionic surfactant to spray mixtures that contain urea herbicides greatly enhances their postemergence activity.

3. Continuous agitation of the spray mixture is essential to keep urea herbicides in suspension in the water carrier.

4. When applying urea herbicides preemergence or postemergence to the weeds in emerged crops, use directed sprays so as not to contact the crop foliage.

5. To avoid crop injury, do not use urea herbicides on light, sandy soils.

6. Applied under dry soil conditions, the urea herbicides will likely provide poor weed control. Moisture is essential to bring these herbicides into contact with the root systems of the weeds.

7. There is no buildup or accumulation of urea herbicides in soils, even when applied at crop selective rates year after year.

8. To avoid injury or death to desired plants, do not use urea herbicides where they may be root-absorbed by trees, shrubs, or herbaceous plants, or on lawns, walks, driveways, tennis courts, or similar areas.

Diuron

9. Diuron is the active ingredient in a water-dispersible granule formulation (Karmex DF) that contains 80% ai w/w. It is used for selective weed control in certain crops, such as established alfalfa, asparagus, field corn, cotton, grain sorghum, fruit and nut crops, peppermint, grapes, berries, tree plantings, and nonselectively in noncrop areas. It may also be mixed with certain other herbicides to control a broader spectrum of weeds.

10. Diuron (Karmex DF) is applied to the crop as a preplant (not incorporated), preemergence, directed postemergence, or as a lay-by treatment, depending on the crop involved and the geographic (regional) location. Used with a surfactant, diuron has postemergence

activity on young, susceptible weeds. Applied as a directed postemergence treatment, do not allow spray or spray drift of diuron to contact crop foliage. Without surfactant, diuron may be applied over-the-top of certain established crops such as alfalfa, asparagus, red clover, sugarcane, and wheat. Used in croplands, the dosage of diuron varies from 0.8 to 2.0 lb ai/A (1–2.5 lb product/A).

11. Diuron (Karmex DF) may be applied preemergence following a preplant-incorporated (PPI) application of trifluralin (Treflan) in cotton.

Fluometuron

12. Fluometuron is available as the active ingredient in two kinds of formulations: liquid (Cotoran 4L), which contains 4 lb ai/gal, and water-dispersible granules (Cotoran DF), which contain 85% ai w/w.

13. Fluometuron (Cotoran) is recommended for use only in cotton. Refer to product labels for specific precautions.

14. Cotoran is most effective when three sequential applications are made during the crop year. These applications are (a) preplant (shallow incorporation) or preemergence; (b) postemergence (directed, semidirected, or over-the-top) when cotton is 3–6 in. tall; and (c) at lay-by. Cotoran may be tank-mixed with other appropriate herbicides for use in cotton.

15. Cotoran is applied in cotton preplant or preemergence at 1.6 lb ai/A, postemergence at 1.0–2.0 lb ai/A, and at lay-by at 1.0 lb ai/A.

Linuron

16. Linuron is the active ingredient in two kinds of formulations; liquid (Lorox L), which contains 4 lb ai/gal and water-dispersible granules (Lorox DF), which contain 50% ai w/w.

17. Linuron (Lorox) is used for the preemergence or postemergence control of annual grass and broadleaf weeds in certain crops (noted in item 18) and for short-term annual weed control in noncrop areas. The dosage rate varies from 0.5 to 1.5 lb ai/A, depending on the crop involved and timing of application.

18. Crops that may be treated with linuron (Lorox) include asparagus (newly seeded or newly transplanted crowns and established), bulbs (tulip, canna lily, daffodil, and dutch iris only), carrots, celery, corn (field and sweet), cotton, parsnips, potatoes, sorghum, and soybeans.

Siduron

19. Siduron is available as the active ingredient in a wettable-powder formulation (Tupersan) that contains 50% ai w/w.

20. Tupersan is used for the preemergence control of annual grass weeds in turfgrass areas (golf course fairways, lawns, parks, roadsides, and turf grasses grown for seed or sod production). It may be used on newly seeded turfgrass areas, as well as on established turfgrass. It is highly effective in the control of both smooth and hairy crabgrass.

21. It is recommended that Tupersan be applied in at least 100 gal water/A (2.5 gal/1000 ft^2). In new spring plantings, apply Tupersan at the rate of 2–6 lb ai/A (0.75–2.25 oz ai/1000 ft^2), equivalent to 4–12 lb product/A (1.5–4.5 oz product/1000 ft^2). Use lower rates on light, sandy soils and the higher rates on soils high in clay and/or organic matter.

22. Tupersan may be used on the following bentgrass strains: Penncross, Seaside, Highland, Astoria, Nimisila, C-1, C-7, and C-19. Do not use on other bentgrass strains, or injury may result.

23. Tupersan does not control broadleaf weeds.

24. Written or oral warnings must be given to workers who are expected to be in a treated area or in an area about to be treated with Tupersan. Refer to the product label for the specific wording of this warning.

Tebuthiuron

25. Tebuthiuron is available as the active ingredient in a pellet formulation (Spike 20P) that contains 20% ai w/w.

26. Spike 20P is used for the control of brush and woody plants. It is an extremely active herbicide that will kill trees, shrubs, and other forms of vegetation whose roots extend into the treated area. Spike 20P may seriously injure desirable forage legumes such as lespedeza or clover.

27. Spike 20P may cause temporary herbicidal symptoms to appear on perennial grasses. Dormant-season application is recommended to minimize the herbicidal effects on desired grasses.

28. Spike 20P is a root-active, soil-applied herbicide, and sufficient rainfall is required to move the herbicide into the root zone. The time required to control trees and brush with Spike 20P may vary from 1 to 3 years and is dependent on the amount of rainfall, soil

texture, soil/rooting depth, and susceptibility of the targeted species.

29. The broadcast rate of Spike 20P in pasture and rangeland is 3.75–20 lb product/A; in noncrop areas, the rate is 3.75–30 lb product/A. For individual plant treatment, apply Spike 20P around the base of the stem at a rate 0.25 oz per 1–2 in. of stem diameter or, for clumps or stands of multiple stems, apply 0.25 oz product per 22 ft^2.

30

NONFAMILY HERBICIDES

■ INTRODUCTION

Herbicides not having a common chemical nucleus to justify their inclusion in a herbicide family are grouped as nonfamily herbicides. Herbicides included in this group are listed below, and their respective characteristics are given in the following discussions.

Amitrole	Ethofumesate
Asulam	Fluridone
Bensulide	Glyphosate
Bentazon	Isoxaben
Clomazone	Pyridate
Diethatyl	Sulfosate
Difenzoquat	Inorganic herbicides
Dithiopyr	

The arrangement of this chapter is somewhat different from the others in Section II. For clarity, each herbicide has its own "Information of Practical Importance" list, as opposed to one large list covering an entire family.

■ AMITROLE

The chemical name of amitrole is *3-amino-1,2,4-triazole*; it is a Restricted Use Pesticide because it can cause cancer in laboratory animals. Its chemical structure is:

amitrole
(Amizol, Amitrol-T)

Principal Use

Amitrole is principally used as a nonselective, systemic herbicide applied postemergence to control annual grasses, annual and perennial broadleaf weeds, and certain woody plants and vines in industrial and other noncrop areas.

Mode of Action

Amitrole causes bleaching in plants, and its primary mode of action is the inhibition of cartenoid formation. Its site of action is the thylakoid membrane in chloroplasts.

Translocation

Amitrole is readily foliar-absorbed and translocated throughout the plant, accumulating in active meristem tissue such as buds and young tissue.

Selectivity

Herbicide placement and dosage are the primary factors contributing to amitrole selectivity.

Information of Practical Importance

1. Amitrole is the active ingredient in a water-soluble liquid formulation (Amitrol-T) that contains 2 lb ai/gal, and in a water-soluble powder formulation (Amizol) that contains 90% ai w/w.

2. Amitrol-T is used as a spray on green, actively growing succulent and woody plants at a rate of 1–2 gal product/A mixed with 100 gal water. For best results, all leaves, stems, and suckers (when present) must be thoroughly wet to the ground line.

3. Amizol is used at 1–4 lb product/A mixed with 100 gal water. Amizol is used in combination with other herbicides (primarily triazine, urea, or uracil herbicides) that are phytotoxic to underground plant parts (roots, rhizomes, reproductive parts) and have a prolonged period of residual activity in the soil.

4. Both formulations of amitrole are used in industrial and other noncrop areas. They control annual and perennial broadleaved plants and suppress perennial grasses. They are especially effective in controlling quackgrass, Canada thistle, poison ivy and poison oak, honeysuckle (which may require a second application), kudzu, and certain woody plants. The control of quackgrass, bermudagrass, and stoloniferous bent grasses is increased by tank-mixing amitrole with ammonium thiocyanate.

5. Plant tissue formed after treatment with amitrole is chlorotic (white) and often completely devoid of chorophyll. When sublethal amounts of amitrole are used, chlorotic tissues may, in time, regain their normal green color. When death occurs, the regions of active growth die first, with necrosis progressing downward along the stem.

6. Amitrole quickly decomposes in warm, moist soils, with an application of 4 lb ai/A decomposing in about 7 days. Most of the herbicide spray is retained on the plant foliage, with a minor portion reaching the soil surface.

■ ASULAM

Asulam is a sulfonylcarbamate, with a sulfonyl group bonding a phenyl ring to the nitrogen of the carbamate nucleus. The chemical structures of the methyl ester of phenylcarbamic acid and asulam are shown in Figure 30.1. The chemical name of asulam is *methyl[(4-aminophenyl)sulfonyl]carbamate*.

It is of interest to compare the similarity in chemical structure of asulam with that of chlorsulfuron (Chapter 25) and their dissimilarity in herbicidal activity (where asulam controls grass weeds and chlorsulfuron controls broadleaf weeds).

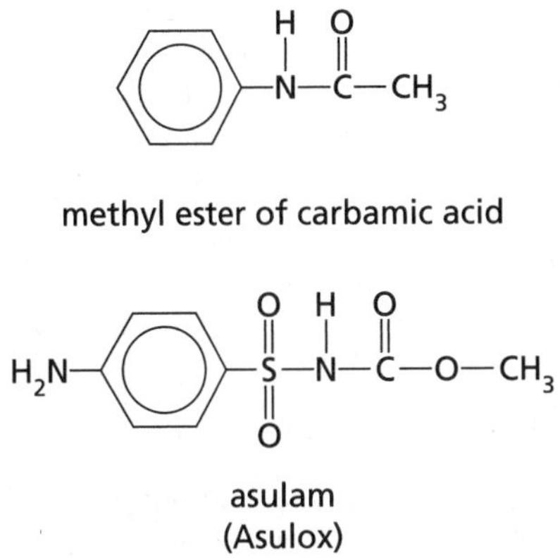

methyl ester of carbamic acid

asulam
(Asulox)

FIGURE 30.1 ■ Chemical structures of the methyl ester of carbamic acid and asulam.

Principal Use

The principal use of asulam is the postemergence control of annual and perennial grasses in certain crops and in noncrop areas. Asulam is one of the best herbicides for the postemergence control of brackenfern.

Mode of Action

Asulam inhibits mitosis and thereby interferes with cell division and expansion.

Translocation

Asulam is readily absorbed by roots and foliage and translocated to other plant parts, where it accumulates in the meristems.

Selectivity

The selective action of asulam is due to the rapid metabolic degradation of the herbicide by tolerant plants and the failure of susceptible plants to do the same.

Information of Practical Importance

1. Asulam is the active ingredient in a water-soluble liquid formulation (Asulox) that contains 3.34 lb ai/gal.

2. Asulox is used to selectively control annual and perennial grasses in sugarcane, Christmas tree plantings, warm-season turfgrasses (St. Augustinegrass, bermudagrass), ornamentals (junipers and yews), and noncrop areas. In croplands, it is applied at 2–3.3 lb ai/A (0.6–1 gal product/A). In noncrop areas, Asulox is applied at 3.3–6.7 lb ai/A (1–2 gal product/A).

3. Weeds controlled by Asulox include barnyardgrass, crabgrass spp., foxtail spp., goosegrass, sandbur spp., itchgrass (Raoulgrass), johnsongrass, paragrass, alexandergrass, and western brackenfern.

4. Asulam is readily leached in soils, and its soil persistence is apparently less than 4 weeks.

■ BENSULIDE

Bensulide is a benzenesulfonamide herbicide. Its chemical name is S-(O,O-*diisopropylphosphorodithioate*)ester of N-(*2-mercaptoethyl) benzene sulfonamide*. The chemical structure of bensulide is:

Bensulide (Bensumec, Prefar)

Principal Use

Bensulide is used for the preemergence control of annual grass and broadleaf weeds in vegetable crops, turfgrass, ornamentals, and ground covers.

Mode of Action

Bensulide inhibits root growth. Its exact mode of action is not known.

Translocation

Bensulide undergoes little or no translocation in plants. It is adsorbed to the root surfaces and a small amount is absorbed by the roots.

Selectivity

The mechanism by which selectivity is achieved is not known.

Information of Practical Importance

1. Bensulide is the active ingredient in a liquid flowable formulation (Bensumec 4LF) and an emulsifiable concentrate (Prefar 4EC), each containing 4 lb ai/gal.

2. Bensulide does not control emerged weed seedlings or established plants.

3. Bensulide is recommended for use on mineral soils only. It is strongly adsorbed to organic colloids and thereby inactivated in soils of high organic matter.

4. Bensulide leaches very little in mineral or organic soils.

5. Bensumec 4LF is recommended for use in established turfgrass in home lawns, parks, and golf courses. *Precationary statement:* Do not apply peat moss or manure as a top dressing to the turfgrass before applying Bensumec, as the herbicide will be strongly adsorbed to the organic matter and inactivated.

6. Bensumec controls annual grass weeds, such as annual bluegrass, barnyardgrass, crabgrass (hairy, smooth), fall panicum, foxtails, and goosegrass, and annual broadleaf weeds, such as henbit, common lambsquarters, redroot pigweed, and shepherd's-purse.

7. Bensumec must be applied prior to emergence of weed seedlings from the soil; if seedling growth is visible above the soil, it is too late to apply Bensumec.

8. Crabgrass will emerge from the soil 4 to 6 weeks before it is visible above the grass sod.

9. In general, Besumec should be applied in March for summer weed control and in August or September for winter weed control.

10. Bensumec can be applied to dichondra lawns at the time of seeding or at any time thereafter.

11. Bensumec 4LF is applied at 7.5–10 lb ai/A (1.875–2.5 gal product/A) or 2.8–4.7 oz ai/1000 sq ft (5.6–9.4 fl oz product/1000 sq ft. Use the lower rate for crabgrass control and the higher rate for control of annual bluegrass and other annual weeds. Sequential or repeat treatments may be used.

12. Prefar is applied preplant incorporated or preemergence for control of annual weeds in vegetable crops such as carrots, cole crops, cucurbits, endive, lettuce, onions, parsnips, peppers, and tomatoes. Refer to label for state restrictions.

13. Prefar is applied at 3–6 lb ai/A (3–6 qt product/A). The rate used varies with the crop involved.

■ BENTAZON

The chemical name of bentazon is *3-(1-methylethyl)-1H-2,1,3-benzothiadiazin-4(3H)-one 2,2-dioxide*, and its chemical structure is:

bentazon (Basagran)

Principal Use

The principal use of bentazon is the postemergence control of broadleaf weeds and yellow nutsedge in certain grass and broadleaf crops. It does not control grasses.

Mode of Action

Bentazon inhibits photosynthesis by blocking electron transport *via* photosystem II (PS II).

Translocation

Bentazon undergoes little or no translocation *via* the photosynthate stream following foliar absorption. It is rapidly translocated *via* the transpiration stream to the leaves following root absorption.

Selectivity

Bentazon is rapidly degraded metabolically to nonphytotoxic metabolites in tolerant plants, while susceptible plants are unable to do this.

Information of Practical Importance

1. Bentazon is the active ingredient in a water-soluble concentrate formulation (Basagran) that contains 4 lb ae/gal. It is formulated as the sodium salt.

2. Basagran is applied postemergence to the weeds. It is effective mainly through foliar contact.

3. When Basagran is applied in tank mix with any one of the various graminicide herbicides (e.g., diclofop-methyl, fluazifop-butyl, haloxyfop-methyl, and sethoxydim), the phytotoxicity of the graminicide is reduced. The activity of Basagran is not affected by the tank mix.

4. Basagran is applied postemergence at 0.5–1 lb ai/A (1–2 pt product/A), depending on the targeted species and the stage of growth.

5. Soybeans are tolerant to Basagran at all growth stages. When the soybeans are in the one- to four-trifoliate leaf stage, the first flush of weeds are generally in the proper stage to treat with Basagran.

6. A nonphytotoxic oil concentrate must be added to the Basagran spray mixtures prior to application. Urea ammonium nitrate (UAN) solution or ammonium sulfate (AMS) may be added to the spray mixture in place of the oil concentrate for improved control of velvetleaf, cocklebur, common sunflower, Pennsylvania smartweed, devilsclaw, venice mallow, and wild mustard.

7. Certain weed species (such as common lambsquarter and common ragweed) may not be controlled when UAN or AMS is used in place of an oil concentrate. When such weeds are present, along with those better controlled with UAN or AMS, then both an oil concentrate and UAN or AMS should be added to the spray mixture. When applied by ground spray equipment, the oil concentrate is added at 1.25% v/v (maximum of 2 pt/A), UAN is added at 0.5–1 gal/A, and AMS at 2.5 lb/A. UAN is a liquid nitrogen solution, while AMS is a dry, granular, nitrogen-source fertilizer.

8. For best results in yellow nutsedge control, make two sequential applications of Basagran at 1.5–2 pt product/A when plants are 6–8 in. high. The second application may be made 7–10 days after the first.

9. Bentazon is one of two active ingredients in two premix liquid formulations (Laddock, Prompt), each containing 1.66 lb/gal bentazon and 1.66 lb/gal atrazine. Laddock and Prompt are Restricted Use Pesticides because of the presence of atrazine in their formulations. Laddock is used postemergence in corn (all types) and sorghum before these crops reach 12 in. in height. Prompt is used only in the warm-season turfgrasses St. Augustinegrass, zoysiagrass, and centipedegrass for the postemergence control of young, actively growing broadleaf weeds.

■ CLOMAZONE

The chemical name for clomazone is *2-[(2-chlorophenyl)-methyl]-4,4-dimethyl-3-isoxazolidinone*. The chemical structure of clomazone is:

clomazone (Command)

Principal Use

Clomazone is applied to the soil and incorporated for preemergence control of annual grass and broadleaf weeds in soybeans, green peas, and pumpkins.

Mode of Action

Clomazone causes bleaching in susceptible plants. Its primary mode of action is inhibition of carotenoids. Seedlings of susceptible plants emerge from the soil devoid of chlorophyll and white in color, with death occurring shortly thereafter. Clomazone inhibits the biosynthesis of carotenoids and chlorophyll in susceptible plant species by inhibiting the enzyme *phytoene desaturase* in the carotenogenic pathway.

Translocation

Clomazone is readily absorbed by roots and underground shoots growing in treated soil. Following absorption, it is translocated to the leaves *via* the transpiration stream, accumulating in the leaves. Clomazone is foliar-absorbed, as demonstrated by the injuries it causes through spray and vapor drift. Clomazone is not translocated in the photosynthate stream from the leaves.

Selectivity

The primary factors contributing to clomazone selectivity are not known. Soybean plants are totally tolerant of clomazone at all stages of growth. Velvetleaf, a primary weed in soybeans grown in the Midwest, is highly susceptible to clomazone.

Information of Practical Importance

1. Clomazone is the active ingredient in an emulsifiable-concentrate formulation (Command 4EC) that contains 4 lb ai/gal (0.5 kg ai/L).

2. Command is applied preplant-surface, preplant-incorporated, or preemergence for selective weed control in soybeans at 0.5–1.25 lb ai/A (1–1.5 pt product/A). It is also applied preplant-incorporated or preemergence in green peas and pumpkins at a rate of 0.5 lb ai/A (1 pt product/A). If the initial stand of the pea or pumpkin crop fails, the crop can be replanted in fields treated with Command alone. Do not re-treat with a second application of Command.

3. Command presents three potential problems for the user: soil carryover, spray- and vapor-drift injury, and its inability to control all problem broadleaf weeds.

4. Wheat, a common rotational crop with soybeans, cannot be seeded in Command-treated soil until 12 months later, or crop injury will ensue. Other rotational crops require a waiting period of a minimum of 9 months after application. Under very dry conditions, the waiting period before planting can be even longer.

5. Spray and vapor drift of Command both pose a real hazard to susceptible plants located within several hundred yards of the treated area; it is particularly apparent because it turns affected plants white. Directions on the product label suggest a buffer zone of 1000 ft between point of application and sensitive vegetation. To reduce or avoid vapor drift, Command must be shallowly soil-incorporated immediately after application, even when applied as a preemergence treatment to the crop.

6. Command effectively controls many grass and broadleaf weeds. However, it does not effectively control certain problem broadleaf weeds (such as cocklebur, ivyleaf morningglory, pigweeds) or weeds that have developed resistance to the triazine herbicides (e.g., common groundsel, kochia, and velvetleaf). To control these resistant broadleaf species, it is necessary to use a tank mix of Command and a suitable herbicide at the time of application or as a sequential postemergence treatment.

7. Command is adsorbed to organic colloids in the soil, and it is not readily leached.

8. Clomazone is one of two active ingredients in a premixemulsifiable-concentrate formulation (Commence EC) that contains 2.25 lb clomazone/gal and 3.0 lb trifluralin/gal. Commence EC is used preplant-incorporated in soybeans.

■ DIETHATYL

The chemical name of diethatyl is N-*(chloroacetyl)*-N-*(2,6-diethylphenyl)glycine*. However, the ethyl ester of diethatyl is the herbicidal form, rather than the parent acid, and diethatyl ethyl is the active ingredient in the commercial formulation of this herbicide. The chemical structure of the ethyl ester of diethatyl is:

diethatyl ethyl (Antor 4ES)

Principal Use

Diethatyl ethyl is used preplant or preemergence to control annual grass and certain broadleaf weeds in sugar beets, spinach, red beets, and in bermudagrass (grown for seed only).

Mode of Action

The primary mode of action of diethatyl ethyl is not known.

Translocation

Diethatyl ethyl is almost completely metabolized in the plant within 96 hours. Its metabolites are translocated throughout the plant. With grass seedlings, the coleoptile and shoot are the principal sites of absorption. With broadleaf seedlings, the roots are the principal site of absorption.

Selectivity

The primary factors contributing to the selectivity of diethatyl ethyl are not known. To be effective, diethatyl ethyl must be dispersed in the weed seed germination zone of the soil profile, killing the seedling shortly after emergence from the seed.

Information of Practical Importance

1. Do not confuse diethatyl with dietholate. *Dietholate* is a nonherbicidal chemical that is used as a herbicide extender in combination with certain herbicides (e.g., EPTC).

2. Diethatyl ethyl is the active ingredient in an emulsifiable solution formulation (Antor ES) that contains 4 lb ai/gal. Note that the active ingredient of Antor is based on the ethyl ester of diethatyl, not on its acid form.

3. Antor is used preplant-incorporated in sugar beets. Where sugar beets are grown in beds, apply Antor 4ES after bedding and incorporate. Irrigate until tops of beds are thoroughly wetted.

4. Antor may be applied preemergence after seeding sugar beets. It is important that the top 2 in. of soil be thoroughly wetted after application to move the herbicide into the weed seed germination zone.

5. Antor is used in sugar beets at 3–6 lb ai/A (3–6 qt product/A), with the lowest rate on light (sandy) soils and the highest rate on heavy (clay) soils.

6. Antor can be tank-mixed or applied sequentially with other herbicides registered for use in sugar beets.

7. Antor is applied preplant-incorporated or preemergence in red beets or spinach at 3–4 lb ai/A (3–4 qt product/A). To be effective, the upper 2-in. soil layer must be thoroughly wetted by rainfall or irrigation.

8. The soil persistence of Antor is 6–10 weeks, depending on soil texture and moisture.

9. Diethatyl is related chemically to the herbicides glyphosate and sulfosate *via* their common *glycine* nucleus (Chapter 30).

■ DIFENZOQUAT

The chemical name of difenzoquat is *1,2-dimethyl-3,5-diphenyl-1H-pyrazolium*. Difenzoquat is formulated as the methyl sulfate salt of difenzoquat, which dissociates into its respective ions in the aqueous spray solution. The chemical structure of difenzoquat methyl sulfate is:

difenzoquat methyl sulfate (Avenge)

Principal Use

Difenzoquat is used in the selective, postemergence control of wild oat in barley and wheat. Difenzoquat does not control broadleaf weeds.

Mode of Action

The mode of action of difenzoquat is inhibition of ATP production, inhibition of phosphorus incorporation into phospholipids and DNA, and mild inhibition of photosynthesis.

Translocation

Difenzoquat is rapidly absorbed by the leaves. It is not translocated from the leaves to other plant parts.

Selectivity

The primary factors contributing to the selectivity of difenzoquat are not known. The difenzoquat cation is stable in plants.

Information of Practical Importance

1. Difenzoquat is the active ingredient in a water-miscible liquid formulation (Avenge) that contains 2 lb difenzoquat cation /gal.

2. Avenge is applied alone, or in tank mix with Assert, postemergence to wild oats when the majority of the plants are in the three- to five-true-leaf stage (beginning of tillering). The timing of this application frequently coincides with barley in the two- to seven-leaf stage, spring-seeded wheat in the five- to six-leaf stage, and fall-seeded wheat in the four-leaf to tillering stage of growth.

3. Avenge is applied at 2.5–4 pt product/A (0.62–1 lb difenzoquat cation/A), depending on density of wild oat infestation.

4. Avenge can be tank-mixed with many broadleaf herbicides registered for use in wheat and/or barley such as 2,4-D, MCPA, bromoxynil, Curtail, Curtail M, Express, Glean, Harmony, and Harmony Extra. Select a broadleaf herbicide that coincides with Avenge in timing of application.

5. Add a nonionic surfactant to the spray mixture, either with Avenge alone or in tank mixes.

6. Difenzoquat has no soil activity as it is strongly adsorbed to soil colloids.

■ DITHIOPYR

Dithiopyr is a substituted pyridine herbicide. However, it is not a growth regulator–type herbicide, as are the pyridines clopy-

ralid, picloram, and triclopyr. The chemical name of dithiopyr is *S,S-dimethyl 2-(difluoromethyl-4-(2-methylpropyl)-6-(trifluoromethyl)-3,5-pyridinedicarbothioate.* The chemical structure of dithiopyr is:

dithiopyr
(Dimension)

Principal Use

The principal use of dithiopyr is the preemergence control of annual grass and broadleaf weeds in established cool- and warm-season turfgrass lawns and ornamental turf. Used early postemergence, it controls both smooth and large crabgrass.

Mode of Action

Dithiopyr inhibits mitosis. It may interact with a microtubule-associated protein and/or microtubule-organizing centers, altering microtubule polymerization and stability rather than interacting with the dimeric protein *tubulin.*

Translocation

Dithiopyr is primarily used preemergence, and any translocation that occurs is local and short in distance.

Selectivity

Factors contributing to the selectivity of dithiopyr are not known.

Information of Practical Importance

1. Dithiopyr is absorbed through the crowns, roots, and shoots of plants; it inhibits cell division and growth.

2. The major sites of physiological activity of dithiopyr are the meristems of roots and shoots of susceptible plants.

3. Dithiopyr is the active ingredient in an emulsifiable-concentrate formulation (Dimension) that contains 1.0 lb ai/gal.

4. Dimension can be used on seeded, sodded, or sprigged turfgrass that is well established. Dimension does not adversely affect the root development of established turfgrasses. It is used on most cool- and warm-season turfgrass species. Dimension may injure Chewings fine fescue.

5. This herbicide is applied as a preemergence treatment in the spring or fall prior to weed seed germination. It may be applied early postemergence to the weeds in the spring/summer. It is used preemergence in the fall to control annual bluegrass.

6. Applied preemergence, Dimension selectively controls many annual grass weeds in turfgrass, such as barnyardgrass and foxtail species, and many annual broadleaf weeds, such as common chickweed, corn speedwell, henbit, prostrate spurge, shepherd's-purse, and yellow woodsorrel.

7. Dimension controls both smooth and large crabgrass when it is applied preemergence or early postemergence prior to the crabgrass reaching the tillering stage. To control tillered (up to three tillers) crabgrass, apply Dimension in a tank mix with fenoxyprop (Acclaim) or MSMA.

8. Dimension is applied at 0.25–0.5 lb ai/A (1–2 qt product/A), equivalent to 0.75–1.5 fl oz product per 1000 ft^2, depending on the weed species involved. Do not apply more than 2 qt of Dimension per acre per year.

9. Do not apply Dimension to desired flowers, vegetables, shrubs, or trees, keeping in mind that it does not injure established ornamentals when applied according to the directions. Do not use clippings from treated turfgrass for mulching vegetables or fruit trees.

■ ETHOFUMESATE

The chemical name of ethofumesate is (±)-*2-ethoxy-2,3 dihydro-3,3-dimethyl-5-benzofuranyl methanesulfonate*. Its chemical structure is:

ethofumesate (Nortron, Progress)

Principal Use

Ethofumesate is a selective herbicide used in the preemergence or early postemergence control of annual grass and broadleaf weeds.

Mode of Action

The biochemical mode of action of ethofumesate is not known.

Translocation

Ethofumesate is translocated to the leaves *via* the transpiration stream following absorption by the root and emerging shoot. It is not translocated from the leaves to other plant parts. It is not readily foliar-absorbed after the plant has formed a mature cuticle.

Selectivity

The selectivity of ethofumesate appears to be from reduced translocation and rapid metabolism in tolerant plants, as compared to that of susceptible plant species. Differences in root absorption of ethofumesate do not appear to be a factor contributing to selectivity.

Information of Practical Importance

1. Sugar beets are highly tolerant to ethofumesate, as is common ragweed. In contrast, redroot pigweed and common lambsquarters are highly susceptible to this herbicide.

2. In plants exposed to ethofumesate when it is applied preemergence, the subsequent formation and deposition of epicuticular wax on leaf surfaces of the growing plant are severely decreased, particularly the deposition of alkane and *sec*-ketones. However, the percentage of long-chain waxy esters is increased. Foliar absorption of subsequently applied postemergence herbicides is enhanced.

3. Ethofumesate is the active ingredient in an emulsifiable-concentrate formulation (Nortron, Progress) that contains 1.5 lb ai/gal.

4. Nortron is applied preplant-incorporated or preemergence in sugar beets at a rate of 1.1–3.75 lb ai/A, depending on the soil type and whether it is tank-mixed with Antor, Eptam, or Pyramin.

5. Nortron alone is not recommended for postemergence use. However, it can be applied postemergence to sugar beets in tank mix, or sequentially, with Betanex or Betamix after the beets have four fully expanded true leaves. These tank mixes broaden and enhance the selective control of weeds in sugar beets. The mixtures are usually applied as a band treatment over the crop row.

6. Nortron is used preemergence to new seedings of Italian ryegrass or postemergence to established stands

of perennial ryegrass, tall fescue, or bentgrass grown for seed in Oregon and Washington. It is applied at 0.75–1.5 lb ai/A, depending on the targeted species.

7. Progress is used preemergence or early postemergence in newly established or mature perennial ryegrass turf, and Kentucky bluegrass, fairway-height bentgrass, or dormant bermudagrass overseeded with perennial ryegrass. It is applied at 0.5–2 lb ai/A, depending on targeted species.

8. In dormant bermudagrass turf (alone or overseeded to perennial ryegrass), Progress is applied preemergence to the weeds to control annual bluegrass, large crabgrass, barnyardgrass, green foxtail, yellow foxtail, canarygrass, common chickweed, common purslane, and redroot pigweed. It may be applied postemergence to control annual bluegrass and common chickweed.

■ FLURIDONE

The chemical name of fluridone is *1-methyl-3-phenyl-5-[3-(trifluoromethyl)phenyl]-4(1H)-pyridinone,* and its chemical structure is:

fluridone (Sonar)

Principal Use

Fluridone is a broad-spectrum aquatic herbicide used to control submersed and emersed vascular plants in ponds, lakes, and drainage canals.

Mode of Action

Fluridone inhibits carotenoid synthesis, which ultimately results in foliar albinism.

Translocation

Foliar absorption of fluridone from the surrounding water is the primary site of entry for aquatic weeds, as little or no translocation appears to take place even when root-absorbed.

Selectivity

Factors responsible for the selectivity of fluridone are not known. Fluridone has little effect on algae, and most floating aquatic weeds are only partially affected. It provides only partial control of cattails.

Information of Practical Importance

1. Fluridone is the active ingredient in a water-soluble formulation (Sonar A.S.) that contains 4 lb ai/gal, and in a pellet formulation (Sonar 5P) that contains 5% ai w/w.

2. Fluridone is applied at 0.5–1.5 lb ai/A to the water surface. It should be applied when there is little water movement and prior to or as weeds begin active growth.

3. Symptoms of injury from fluridone include albinism (whitening) of young leaves, retarded growth, and leaf necrosis. Its average half-life is about 3 weeks in pond water and about 3 months in the surrounding soil (hydrosoil).

■ GLYPHOSATE AND SULFOSATE

The chemical name of glyphosate is N-(*phosphonomethyl)-glycine,* and it is formulated as its isopropylamine salt. Sulfosate is the sulfonium salt of glyphosate, and it is formulated as this salt. Following foliar absorption, the salts of these herbicides ionize, and the anion (common to both glyphosate and sulfosate) is the herbicidally active portion. The chemical structures of glyphosate and sulfosate are:

glyphosate
(Ranger, Rodeo, Roundup)

sulfosate
(Touchdown)

Principal Use

Glyphosate is principally used in the nonselective postemergence control of annual, biennial, and perennial grasses and broadleaf weeds, woody brush and trees, and aquatic weeds.

Sulfosate is used as a nonselective, postemergence herbicide for control of annual and perennial grasses and broadleaf weeds in nonbearing groves, orchards, and vineyards, and in noncrop areas.

Mode of Action

Glyphosate and sulfosate inhibit the shikimic acid pathway by inhibiting 5-enolpyruvylshikimate-3-phosphate (EPSP) synthase. They also inhibit the synthesis of δ-aminolevulinic acid (ALA), thereby blocking porphyrin ring synthesis and, subsequently, all compounds normally containing porphyrin rings such as chlorophyll, cytochromes, and peroxidases.

Translocation

Following foliar application, glyphosate and sulfosate are readily absorbed and translocated throughout aerial and underground plant parts.

Selectivity

Glyphosate and sulfosate are both nonselective, postemergence herbicides. They are strongly adsorbed to soil colloids, and have little or no soil activity. Selectivity is achieved by use of directed or shielded sprays to avoid contact with the foliage and green stems or bark of desired plants, and by applications made prior to crop planting or emergence.

Information of Practical Importance

1. Gyphosate is the active ingredient in three water-soluble formulations: Ranger, which contains 2 lb ae/gal with surfactant; Rodeo, which contains 4 lb ae/gal without surfactant; and Roundup, which contains 3 lb ae/gal without surfactant.

2. Sulfosate is the active ingredient in a water-soluble liquid formulation (Touchdown) that contains 6 lb ai/gal.

3. Mistakes in calculations based on the active ingredient can easily occur by inadvertently using the pounds per gallon (or kilograms per liter) of the salt form in the respective formulation rather than the pounds per gallon of the acid form, or vice versa. The respective amounts of glyphosate's parent acid in Ranger, Rodeo, and Roundup are 2.0, 4.0, and 3.0 lb ae/gal, and those of glyphosate's isopropylamine salt are 2.7, 5.4, and

4 lb/gal. Such a mistake in calculations could lead to the application of too little or too much herbicide, which would result in poor weed control or excessive cost. In general, calculations based on active ingredient use the amount of glyphosate acid (acid equivalent or ae) in the formulation, unless otherwise specified.

4. Used in croplands, glyphosate is applied preplant or preemergence to the crop and postemergence to weeds (*always* prior to crop emergence). However, glyphosate may be applied over-the-top of mature cotton shortly before harvest to control perennial weeds such as johnsongrass, field bindweed, and Texas blueweed.

5. The visual effects of glyphosate or sulfosate following application include a gradual wilting or yellowing of the plant, which advances to complete browning of aboveground vegetation, and deterioration of affected underground plant parts. Visual symptoms are usually apparent within 2–4 days after application, but they may not be apparent for 7 days or more. Cool, cloudy weather following glyphosate application can slow its activity.

6. *Precautionary statement:* Spray solutions that contain glyphosate or sulfosate react with galvanized or unlined steel containers or spray tanks and produce hydrogen gas, which is highly combustible. If ignited by open flame, spark, welder's torch, lighted cigarette, or other ignition source, this gas can flash or explode and cause serious personal injury. Do not mix or apply these spray solutions in galvanized steel or unlined steel (except stainless steel) containers or spray tanks. Spray solutions of glyphosate or sulfosate should be mixed, stored, and applied only in stainless steel, aluminum, fiberglass, or plastic (or plastic-lined steel) containers.

Ranger

7. Ranger is used in the postemergence control of annual and perennial weeds, such as quackgrass, and in annual cropping systems, such as pastures and sods. It is applied at a rate of 0.75 lb ae/A (3 pt product/A) when the weeds have reached the three- to four-leaf stage and are at least 6–8 in. high; in sods established 3 years or more, use 5–6 pt product/A.

8. Ranger may be applied in the following crops prior to their emergence from the soil: alfalfa, barley, corn (all), forage grasses, forage legumes, oats, wheat.

9. Ranger may be tank-mixed with 2,4-D or dicamba (Banvel) for enhanced control of forage legumes. Do not add additional surfactant to the spray mixture.

Rodeo

10. Rodeo is used in aquatic areas to control emerged weeds (standing above the water surface) in all bodies of fresh and brackish water; these include flowing, nonflowing, or transient bodies such as lakes, rivers, ponds, estuaries, rice levees, seeps, irrigation and drainage ditches, canals, and reservoirs. There is no restriction on the use of the treated water for irrigation, recreation, or domestic purposes. However, do not apply upstream to flowing water within 0.5 mi (0.8 km) of potable water intake or within 0.5 mi of potable water intake in standing water such as ponds, lakes, or reservoirs.

11. Rodeo does not control plants that are submerged or that have a majority of their foliage underwater. Floating mats of aquatic weeds may require re-treatment. Applications to flowing water must be made while traveling upstream (in a direction opposite to the flow of water) to prevent the herbicide from becoming too concentrated. When applying Rodeo along banksides, do not allow the application to overlap more than 1 ft into open water.

12. For annual weed control, Rodeo is used at 1.5–2.5 pt product/100 gal water, applied to thoroughly wet the foliage. For perennial weed control (terrestrial or aquatic), Rodeo is used at 4.5–7.5 pt/100 gal water, applied to thoroughly wet the foliage.

13. Rodeo contains no surfactant, and a suitable nonionic surfactant must be added to the aqueous spray mixture in the amount of 2 qt nonionic surfactant/100 gal spray solution prior to use.

Roundup

14. Roundup is applied postemergence to control annual and perennial grass and broadleaf weeds. A wide variety of agricultural situations exist in which it may be used, and the reader is referred to the product label for this information.

15. Roundup contains no surfactant, and a suitable nonionic surfactant may be be added to the spray solution in the amount of 2 qt nonionic surfactant/100 gal spray solution.[1616]

16. Roundup may be used with special application equipment (shielded sprayers, recirculating sprayers, and rotary-wiper and rope-wick applicators) that allows the herbicide to be applied to the weed foliage without contacting the crop foliage. In general, this equipment is used to control weeds that have grown taller than the crop plants such as johnsongrass in cotton or soybeans.[1717]

17. Roundup may be applied with hand-held sprayers for applications made to wet the foliage of targeted vegetation. Do not spray to the point of run-off. Use an aqueous spray solution containing 0.5% Roundup and 1% nonionic surfactant and apply to weeds less than 6 in. (15 cm) in height or runner length. Apply prior to seedhead formation in grasses or bud formation in broadleaf weeds. To prepare a 0.5% solution, mix 4 teaspoons (30 ml) Roundup with 1 gal water and add 4 teaspoons nonionic surfactant and mix well. For larger or smaller volumes of spray solution, increase or decrease ingredients proportionally.

Touchdown

18. Touchdown is applied postemergence to young, actively growing weeds. Weeds 6 in. in height are easiest to control. The exceptions to this are johnsongrass and bermudagrass, which should be allowed to reach the seedhead stage before herbicide application.

19. Touchdown may be tank-mixed with 2,4-D or Banvel for greater control of certain perennial weeds, such as Canada thistle and field bindweed, and certain annual broadleaf weeds, such as common ragweed and morningglory spp. Touchdown may be tank-mixed with certain herbicides that provide residual activity in the soil. Refer to the product label for the specific herbicides.[20]

20. To improve foliar wetting, a nonionic surfactant (2 qt/100 gal water) should always be added to the spray solution, which can contain Touchdown alone or in combination with other herbicides. Weed control can also be improved by adding dry ammonium sulfate to the spray solution at 2% by weight or 17 lb/100 gal water.

21. The rate of Touchdown applied varies widely (3 fl oz to as much as 42 fl oz product/A) depending on targeted species.

■ ISOXABEN

The chemical name of isoxaben is N-[3-(1-ethyl-1-methylpropyl)-5-isoxazolyl]-2,6-dimethoxybenzamide, and its chemical structure is:

isoxaben (Gallery)

Principal Use

Isoxaben is used in the preemergence control of broadleaf weeds in established turfgrasses, certain other crops, and in noncrop areas.

Mode of Action

Isoxaben inhibits the biosynthesis of cellulose from glucose. It also disrupts root and hypocotyl development in susceptible broadleaf weeds; they are killed prior to emergence from the soil.

Translocation

Isoxaben is translocated into the stems and leaves of cereal plants following absorption from the soil. However, as it is applied preemergence to the weeds, translocation from leaves is not a factor in its activity.

Selectivity

Seedling broadleaf weeds are generally susceptible to isoxaben. The resistance of some dicot weed species (e.g., catchweed bedstraw, redroot pigweed, and velvetleaf) to isoxaben appears to be decreased sensitivity at the target site. Grass seedlings are partially resistant, depending on the application rate. It has been suggested that monocot resistance to isoxaben results from insensitivity at the target site. Established broadleaf and grass plants are tolerant of isoxaben.

Information of Practical Importance

1. Isoxaben is the active ingredient in a dry-flowable formulation (Gallery) that contains 75% ai w/w.
2. Gallery is used in established turfgrasses, ornamentals, nonbearing fruit and nut crops, nonbearing vineyards, and noncrop areas.

3. It is applied in late summer to early fall or in early spring, prior to weed seed germination. It does not control emerged or established weeds. Weeds that have emerged at time of treatment should be controlled by tillage or with postemergence herbicides.
4. Gallery is applied at 0.5–1.0 lb ai/A (0.66–1.33 lb product/A), depending on the targeted species. At the higher recommended rates, Gallery suppresses or partially controls certain annual grass weeds.
5. To be effective, Gallery must be moved into the weed seed germination zone by rainfall or irrigation (0.5 in. or more). Gallery is stable on the soil surface for up to 3 weeks. In field soils, its half-life persistence is about 6 months.
6. In Arizona, Gallery can be used in established ornamental lawns and in noncrop areas. However, it cannot be used on any plants grown for commercial production, such as turf sod farms, ornamentals and nursery stock grown for resale, and nonbearing fruit and nut trees and vineyards.
7. Gallery can be tank-mixed with other herbicides registered for use in established turfgrasses and the crop and noncrop areas noted above.

■ PYRIDATE

The chemical name for pyridate is O((*6-chloro-3-phenyl*)-*4-pyridazinyl*)-S-*octyl carbonothioate*. The chemical structure of pyridate is:

pyridate (Lentagran, Tough)

Principal Use

Pyridate is used as an early postemergence herbicide to control annual broadleaf weeds in cabbage, field corn, peanuts, and winter wheat.

Mode of Action

Pyridate inhibits photosynthesis by blocking electron transport *via* photosystem II (PS II).

Translocation

Rapidly foliar-absorbed, pyridate does not translocate in plants.

Selectivity

The exact factors contributing to the selectivity of pyridate are not known.

Information of Practical Importance

1. Pyridate is the active ingredient in a wettable-powder formulation (Lentagran) that contains 45% ai w/w, and in an emulsifiable-concentrate formulation (Tough) that contains 3.75 lb ai/gal.

2. Lentagran is used only in cabbage at 0.9 lb ai/A (2 lb product/A), applied early postemergence to grass and broadleaf weeds in the one-true-leaf stage and followed with a cultivation in 10–14 days.

3. Tough is used in field corn and peanuts at 0.9–1.4 lb ai/A (2–3 pt product/A). It may be applied in combination with a triazine herbicide when the broadleaf weeds are in the one- to four-leaf stage.

4. When Tough is combined with a triazine herbicide, the requisite amount of triazine may be reduced.

5. Tough may be applied at-cracking or postemergence to peanuts as a sequential treatment to other herbicides, such as Balan, Dual, Lasso, and Prowl, that were applied preplant, preemergence, or at-cracking. Apply Tough to broadleaf weeds in the two- to four-leaf stage (2–3 in. high).

6. Pyridate has no soil activity and presents no soil-residue problems.

■ INORGANIC HERBICIDES

Most herbicides are organic chemicals, but some inorganic herbicides do exist; the primary ones are sodium chlorate ($NaClO_3$) and sodium metaborate tetrahydrate ($Na_2B_4O_7$). They are nonselective contact herbicides. In general, these inorganic herbicides are used as nonselective soil sterilants to control unwanted vegetation, especially perennial weeds. Soil applications are best for a sterilant effect. These herbicides control annual and perennial grass and broadleaf weeds, trees, and stumps. The phytotoxic portion of sodium metaborate tetrahydrate is boron trioxide (B_2O_3).

Chlorate is a strong oxidizing agent and, if not properly combined with a fire retardant, can readily support combustion. Sodium chlorate presents a fire hazard from spontaneous combustion when it is dry and in contact with organic matter such as cotton clothing and dead, dry vegetation. Moist sodium chlorate will not burn, and sodium chlorate must be mixed and used with a water-soluble, fire-retardant chemical such as sodium metaborate, soda ash, magnesium chloride, or urea. All clothing contaminated with chlorate materials presents a fire hazard to the wearer, and it should be kept wet, changed immediately, and washed before reuse.

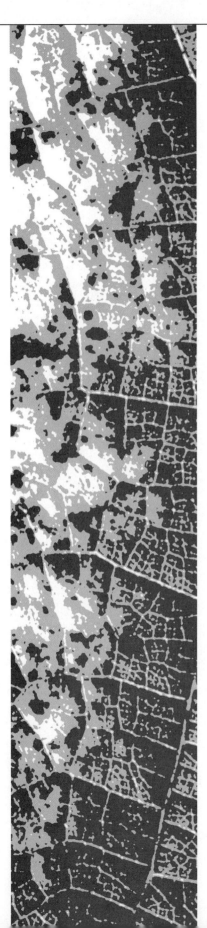

31 WEED CONTROL IN SELECTED CROPS: An Introduction

■ INTRODUCTION

Soils devoted to agronomic and horticultural crops in the United States and Canada differ widely in topography, parent material, soil texture, soil pH, organic matter content, and climatic conditions. It is impractical to attempt to cover in this text *all* aspects of weed control in even a few major crops because of this wide diversity.

■ WEED AND WEED-CONTROL LITERATURE

It is assumed that the scientific literature, herbicide product labels, state and regional weed-control guides, and other suitable publications can provide the reader with pertinent information on local weeds, weed-control practices, and herbicide recommendations for the crops of interest. These publications are invaluable in broadening one's perspective and in accessing current information. The scientific and popular literature relative to weeds and weed control is multitudinous. Some excellent scientific publications include *Weed Science*, *Reviews of Weed Science*, *Weed Technology*, *Weed Research* (British), regional *Weed Science Proceedings and Research Reports* (Northeast, North

Central, Southern, and Western), *Canadian Journal of Plant Science*, and *Canadian Weed Control Conference Proceedings* (East and West).

■ KEEPING UP TO DATE

One of the problems in any attempt to provide up-to-date information on herbicides is that products are constantly being changed or withdrawn, and new ones are continually being offered. The common chemical name or trade names may be changed, even though the active ingredient remains the same. In addition, users tend to assume that only specific chemical companies produce and market certain herbicides, but other companies often buy the rights to those herbicides, thereby making it more difficult to obtain the desired product. And, of course, a company may undergo a name change. Each change adds to the confusion of keeping up-to-date.

■ USE OF TRADE NAMES

In the following chapters, trade names have been freely used to enable the reader to associate a particular active ingredient with its commercial product(s). It is impractical to include in the text every available trade-named product. The omission of a particular product or trade name is not intended to reflect adversely, or to show bias against, any product or trade name not mentioned.

■ WEED-CONTROL PRACTICES

Cultivation is a traditional means of controlling weeds in agronomic and horticultural crops. Today, a combination of practices are utilized for a successful weed-control program. These practices include cultivation, applications of herbicides, crop rotation, and other weed-control and crop-management procedures. They vary from place to place, from region to region, and are continually being adapted to local crops and growing conditions. The practices chosen for weed control are determined by the crop involved, weather conditions, effectiveness, convenience, available equipment, and comparative costs to the grower.

■ HERBICIDES

In the following chapters, weed control in various crops, turfgrass, and pastures and rangelands will emphasize the use of herbicides, with the understanding that cultivation and other weed-control measures are practiced in conjunction with chemical practices. Even this presentation is limited to selected herbicides, as there is a multitude of available herbicides registered for use in the major crops grown in the United States and Canada. For example, in 1992, there were 17 active ingredients (herbicides) and 27 herbicide tank mixes registered for use in cotton; 22 active ingredients, 12 premix products, and more than 100 tank mixes registered for field corn; 23 active ingredients, 16 premix products, and more than 100 tank mixes registered for soybeans; and 17 active ingredients, 5 premix products, and 50 or more tank mixes registered for winter wheat. In addition, each of the active ingredients can be available in more than one kind of formulation and then applied at various recommended dosages.

Newly introduced herbicide premixes have proliferated since 1985, with 98 premixes listed in *Weed Technology*, Vol. 6, No. 4, 1992. A premix is a manufacturer-formulated product that contains two or more herbicides.

To use herbicides effectively, it is necessary to identify the problem weeds and to choose a herbicide (or herbicides) that, when properly applied, will selectively control these weeds with little or no crop injury. Mapping a field to show locations and densities of problem weed species aids in the selection of the proper herbicide and also in its application. Predominant weed species change over the years as the environment and weed-control practices change.

■ WEED CONTROL IN CROPLANDS

The objectives of weed control in croplands are (1) to remove or suppress the growth of weeds that are detrimental to crop production, and (2) to prevent their setting seed—seed that would otherwise be added to the weed seed bank. Weed control can be achieved through a variety of methods, such as cultivation and/or herbicides in conjunction with crop rotation, as was discussed in Chapter 3.

The weed-control practices used will depend on the specific crop and weed species involved. Such practices will differ for a row crop (such as corn, cotton, lettuce) and a broadcast crop (such as alfalfa, small grains, and pastures) and will be affected by the characteristic growth of the crop plants. Experience has shown that using a combination of weed-control practices is often more effective than relying on one method alone.

Weed control can be practiced to contain or eliminate existing weeds in the short term or to contain or eliminate perennial weeds and other persistent weeds in the long term. The control of annual weeds involves early control and prevention of seed set by surviving weeds. Control of established perennial weeds may well extend over a period of years.

Perennial weeds, such as field bindweed, quackgrass, and yellow nutsedge, pose a long-term control problem, due to their growth habits. In general, perennial weeds survive year after year because of their underground propagules, which are capable of regenerating even from fragments left by the cutting action of cultivating tools. Cultivation equipment tends to spread these propagules from one place to another.

Herbicides are an excellent tool to control weeds in crop and noncrop areas. However, care must be taken that the proper herbicide for the crop and weed species involved is chosen and then correctly applied. Each herbicide has its own unique properties: some can be used selectively (safely) only in certain specified crops and are effective only against certain weed species or groups of weeds. The effectiveness and selectivity of a particular herbicide is often dependent on application timing and placement.

32 FIELD CORN

■ INTRODUCTION

In 1992, 71.4 million A (28.9 million ha) were planted to field corn in the 17 major corn-producing states in the United States, and 96% of this acreage was treated with herbicides (Table 32.1). Over 90% of this crop is used for animal feed, with most of the remainder used in food products for humans and in ethanol production. In 1991, weeds caused an estimated loss of $1.04 billion in corn, even though herbicides were used. Under the same conditions, but without herbicides, the estimated loss would have been $4.74 billion.

In 1992, 2.7 million A (1.1 million ha) were planted to field corn (grain and silage) in Canada. The average yield loss in grain and silage corn due to weeds in the eastern provinces (Quebec and Ontario) was 7 and 5%, respectively, while that in the western provinces (Manitoba, Alberta, and British Columbia), was 8 and 9%, respectively. The loss due to weeds in eastern Canada was valued at $44 million for grain corn and $6 million for silage corn, while that in western Canada was valued at $1.55 million and $2.46 million, respectively.

Principal weeds causing losses in field corn grown in the Corn Belt of the United States are given in Table 32.2. Although annual grass and broadleaf weeds present a continual problem to the corn grower, most can be adequately controlled through the judicious use of available herbicides. However, the perennial weeds Canada thistle, field bindweed, johnsongrass, quackgrass, purple nutsedge, and yellow nutsedge continue to be problems.

Cultural practices associated with corn vary from region to region and are influenced by soil textures, weather conditions, available equipment, and economics. In some areas, overhead (center-pivot) or furrow irrigation may be used. In other areas, moisture is provided only by rainfall, which may be erratic. Crop rotations may vary, with the principal crops adapted to a particular area following one another in rotation, such as corn–soybeans or corn–wheat.

Herbicides are the primary means of weed control in field corn, but some tillage is still being practiced. Corn is a crop that can be grown with much or little tillage. Once the seedbed is prepared, the main reason to cultivate is weed control. Where herbicides adequately control weeds, there is little reason to cultivate. Reduced-tillage systems for corn are gaining favor in some corn-producing areas, particularly as a means to reduce soil erosion. With no-tillage systems, however, herbicides are the only means of weed control. The critical weed-free period for corn is during the first 4 weeks following crop emergence; weed competition during this period will reduce yields.

TABLE 32.1 ■ Corn acreage planted and treated with herbicides in the 17 major corn-producing states in the United States in 1992.*

STATE	ACRES (MILLIONS)	PERCENT TREATED
Georgia	0.75	77
Illinois	11.20	99
Indiana	6.10	98
Iowa	13.20	98
Kansas	1.85	99
Kentucky	1.42	96
Michigan	2.70	97
Minnesota	7.20	99
Missouri	2.50	97
Nebraska	8.30	92
N. Carolina	1.15	91
Ohio	3.80	98
Pennsylvania	1.38	95
S. Carolina	0.38	92
S. Dakota	3.80	92
Texas	1.75	90
Wisconsin	3.90	94

*Total corn acreage was 71.38 million A, with 96% treated with herbicides.

SOURCE: *Agricultural chemical usage: 1992 field crop summary.* United States Department of Agriculture, National Agricultural Statistics Service, Washington, D.C. p. 7.

TABLE 32.2 ■ Principal problem weeds for corn grown in the Corn Belt.*

COMMON WEEDS**	TROUBLESOME WEEDS**
Common cocklebur	Canada thistle
Common lambsquarters	Common cocklebur
Fall panicum	Common lambsquarters
Giant foxtail	Common milkweed
Jimsonweed	Fall panicum
Morningglories	Giant foxtail
Pennsylvania smartweed	Hemp dogbane
Pigweeds	Johnsongrass
Ragweeds	Morningglories
Shattercane	Pennsylvania smartweed
Velvetleaf	Quackgrass
	Ragweeds
	Shattercane
	Velvetleaf

*Corn Belt states: Illinois, Indiana, Iowa, Missouri, Ohio.
**Weed species present in two or more states in the Corn Belt. Common weeds are those that abundantly infest a significant portion of the corn acreage. Troublesome weeds are those that, despite weed-control efforts, are inadequately controlled and interfere with crop production and/or yield, crop quality, or harvest efficiency.

SOURCE: D. C. Bridges, ed. 1992. *Crop losses due to weeds in the United States.* Champaign, Ill.: Weed Science Society of America.

■ CHEMICAL WEED CONTROL

The principal herbicides used in field corn in 1990, by region and percent acreage treated, are given in Table 32.3. The comparative use of the top five field corn herbicides in 16 major corn-producing states in 1991, 1992, and 1993, expressed as a percent of the acreage planted, is shown in Figure 32.1 on page 268.

The herbicide applicator is cautioned that corn is very sensitive to many of the new soybean herbicides such as chlorimuron-ethyl, clomazone, imazaquin, and imazathapyr. Special care should be taken to thoroughly clean a sprayer that is used first to apply these herbicides to soybeans and then later to apply other herbicides to corn.

■ INFORMATION OF PRACTICAL IMPORTANCE

Metolachlor (Dual 8E or 25G) is an acid amide herbicide (Chapter 12). It is a soil-incorporated herbicide used primarily to control annual grasses. It also controls (or sup-presses) yellow nutsedge, but not purple nutsedge. It does not control emerged weeds. Metolachlor can be applied preplant-surface, preplant-incorporated, preemergence, or at lay-by (postdirected to corn 5–40 in. or 13–102 cm tall) at a rate of 1.5–3 lb ai/A. Metolachlor is available with atrazine in the premix Bicep and with cyanazine in the premix Cycle. Metolachlor can be tank-mixed with other herbicides to broaden the spectrum of weeds controlled.

Alachlor (Lasso 4EC) is an acid amide herbicide (Chapter 12). It is a soil-incorporated herbicide used primarily to control annual grasses. It can be applied preplant-incorporated, preemergence-surface, or early postemergence before the corn height exceeds 5 in. at 2–4 lb ai/A. Alachlor also controls yellow nutsedge and certain annual broadleaf weeds. It does not control emerged weeds. Alachlor can be tank-mixed with other appropriate herbicides, such as atrazine, cyanazine, dicamba, and 2,4-D, to control a broader spectrum of weeds.

Atrazine and *cyanazine* are both triazine herbicides (Chapter 27). Their use and activity is similar, with some differences. Corn has a greater tolerance of atrazine than cyanazine, but cyanazine has a shorter soil persistence than atrazine. Atrazine and cyanazine provide about the same control of annual grass and broadleaf weeds, with cyanazine

TABLE 32.3 ■ **Principal herbicides used in corn in 1990, expressed as the average percentage of the total acres grown in each region.**

HERBICIDE**	REGION*							MEAN
	1	2	3	4	5	6	7	
Alachlor	28	31	17	25	15	29	36	26
Atrazine	75	70	27	37	45	90	81	61
Cyanazine	20	28	10	4	12	4	21	14
2,4-D	11	15	29	14	15	21	8	16
Dicamba	17	24	17	3	15	9	7	13
EPTC	6	22	14	15	15	9	7	13
Metolachlor	28	25	7	14	21	26	43	23

*Regions: 1 = Corn Belt; 2 = North Central; 3 = Mountain; 4 = Pacific Coast; 5 = Southwest; 6 = Southeast; 7 = Northeast.
**A particular herbicide may not be used in every state within a region.

SOURCE: D. C. Bridges, ed. 1992. *Crop losses due to weeds in the United States.* Champaign, Ill.: Weed Science Society of America.

slightly better on certain grass species than atrazine and atrazine better on pigweeds and yellow nutsedge. Atrazine controls quackgrass, a perennial grass; cyanazine does not control quackgrass. Both can be applied preplant-incorporated, preemergence, or early postemergence before the corn is 12 in. (3 cm) high in the case of atrazine and through the four-leaf stage (but before the fifth leaf is visible) with cyanazine. Applied postemergence, broadleaf weeds should be less than 1.5 in. high, and a suitable surfactant and/or crop oil is recommended for best results. Atrazine is applied at 2–3 lb ai/A, and cyanazine at 1.25–4.75 lb ai/A.

Atrazine is often used in combination with the grass-control herbicides alachlor or metolachlor. There are at least 8 premix products containing atrazine as one of two herbicides and at least 16 registered tank mixes in which atrazine can be mixed with other herbicides. In general, mixing atrazine with other herbicides allows the atrazine to be applied at a reduced rate, which favors shorter soil persistence and reduces the groundwater hazard. Cyanazine can also be applied in tank mixes with other herbicides to broaden the spectrum of weeds controlled.

Alachlor, atrazine, and cyanazine are Restricted Use Pesticides. Alachlor is restricted because of its tendency to produce tumors (oncogenicity) in laboratory animals. Atrazine and cyanazine are restricted because of their potential to move through the soil and contaminate groundwater (water stored underground).

Clopyralid (Stinger) is a growth regulator–type herbicide (Chapter 19). It is applied postemergence to control annual and certain perennial broadleaf weeds. Clopyralid can be applied in corn any time from emergence to up to 24 in. (61 cm) tall. Clopyralid is applied at 0.1–0.25 lb ae/A (0.25–0.67 pt product/A).

Dicamba and *2,4-D* are growth regulator–type herbicides (Chapter 19). Dicamba (Banvel 4EC) and 2,4-D (Formula 40, Weedar 64, Weedone LV4) are applied postemergence to control annual and perennial broadleaf weeds, with little or no control of grasses. They can be applied preplant, preemergence, or early postemergence (spike to five-true-leaf corn stage), or late postemergence (corn 8–36 in. or 20–91 cm high) but no later than 15 days prior to tassel emergence. When spraying in corn over 10 in. (25 cm) tall, use drop nozzles (spray nozzles on extended arms so that the spray can be applied under the crop canopy) to avoid spraying into the leaf whorls and possibly injuring the terminal meristem of the corn. Dicamba is applied at 0.25–0.5 lb ae/A, while 2,4-D is applied at 1–2 lb ae/A.

Nicosulfuron (Accent 75DF, Accent SP) is a sulfonylurea herbicide (Chapter 25). It is applied postemergence to control annual and perennial grasses and certain broadleaf weeds 1–4 in. (2.5–10 cm) tall, in some cases up to 12 in. tall. Field corn has excellent tolerance to nicosulfuron, exhibiting a four- to eightfold margin of safety above the recommended rate. Nicosulfuron can be applied broadcast or with drop nozzles to corn up to 24 in. tall and with drop nozzles to corn 24–36 in. tall. A second application can be made 14–28 days after the first to control later-emerging weeds. The use of a nonionic surfactant or crop oil concentrate is required with all applications of nicosulfuron, except in tank mix with Marksman (potassium salt of dicamba and atrazine). Accent SP is packaged in water-soluble packets, each containing 2.67 oz product, enough to

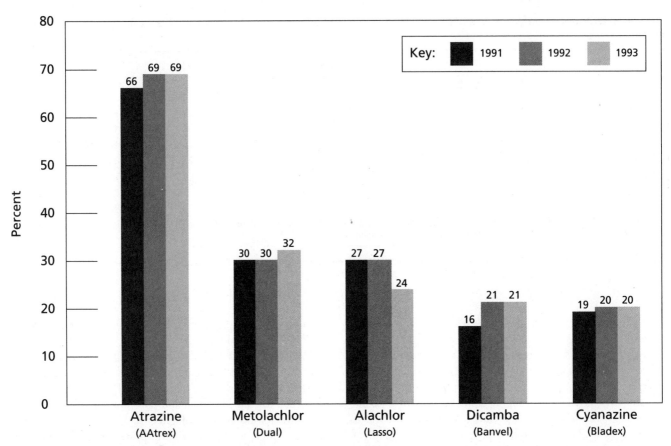

FIGURE 32.1 ■ **Comparative use of the top five herbicides in corn in 1991, 1992, and 1993, based on percent of acreage planted in 17 states in the respective years.**

SOURCE: *Agricultural chemical usage: 1993 field crop summary.* United States Department of Agriculture, National Agricultural Statistics Service, Washington, D.C. p. 5.

treat 4 A at the recommended rate. Nicosulfuron is applied at 0.5 oz ai/A (0.67 oz product/A).

Nicosulfuron can be tank-mixed with atrazine, bromoxynil, or dicamba. To avoid severe crop injury, do not tank-mix with bentazon or apply nicosulfuron to corn previously treated with the insecticide Counter.

Primisulfuron (Beacon) is a sulfonylurea herbicide (Chapter 25). It is applied postemergence to control annual and perennial grass and broadleaf weeds. Primisulfuron can be applied over-the-top, directed, or semidirected to corn 4–20 in. (10–51 cm) tall. A directed split application can be made when corn is between 20 in. tall and before tassel emergence. Beacon is packaged in water-soluble packets that contain 1.52 oz product per packet, enough to treat 2 A of corn. Primisulfuron can be tank-mixed with bromoxynil (Buctril), dicamba (Banvel), or 2,4-D to increase the spectrum of broadleaf weeds controlled.

Pyridate (Tough) is a nonfamily herbicide (Chapter 30). It is a postemergence, contact-type herbicide used to control broadleaf weeds in corn. Field corn is tolerant to pyridate at all stages of growth. For optimal weed control, time the application to the stage of weed growth. Pyridate is most effective when applied to broadleaf seedlings in the one- to four-leaf stage. Rainfall 1–2 hours after application has little effect on the degree of weed control. Tough may be applied in tank mix with atrazine or cyanazine. Pyridate is applied at 0.47–0.93 lb ai/A (1–2 pt product/A).

■ SELECTED REFERENCES

Agricultural chemical usage: 1993 field crop summary. United States Department of Agriculture, National Agricultural Statistics Service, Washington, D.C. p. 5.

Anonymous. 1994. *Weed control guide 1994.* Willoughby, Ohio: Meister Publishing. p. 61–74.

Bridges, D. C., ed. 1992. *Crop losses due to weeds in the United States.* Champaign, Ill.: Weed Science Society of America.

Herbicide product labels.

Swanton, C. J., K. N. Harker, and R. L. Anderson. 1993. Crop losses due to weeds in Canada. *Weed Technol.* 7:537–542.

Tripplet, G. B., Jr. 1985. Principles of weed control for reduced-tillage corn production. In *Weed control in limited-tillage systems,* Monograph No. 2, pp. 26–40. Champaign, Ill: Weed Science Society of America.

33 COTTON

■ INTRODUCTION

In 1991, cotton was grown on 13 million A of land in the United States. The principal type of cotton grown in the United States is the short-staple American Upland (*Gossypian hirsutum*), with the long-staple American Pima (*Gossypium barbadense*) cotton grown to a much lesser extent. The short-staple cotton is generally grown throughout the cotton-growing areas of the United States, while the long-staple American Pima cotton is grown entirely in the irrigated desert areas of the southwestern United States.

Cotton is produced in the United States in four major geographic regions: (1) the *Southeast*, encompassing the states of Alabama, Florida, Georgia, North Carolina, and South Carolina; (2) the *Midsouth*, encompassing Arkansas, Louisiana, Mississippi, Tennessee, and Missouri; (3) the *Southwest*, comprised of Texas, Oklahoma, and New Mexico; and (4) the *West*, comprised of Arizona and California. The cotton production acreage for these regions is shown in Table 33.1.

■ WEEDS

In 1991, weeds caused an estimated loss of $450 million in cotton grown in the United States, even though herbicides were used. Under the same conditions, but without herbicides, the estimated crop loss would have been $2.3 billion.

Representative grass and broadleaf weeds in cotton in the United States are listed in Table 33.2. Nearly 66% of all cotton losses in the United States due to weeds have been caused by five weed species: common cocklebur, johnsongrass, morningglory spp., prickly sida, and yellow nutsedge. Johnsongrass and yellow nutsedge are the most widespread perennial weeds infesting cotton fields in the United States. Of these, johnsongrass is the most competitive with cotton, with its rhizome system aggressively claiming soil space and its foliage topping that of cotton by as much as 2 ft or more.

Weeds are a major problem in the production of cotton, especially during the first 8–10 weeks after planting. If cotton is kept weed-free for the first 8–10 weeks after

TABLE 33.1 ■ Cotton-growing regions in the United States and acreage planted in 1991.*

STATE	ACRES (× 1000)	STATE	ACRES (× 1000)
Southeast		*Midsouth*	
Alabama	410	Arkansas	980
Florida	49	Louisiana	820
Georgia	425	Mississippi	1,310
N. Carolina	457	Tennessee	610
S. Carolina	210	Missouri	327
Virginia	18	**Regional total: 4.05 million A**	
Regional total: 1.57 million A			
Southwest		*West*	
New Mexico	85	Arizona	462
Oklahoma	380	California	1,020
Texas	5,357	**Regional total: 1.48 million A**	
Regional total: 5.82 million A			
*Total cotton acreage in the United States in 1991 was 13 million A.			

SOURCE: *Agricultural statistics 1992.* United States Department of Agriculture, Government Printing Office, Washington, D.C. p. 62.

planting, it is sufficiently competitive to suppress further weed growth. The net effect of weed competition on cotton is a reduction in the yields of cotton lint and seed, usually through a reduced number of bolls per plant. In addition, green weed vegetation collected along with the harvested cotton tends to stain the lint green, resulting in a reduction of lint grade and dockage at marketing.

■ CROP MANAGEMENT

Good crop-management practices—those that create the most favorable environment for the cotton crop—may also create just as favorable conditions for certain weed species. To combat this problem, crop rotation, used along with weed-control programs (both chemical and mechanical), has been successful against specific problem weeds. Such rotational crops should differ in growth habits, cultural requirements, and weed-control practices from those of cotton. A rotational crop is an excellent opportunity to utilize different herbicides from those used selectively in cotton. Cotton rotated with cover crops or fallowing is an effective means of controlling weeds. In addition, cotton planted in 20-in. row spacings, rather than the more conventional 40-in. row spacing, requires a shorter weed-free maintenance period before it becomes competitive with weeds.

TABLE 33.2 ■ Representative weeds in cotton grown in the United States.

GRASS WEEDS	BROADLEAF WEEDS
Annual	*Annual*
Barnyardgrass	Buffalobur
Broadleaf signalgrass	Common cocklebur
Field sandbur	Common lambsquarters
Goosegrass	Common purslane
Junglerice	Common sunflower
Large crabgrass	Florida pusley
Texas panicum	Florida beggarweed
Perennial	Groundcherry, lanceleaf
Bermudagrass	Groundcherry, Wright
Johnsongrass	Hemp sesbania
Sedges (perennial)	Jimsonweed
Purple nutsedge	Morningglories
Yellow nutsedge	Nightshade, black
	Nightshade, hairy
	Pigweeds
	Prickly sida
	Sicklepod
	Spotted spurge
	Spurred anoda
	Tropic croton
	Perennial
	Field bindweed
	Redvine
	Silverleaf nightshade
	Trumpetcreeper

SOURCE: D. C. Bridges, ed. 1992. *Crop losses due to weeds in the United States.* Champaign, Ill: Weed Science Society of America.

■ TILLAGE

Tillage as a means of weed control works the soil to a shallow depth (about 2 in.) in such a manner that young weeds are destroyed and normal crop growth is not disrupted. Tillage is effective in the control of relatively small annual weeds, but it is far less effective against established perennial weeds. Mechanical control of weeds in cotton is achieved with a variety of tractor-mounted tillage tools such as knives, sweeps, rotary hoes, listers (disks), shovels, and various spring-toothed or spring-tined tools. Often, such tools are tractor-mounted to work 4–8 rows at the same time, requiring precision adjustment of the tools on the tool bar to which they are attached. The rotary hoe is a popular and useful tool for breaking the soil crust over emerging cotton seedlings and for destroying young weeds by uprooting them. Tillage has been one of the most important practices in cotton production in the United States, generally involving a total of 8–12 tillage operations per year, varying from state to state, with at least half of these used specifically for weed control.

Hand-hoeing, although expensive, is still an integral part of weed control in cotton production in many areas, supplementing herbicides and tractor-mounted cultivators. Prior to the introduction of herbicides, hand-hoeing accounted for more than half of the total labor requirement in cotton production.

While the effectiveness of tillage in controlling weeds is unquestioned, it does have its disadvantages, such as no residual weed control; the difficulty of controlling weeds between plants in the row; exposure of the soil to possible water erosion; and the termination of mechanical weed control after the cotton crop reaches a size that would be injured by the tillage equipment.

■ CONSERVATION TILLAGE

With the coming development of much-needed technology, cotton grown on erodible soils will probably utilize some form of conservation tillage in the future. At present, reduced-tillage systems have not been as readily adapted to cotton as to other crops. Reduced-tillage implies that tillage operations have been reduced to a minimum without adversely affecting maximum crop yields. In no-tillage production, residues of previous crops and weeds remain on the soil surface as a mulch to reduce wind and water erosion. The critical need for postemergence control of weeds in cotton has resulted in the use of fewer no-tillage production practices in cotton than in any other major crop in the United States.

The potential for producing cotton with reduced- or no-tillage appears greater in the Southern Plains (Oklahoma, Texas, and eastern New Mexico) than in the other cotton-producing regions of the United States. The Delta States (Arkansas, Louisiana, and Mississippi) are the least likely to adopt minimum-tillage practices due to the luxuriant weed growth that results from the region's highly fertile alluvial soils and high rainfall. The success of any conservation tillage program in cotton is dependent on the appropriate use of postemergence herbicides.

■ HERBICIDES

Herbicides are used more extensively to control weeds in cotton than in any other major crop in the United States

TABLE 33.3 ■ Principal herbicides used in cotton in 1991, expressed as the average percentage of total acreage grown in each region.

HERBICIDE*	PERCENT	HERBICIDE*	PERCENT
Southeast		*Mid-South*	
Flumoeturon	63	Fluometuron	82
Trifluralin	50	Trifluralin	56
MSMA	49	MSMA	53
Pendimethalin	33	Norflurazon	50
Cyanazine	30	Cyanazine	33
DSMA	28	Pendimethalin	22
Norflurazon	25	Prometryn	21
Fluazifop-P-ethyl	9	Methazole	16
Regional acreage:		DSMA	15
812,133 A		Fluazifop-P-ethyl	14
West		Paraquat	13
Oxyfluorfen	55	**Regional acreage:**	
Paraquat	51	**2,456,836 A**	
Trifluralin	50	*Southwest*	
Pendimethalin	33	Trifluralin	68
Prometryn	33	Prometryn	20
Glyphosate	14	Pendimethalin	19
Cyanazine	13	Glyphosate	18
Regional acreage:		**Regional acreage:**	
1,465,544 A		**4,526,809 A**	

*A particular herbicide may not be applied in every state within a region.

SOURCE: D. C. Bridges, ed. 1992. *Crop losses due to weeds in the United States.* Champaign, Ill: Weed Science Society of America.

(McWhorter and Bryson, 1992). In 1992, herbicides were applied to 88% of the Upland cotton acreage in the United States. The seven top herbicides used in Upland cotton in 1993 (and the percentage of the crop treated) were: trifluralin (57%), fluometuron (29%), prometryn (25%), MSMA (24%), cyanazine (19%), norflurazon (18%), and pendimethalin (17%). The principal herbicides used in cotton in the United States in 1991 are shown in Table 33.3.

Herbicides are especially useful in controlling weeds in the drill row of cotton. In conventional cotton management systems, herbicides are applied (1) preplant to the crop but postemergence to emerged weeds, (2) preplant soil-incorporated, (3) preemergence, (4) postdirected (before and at lay-by), (5) over-the-top, and (6) spot treatment.

With the introduction and wide use of the dinitroaniline herbicide trifluralin (Treflan) in the 1960s, a noticeable shift in problem weed populations occurred, with a decline in annual grass weeds and an increase in annual broadleaf weeds such as common cocklebur, morningglories, prickly sida, sicklepod, spurred anoda, and tropic croton. There was also a noticeable increase in perennial weeds such as johnsongrass and yellow nutsedge. Over the years, other herbicides have also caused weed shifts, and this has led to the documentation of weeds highly tolerant to specific herbicides; such documentation is often noted on the label of a particular herbicide. To overcome shifts in problem weeds, combinations of weed-control practices are exercised, and herbicide combinations are used as well.

■ INFORMATION OF PRACTICAL IMPORTANCE

Cyanazine (Bladex 4L or 90DF) is a triazine herbicide (Chapter 27). It is used to control annual grass and broadleaf weeds. It can be applied either preplant-surface (California only), preemergence, or postdirected at lay-by. Cyanazine is registered as a preemergence application only in the states of Alabama, Arkansas, Louisiana, and Mississippi. Cyanazine can be applied preemergence in tank mix with norflurazon (Zorial). It can be applied as a postdirected treatment in cotton that has attained a minimum height of 6 in. and where the weeds are not more than 2 in. tall. It can be applied postdirected in tank mix with MSMA after the cotton is 6 in. tall but before the first bloom stage. Apply no more than two applications of this mixture before first bloom. Cyanazine can be applied postdirected at lay-by after the cotton is 12 in. or more in height for preemergence or postemergence control of broadleaf weeds not more than 2 in. tall. Make no more than one preemergence and two postdirected applications of cyanazine in any one year. If cyanazine is not applied preemergence, apply no more than three postdirected applications, including lay-by, in any one year. The use of leaf lifters or spray shields on application equipment is recommended to avoid spraying the cotton foliage. Cyanazine is applied preplant at 1.5–4 lb ai/A, depending on soil texture (lowest rate on coarse soils, highest rate on fine soils). It is applied at 0.5–1.2 lb ai/A preemergence and 0.6–1 lb ai/A directed postemergence.

Fluometuron (Cotoran 4L or 85DF) is a substituted urea herbicide (Chapter 29). It is used to control annual grass and broadleaf weeds. It can be applied preemergence to the soil surface, or lightly incorporated after planting, or as a directed, semidirected, or over-the-top postemergence treatment. When applied preemergence, fluometuron can be tank-mixed with most liquid nitrogen fertilizers. It is most effective when three applications are made as follows: (1) preemergence at 1–2 lb ai/A, using the lowest rate on light silt or sandy soils, low in organic matter; (2) postemergence as a directed, semidirected, or over-the-top treat-

ment when cotton is 3–6 in. high using no more than 1 lb ai/A; and (3) if needed, postemergence (directed, or semidirected), and at lay-by (last normal cultivation), using no more than 1 lb ai/A. Fluometuron may be applied postemergence to cotton 3 in. tall to lay-by but preemergence to weeds, or postemergence to both cotton and weeds; weeds should be 2 in. or less in height. Fluometuron can be tank-mixed with suitable herbicides.

MSMA (Drexel MSMA 6 Plus) is a methanearsonate herbicide (Chapter 21). It is used postemergence to control young, actively growing grass and broadleaf weeds, with best results obtained when applied at temperatures above 70°F. The phytotoxic properties of MSMA are quickly inactivated on contact with the soil. MSMA can be applied preplant to cotton, but postemergence to emerged weeds. Cotton can be planted immediately after the preplant application. MSMA can be applied as a directed postemergence application when cotton is 3 in. high to first bloom for the control of emerged weeds. Do not apply after first bloom. Drexel MSMA 6 Plus contains surfactant; do not add additional surfactant to the spray solution. MSMA is applied at 2 lb ai/A (2.6 pt product/A).

Norflurazon (Zorial Rapid 80DF) is a phenyl pyridazone herbicide (Chapter 23). It is used to control annual grass and broadleaf weeds and purple and yellow nutsedges. Norflurazon can be applied preplant and incorporated not deeper than 2–3 in. within 1 week after application. It may be applied as a split application, with one-half the recommended rate applied preplant-incorporated and the remainder applied preemergence as an overlay to the preplant treatment. Noraflurazon is not registered for use in Arizona, New Mexico, or California. In the mineral soils of these states, noraflurazon persists in the soil more than 1 year. The recommended rate for norflurazon is 1–2 lb ai/A.

Pendimethalin (Prowl 3.3 EC) is a dinitroaniline herbicide (Chapter 17). It is used to control germinating grass and certain broadleaf weeds. Pendimethalin can be applied preplant-incorporated or preemergence no later than 2 days after planting. Following application, pendimethalin must be incorporated 1–2 in. deep, in the weed seed germination zone. Pendimethalin does not control emerged weeds. It is applied at 0.5–1.5 lb ai/A, with the lowest rate on coarse-textured soils and the highest rate on fine-textured soils.

Prometryn (Caparol 4L) is a triazine herbicide (Chapter 27). It is used to control annual grass and broadleaf weeds preemergence or early postemergence. Prometryn can be applied in cotton preplant-incorporated in Arizona, New Mexico, and California. Following application, incorporate up to 4 in. deep to mix herbicide into weed seed germination zone. Prometryn can be applied preemergence in all cotton-producing states except Arizona and California. If necessary, cotton may be replanted in soil previously treated with Caparol; do not re-treat with Caparol. Prometryn can be applied as a directed postemergence treatment to cotton 3 or more in. tall or postdirected lay-by. Avoid contact of spray with cotton foliage. As a lay-by treatment, apply prometryn to cotton at least 12 in. tall and before weeds are 2 in. tall. The application rate of prometryn varies from 1.2 to 2.4 lb ai/A, depending on the soil texture, timing of application, and the state in which applied.

Trifluralin (Treflan 4EC) is a dinitroaniline herbicide (Chapter 17). It is used to control germinating grass and broadleaf weeds. Trifluralin is applied to the soil surface and incorporated 2–3 in. deep in the final seedbed within 24 hours after application. Trifluralin is soil-incorporated to mix the herbicide into the seed germination zone and to avoid herbicide loss by volatility. It may be applied and incorporated before or at planting, immediately after planting, or at lay-by in cotton. When incorporated after planting, take care not to injure or disturb the cotton seed. When applied at lay-by, direct the spray below the cotton canopy onto the soil surface. Applied at twice the normal dosage, it may be used to control rhizome johnsongrass to a commercially acceptable level in all cotton-producing states except Arizona and California. Trifluralin is applied at 0.5–1.25 lb ai/A, depending on soil type, organic matter content, and the state in which the cotton is grown.

■ SELECTED REFERENCES

Agricultural chemical usage: 1993 field crop summary. United States Department of Agriculture, National Agricultural Statistics Service, Washington, D.C. p. 29.

Agricultural statistics 1992. United States Department of Agriculture, Government Printing Office, Washington, D.C. p. 62.

Bridges, D. C., ed. 1992. *Crop losses due to weeds in the United States.* Champaign, Ill: Weed Science Society of America.

Bryson, C. T., and P. E. Keeley. 1992. Reduced-tillage systems. In *Weeds of cotton: Characterization and control,* C. G. McWhorter and J. R. Abernathy, eds., Reference Book Series, No. 2, pp. 323–363. Memphis, Tenn.: The Cotton Foundation.

Buchanan, G. A. 1981. Management of weeds in cotton. In *CRC handbook of pest management in agriculture,* Vol. 3, pp. 215–242. Boca Raton, Fla.: CRC Press.

Buchanan, G. A. 1992. Trends in weed control methods. In *Weeds of cotton: Characterization and control,* C. G. McWhorter and J. R. Abernathy, eds., Reference Book Series, No. 2, pp. 47–72. Memphis, Tenn.: The Cotton Foundation.

Danielson, L. L. 1981. Evaluation of the importance of nonchemical methods of weeds and pest control in crop production. In *CRC handbook of pest management in agriculture,* Vol. 2, pp. 377–381. Boca Raton, Fla.: CRC Press.

McWhorter, C. G., and T. N. Jordan. 1985. Limited tillage in cotton production. In *Weed control in limited-tillage systems,* A. F. Wiese, ed., Monograph Series, No. 2, pp. 61–76. Champaign, Ill.: Weed Science Society of America.

McWhorter, C. G., and C. T. Bryson. 1992. Herbicide use trends in cotton. In *Weeds of cotton: Characterization and control,* C. G. McWhorter and C. T. Bryson, eds., Reference Book Series, No. 2, pp. 233–294. Memphis, Tenn.: The Cotton Foundation.

Murray, D. S., L. M. Verhalen, and R. J. Tyri. 1992. The changing weed problem in cotton. In *Weeds in cotton: Characterization and control,* C. G. McWhorter and C. T. Bryson, eds., Reference Book Series, No. 2, pp. 117–232. Memphis, Tenn.: The Cotton Foundation.

Rudd, P. A. (Senior writer.) 1984. Cotton. (Western Regional Integrated Pest Management Project), University of California, Oakland: Division of Agriculture and Natural Resources.

Snipes, E. S., et al. 1992. Cotton (*Gossypium hirsutum*) yield response to cultivation timing and frequency. *Weed Technol.* 6:31–35.

Timmons, F. L. 1970. A history of weed control in the United States and Canada. *Weed Sci.* 18:294,307.

Unger, P. W., and T. M. McCalla. 1981. Conservation tillage systems. *Ad. Agron.* 33:1–44.

Zimdahl, R. L. 1981. Extent of mechanical, cultural, and other nonchemical methods of weed control. In *CRC handbook of pest management in agriculture,* Vol. 2, pp. 79–83. Boca Raton, Fla.: CRC Press.

34 PEANUTS

■ INTRODUCTION

In 1991, peanuts were planted on 2 million A in the United States (Table 34.1). In the same year, weeds caused an estimated loss of $127 million in peanuts, even though herbicides were used. Under the same conditions, but without herbicides, the estimated loss would have been $791 million. The principal problem weeds for peanuts grown in the southeastern United States are given in Table 34.2 on page 276.

There are three major peanut-producing regions in the United States: the Southeast, the Virginia–Carolina area, and the Southwest (Table 34.1). For commercial purposes, peanuts are grouped into four types: (1) the large-seed Virginia with both bunch- and runner-type plants, (2) the true Runners, (3) the Spanish, and (4) the Valencia. The latter two are erect or bunch types. Bunch peanuts have erect branches and produce most of their pods from a cluster of pegs developing in the soil at the base of the plant. Runners have prostrate branches from which most of the pegs and pods develop. The large-seed Virginia peanut is grown almost entirely in the Virginia–Carolina region. The Runner, Virginia, and Spanish types are grown in the Southeast region. The Spanish type is grown in Oklahoma and Texas. The Valencia peanut is grown almost entirely in

New Mexico. The Virginia and Valencia types are used largely for roasting in the shell and for specialty products. The bulk of the Spanish and Runner peanuts are used in making peanut butter and confections. Most of the peanut crop is processed for oil and meal. The peanut kernel (nut) contains 43–50% oil and 25–30% protein.

■ WEEDS

The most common and troublesome grass weed in peanuts grown in the Southeast is Texas panicum, and the three most common and troublesome broadleaf weeds are sicklepod, Florida beggarweed, and morningglories. The presence of Texas panicum not only reduces peanut yield, but also reduces harvesting efficiency as peanut pods become embedded in its extensive root system. Consequently, complete control of Texas panicum is desirable. Wild poinsettia is a newer weed that is of major economic importance in peanuts grown in the Southeast.

■ WEED CONTROL

Precision cultivation for weed control in peanuts is recommended using flat sweeps set to run shallow between rows.

TABLE 34.1 ■ Peanut-growing regions in the United States and acreage planted in 1991.*

STATE	ACRES (× 1000)	STATE	ACRES (× 1000)
Southeast		*Southwest*	
Alabama	280	New Mexico	22.7
Georgia	900	Oklahoma	110
Florida	123	Texas	330
S. Carolina	14.5	**Regional total: 462,700 A**	
Regional total: 1,317,500 A			
Virginia and North Carolina			
N. Carolina	164		
Virginia	98		
Regional total: 262,000 A			
*Total peanut acreage in the United States in 1991 was 2,042,200 A.			

SOURCE: *Agricultural statistics 1992.* United States Department of Agriculture, Government Printing Office, Washington, D.C. p 118.

TABLE 34.2 ■ Principal problem weeds in peanuts in the southeastern United States.*

COMMON WEEDS**	TROUBLESOME WEEDS†
Amaranthus spp. (pigweeds)	*Amaranthus* spp. (pigweeds)
Common cocklebur	Bristly starbur
Crabgrasses	Common cocklebur
Florida beggarweed	Florida beggarweed
Goosegrass	Morningglories
Morningglories	Nutsedge, purple
Nutsedge, purple	Nutsedge, yellow
Nutsedge, yellow	Texas panicum
Sicklepod	
Signalgrasses	
Texas panicum	

*Weed species present in three or more states in the Southeast.
**Common weeds are those that abundantly infest a significant portion of the peanut acreage.
†Troublesome weeds are those that, despite weed-control efforts, are inadequately controlled and interfere with crop production and/or yield, crop quality, or harvest efficiency.

SOURCE: Anonymous. 1992. *Crop losses due to weeds in the United States.* Champaign, Ill.: Weed Science Society of America.

Positive depth and lateral control of tillage equipment is necessary to avoid injuring the peanuts. The peanut plant is especially susceptible to injury during pegging (runner formation and setting of the underground nut). The use of fenders or shields on the tillage equipment to prevent soil movement onto the plants at any time during the season is a desirable practice. Rolling cultivators can be used effectively. When using gang cultivators such as the Lilliston, the gangs should be set for minimum lateral movement of the soil.

Standard weed-control practices involve thorough seedbed tillage and soil incorporation of dinitroaniline herbicides such as benefin, pendimethalin, or ethalfluralin. Texas panicum and other annual grasses and many annual broadleaf weeds can be controlled with these herbicides. Texas panicum and other grasses can be controlled postemergence with fenoxaprop or sethoxydim. Early to midseason control of common cocklebur, sicklepod, and other annual broadleaf weeds is achieved postemergence with acifluorfen, 2,4-DB, or pyridate. The principal herbicides used in peanuts in 1991 in the Southeast and Southwest are given in Table 34.3.

■ INFORMATION OF PRACTICAL IMPORTANCE

Preplant and Preemergence

Alachlor (Lasso 4EC) is an acid amide herbicide (Chapter 12) and a Restricted Use Pesticide due to its oncogenicity. It is used to control grass and broadleaf weeds. It can be applied within 7 days before planting and incorporated 1–2 inches deep, or preemergence to the soil surface after planting but prior to crop and weed emergence. Alachlor may be applied preemergence at the ground-cracking stage of peanuts. For best results, rainfall or overhead sprinkler irrigation within 7 days after application, and prior to weed emergence, is needed to move the herbicide into the weed germination zone. To broaden the spectrum of weeds controlled, alachlor can be applied preplant-incorporated in tank mix with benefin, pendimethalin, or ethalfluralin. Alachlor is applied at 3–4 lb ai/A.

Metolachlor (Dual 8EC) is an acid amide herbicide (Chapter 12). It can be applied preplant-incorporated, postplant-incorporated, preemergence, or at lay-by in peanuts. When applied postplant, incorporate metolachlor shallowly, above the crop seed, or seed may be damaged. Multiple applications of metolachlor can be made in all states but those in the Southwest. Metolachlor can be applied preplant-incorporated in tank mix with benefin within 14 days prior to planting. A metolachlor application can follow preplant-incorporated treatments with benefin, ethalfluralin, trifluralin (in Spanish peanuts in Oklahoma and Texas only), and vernolate. Metolachlor is applied at 1.5–3 lb ai/A.

TABLE 34.3 ■ Principal herbicides used in peanuts in 1991, expressed as the average percentage of total acreage grown in each region.

| HERBICIDE* | AVERAGE PERCENT OF TOTAL ACREAGE | | |
	Southeast Region	Southwest Region	Mean
Acifluorfen (Blazer)	29	12	20.5
Alachlor (Judge, Lasso)	62	8	35.0
Benefin (Balan)	44	11	27.5
Bentazon (Basagran, others)	26	6	16.0
Chlorimuron (Classic)	19	0	9.5
2,4-DB (Butoxone)	68	48	58.0
Ethalflurafin (Sonalan)	20	2	11.0
Metolachlor (Dual)	29	20	24.5
Paraquat (Starfire)	61	1	31.0
Pendimethalin (Prowl)	24	35	29.5
Sethoxydim (Poast Plus)	15	8	11.5
Trifluralin (Treflan, others)	0	43	21.5
Vernolate (Vernam)	22	6	14.0

*A particular herbicide may not be used in every state within a region.

SOURCE: Bridges, D. C., ed. 1992. *Crop losses due to weeds in the United States.* Champaign, Ill.: Weed Science Society of America.

Benefin, ethalfluralin, pendimethalin, and *trifluralin* are dinitroaniline herbicides (Chapter 17). These herbicides provide residual control of grass and broadleaf weeds as they germinate. They do not control emerged weeds. They are soil-incorporated to move them into the seed germination zone and to prevent their loss by volatility from the soil surface.

■ *Benefin* (Balan 1.5EC) must be incorporated 2–3 in. deep (to reduce loss by volatility) within 4 hours after application in the western United States and within 8 hours in the eastern United States. Benefin is incorporated to a depth of 3 in. for more effective Texas panicum control. It can be applied immediately before planting to as much as several weeks before planting. Benefin may be applied preplant-incorporated in tank mix with metolachlor or vernolate. Benefin is applied at 1.1–1.5 lb ai/A.

■ *Ethalfluralin* (Sonalan 3EC) must be incorporated 2–3 in. deep within 2 days after application. It can be applied and incorporated up to 3 weeks prior to planting. Ethalfluralin can be applied preplant in tank mix with alachlor, metolachlor, or vernolate. Ethalfluralin is applied at 0.5–1.1 lb ai/A.

■ *Pendimethalin* (Prowl 4EC) must be incorporated 1–2 in. deep within 7 days after application. Pendimethalin

can be applied immediately before planting or up to 60 days before planting. It can be applied preplant-incorporated in tank mix with metolachlor or vernolate. Pendimethalin is applied at 0.5–1.0 lb ai/A.

■ *Trifluralin* (Treflan 4EC, others) can be applied before planting, at planting, or immediately after planting. Trifluralin must be incorporated 2–3 in. deep within 24 hours after application. When soil-incorporating after planting, take care not to disturb the crop seed. A tank mix of trifluralin and vernolate can be applied up to 10 days prior to planting peanuts; incorporate immediately after application. Use trifluralin or the tank mix only on Spanish peanuts grown in Texas and Oklahoma. Trifluralin is applied at 2–2.5 lb ai/A.

Imazethapyr (Pursuit 2EC) is an imidazolinone herbicide (Chapter 20). It is used to control grass and broadleaf weeds. It can be applied preplant-incorporated, preemergence, ground-cracking, or postemergence to peanuts. It may be applied as a split application (sequential) with one-half the recommended rate applied preplant or preemergence and the remaining half applied at ground-cracking or postemergence. Imazethapyr is applied at 0.25 lb ai/A.

Vernolate (Vernam 7EC) is a thiocarbamate herbicide (Chapter 26). It is used to control grass and broadleaf weeds. It is applied at planting and immediately incorporated 2–3 in. deep with tandem disks or power-driven equipment. It may also be applied with injector equipment spaced 3 in. apart, with six injectors per 18-in. band. To avoid crop injury, do not plant deeper than 2 in. into treated soil. Do not re-treat if replanting is necessary. Vernolate is applied at 2–2.5 lb ai/A.

Postemergence

Acifluorfen (Blazer 2WS) is a diphenyl ether herbicide (Chapter 18). It is used to control grass and broadleaf weeds. It can be applied over-the-top to peanuts as a contact spray to young, actively growing grass and broadleaf weeds. It can be tank-mixed with bentazon, 2,4-DB, alachlor, metolachlor, and sethoxydim. Add 1 pt of nonionic surfactant per 100 gal of spray mixture when applying acifluorfen alone and in all tank mixes except those with sethoxydim. In tank mixes with sethoxydim, add 2 pt of oil concentrate to spray mixture per acre. Acifluorfen is applied at 0.25–0.38 lb ai/A.

Bentazon (Basagran 4WS) is a nonfamily herbicide (Chapter 30). It is used to control broadleaf weeds and yellow nutsedge. It can be applied over-the-top of peanuts. Peanut plants are tolerant to bentazon at all stages of growth. Used alone, add 1 pt of a nonionic surfactant per 100 gal of spray mixture. Basagran may be tank-mixed with 2,4-DB. Basagran does not control grasses, and it has no residual activity. Basagran is applied at 0.5–1 lb ai/A.

The *premix Storm 4EC,* which contains *acifluorfen + bentazon,* is applied to peanuts from cracking through the expanded, tetrafoliate, two-leaf stage, a stage favoring the control of young, actively growing broadleaf weeds. Storm is effective mainly through contact action. It does not control grasses. Spray additives are needed with Storm: add either a nonionic surfactant, an oil concentrate, or urea ammonium nitrate (UAN) to the spray mixture as recommended on the product label. Storm is applied at 0.75 lb ai/A.

2,4-DB (Butoxone 1.75EC) is a growth regulator–type herbicide (Chapter 19). It is used to control broadleaf weeds. It does not control grasses. It can be applied over-the-top to peanuts up to the bloom stage. Avoid spray drift to susceptible broadleaf plants. 2,4-DB has no residual activity. It is applied at 0.2–0.4 lb ae/A.

Fenoxaprop-ethyl (Bugle 0.8EC) is an aryloxyphenoxypropionate herbicide (Chapter 13). It is used to control grasses. It does not control broadleaf weeds or sedges. Fenoxaprop-ethyl can be applied over-the-top to peanuts. It is applied at 0.05–0.15 lb ai/A.

Pyridate (Tough 3.75EC) is a nonfamily herbicide (Chapter 30). It is used to control broadleaf weeds, purple nutsedge, and yellow nutsedge. It can be applied over-the-top to peanuts. Peanut plants are tolerant of pyridate at all growth stages. Pyridate does not control grasses. It is applied at 0.9–1.4 lb ai/A.

Paraquat (Starfire 1.5L) is a bipyridinium herbicide (Chapter 15). It is a nonselective, contact-type, postemergence herbicide. Do not apply paraquat as an over-the-top spray to peanuts, as the plants will be killed. Paraquat can be applied in peanuts from ground-cracking to 28 days after cracking. Paraquat is applied to young, actively growing weeds using shielded sprays to prevent the spray mixture from contacting the peanut foliage. If paraquat makes contact with the peanut foliage, the contaminated part will turn brown and crinkled, but the crop will recover and develop normally. Paraquat is applied at 0.13 lb ai/A.

Precautionary statement: Paraquat is a Restricted Use Pesticide. It is fatal to humans if swallowed, inhaled, or absorbed through the skin. Paraquat reacts with aluminum to produce hydrogen gas. Do not mix or store this product in containers or equipment made of aluminum or that have aluminum fittings.

■ SELECTED REFERENCES

Agricultural statistics 1992. United States Department of Agriculture, Government Printing Office, Washington, D.C. p. 118.
Anonymous. 1992. *Weed control guidelines for Mississippi 1992,* pp. 102–109. Mississippi State, Miss.: Mississippi State University.
Anonymous. 1993. New herbicides for peanuts. *Progressive Farmer.* April Issue, p. 32.
Anonymous. 1994. *Weed control guide 1994,* pp. 153–157. Willoughby, Ohio: Meister Publishing.
Bridges, D. C., ed. 1992. *Crop losses due to weeds in the United States.* Champaign, Ill: Weed Science Society of America.
Bridges, D. C., B. J. Brecke, and J. C. Barbour. 1992. Wild poinsettia (*Euphorbia heterophylla*) interference with peanut (*Arachis hypogaea*). *Weed Sci.* 40:37–42.
Herbicide product labels.
Leidner, J. 1990. How a peanut plant grows. *Prog. Farmer.* February Issue, pp. 35–49.
Wilcut, J. W., A. C. York, and G. R. Wehtje, 1994. The control and interaction of weeds in peanuts (*Arachis hypogaea*). *Rev. Weed Sci.* 6:177–205.

35

RICE

■ INTRODUCTION

In 1991, rice was planted on 2.9 million A in the United States (Table 35.1). In the same year, weeds caused an estimated loss of $59 million in rice grown with the best available weed-control practices, including herbicides. Under the same conditions, but without herbicides, weeds would have caused an estimated loss of $690 million. The principal weeds infesting the rice crop in the southeastern United States are given in Table 35.2. In 1992, herbicides were applied to 97% of the rice acreage in Arkansas and Louisiana, the two principal rice-producing states.

Proper water management is the most important factor in successful weed control in rice, and a well-leveled field is essential. A permanent flood restricts oxygen diffusion into the soil, resulting in reduced weed seed germination and growth. Lowering or draining the floodwater favors weed seed germination and growth. A continuous water depth of 3–6 in. is necessary for the effectiveness of most currently used herbicides.

The principal herbicides used in rice in 1991, by region and state, are given in Table 35.3. These same herbicides were registered for use in rice in 1994.

■ INFORMATION OF PRACTICAL IMPORTANCE

Bensulfuron-methyl (Londax 60 DF) is a sulfonylurea herbicide (Chapter 25). It is used postemergence to control most broadleaf weeds and sedges. It does not control grasses. Bensulfuron-methyl is applied when rice is in the one- to three-leaf stage, and when weeds are actively growing and are in less than the three-leaf stage of growth. It is applied directly to standing water, after establishment of permanent flood, and the floodwater is retained for a minimum of 7 days after application. Bensulfuron-methyl is applied at 0.6–1 oz ai/A (1–1.6 oz product/A).

Bentazon (Basagran 4WS) is a nonfamily herbicide (Chapter 30). It is used postemergence to control annual broadleaf weeds and yellow nutsedge. It does not control grasses. Bentazon has no adverse effect on rice when used as recommended, and it may be used on the first and second (ratoon) crops. If grasses are a problem, use bentazon in tank mix with propanil. Applications of bentazon must be made when there is no water on the field and 24 hours or more prior to flooding. Where continuous flooding is the practice, treatment should be made to weed foliage

TABLE 35.1 ■ Rice-growing regions in the United States and acreage planted in 1991.

STATE	ACRES (× 1000)	STATE	ACRES (× 1000)
Corn Belt		*Southeast*	
Missouri	97	Arkansas	1,300
Regional total: 97,000 A		Louisiana	560
Pacific Coast		Mississippi	225
California	330	**Regional total: 2,085,000 A**	
Regional total: 330,000 A			
Southwest			
Texas	345		
Regional total: 345,000 A			
*Total rice acreage in 1991 was 2,857,000 A.			

SOURCE: *Agricultural statistics 1992.* United States Department of Agriculture, Government Printing Office, Washington, D.C. p. 22.

TABLE 35.2 ■ Principal problem weeds for rice in the southeastern United States.*

COMMON WEEDS**	TROUBLESOME WEEDS†
Barnyardgrass	Barnyardgrass
Ducksalad	Broadleaf signalgrass
Eclipta	Ducksalad
Hemp sesbania	Hemp sesbania
Jointvetches	Jointvetches
Nutsedge, yellow	Nutsedge, yellow
Red rice	Red rice
Sprangletops	Sprangletops

*Southeastern states: Arkansas, Louisiana, Mississippi.
**Weed species present in two or more states in the Southeast. Common weeds abundantly infest a significant portion of the rice acreage.
†Troublesome weeds are inadequately controlled despite weed-control efforts, and interfere with crop production and/or yield, crop quality, and harvest efficiency.

SOURCE: Anonymous. 1992. *Crop losses due to weeds in the United States.* Champaign, Ill.: Weed Science Society of America.

TABLE 35.3 ■ Principal herbicides used in rice in 1991, expressed as the average percentage of total rice acreage in each region.

HERBICIDE**	REGION* 1	2	3	4	MEAN
Bentazon	10	34	28	5	19
MCPA	0	62	3	4	17
Molinate	7	74	50	37	42
Propanil	94	5	95	83	69
Thiobencarb	20	24	50	11	26

*Regions: 1 = Corn Belt; 2 = Pacific Coast; 3 = Southwest; 4 = Southeast.
**A particular herbicide may not be used in every state within a region.

SOURCE: Anonymous. 1992. *Crop losses due to weeds in the United States.* Champaign, Ill.: Weed Science Society of America.

Molinate (Ordram 8E or 10G) is a thiocarbamate herbicide (Chapter 26). A premix (Arrosolo) contains molinate and propanil as its active ingredients, with 3 lb of each herbicide per gallon. Molinate is used specifically to control barnyardgrass, with partial control of other early-season grasses and nutsedge. It may be applied preplant-incorporated prior to flooding, or postemergence when barnyardgrass is at least two-thirds submerged in the flood-water. Directions for use of Ordram 8E vary for dry- and water-seeded rice, as noted in the following paragraphs. Directions for use of Ordram 10G are also given.

■ In *water-seeded rice* in all states, Ordram 8E may be applied preplant and soil-incorporated immediately to a depth of 1–2 in. Flood the field for seeding and hold for minimum of 4–6 days after seeding. The application rate is 3 lb ai/A (3 pt product/A).

■ In *dry-seeded rice* in all states except California, Ordram 8E may be applied preplant and soil-incorporated to a depth of 1–2 in. Applied in liquid fertilizer, incorporation may be delayed up to 2 hours after application. The application rate is 3–4 lb ai/A (3–4 pt product/A).

■ In *dry- or water-seeded rice* in all states except California, apply Ordram 8E directly to flooded field by air only. At time of treatment, barnyardgrass must be 2–5 in. tall and at least two-thirds submerged. The application rate is 3 lb ai/A (3 pt product/A).

■ In *dry-seeded rice* in all states except California, apply Ordram 8E postemergence to rice and to barnyardgrass after flooding. Rice must be established at time of application. Barnyardgrass must be at least two-thirds

standing above the water surface. Control of submerged weeds is poor. The application rate is 0.75–1 lb ai/A (1.5–2 pt product/A), depending on the weeds' stages of growth.

Propanil (Cedar Propanil 4EC, Wham 80DF) is an acid amide herbicide (Chapter 12). It is used postemergence to control annual grass and broadleaf weeds in dry- or water-seeded rice. Weed foliage must not be covered with water at time of treatment. Rice treated after the four-leaf stage may be visibly injured by propanil. Propanil is applied at 3–6 lb ai/A (3–6 qt 4EC/A or 3.8–7.5 lb 80DF/A).

submerged at time of treatment. The application rate is 3 lb ai/A (3 pt product/A).

- In *drilled, water-seeded rice* in California, apply Ordram 8E by air to flooded fields for postemergence control of barnyardgrass. Barnyardgrass should be less than 5 in. tall and at least two-thirds submerged at time of treatment. Rice should be in the seedling stage. Hold floodwater a minimum of 4 days after application. The application rate is 3–5 lb ai/A (3–5 pt product/A).

- *Ordram 10G* is used in California. It may be applied preplant-incorporated (preflood) or postemergence (postflood) to barnyardgrass at least two-thirds submerged. Either treatment may be applied by air or ground equipment. The application rate is 3–5 lb ai/A (30–50 lb Ordram 10G/A).

Thiobencarb (Bolero 8EC or 10G) is a thiocarbamate herbicide (Chapter 26). Bolero 8EC is applied to dry-seeded rice 1–5 days before emergence. It is applied early postemergence to barnyardgrass, sprangletop, and aquatic weeds. The soil should be wet and barnyardgrass at not more than the two-leaf stage at time of application. The application rate for Bolero 8EC is 4 lb ai/A (2 qt product/A). Bolero 8EC may be tank-mixed with propanil.

Bolero 10G is applied by air to flooded rice in the expanded two-leaf stage. The floodwater level should be maintained 3–4 days after application, with no soil exposed. Rice smaller than the two-leaf stage will be killed or injured. Grass weeds should be in the two-leaf stage or smaller and aquatic weeds less than 0.5 in. tall. Weeds larger than specified will not be controlled. Do not drain field for a minimum of 4 days after application. In California, the rate of application is 4 lb ai/A (40 lb Bolero 10G/A). In all other states, the application rate is 3–4 lb ai/A (30–40 lb Bolero 10G/A).

■ SELECTED REFERENCES

Agricultural chemical usage: 1993 field crop summary. United States Department of Agriculture, National Agricultural Statistics Service, Washington, D.C. p. 64.

Agricultural statistics 1992. United States Department of Agriculture, Government Printing Office, Washington, D.C. p. 22.

Anonymous. 1992. *Weed control guidelines for Mississippi.* Mississippi State, Miss.: Mississippi State University. pp. 83–90.

Anonymous. 1994. *Weed control guide 1994.* Willoughby, Ohio: Meister Publishing. pp. 157–160.

Bridges, D. C., ed. 1992. *Crop losses due to weeds in the United States.* Champaign, Ill.: Weed Science Society of America.

36 GRAIN SORGHUM

■ INTRODUCTION

In 1991, grain sorghum (often called "milo") was grown on 11 million A in the United States, with about one-half of the crop grown in Kansas and Nebraska and one-third in Texas. The acreages for this crop are given in Table 36.1. Grain sorghum is grown in the United States primarily as a source of high-protein animal feed. It is second in importance only to corn for this purpose. In 1991, weeds caused an estimated crop loss of $103 million, even though herbicides were used. If herbicides had not been used, the estimated crop loss would have been $452 million.

Grain sorghum is grown as an annual, although in southern California it may be perennial. Grain sorghum is well adapted to regions of limited rainfall, 17–25 in. It is also highly productive when grown under irrigation. Grain sorghum is grown on flat and raised seedbeds. High-yielding dwarf hybrids, about 3 ft tall, have been developed for combine harvesting.

■ WEEDS

The first 4 weeks after planting is the critical weed-free period for sorghum. During this period the crop must be kept weed-free to avoid reduced crop yield due to weed competition.

Shattercane, an annual, weedy, forage-type sorghum, is one of the most troublesome weeds in grain sorghum. It is so closely related to grain sorghum that it shares the same scientific name. It can grow 3–10 ft tall and produce over 1000 seeds per panicle. Its seed can remain viable in soil for up to 13 years. Shattercane seeds are deciduous and frequently drop prior to crop harvest. The principal weeds in grain sorghum are listed in Table 36.2.

■ WEED CONTROL

An integrated system of cultural, mechanical, and chemical practices comprises successful weed control in grain sorghum. These practices need not provide season-long weed control, but they must be timely. Crop rotation is the best cultural practice to control weeds in grain sorghum.

The most common crop rotation in the drier parts of Kansas and Nebraska is winter wheat–sorghum–fallow. Weeds that infest winter wheat are winter annuals and those that infest sorghum are summer annuals. When summer annuals become problem weeds in sorghum, rotating to winter wheat for a few years eliminates the problem.

TABLE 36.1 ■ Grain sorghum–growing regions in the United States and acreage planted in 1991.*

STATE	ACRES (× 1000)
Corn Belt	
Illinois	180
Missouri	550
Regional total: 730,000 A	
Great Plains	
Kansas	3,400
Nebraska	1,450
S. Dakota	500
Regional total: 5,350,000 A	
Southwest	
New Mexico	180
Oklahoma	350
Texas	3,200
Regional total: 3,730,000 A	
Southeast	
Alabama	30
Arkansas	290
Georgia	90
Kentucky	32
Louisiana	205
Mississippi	85
N. Carolina	45
S. Carolina	32
Tennessee	75
Regional total: 884,000 A	
Mountain	
Colorado	320
Regional total: 320,000 A	
*Total grain sorghum acreage planted in 1991 was 11,014,000 A.	

SOURCE: *Agricultural statistics 1992.* United States Department of Agriculture, Government Printing Office, Washington, D.C. p. 52.

TABLE 36.2 ■ Principal weeds causing losses in grain sorghum in the Northern Plains region.*

COMMON WEEDS**	TROUBLESOME WEEDS†
Common cocklebur	Common milkweed
Common sunflower	Field bindweed
Field bindweed	Hemp dogbane
Foxtails	Sandburs
Kochia	Shattercane
Pigweeds	Velvetleaf
Sandburs	
Shattercane	
Velvetleaf	

*Northern Plains states: Kansas, Nebraska, South Dakota.
**Weed species present in two or more states in the Northern Plains region. Common weeds are those that abundantly infest a significant portion of the sorghum acreage.
†Troublesome weeds are those that, despite weed-control efforts, are inadequately controlled and interfere with crop production and/or yield, crop quality, or harvest efficiency.

SOURCE: D. C. Bridges, ed. 1992. *Crop losses due to weeds in the United States.* Champaign, Ill.: Weed Science Society of America.

After a few years of winter wheat, the seeds of summer annuals in the soil are reduced, and grain sorghum can be grown again. Tillage in fallow between wheat crops prevents summer annual weeds from going to seed and also soil moisture losses by transpiration.

Mechanical Weed Control

In the drier part of Kansas and Nebraska, primary tillage is done with subsurface tillage tools or by one-way disk plowing. Subsequent cultivation is with field cultivators, rod weeders, or disks. In the wetter parts of these states, moldboard plowing is usually the first tillage operation after harvest. This is followed by disking or cultivation with small sweeps. Moldboard plowing leaves the soil exposed to wind and water erosion.

Success with mechanical weed control in grain sorghum is timeliness. Sorghum seedlings are not vigorous, and they are at the mercy of weed seedlings. Cultivation is done with small sweeps, rotary hoes, rolling cultivators, and various types of harrows. Small weeds are easily uprooted or buried with soil by cultivation. Large weeds are difficult to kill by cultivation.

Reduced-Tillage

Over the past 25 years, reduced-tillage (see Chapter 4), also called *chemical fallow,* in grain sorghum production has begun to replace conventional tillage during the fallow period between the harvest of one crop and the planting of the next. With this system, herbicides (rather than tillage) are used to kill weeds. The abundant crop residue on the soil surface greatly reduces the potential for wind and water erosion. Often, water is conserved and crop yields increased. Another perceived advantage is a reduction in production costs. The key to success with reduced-tillage is obtaining adequate weed control with herbicides.

Atrazine in tank mix with either 2,4-D, dicamba, paraquat, or glyphosate is an economical and effective way to control weeds in winter wheat stubble during the 8- to 11-month fallow prior to planting sorghum. The number of

TABLE 36.3 ■ Principal herbicides used in sorghum in 1990 expressed as the average percentage of total acreage grown in each region.

HERBICIDE**	REGION* 1	2	3	4	5	MEAN
Alachlor	55	7	4	15	20	20
Atrazine	95	52	11	0	76	17
2,4-D	2	12	16	10	24	13
Dicamba	7	0	9	5	7	6
Glyphosate	1	3	5	25	6	6
Paraquat	2	0	3	15	10	6
Propachlor	8	22	0	25	1	11

*Regions: 1 = Corn Belt; 2 = Northern Plains; 3 = Southwest; 4 = Mountain; 5 = Southeast.
**A particular herbicide may not be used in every state within a region.

SOURCE: D. C. Bridges, ed. 1992. *Crop losses due to weeds in the United States.* Champaign, Ill.: Weed Science Society of America.

tillage operations during fallow can be reduced by applying residual herbicides such as chlorsulfuron, cyanazine, or metribuzin. In the southeastern United States, a mixture of paraquat or glyphosate with atrazine or propazine applied in wheat stubble prior to planting sorghum will control weeds in the stubble. A satisfactory reduced-tillage system for furrow-irrigated, continuous-grain sorghum has yet to be developed.

Chemical Weed Control

Herbicides have been widely used for weed control in sorghum because they give effective and economical control of early weeds. The principal herbicides used in grain sorghum in 1990 are shown in Table 36.3.

■ INFORMATION OF PRACTICAL IMPORTANCE

Alachlor (Lasso 4EC) is an acid amide herbicide (Chapter 12) and a Restricted Use Pesticide because of its oncogenicity. It is used to control annual grass and broadleaf weeds and yellow nutsedge. Alachlor can be applied preplant-incorporated or preemergence to grain sorghum. Applied preplant, soil-incorporate 1–2 in. deep prior to planting. It may be applied in water or a sprayable fluid fertilizer solution. It is applied preplant-incorporated at 2–3 lb ai/A (2–3 qt product/A), and preemergence at 1.5–2.5 lb ai/A (1.5–2.5 qt product/A). Alachlor can be tank-mixed with atrazine to broaden the spectrum of weeds controlled. Alachlor can be used in conservation or minimum-tillage systems in tank mixes with atrazine, glyphosate, or paraquat.

Precautionary statement: Preplant-incorporated and preemergence-surface applications of alachlor must be made *only* to grain sorghum that has been properly treated with the seed protectant or safener *Screen,* which contains flurazole as its active ingredient (see Chapter 8 for discussion of safeners).

Metolachlor (Dual 8E) is an acid amide herbicide (Chapter 12). It is used to control annual grasses and certain broadleaf weeds, and to control or suppress yellow nutsedge. It does not control emerged weeds. Metolachlor can be applied preplant-surface, preplant-incorporated, or preemergence in grain sorghum. It can be applied preplant-surface as a split application 30–45 days prior to planting, with two-thirds of the broadcast rate applied initially and the remaining one-third at planting. If applied less than 30 days prior to planting, a split or single application may be made. The broadcast rate for this treatment is 2.0–2.5 lb ai/A, depending on the soil type and time of application. Metolachlor can be applied preplant and incorporated into the top 2 in. of soil within 14 days prior to planting. It can be applied preemergence directly behind the planter or after planting but before weeds or crop emerges. The broadcast rate for metolachlor applied preplant-incorporated or preemergence is 1.5–2.5 lb ai/A, depending on the soil type.

Caution: Apply metolachlor alone or in tank mixtures only when the seed has been properly treated by the seed company with the safener *Concep* (see Chapter 8 for a discussion of safeners).

Atrazine (AAtrex 4L and Nine-O) is a triazine herbicide (Chapter 27) and a Restricted Use Pesticide due to ground and surface water concerns. It is used to control grass and broadleaf weeds. Atrazine can be applied preplant-surface, preplant-incorporated, preemergence during or shortly after planting, or postemergence before weeds exceed 1.5 in. in height and before sorghum exceeds 12 in. in height. Atrazine is applied preplant or preemergence at 1.6–2 lb ai/A. It is applied postemergence at 1.2 lb ai/A, with an emusifiable oil or oil concentrate, for control of annual broadleaf weeds. Atrazine may be used alone or in tank mix with paraquat for chemical fallow in a wheat–sorghum–fallow rotation.

2,4-D (Formula 40, Weedar 64, Weedone LV4) is a growth regulator–type herbicide (Chapter 19). It is used postemergence in grain sorghum to control broadleaf weeds. 2,4-D can be applied when the sorghum plants are 6–15 in. tall. To avoid crop injury, use drop nozzles and keep spray off crop foliage and out of the leaf whorl if sorghum is taller than 8 in. To avoid yield reductions, do not treat grain sorghum during the boot, flowering, or dough stage. 2,4-D is applied at 0.35–0.53 lb ae/A.

Dicamba (Banvel 4WS) is a growth regulator–type herbicide (Chapter 19). It is used postemergence in grain sorghum to control broadleaf weeds. Dicamba can be applied to emerged weeds at least 15 days prior to planting. It can be applied postemergence to sorghum from the two-leaf stage up to 15 in. tall. For best results, apply dicamba when sorghum is in the three- to five-leaf stage and weeds are less than 3 in. tall. To avoid crop injury, use drop nozzles when sorghum is taller than 8 in., keeping the spray off the crop foliage and out of the leaf whorl. Dicamba can be tank-mixed with atrazine or bromoxynil for improved control of broadleaf weeds. Dicamba is applied at 0.25 lb ae/A.

Glyphosate (Roundup 3WS) is a nonfamily herbicide (Chapter 30). Glyphosate is a nonselective, translocated herbicide applied preemergence to grain sorghum to control emerged weeds. Emerged grain sorghum plants are not tolerant of glyphosate. Glyphosate is applied at 0.25–1.5 lb ai/A or 0.19–1.1 lb ae/A (0.5–3 pt Roundup/A), depending on the target weeds.

Paraquat (Gramoxone Extra 2.5EC) is a bipyridinium herbicide (Chapter 15). It is a postemergence, nonselective, contact-type herbicide. It can be applied preplant or preemergence to grain sorghum to control emerged weeds. Sorghum plants are not tolerant of paraquat; do not overspray. Paraquat is applied at 0.5–1 lb ai/A (2–3 pt product/A), depending on the weeds present.

Caution: Paraquat is a Restricted Use Pesticide. It is fatal to humans if swallowed, inhaled, or absorbed through the skin. Paraquat reacts with aluminum to produce hydrogen gas. Do not mix or store this product in containers or equipment made of aluminum or that have aluminum fittings.

■ SELECTED REFERENCES

Agricultural statistics 1992. United States Department of Agriculture, Government Printing Office, Washington, D.C. p. 52.

Anonymous. 1994. *Weed control guide 1994.* Willoughby, Ohio: Meister Publishing. pp. 161–167.

Bridges, D. C., ed. 1992. *Crop losses due to weeds in the United States.* Champaign, Ill.: Weed Science Society of America.

Wiese, A. F. 1981. Pest management systems for sorghum weeds. In *CRC handbook of pest management in agriculture,* Vol. III, pp. 575–586. D. Pimentel, ed. Boca Raton, Fla.: CRC Press.

Wiese, A. F., P. W. Unger, and R. R. Allen. 1985. Reduced tillage in sorghum. In *Weed control in limited-tillage systems,* Monograph No. 2, pp. 51–60. A. F. Wiese, ed. Champaign, Ill.: Weed Science Society of America.

37 SOYBEANS

■ INTRODUCTION

In 1991, 59.1 million A of soybeans were grown in 29 states in the United States (Table 37.1). Of this acreage, 97% was treated with herbicides. In this same year, weeds caused an estimated loss of $800 million in soybean yields, even though herbicides were used. If herbicides had not been used, the estimated loss would have been $3.9 billion.

In 1992, 1.4 million A (568,000 ha) of soybeans were harvested in Canada, all in the eastern province of Ontario. Weeds reduced the soybean crop by 10% and caused a monetary loss of $31.8 million.

Weed species common to soybeans grown in the United States are given in Table 37.2. Velvetleaf and volunteer corn are two of the worst weeds in soybeans grown in the Corn Belt. Sicklepod is one of the most serious soybean weeds in the warm, humid southeastern United States.

Soybean is an annual leguminous plant that attains a height of 2–4 ft, depending on the variety. It is a major oil seed crop, grown for its seed, oil, and meal. Soybean seed and oil are used primarily for food, with over 90% of the oil used in edible products such as margarine, shortening, salad and cooking oils, and mayonnaise. In 1991, 7 million tons of soybean oil were produced in the United States.

The soybean meal is a high-protein source of animal feed, with about 98% of the meal used for this purpose.

■ CROP ROTATIONS

In the Midwest, soybeans are commonly rotated with corn, small grains, and forages. Throughout the southern United States, soybeans are commonly rotated with cotton, corn, wheat, grain sorghum, Italian ryegrass, and rice. This rotation can take place as a double-cropping program immediately following soybean harvest, or rotated with soybeans the following year.

■ WEED CONTROL

Soybean yields are reduced only slightly when weeds are present for the first 4–6 weeks, but the rate of yield loss increases rapidly when weeds are present longer. When weeds are present during the first 12 weeks, the yield loss is nearly as great as when weeds compete for the entire season. Late-emerging weeds can interfere with harvesting and cause market dockage, resulting in diminished profits.

TABLE 37.1 ■ Principal soybean-growing regions in the United States and acreage planted to soybeans in 1991.*

STATE	ACRES (× 1000)
Corn Belt	
Illinois	9,200
Indiana	4,450
Iowa	8,700
Missouri	4,500
Ohio	3,800
Regional total: 30.65 million A	
North Central	
Kansas	1,950
Michigan	1,400
Minnesota	5,500
Nebraska	2,500
N. Dakota	635
S. Dakota	2,200
Wisconsin	570
Regional total: 14.81 million A	
Southwest	
Oklahoma	260
Texas	180
Regional total: 0.44 million A	
Southeast	
Alabama	360
Arkansas	3,250
Florida	45
Georgia	600
Kentucky	1,150
Louisiana	1,030
Mississippi	1,900
N. Carolina	1,350
S. Carolina	650
Tennessee	1,100
Virginia	530
Regional total: 11.97 million A	
Northeast	
Delaware	255
Maryland	510
New Jersey	125
Pennsylvania	310
Regional total: 1.2 million A	

*Total acres planted in soybeans was 59.07 million.

SOURCE: *Agricultural statistics 1992.* United States Department of Agriculture, Government Printing Office, Washington, D.C. p. 123.

TABLE 37.2 ■ Representative problem weeds for soybeans in the United States.

Common Weeds*	Troublesome Weeds**
Grasses	**Grasses**
Crabgrass, large	Cupgrass, woolly
Foxtail, giant	Foxtail, giant
Foxtail, green	Johnsongrass
Foxtail, yellow	Quackgrass
Johnsongrass	Shattercane
Quackgrass	**Broadleafs**
Shattercane	Burcucumber
Broadleafs	Canada thistle
Cocklebur, common	Cocklebur, common
Horseweed	Dogbane, hemp
Jimsonweed	Jimsonweed
Lambsquarters, common	Lambsquarters, common
Morningglories	Milkweed, common
Morningglory, ivyleaf	Morningglory, ivyleaf
Pigweeds	Morningglory, tall
Pigweed, redroot	Nightshade, Eastern black
Ragweed, common	Pigweeds
Ragweed, giant	Ragweed, giant
Smartweed, Pennsylvania	Smartweed, Pennsylvania
Sunflower, common	Sunflower, common
Velvetleaf	Thistles
Waterhemp, common	Velvetleaf
	Waterhemp, common

*Common weeds infest a significant portion of the soybean acreage, but they are adequately controlled by available control practices.
**Troublesome weeds infest the soybean crop, or portions thereof, and they are inadequately controlled.

SOURCE: D. C. Bridges, ed. 1992. *Crop losses due to weeds in the United States.* Champaign, Ill.: Weed Science Society of America.

will intercept all of the available sunlight, making it more difficult for weeds to become established and compete with the crop.

Weeds are a major problem in narrow-row soybeans. Cultivation becomes very difficult when row widths are less than 20 in. because large tractor tires don't fit between the rows. Rows narrower than 20 in. are difficult to cultivate even when skip-row patterns are used.

Herbicides are often the only practical means of controlling weeds in narrow-row soybeans. Preplant-incorporated herbicides should be used for early control of grass weeds, followed with one or two over-the-top treatments with a suitable postemergence herbicide(s). The principal herbicides used in soybeans in 1991 in the United States are shown in Table 37.3. Comparative use of the top five herbicides in soybeans in 1991, 1992, and 1993, based on the percentage of acreage planted, is shown in Figure 37.1.

Soybeans planted in narrow rows, about 20 in. (50 cm) apart, require about 47 days for the foliar canopy to develop and provide full shading of the soil. Soybeans planted in rows spaced 30 in. apart require about 58 days for full shading of the soil. When the soil is fully shaded, the crop

TABLE 37.3 ■ Principal herbicides used in soybeans in 1991, expressed as the average percentage of total acreage grown in each region.

HERBICIDE**	REGIONS 1	2	3	4	5	MEAN
Acifluorfen	7	10	14	12	18	12
Alachlor	16	15	22	34	26	23
Bentazon	18	22	19	8	10	15
Chlorimuron-ethyl	33	8	20	24	3	18
Imazaquin	18	10	12	13	24	15
Linuron	7	17	11	60	4	20
Metolachlor	14	19	12	44	25	23
Metribuzin	26	16	26	12	10	18
Paraquat	3	0	19	21	11	11
Pendimethalin	13	13	17	14	13	14
Sethoxydim	4	2	8	2	5	26
Trifluralin	31	39	34	5	23	26

*Regions: 1 = Corn Belt; 2 = North Central; 3 = Southeast; 4 = Northeast; 5 = Southwest.
**A particular herbicide may not be used in every state within a region.

SOURCE: D. C. Bridges, ed. 1992. *Crop losses due to weeds in the United States.* Champaign, Ill.: Weed Science Society of America.

No-Tillage

The four basic tillage systems in soybean production are the moldboard plow, the chisel plow, the disk, and the no-tillage system. The moldboard plow is considered the conventional tillage system, and the others are considered conservation tillage systems.

The development of effective herbicides and suitable planters has generated considerable interest in no-tillage soybeans. No-tillage was used on 12 million A of full-season soybeans and 3 million A of double-crop soybeans grown in the United States in 1993; this was, respectively, 22 and 56% of the total soybean acreage. In Ohio, the acreage of full-season, no-till soybeans increased from 3000 A in 1982 to more than 800,000 A in 1988, and by year 2000, about two-thirds of the crop is expected to be grown without tillage.

Weed control is one of the greatest problems encountered with no-tillage soybeans, especially the control of perennial weeds. Weed control in no-tillage soybeans depends almost entirely on foliar- and surface-applied herbicides, as conventional seedbed preparation is eliminated and cultivation impractical. A successful weed-control program in no-tillage soybeans usually includes a foliar-active herbicide to control weeds existing at or before seeding, a soil-residual herbicide to control later-emerging weeds, and one or more postemergence herbicides to control weed escapes.

Mechanical Control

A good approach to weed control in soybeans is to combine cultural practices with mechanical and chemical weed-control measures. Cultural practices include proper planting date, seed rate and depth, row width; the use of weed-free, high-quality seed of adapted varieties; proper liming and soil fertility; and crop rotation. The most commonly used mechanical practices are row cultivation and rotary hoeing.

The primary purpose of row cultivation in soybeans is to control weeds between rows. To avoid crop injury, the cultivation shovels should not be set more than 1–2 in. (2.5–5 cm) deep and should be placed no closer to the row than 6 in. (15 cm), or two-thirds the height of the crop, whichever is greater. Shovels may be adjusted to throw soil into the crop row to smother small weeds in the row, but this approach also builds ridges in the row that interfere with the combine header during harvest. On the other hand, sweeps and rotary hoeing do not build ridges during cultivation, and sweeps permit closer cultivation than do shovels. Shallow sweep cultivation minimizes soybean root-

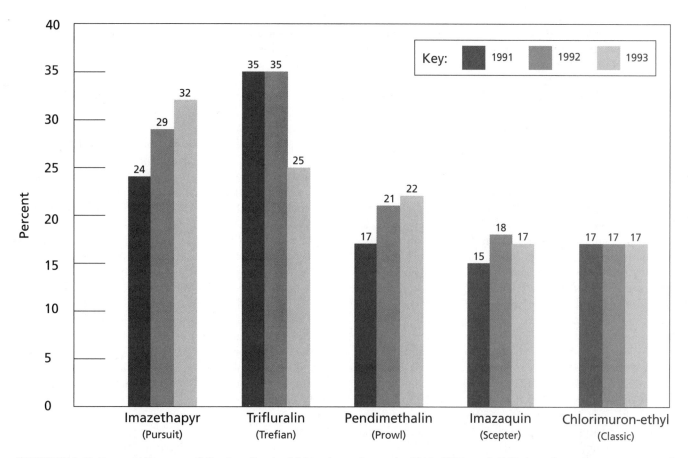

FIGURE 37.1 ■ **Comparative use of the top five herbicides in soybeans in 1991, 1992, and 1993, based on the percentage of acreage planted in 13 states in the respective years.**

SOURCE: *Agricultural chemical usage: 1993 field crop summary.* United States Department of Agriculture, National Agricultural Statistics Service, Washington, D.C. p. 5.

pruning and avoids disturbing weed seeds at deeper levels. Deep cultivation, more than 2 in., will prune roots excessively, resulting in reduced yields. Cultivation produces its best results when the weeds are small.

The rotary hoe (often called a "Lilliston") is particularly effective in controlling small weeds, with best results obtained when used about the time the weeds are emerging. It should be operated at 6–8 mph (10–13 kph) and weighted sufficiently to work the soil to a depth of about 2 in. Rotary hoeing aids soybean emergence by breaking crusted soil. The rotary hoe can be used until soybeans are 4 in. tall, but particular care must be taken when soybeans are in the "crook" stage, as they are brittle (turgid) and highly susceptible to injury at this time. The rotary hoe is most effective when used in combination with sweep cultivation and herbicides. Sweeps are more effective than

the rotary hoe in controlling established weeds. The rotary hoe is an effective tool for incorporating preplant and preemergence herbicides in soybeans. As tillage is reduced in soybeans, an increased demand is placed on herbicides to provide weed control throughout the growing season.

■ INFORMATION OF PRACTICAL IMPORTANCE

Pendimethalin (Prowl 4EC) is a dinitroaniline herbicide (Chapter 17). It is used to control most annual grass and certain broadleaf weeds as they germinate. It is applied preplant-incorporated or preemergence at planting or up to 2 days after planting soybeans. Preemergence treatments are most effective when shallowly incorporated, using a rotary hoe or similar tool, or when adequate rainfall or sprinkler irrigation is received within 7 days after application. Pendimethalin is available in a premix (Squadron) with imazaquin. The broadcast rate of pendimethalin is 0.5–1.5 lb ai/A.

Trifluralin (Treflan 4EC) is a dinitroaniline herbicide (Chapter 17). Trifluralin is a preplant-incorporated herbicide that provides long-lasting control of many annual grass and broadleaf weeds as they germinate. To reduce herbicide loss from the soil surface by volatility and to place it in the seed germination zone, trifluralin must be incorporated into the top 2–3 in. (5–8 cm) of the final seedbed. For uniform mixing into the soil, it is best to incorporate twice, the second pass over the soil with tillage equipment made at a right angle (if possible) to the first. Treflan may be applied in the fall or spring. In the fall, apply and incorporate any time between October 15 and December 31. The ground may be left flat or bedded-up over winter. In the spring, apply and incorporate before or after bedding. Trifluralin is available in various premixes, and it can be tank-mixed with suitable herbicides to broaden the spectrum of weeds controlled. The broadcast rate of Treflan is 0.5–1.25 lb ai/A (1–2.5 pt product/A).

Imazethapyr (Pursuit 2EC) is an imidazolinone herbicide (Chapter 20). It is used to control annual grass and broadleaf weeds. It can be applied early preplant, preplant-incorporated, preemergence, or early postemergence to soybeans. Applied postemergence, weeds should be actively growing and not more than 3 in. (8 cm) in height. With postemergence applications, use a nonionic surfactant or crop oil concentrate and a recommended nitrogen-based liquid fertilizer solution for best results. Imazethapyr can be tank-mixed with many grass and broadleaf herbicides, or they may be applied as sequential treatments. Imazethapyr is available in premix with pendimethalin (Pursuit Plus) and trifluralin (Passport). Imazethapyr is applied at 1.0 oz ae/A (4 fl oz or 0.25 pt product/A).

Imazaquin (Scepter 1.5EC or 70 DG) is an imidazolinone herbicide (Chapter 20). It is a broad-spectrum herbicide that may be applied preplant-incorporated, preemergence, or postemergence in soybeans. Following its application, weeds stay green and appear unaffected for as long as 2 weeks. However, growth has stopped, and the weeds soon die or those that remain are not competitive with the crop.

Applied preplant, incorporate imazaquin 1–2 in. immediately before planting or up to 30 days before planting. Imazaquin may be applied preemergence before, during, or after planting but before crop emerges. For best results, incorporate shallowly if adequate rainfall or sprinkler irrigation is received within 7 days after application. Applied postemergence, imazaquin is absorbed by both the foliage and roots, and susceptible weeds stop growth and either die or are not competitive. Postemergence applications should be made when soybeans are small so that the foliar canopy will not intercept and prevent imazaquin from reaching the soil surface. Use a nonionic surfactant or a crop oil concentrate with postemergence applications.

Imazaquin is available in premixes with acifluorfen (Scepter O.T.), pendimethalin (Squadron), and trifluralin (Tri-Scept). Imazaquin can be tank-mixed with suitable herbicides to broaden the spectrum of weeds controlled. Imazaquin is applied preplant or preemergence at 2 oz ae/A (0.67 pt Scepter 4EC/A or 2.86 oz Scepter 70 DG/A). The postemergence rate is 1–2 oz ae/A (0.33–0.67 pt Scepter 4EC/A or 1.43–2.86 oz Scepter 70 DG/A).

Chlorimuron-ethyl (Classic 25 DG) is a sulfonylurea herbicide (Chapter 25). It is used postemergence to control broadleaf weeds and yellow nutsedge. It is especially effective against common cocklebur and sicklepod. Chlorimuron-ethyl may be applied anytime after the first trifoliate soybean leaves have opened but not later than 60 days before soybean maturity. For best results, always add a nonionic surfactant or crop oil concentrate to the spray mixture. To control velvetleaf, a nonionic surfactant or crop oil concentrate and a high-quality liquid nitrogen fertilizer should be added to the spray mixture. Cultivating approximately 14 days after application will help control suppressed weeds.

Chlorimuron-ethyl is available in premixes with linuron (Gemini and New Lorox Plus), metribuzin (Canopy and Preview), and thifensulfuron (Concert). It can be tank-mixed with suitable herbicides to broaden the spectrum of weeds controlled. Chlorimuron-ethyl is applied at 0.125–0.188 oz ai/A (0.5–0.75 oz product/A).

Clomazone (Command) is a nonfamily herbicide (Chapter 30). It is a broad-spectrum herbicide used to control annual grass and broadleaf weeds. It may be applied preplant-surface, preplant-incorporated, or preemergence to soybeans. *Soybeans are totally tolerant to clomazone.* Clomazone is a pigment inhibitor, and susceptible plants turn white. There are state restrictions relative to the use of Command. It is applied at a rate of 0.5–1.25 lb ai/A (1–2.5 pt product/A). Command may be tank-mixed with other soybean herbicides.

Clomazone is extolled as pure death to velvetleaf, consistently providing 100% control. However, there are three potential problems associated with its use: (1) a soil-residual problem, which prevents recropping to small grains within 1 year after application; (2) vapor drift to nearby plants, turning them white; and (3) failure to control some important weeds such as common cocklebur, ivyleaf morning-glory, and pigweeds.

■ SELECTED REFERENCES

Agricultural chemical usage: 1993 field crop summary. United States Department of Agriculture, National Agricultural Statistics Service, Washington, D.C. p. 5.

Agricultural statistics 1992. United States Department of Agriculture, Government Printing Office, Washington, D.C. p. 123.

Anonymous. 1982. *Illinois growers' guide to superior soybean production,* Circular 1200. Urbana, Ill.: Cooperative Extension Service, University of Illinois.

Anonymous. 1989. *Ohio agronomy guide,* 12th ed. Columbus, Ohio: Cooperative Extension Service, The Ohio State University.

Anonymous. 1994. *Weed control guide 1994.* Willoughby, Ohio: Meister Publishing.

Bozsa, R. C., L. R. Oliver, and T. L. Driver. 1989. Intraspecific and interspecific sicklepod (*Cassia obtusifolia*) interference. *Weed Sci.* 37:670–673.

Bridges, D. C., ed. 1992. *Crop losses due to weeds in the United States.* Champaign, Ill.: Weed Science Society of America.

Fellows, G. M., and F. W. Roeth. 1992. Shattercane (*Sorghum bicolor*) interference in soybean (*Glycine max*). *Weed Sci.* 40:68–73.

Horn, P. W., and O. C. Burnside. 1985. Soybean growth as influenced by planting date, cultivation, and weed removal. *Agron. J.* 77:793–795.

Lewis, W. M. 1985. Weed control in reduced-tillage soybean production. In *Weed control in limited-tillage systems,* A. F. Wiese, ed., Monograph No. 2 pp. 41–50. Champaign, Ill.: Weed Science Society of America.

Spenser, N. R. 1984. Velvetleaf, *Abutilon theophrasti* (Malvaceae), history and economics in the United States. *Econ. Bot.* 38:407–416.

Stoller, E. W., et al. 1987. Weed interference in soybeans (*Glycine max*). *Rev. Weed Sci.* 3:155–181.

Swanton, C. J., K. N. Harker, and R. L. Anderson. 1993. Crop losses due to weeds in Canada. *Weed Technol.* 7:537–542.

Thurlow, D. L., and G. A. Buchanan. 1972. Competition of sicklepod with soybeans. *Weed Sci.* 20:379–384.

Wixson, M. B., and D. R. Shaw. 1992. Effects of soil-applied AC 263,322 on crops rotated with soybean (*Glycine max*). *Weed Technol.* 6:276–279.

38 SUGAR BEETS

■ INTRODUCTION

In the United States in 1991, sugar beets were grown on 1.4 million A in 12 states (Table 38.1). Over 90% of this acreage was treated with at least one herbicide. In the same year, weeds caused an estimated crop loss valued at $62 million, even though herbicides were used. If herbicides had not been used, the estimated loss would have been $357 million. In 1989, sugar beets were produced on 59.3 thousand A in western Canada, with a yield of 2 million tons of beets. In the same year, weeds caused an average 7% crop loss, valued at $2.1 million.

In 1991, 28 million tons of sugar beets were produced in the United States, yielding 3.5 million tons of refined sugar. About one-half of the crop is produced with irrigation. An average sugar beet weighs about 2 lb, 15% of which is sugar. Beet yields average about 20 tons/A, varying from 15.5 to 29 tons/A. An acre of land yielding 20 tons of beets will produce about 6000 lb of sugar. By-products of the sugar beet crop are the beet tops, pulp, and molasses, which are utilized as animal feed.

■ WEEDS

Sugar beets are poor competitors with weeds. Weeds in sugar beets are problems during (1) planting to stand-thinning, and (2) stand-thinning through last cultivation (lay-by). Weeds are most difficult to control during the first of these periods, as beet seedlings have low tolerance to herbicides and cultivators easily cover the seedlings with soil. During the second period, the plants are larger and can better tolerate cultivation and herbicides that could not be used earlier. Weeds that emerge after sugar beets reach the six- to eight-leaf stage can be controlled with cultivation or postemergence herbicides. Weeds within the crop row are major problems. These can be controlled during thinning, cultivation with flexible tines and rotary hoes, and with herbicides. Weeds that escape these methods of control are usually removed by hand, if their numbers warrant the added cost. Sugar beet yields are reduced about 8% if only two kochia plants per 100 ft of row remain after thinning. After lay-by, if the stand is uniform and vigorous, the sugar beet plants are large enough to suppress

TABLE 38.1 ■ Sugar beet acreage planted in the United States in 1991.*

AREA	ACRES (× 1000)
Corn Belt	
Ohio	20.3
Regional total: 20,300 A	
Mountain	
Colorado	39.4
Idaho	196
Montana	56.6
Wyoming	69
Regional total: 361,000 A	
Southwest	
Texas	41.5
Regional total: 41,500 A	
North Central	
Michigan	171
Minnesota	369
Nebraska	82
North Dakota	195
Regional total: 817,000 A	
Pacific Coast	
California	160
Oregon	18.9
Regional total 178,900 A	
Other	23

*Total sugar beet acreage planted in the United States in 1991 was 1,421,000 A.

SOURCE: *Agricultural statistics 1992.* United States Department of Agriculture, Government Printing Office, Washington, D.C. p. 74.

TABLE 38.2 ■ Representative weeds in sugar beet–growing regions of the United States and Canada.*

Corn Belt	North Central
Foxtail, green	Canada thistle
Foxtail, yellow	Foxtail, green
Lambsquarters, common	Kochia
Pigweed, redroot	Lambsquarters, common
Velvetleaf	Nightshade, hairy
	Pigweed, redroot
Mountain & State of Oregon	Ragweed spp.
Barnyardgrass	Smartweed spp.
Foxtail, green	Sunflower, common
Foxtail, yellow	Velvetleaf
Kochia	Wild buckwheat
Lambsquarters, common	Wild oats
Nightshade, hairy	
Pigweed, redroot	*California*
Wild oats	Barnyardgrass
Southwest	Cereals, volunteer
Barnyardgrass	Chickweeds, common
Johnsongrass	Groundcherry spp.
Kochia	Knotweed, prostrate
Nutsedge, yellow	Lambsquarters, common
Pigweed, prostrate	Mallow spp.
Pigweed, redroot	Mustard spp.
	Nettle, burning
Canadian Provinces	Nutsedge, yellow
Canada thistle	Pigweed spp.
Foxtail, green	Purslane, common
Kochia	Ryegrass, Italian
Ladysthumb	Shepherd's-purse
Lambsquarters, common	
Mallow, common	
Pigweed, redroot	
Ragweed, common	
Wild buckwheat	
Wild oats	

*All species may not be a problem in every state within a region or in every province. The Canadian provinces are Alberta, Manitoba, and Quebec.

SOURCE: Abstracted from E. E. Schweizer and A. G. Deuter, 1987, *Rev. Weed. Sci.* 3:113–133.

newly emerging weeds. However, growers may obtain good early-season weed control only to have annual grass and broadleaf weeds appear later.

Sugar beet growers in California must contend with weeds all year, while growers in other areas have weed problems only in the spring and summer. The principal weeds in sugar beets grown in the United States and Canada are given in Table 38.2.

■ WEED CONTROL

With the development of monogerm sugar beet seed and effective herbicides, labor costs in sugar beet production have decreased greatly. Prior to these developments, hand labor and mechanical cultivation were the only means by which weeds were controlled. Hand labor for weeding and beet thinning is still widely used, in spite of new developments in mechanical thinning and planting to a stand. Hand labor for weeding and thinning will continue to be utilized at least into the next decade, depending on the availability of labor and cost effectiveness. In 1987, hand labor for thinning sugar beet plants and removing weeds at the same time averaged $33/A; a second weeding cost $22/A.

Weed control by mechanical means is achieved with harrows, rotary hoes, and row-crop cultivators from crop emergence to lay-by. Cultivation needs to begin soon after

TABLE 38.3 ■ **Principal herbicides used in sugar beets in 1990, expressed as the average percentage of total acreage grown in each region.**

| HERBICIDE** | REGION* | | | | | MEAN |
	1	2	3	4	5	
Cycloate	12	30	51	45	0	28
Desmedipham	67	79	30	58	10	49
Diethatyl	0	27	14	13	33	17
Ethofumesate	25	19	25	22	60	30
Phenmedipham	67	76	30	50	50	55
Chloridazon, formerly pyrazon	68	23	3	10	0	21
Sethoxydim	0	40	17	8	20	17
Trifluralin	0	4	7	58	80	30

*Regions: 1 = Corn Belt; 2 = North Central; 3 = Mountain; 4 = Pacific Coast; 5 = Southwest.
**A particular herbicide may not be used in every state within a region.

SOURCE: D. C. Bridges, ed. 1992. *Crop losses due to weeds in the United States.* Champaign, Ill.: Weed Science Society of America.

beet emergence, taking care not to throw soil over the seedling beets. Chemical weed control is obtained with a variety of preplant, preemergence, and postemergence herbicides. For season-long weed control, a preplant or preemergence herbicide must be followed with one or more applications of a postemergence herbicide. The principal herbicides used in sugar beets in 1990 are given in Table 38.3.

■ INFORMATION OF PRACTICAL IMPORTANCE

Preplant

Cycloate (Ro-Neet 6E) is a thiocarbamate herbicide (Chapter 26). It is used to control grass and broadleaf weeds prior to emergence. It may cause crop injury when applied on very light sandy soil. Cycloate is applied preplant to the seedbed and immediately incorporated to a depth of 2–3 in. to prevent loss by volatility. It can be applied in tank mix with a compatible liquid fertilizer. Cycloate is applied broadcast at 3–4 lb ai/A (2–2.67 qt product/A).

Diethatyl (Antor 4ES) is a nonfamily herbicide (Chapter 30). It is used to control grass and broadleaf weeds prior to emergence. It is applied preplant to the seedbed. It need not be incorporated immediately following application, but it must be incorporated 1–2 in. deep prior to weed seed germination. Diethatyl is applied broadcast at 3–6 lb ai/A (3–6 qt product/A), depending on soil type.

Ethofumesate (Nortron 1.5EC) is a nonfamily herbicide (Chapter 30). It is used to control grass and broadleaf

weeds prior to emergence. Nortron EC is applied to the seedbed and soil-incorporated 1–2 in. deep. Deeper incorporation may reduce effectiveness. Ethofumesate is applied broadcast at 1.1–3.8 lb ai/A (0.75–2.5 gal product/A), depending on soil type.

Preemergence

Chloridazon, formerly *pyrazon* (Pyramin 68DF), is a phenyl pyridazinone herbicide (Chapter 23). It is used to control annual broadleaf weeds. Chloridazon is applied to the surface of the seedbed after planting and prior to sugar beet or weed emergence. Where sprinkler irrigation is used, plant sugar beet seed and preirrigate the seedbed to field capacity prior to applying chloridazon. Chloridazon is applied at 3.1–3.7 lb ai/A (4.6–5.4 lb product/A).

Diethatyl (Antor 4ES) is a nonfamily herbicide (Chapter 30). It is used to control grass and broadleaf weeds. Diethatyl is applied to the surface of the seedbed at time of planting or shortly thereafter, but prior to weed seed germination. For best results, diethatyl must be moved into the weed seed germination zone (upper 2 in.) by irrigation or rainfall. Diethatyl is applied at 1–6 lb ai/A (3–6 qt product/A), depending on soil type.

Postemergence

Desmedipham (Betanex 1.3EC) is a phenylcarbamate herbicide (Chapter 22). It is used postemergence to control small (two-true-leaf stage) annual broadleaf weeds and certain grasses. Desmedipham is selective in sugar beets past the two-true-leaf stage. The beets may be severely injured if treated before this stage. Desmedipham is applied broadcast at 0.73–1.2 lb ai/A (4.5–7.5 pt product/A).

If necessary, a second application of desmedipham can be applied at the same rate as the first application when the beets have at least four leaves. Allow at least 7 days between the first and second applications. Except in California, Betanex may be applied as a split application to sugar beets at any stage of growth to control early-germinating weeds; apply 0.33–0.5 lb ai/A (2–3 pt product/A) in each of two applications. The first spray must be applied when the earliest emerging weeds are in the cotyledonary stage.

Betamix 1.3EC is a premix formulation that contains equal amounts of desmedipham and phenmedipham, both phenylcarbamate herbicides (Chapter 22). The herbicidal characteristics of Betamix are similar to desmedipham. *However, due to the reduced rate of desmedipham in the premix, grasses are not controlled, although use of the premix does broaden the spectrum of broadleaf weeds controlled.* A repeat application, or sequential applications, of Betamix can be made as set forth for desmedipham. Betamix is applied broadcast at 0.73–1.2 lb ai/A (4.5–7.5 pt product/A).

Sethoxydim (Poast 1.5EC) is a cyclohexanedione herbicide (Chapter 16). It is a systemic, postemergence herbicide used to control annual and perennial grasses. It does not control broadleaf weeds or sedges. *Sugar beets are tolerant of sethoxydim at all growth stages.* Sethoxydim is applied in 5–20 gal water/A (10 gal water/A is optimal), with the product Dash or a nonphytotoxic oil concentrate added to the spray mixture as recommended. Sethoxydim is applied broadcast at 0.1–0.5 lb ai/A (0.5–2.5 pt product/A).

Trifluralin (Treflan 4EC) is a dinitroaniline herbicide (Chapter 17). Trifluralin is applied and incorporated as a broadcast, over-the-top spray to clean-tilled soil when the sugar beets are 2–6 in. tall. To reduce possible root girdling, exposed beet roots should be covered with soil before trifluralin is applied. Set incorporation machinery to throw treated soil toward the plants in the row, taking care not to damage the sugar beet taproot. A tine-toothed harrow can be used to incorporate trifluralin, with two passes (one at right angle to the other) made across the field. Set the harrow to cut 1–2 in. deep, taking care not to damage the sugar beet taproot. Trifluralin is applied broadcast at 0.5–0.75 lb ai/A, depending on soil type.

■ SELECTED REFERENCES

Agricultural statistics 1992. United States Department of Agriculture, Government Printing Office, Washington, D.C. p. 74.

Anonymous. 1994. *Weed control guide 1994.* Willoughby, Ohio: Meister Publishing. pp. 167–170.

Bridges, D. C., ed. 1992. *Crop losses due to weeds in the United States.* Herbicide product labels. Champaign, Ill.: Weed Science Society of America.

Schweizer, E. E., and A. G. Dexter. 1987. Weed control in sugar beets (*Beta vulgaris*) in North America. *Rev. Weed Sci.* 3:113–133.

39 WHEAT

■ INTRODUCTION

In the United States in 1991, 69.9 million A (28.3 million ha) were planted to wheat (Table 39.1). Of this acreage, 73% was winter wheat, 5% durum wheat, and 22% other spring wheats. In this same year, weeds caused an estimated crop loss of $396 million, even though herbicides were used. If herbicides had not been used, the estimated loss would have been $1.25 billion. In 1992, herbicides were used on 33% of the winter wheat acreage, 93% of the durum wheat, and 87% of the other spring wheats. The regions of wheat production in the United States, the total wheat acreage in each region, and the types of wheat grown are shown in Table 39.1. Acreages for 1991 in each state are shown in Table 39.2, and Table 39.3 on page 298 lists the uses for the various types of wheat.

In Canada in 1991, 36 million A (14.6 million ha) were seeded to wheat, including hard red winter, hard red spring, soft white spring, and durum wheats. In this same year, weeds caused an estimated 10% loss in wheat yields averaged for the eastern provinces and 9% loss averaged for the western provinces, equivalent to a monetary loss of $10.5 million and $281 million, respectively.

Wheat is grown in the United States at elevations ranging from about 330 ft to more than 3000 ft above sea level. It is grown on soils that vary in pH from 4.8 to 8.2, with a pH of 6.0–6.5 considered optimum. Soils vary in texture from sandy loams to clay loams. Soil organic matter content varies from less than 1% to more than 3%. Wheat is grown under conditions of abundant rainfall, minimal rainfall (dryland), and irrigation, depending on local conditions.

■ WEEDS

With so much diversity in soil, topography, geography, and climatic conditions in spring and winter wheat croplands of North America, it is not surprising that the weed species common to these crops vary greatly. Often more than 50 species are listed as problem weeds for wheat in a given region; many of these are unique to a region, while others are common throughout the wheat-producing regions. In general, weeds plague spring-planted wheat more than winter-planted wheat.

The principal weeds plaguing winter wheat are winter annual grass and broadleaf species. They have a growing season similar to winter wheat. They germinate and emerge from the soil along with the wheat seedlings, withstand adverse weather conditions, and put on new growth in the spring. Winter wheat is a competitive crop, and it tends to

TABLE 39.1 ■ Wheat-growing regions of the United States, with total wheat acreage and type(s) grown in each region in 1991.

REGION*	TOTAL ACRES** (MILLIONS)	WHEAT TYPE
Northern Great Plains	20.7	Hard red spring Durum†
Central Great Plains	17.1	Hard red winter
Pacific Northwest	5.9	Soft white winter Hard red winter
Midwest	6.0	Soft white winter Soft red winter
Northeast	0.6	Soft white winter Soft red winter
Southeast	4.7	Soft red winter
Southwest	14.2	Hard red winter
West	0.7	Soft white winter Soft white spring Hard red winter Hard white winter Durum

*Refer to Table 39.2 for a breakdown of the various regions.
**Includes all types of wheat.
†North Dakota produced 89% of the durum wheat grown in the United States in 1991.

SOURCE: Adapted from *Agricultural statistics, 1992.* United States Department of Agriculture, Government Printing Office, Washington, D.C. p. 5.

TABLE 39.2 ■ Wheat-growing regions in the United States and total acreage planted to wheat in 1991.*

STATE	ACRES (× 1000)	STATE	ACRES (× 1000)
Northern Great Plains		**Central Great Plains**	
N. Dakota	10,000	Colorado	2,640
S. Dakota	3,370	Kansas	11,800
Minnesota	2,190	Iowa	80
Montana	5,130	Nebraska	2,350
Regional total: 20.69 million A		Wyoming	220
Midwest		Regional total: 17.09 million A	
Illinois	1,650	**Northeast**	
Indiana	850	Delaware	70
Michigan	570	Maryland	210
Missouri	1,650	New Jersey	40
Ohio	1,150	New York	120
Wisconsin	150	Pennsylvania	180
Regional total: 6.02 million A		Regional total: 0.62 million A	
Pacific Northwest		**Southeast**	
Idaho	1,340	Alabama	170
Oregon	900	Arkansas	1,100
Washington	3,700	Florida	50
Regional total: 5.94 million A		Georgia	500
Southwest		Kentucky	640
Arizona	70	Louisiana	300
Oklahoma	7,400	Mississippi	350
New Mexico	550	N. Carolina	550
Texas	6,200	S. Carolina	300
Regional total: 14.22 million A		Tennessee	440
West		Virginia	280
California	480	W. Virginia	10
Nevada	10	Regional total: 4.69 million A	
Utah	170		
Regional total: 0.66 million A			

*Total: 69.9 million A.

SOURCE: *Agricultural statistics 1992.* United States Department of Agriculture, Government Printing Office, Washington, D.C. p. 5.

suppress dense stands of many weeds. However, if the wheat stand becomes thin due to winter kill, disease, or other reasons, weeds emerge wherever there are bare areas. Small weeds are commonly present under the winter wheat foliar canopy, and they develop rapidly when harvesting removes this canopy. Unless soon controlled, these weeds will mature and set an abundance of seed that will ensure their presence in succeeding crops.

The presence of tall, green weed vegetation at harvest time (such as Canada thistle, common sunflower, field bindweed, kochia, and wild buckwheat) may disrupt harvesting. Figures 39.1 and 39.2 on page 299 show a wheat field so heavily infested with kochia that the crop could not be harvested with a combine.

Wild oat is usually not a problem in fall-seeded wheat, as it has little resistance to frost. Winter wild oat is a problem in fall-seeded wheat, especially in the southern Great Plains and eastern Washington. However, the presence of wild oat seed in the harvested grains of winter and spring wheats can cause dockage when the wheat is marketed.

Wild garlic and wild onion are not competitive with wheat, but the presence of their aerial bulblets in harvested grain causes dockage at market time. Wheat is graded "garlicky" when two or more green aerial bulblets of either species or an equivalent of dry or partially dry bulblets are present in 1000 g of wheat seed. These bulblets impart an odor and flavor to the wheat and wheat products that persist even after the bulblets are removed. When crushed in wheat milling, the bulblets contaminate the milling equipment, forcing the mill to shutdown for cleaning.

Representative weeds in wheat grown in the United States are listed in Table 39.4. A particular species can be a problem in one wheat-growing region but not in another,

TABLE 39.3 ■ Major types of wheat grown in the United States and their respective uses.

WHEAT TYPE		
Soft red winter	Hard red winter	Durum
Soft white winter	Hard red spring	
Soft white spring	Hard white winter	
	Hard white spring	
USE		
Biscuits and muffins	Bread and rolls	Macaroni
Cakes and cookies	Baby foods	Puffed breakfast
Crackers	Hot breakfast	cereals
Doughnuts	cereals	Spaghetti
Ice cream cones	Farina	Wheat germ
Noodles	Wheat germ	
Pancakes and waffles		
Pie crusts and pizza		
Breakfast cereals (flaked, puffed, granola)		
Gravy and soup thickeners		
Wheat germ		

SOURCE: L. W. Briggle. 1980. Origin and botany of wheat. In *Wheat, a technical monograph*, pp. 6–13. Greensboro, N.C.: Ciba.

TABLE 39.4 ■ Representative problem weeds in wheat grown in the United States.

GRASS WEEDS	BROADLEAF WEEDS
Annual	*Annual*
Canarygrass	Common lambsquarters
Cheat	Common ragweed
Downy brome	Common sunflower
Foxtail, green	Field pennycress
Foxtail, yellow	Flixweed
Hairy chess	Henbit
Italian ryegrass	Kochia
Japanese brome	Prickly lettuce
Jointed goatgrass	Redroot pigweed
Persian darnel	Russian thistle
Volunteer rye	Tumble pigweed
Volunteer winter wheat	Wild buckwheat
Wild oat	Wild mustard
Winter wild oat	Wild radish
Perennial	*Perennial*
Bentgrass	Canada thistle
Johnsongrass	Curly dock
Quackgrass	Field bindweed
Perennial monocots, but not grasses	
Wild garlic	
Wild onion	

SOURCE: D. C. Bridges, ed. 1992. *Crop losses due to weeds in the United States*. Champaign, Ill.: Weed Science Society of America.

and others may be common to most regions. Weed species can also differ in growing wheat, wheat stubble after harvest, and in land lying fallow.

■ WEED CONTROL

Preventive Weed Control

Preventive weed control should begin soon after wheat harvest, during the fallow period between crops, to control weeds present in the wheat understory. If not controlled, these weeds undergo rapid growth when released from the shading effect of the wheat canopy and then mature and set an abundance of seed. The weed species may be winter annuals common to the wheat crop and summer annuals common to spring-planted crops, as well as established perennials.

The weeds are usually controlled with primary tillage, and later-emerging weeds are controlled with secondary tillage tools such as large sweeps, tandem disks, and rod weeders, and/or herbicides. Usually, tillage and herbicides are used jointly in a preventive weed-control program.

Three large tractor-mounted sweeps and a rodweeder are pictured respectively in Figures 39.3 and 39.4 on page 300.

Cleaning the combine after harvesting a wheat field that has serious weed problems is a good practice in order to avoid spreading the weed propagules from field to field. Harvesting infested portions of a field separately can also be effective. Custom combines should be thoroughly cleaned before moving onto fields of the next customer. *The combine harvester has been described as the perfect seed-dispersal device for late-maturing weed species.*

Crop Rotation

Where wheat is grown in rotation with other crops, such as corn, soybeans, or hay, weed problems in the wheat may be reduced by controlling weeds in the rotational crops. Weed-free corn or soybean fields reduce or eliminate the amount of weed seed that is produced and that will persist to germinate and grow in the following crop. Annual weeds

FIGURE 39.1 ■ **A wheat field in southcentral New Mexico heavily infested with kochia.**

SOURCE: Photo by the author.

FIGURE 39.2 ■ **Same wheat field as shown in Figure 39.1 after attempt to harvest with a grain combine.** Heavy infestation with kochia jammed the combine, preventing normal harvest. The field was eventually mowed.

SOURCE: Photo by the author.

may be controlled in rotational crops by cultivation and herbicides. Quackgrass, a perennial, can be controlled in corn with atrazine. Johnsongrass can be controlled prior to planting soybeans by thoroughly disking the soil to cut its rhizomes into segments to encourage sprouting from adventitious buds, and then soil-incorporate trifluralin at double the normal rate. Sethoxydim may be applied postemergence to control perennial grasses in soybeans. Glyphosate,

FIGURE 39.3 ■ Three large tractor-mounted sweeps. These are pulled through the soil at a depth of about 4 in. to kill weeds in fallow land.

SOURCE: Photo by the author.

fluazifop-P-butyl, and sethoxydim may be used post-emergence to control quackgrass and johnsongrass and annual grasses following the wheat harvest. Perennial broad-leaf weeds, such as common milkweed, curly dock, and hemp dogbane, can be controlled with 2,4-D or dicamba during the fallow following the wheat harvest.

Chemical Weed Control

Since the introduction of 2,4-D in 1946 and MCPA shortly thereafter, the use of herbicides as a means of selectively controlling weeds in wheat has become a common practice. Herbicides registered for use in wheat in 1994 are listed in Table 39.5, and the comparative use of these herbicides in winter, durum, and other spring wheats in 1992 is shown in Table 39.6. Herbicides registered in 1992 for use during the wheat fallow period are listed in Table 39.7. Weed responses to four broadleaf herbicides are shown in Table 39.8. The comparative use of the top

FIGURE 39.4 ■ A rodweeder in operation. Rodweeders function through a ground-driven, 1-in. diameter, round or square rod that is geared to rotate in the *opposite* direction to that of the tractor to which it is mounted. It is generally set to operate at a depth of 1–2 in. below the soil surface. Rod-weeders are used primarily to control weeds in wheatlands during fallow periods after plowing. However, the rodweeder pictured here is being adjusted to test its effectiveness in incorporating a surface-applied herbicide on onion seedbeds. It is used commercially to lift onion bulbs at harvest.

SOURCE: Photo by the author.

TABLE 39.5 ■ Herbicides registered for use in wheat in the United States in 1994.*

COMMON NAME	TRADE NAME
Grass Weed Control	
Preplant-incorporated or preemergence-incorporated	
Triallate	Far-go
Triallate + trifluralin	Buckle
Preemergence	
Chlorsulfuron	Glean
Triasulfuron	Amber
Postemergence	
Diclofop	Hoelon
Difenzoquat	Avenge
Imazamethabenz	Assert
Metribuzin	Sencor
Broadleaf Weed Control	
Preemergence or Postemergence	
Chlorsulfuron	Glean
Chlorsulfuron + metsulfuron	Finesse
Diuron	Karmex
Postemergence	
2,4-D	several
MCPA	several
Bromoxynil	Buctril
Bromoxynil + MCPA	Bronate
Clopyralid	Stinger
Clopyralid + 2,4-D	Curtail
Clopyralid + MCPA	Curtail M
Dicamba	Banvel
Imazamethabenz	Assert
Metribuzin	Sencor
Metsulfuron	Ally
Picloram	Tordon
Pyridate	Tough
Triasulfuron	Amber
Tribenuron	Express
Tribenuron + thifensulfuron	Harmony Extra

*Many herbicide premixes and tank mixes are registered for use in wheat.

SOURCE: R.T. Meister, ed. 1994. *Weed control manual.* Willoughby, Ohio: Meister Publishing pp. 114–128.

herbicides in durum, spring (other than durum), and winter wheats in 1993, based on the percentage of acreage planted, is shown in Figure 39.5 on page 303.

TABLE 39.6 ■ Comparative use of herbicides in wheat in the United States in 1992.*

HERBICIDE**	PERCENT	HERBICIDE**	PERCENT
Winter Wheat			
2,4-D	14	Bromoxynil	2
Chlorsulfuron	14	Diclofop	1
Metsulfuron	7	Diuron	1
Dicamba	4	Glyphosate	1
MCPA	3	Metribuzin	1
Tribenuron	3	Triallate	1
Durum Wheat			
2,4-D	43	Tribenuron	10
MCPA	38	Diclofop	7
Dicamba	34	Triallate	7
Trifluralin	31		
Spring Wheats (other than durum)			
2,4-D	52	Metsulfuron	6
MCPA	37	Triallate	4
Dicamba	29	Trifluralin	4
Tribenuron	13	Diclofop	2
Bromoxynil	10	Imazamethabenz	2
Fenoxaprop	8		

*Expected as as a percentage of the wheat acreage treated with the respective herbicides in the 42 states growing wheat in 1992.

**Not all herbicides applied in every state.

SOURCE: *Agricultural chemical usage: 1992 field crop summary.* United States Department of Agriculture, National Agricultural Statistics Service, Washington, D.C. pp. 93, 105, 107.

Winter Wheat's Critical Growth Stage

The stage at which winter wheat is treated with growth regulator–type herbicides, such as 2,4-D, MCPA, clopyralid, and dicamba, is critical if crop damage is to be avoided. This growth stage has been described for many years as "after fully tillered but before jointing," but the criterion of "fully tillered" has proven to be an inadequate guide with respect to gauging the safest growth stage of the wheat. This is a time when the embryonic spike (the plant part most adversely affected by growth regulator–type herbicides) has reached a stage of development most tolerant to these herbicides.

A more suitable criterion is based on measurement of the length of the longest leaf sheath on the main shoot of young winter wheat plants, measured from the soil surface to the uppermost ligule (or base of the uppermost leaf blade). When the length of this leaf sheath exceeds 2 in. (5 cm), measured from the soil surface, but is less than 4 in.

(10 cm), the embryonic spike of winter wheat has reached the stage in its development that is the most tolerant to applications of growth regulator–type herbicides. This stage

is often reached when the wheat plant has six leaves unfolded on the main shoot (Figure 39.6 on page 304). However, it is more accurate to measure the length of the longest leaf sheath on the main shoot than to count the number of leaves present (Tottman, 1976).

An objective determination of "jointing" can be made by counting the nodes that can be seen or felt aboveground within the leaf sheath surrounding the main stem. The first node (beginning of jointing) can usually be found by gently squeezing the leaf sheath of the main shoot above its base between thumb and finger when the leaf sheath measures about 4–4.5 in. (10–12 cm) from the soil surface to the uppermost ligule (or base of the uppermost leaf blade).

Growth regulator–type herbicides applied to winter wheat before the leaf sheath of the main shoot is 2 in. long or after the first node is detectable within the leaf sheath of the main shoot can disrupt the formation of normal seed-heads (spikes), resulting in deformed heads and shriveled (unfilled) grain at harvest. The time span in which these herbicides can be safely applied in the spring to winter wheat is relatively short—about 2 weeks.

TABLE 39.7 ■ Herbicides that can be used during wheat fallow.*

COMMON NAME	TRADE NAME
Chlorsulfuron	Glean
Cyanazine	Bladex
2,4-D	several
Dicamba	Banvel
Glyphosate	Ranger, Roundup
Metsulfuron	Ally
Metribuzin	Lexone
Paraquat	Cyclone, Gramoxone Extra
Trifluralin	Treflan, others
*Various mixtures and premixes may be used.	

SOURCE: Herbicide product labels.

TABLE 39.8 ■ Responses of weeds less than 4 in. tall to four postemergence broadleaf herbicides in irrigated wheat.*

| WEEDS | HERBICIDE | | | |
	2,4-D	MCPA	Bromoxynil	Dicamba
Blue mustard	G-E	F	F	F
Field pennycress	E	E	G	P
Flixweed	E	G	G	G
Kochia	E	F-G	F	F
Lambsquarters	E	E	G	E
Mayweed	P	P	F	G
Pineappleweed	P	P	F	G
Prickly lettuce	E	G	P	–
Prostrate knotweed	P	P	F	E
Prostrate pigweed	E	E	F	E
Puncturevine	G-E	F-G	E	–
Common purslane	P	P	F	F-G
Redroot pigweed	E	E	F	E
Russian thistle	G	P-F	F-G	E
Shepherd's-purse	E	G	G	F
Sunflower	E	G	E	E
Tansy mustard	G	G	G	G
Tumble mustard	E	G	G	–
Wild buckwheat	P	P	E	E
*Key: Weed control: E = 90–100%; G = 75–90%; F = 50–75%; P = 0–50%.				

SOURCE: R. Parker. 1981. Weed control in irrigated wheat. In *Central Washington weed control guide.* Extension bulletin 0760, p. 4. U.S. Department of Agriculture, Cooperative Extension Service, Washington, D.C.

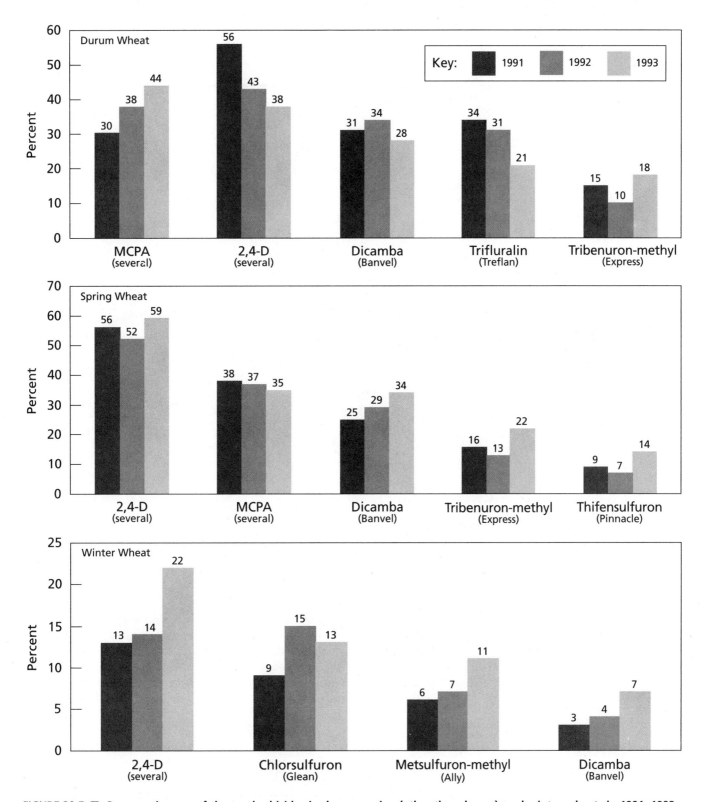

FIGURE 39.5 ■ Comparative use of the top herbicides in durum, spring (other than durum), and winter wheats in 1991, 1992, and 1993, based on the percentage of acreage planted in 15 states in the respective years.

SOURCE: *Agricultural chemical usage: 1993 field crop summary.* United States Department of Agriculture, National Agriculture Statistics Service, Washington, D.C. pp. 82, 96.

Leaf sheath
length
(5 cm+)

FIGURE 39.6 ■ **A young winter wheat plant at the "safe" growth stage for growth regulator-type herbicide application.** This plant has six leaves unfolded (darkened leaves), with center pseudostem erect (darkened), and four tillers (shown with two tillers on each side of main stem and leaves not darkened).

SOURCE: D. R. Tottman. 1976. Spray timing and the identification of cereal growth stages. (Proceedings of British Crop Protection Conference.) *Weeds* 13:791–800.

Wheat growth stages and the proper application timing for the broadleaf herbicides 2,4-D, bromoxynil, MCPA, and dicamba are shown in Figure 39.7. Growth stages of wild oat and the timing of postemergence applications of diclofop and difenzoquat are shown in Figure 39.8 on page 306.

■ INFORMATION OF PRACTICAL IMPORTANCE

Grass Control

Triallate (Far-Go) is a thiocarbamate herbicide (Chapter 26). It can be applied preplant or preemergence in fall- and spring-seeded wheat and shallowly incorporated using a spike-tooth or spring-tooth harrow, with a second incorpo-

ration at right angles. If applied preemergence, the cultivation depth should be adjusted so as not to disturb the crop seed. Triallate is used to control *Bromus* spp., wild oats, and Italian ryegrass (in Oregon only). Triallate may be tank-mixed with trifluralin to control green foxtail and yellow foxtail, as well as wild oats. A premix of triallate + trifluralin (Buckle) is available.

Diclofop (Hoelon) is an aryloxyphenoxypropionate herbicide (Chapter 13) and a Restricted Use Pesticide, due to its toxicity to fish. It can be applied preplant-incorporated (PPI), preemergence (at planting), or postemergence to wheat to control annual grasses such as winter annual bromes, Italian ryegrass (preemergence), and wild oats, Italian ryegrass, barnyardgrass, foxtail spp., and other annual grass weeds (postemergence). For best results, apply diclofop when wild oats and other annual grass weeds

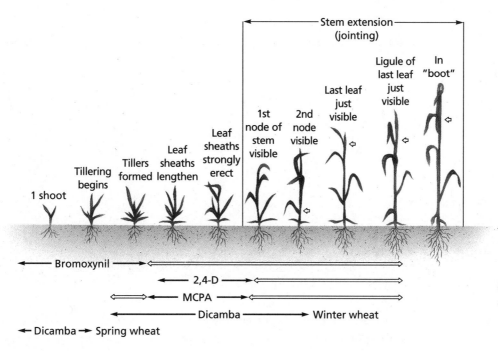

FIGURE 39.7 ■ Wheat growth stages and proper application timing for the broadleaf herbicides bromoxynil, 2,4-D, MCPA, and dicamba.

SOURCE: R. Parker. 1981. Weed control in irrigated wheat. In *Central Washington weed control guide.* Extension bulletin 0760, p. 6. U.S. Dept. of Agriculture, Cooperative Extension Service, Washington, D.C.

are in the two- to four-leaf stage. It can be safely applied to wheat in the one- to four-leaf stage of growth. A tank mix of diclofop and the broadleaf herbicide bromoxynil can be used postemergence to the wheat and weeds. However, do not tank-mix diclofop with the broadleaf herbicides 2,4-D, MCPA, clopyralid, dicamba, or picloram, as the activity of diclofop will be greatly reduced. The activity of the broadleaf herbicides is not affected.

Difenzoquat (Avenge) is a nonfamily herbicide (Chapter 30). It is used to control wild oats when the majority of the plants are in the three- to five-leaf stage (tillering). This stage frequently coincides with fall-seeded wheat in the four-leaf to tillered stage and with spring-seeded wheat in the five- to six-leaf stage. Difenzoquat can be tank-mixed with most broadleaf herbicides used in wheat.

Broadleaf Weed Control

The growth regulator–type herbicides [2,4-D (Weedar 64, Weedone LV4, others); MCPA (Gordon MCPA Amine, Rhomene, others); clopyralid (Stinger); dicamba (Banvel); and picloram (Tordon)] are widely used for broadleaf weed

control in wheat. Of these, *2,4-D* is the most commonly used in wheat grown in the United States and western Canada, and *MCPA* is the next most commonly used.

The growth regulator–type herbicides are recommended for postemergence use after the wheat is well tillered and up to the jointing stage. If applied before tillering, the wheat plants are injured, resulting in reduced yield; if applied after jointing, the stems become brittle and break easily.

Various mixtures of the growth regulator herbicides can broaden the weed spectrum controlled, but only certain mixtures, such as 2,4-D or MCPA with dicamba, or picloram as a tank mix or formulated mixture (premix), are registered. MCPA is used in some mixtures rather than 2,4-D because wheat is more tolerant of MCPA than 2,4-D at early growth stages, the stages when these mixtures are usually applied. These herbicides cannot be used on wheat underseeded to legumes, as severe injury to the legumes will result.

Chlorosulfuron (Glean), *metsulfuron* (Ally), and *triasulfuron* (Amber) are sulfonylurea herbicides (Chapter 25). They have long-term soil persistence, which can result in

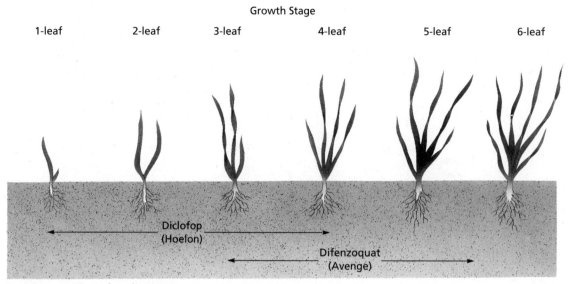

FIGURE 39.8 ■ **Growth stages of wild oat and the timing of postemergence applications of diclofop and difenzoquat.**

SOURCE: R. Parker. 1981. Weed control in irrigated wheat. In *Central Washington weed control guide*. Extension Bulletin 0760, p. 5. U.S. Dept. of Agriculture, Cooperative Extension Service, Washington, D.C.

severe injury to rotational crops other than small grains. *Tribenuron* (Express) is a sulfonylurea herbicide whose soil persistence is shorter than that of chlorsulfuron. *Metribuzin* (Sencor), a triazine herbicide (Chapter 27) also poses a hazard to crop rotation because of its persistence in the soil. Chlorsulfuron, thifensulfuron + tribenuron (Harmony Extra), chlorsulfuron + metsulfuron (Finesse), and metribuzin are being used successfully in land devoted to continuous wheat.

Chemical Fallow

A number of herbicides may be used in fallow land following the harvest of wheat or barley. The primary concern as to which herbicide(s) to use are the soil residues of the herbicide(s) that may still be present to cause injury to the crop planted in the fall or in the following year. The postemergence herbicides glyphosate (Roundup, others) and paraquat (Gramoxone Extra) are recommended for this use, as are 2,4-D, MCPA, dicamba, and bromoxynil and certain soil-residual herbicides such as atrazine (AAtrex, others), cyanazine (Bladex), chlorsulfuron, and metribuzin. Also, certain combinations of these herbicides may be applied.

A balanced fallow program utilizes timely tillage in conjunction with appropriate herbicides.

Spray Drift

The drift of spray droplets of herbicides used in wheat (such as 2,4-D, dicamba, glyphosate, chlorsulfuron, and paraquat) onto adjacent susceptible crops or croplands causes substantial crop loss or injury each year, especially when applied by aircraft. Obviously, greater care should be taken to avoid spray drift. Factors contributing to spray drift are spray droplet size (the smaller the size, the greater the potential for drift); wind speed and direction (wind speeds greater than about 5 mph or 8 kmph favor drift, and spray drift follows direction of air movement); and presence of an air temperature inversion (cold air over warm air). The phenoxy herbicides, in particular, should not be sprayed when an air temperature inversion is present, especially not by aircraft. A number of states have restrictions on the use of phenoxy herbicides on wheat where susceptible crops such as cotton and grapes may be injured by these herbicides.

■ REGIONAL WEEDS AND WEED CONTROL

Southeast

In the Southeast (Tables 39.1 and 39.2), where soft red winter wheat is grown almost exclusively, wheat is most commonly grown as a double crop with soybeans, and in some cases, with cotton, grain sorghum, peanuts, or sunflowers. Corn, cotton, soybean, and grain sorghum are the most common crops to precede wheat in a crop rotation.

Typically, winter wheat is planted in the northern portion of the Southeast in early September to late October and from late October to late November in the southern portion. Primary and secondary tillage prepare the seedbed and provide early weed control. A few producers plant no-till and must use special no-till grain drills to plant their wheat. Regardless of the type of seedbed preparation, no other tillage operation is made until after harvest.

Although few herbicides are used in winter wheat in the Southeast, the ones most commonly used are 2,4-D and Harmony Extra. Occasionally, MCPA, Buctril, or Banvel are used for special problem weeds such as wild radish or wild garlic. Metribuzin is used in certain states, but only after wheat is fully tillered, and with restrictions as to wheat varieties.

Warm-season weed species that are found in summer crops of the Southeast can also be found in wheat fields following harvest. The most common weeds at wheat harvest are common ragweed, common lambsquarters, and horseweed. These weeds will usually dominate the field if nothing is done to control them, and they are of most concern when a summer crop is planted no-till into wheat stubble. A standard practice in the region is to use a nonselective, foliar-active herbicide such as paraquat or glyphosate, applied postemergence to the weeds but pre-plant or preemergence to the crop.

Italian ryegrass is the most serious annual grass weed in winter wheat in the Southeast, with moderate to heavy infestations that cause serious yield reductions, slow harvest, and contribute to the foreign matter in the harvested grain. Italian ryegrass can be controlled with diclofop applied preemergence or postemergence (before it reaches the four-leaf stage). Mixing diclofop with 2,4-D, MCPA, bromoxynil, or dicamba will reduce the activity of diclofop on Italian ryegrass and *Bromus* spp.

If the wheat harvest is delayed, or the rotational crop planting is delayed after harvest, various annual grasses, such as large crabgrass, fall panicum, or Texas panicum, may become problem weeds.

There are few perennial weeds that cause problems in winter wheat in the Southeast. Curly dock may pose problems in minimum-tillage systems, and johnsongrass may develop early enough in the spring to cause yield losses and harvesting problems.

Midwest

In the Midwest (Tables 39.1 and 39.2), winter wheat is generally grown in rotation with corn and soybeans on farms where grains are the principal product. On farms that raise cattle, hay crops such as alfalfa and red clover may be included in the rotations to provide for forage needs.

If weeds have been well controlled in soybeans, only a light disking is necessary to prepare a seedbed for wheat. Where silage corn has been harvested, wheat is seeded after one or more diskings. With the introduction of no-till drills, there has been an increased interest in drilling directly into soybean and silage fields following their harvest. Corn grown for grain is usually harvested too late for winter wheat to be planted in rotation, unless short-season corn varieties are grown. There is considerable interest in double-cropping wheat and soybeans where growing conditions are consistently long enough to mature soybeans planted in late June to mid-July, such as in the southern Corn Belt.

Winter wheat grows rapidly and is a highly competitive crop, and most Midwest farmers do not routinely plan a weed-control program in anticipation of serious weed infestations. A number of spring annual weeds are present in the winter wheat understory, but, since they are shaded by the wheat, they do not adversely affect yields. However, after wheat harvest and the young plants are exposed to light, they grow rapidly. If they are not killed, they mature and produce an abundance of seed to reinfest the soil. During seasons of normal rainfall before or after harvest, weeds such as common ragweed, common lambsquarters, and Pennsylvania smartweed may be present yet still small enough that the combine passes over them. When abnormally heavy rains occur before or during harvest, weed growth is enhanced, and the weeds will be tall enough to be cut with the grain at harvest.

Annual spring weeds that are commonly present in winter wheat are foxtails (giant, green, yellow), common ragweed, giant ragweed, Pennsylvania smartweed, redroot pigweed, among others.

A large proportion of winter annual weeds germinate as soon as adequate rainfall occurs in late August or early September. Tillage prior to planting wheat in October or November, after weed emergence, will kill the seedlings. Some winter annual weeds germinate after the wheat is seeded, but these are usually not a problem.

Northern Great Plains

Most of the hard red spring wheat and durum wheat (a spring wheat) grown in the United States are produced in the Northern Great Plains region (Tables 39.1 and 39.2). Durum wheat is a specialty wheat used mainly by the pasta industry. It is tetraploid, while most other wheats are hexaploid.

Kochia, wild buckwheat, wild oats, and wild mustard are problem weeds in early-planted spring wheat, whereas later plantings favor the growth of green and yellow foxtails, common lambsquarters, and pigweed spp.

In Montana, wild oats, green foxtail, and Persian darnel are major grass weeds. In North Dakota, South Dakota, and Minnesota, wild oats, green foxtail, and yellow foxtail are major grass weeds.

In Montana, kochia, Russian thistle, wild mustard, common lambsquarters, field pennycress, and haresear mustard are major broadleaf weeds. In North Dakota, listed in decreasing order of frequency, wild buckwheat, kochia, Russian thistle, redroot pigweed, field bindweed, common lambsquarters, wild sunflower, and common purslane are problem weeds. In South Dakota, redroot pigweed, wild buckwheat, common lambsquarters, wild mustard, Russian thistle, and kochia are major weeds. In Minnesota, wild buckwheat, pigweed spp., smartweed spp., common lambsquarters, and wild mustard are serious problem weeds.

If soil moisture is adequate, temperature requirements for wheat and weed seed germination often dictate the sequence of weed problems that develop in spring wheat. The base temperature for germinating spring wheat is 37–40°F (2–5°C). The minimum germination temperature for a number of weed seeds is 40°F, including wild oats, wild buckwheat, and wild mustard. Kochia germinates at a base temperature of 35°F (2°C), whereas redroot pigweed, green foxtail, and yellow foxtail germinate at about 50°F (10°C), emerging after wheat planting.

Triallate is the most commonly used herbicide for wild oat control, with lesser amounts of difenzoquat and diclofop used. Green and yellow foxtails are controlled with trifluralin, usually applied preemergence-incorporated, and diclofop applied postemergence.

Wild buckwheat is a phenoxy-tolerant weed species, but it can be controlled with dicamba or picloram applied when the plants are small and the wheat is in the three-leaf stage. Mixtures of bromoxynil and dicamba enhance control of kochia. Better control of wild mustard can be obtained with a mixture of 2,4-D or MCPA with bromoxynil, dicamba, or picloram. A mixture of MCPA and bromoxynil effectively controls common sunflower.

Central Great Plains

Hard red winter wheat is the kind of wheat grown in the Central Great Plains (Tables 39.1 and 39.2). Typical crop rotations in the Central Great Plains are winter wheat–corn–fallow, winter wheat–sorghum–fallow, winter wheat–fallow, winter wheat–soybeans, or continuous wheat.

The predominant winter annual grass weeds infesting winter wheat fields are downy brome, hairy chess, Japanese brome, cheat, and jointed goatgrass. These grass weeds have growth habits similar to those of winter wheat. They are extremely difficult to control in continuous winter wheat or winter wheat–fallow rotation. Wild oat is absent from winter wheat fields in the Central Great Plains except in Colorado. Volunteer winter wheat is the most troublesome weed during the fallow period following wheat harvest.

In general, herbicides are not being used to control grass weeds in winter wheat, apparently due to a lack of effective, selective herbicides for this use, and there is a definite need for a selective herbicide to control jointed goatgrass and *Bromus* spp. Metribuzin is registered for use in wheat to control *Bromus* spp., but its use is restricted to only a few wheat cultivars. Difenzoquat can be used to control wild oats, and diclofop can be used for control of wild oats, barnyardgrass, foxtail spp., and Italian ryegrass. Moderate control of downy brome has been obtained with diclofop and triallate.

Trifluralin is not registered for use in the Central Great Plains. Soil incorporation of herbicides such as triallate and trifluralin tends to bury plant residues needed on the soil surface to protect against soil erosion. Winter wheat varieties with short coleoptiles may be injured by preemergence incorporation.

Most herbicides used in the Central Great Plains to control broadleaf weeds in winter wheat are applied postemergence when the wheat is tillering in the spring. These herbicides include 2,4-D, bromoxynil, chlorsulfuron, dicamba, metsulfuron, triasulfuron, diuron, MCPA, and picloram. The safest time to apply 2,4-D, bromoxynil, dicamba, MCPA, and picloram to wheat is in the tillering stage, and applications made during the boot stage are the most injurious. Most herbicides applied to wheat in the two- to four-leaf stage reduce wheat height and grain yields more than when applied to fully tillered plants.

Aerial application of 2,4-D, glyphosate, or glyphosate + 2,4-D, used as harvest aids when tall broadleaf weeds threaten grain harvesting, should be made 7–10 days prior to wheat harvest when the wheat is in the hard dough stage. Later applications do not allow sufficient time for the weeds to become dry and brittle before harvest. When 2,4-D is used as a harvest aid, the wheat should not be green in the nodes when sprayed, or stem breakage may occur.

The most common winter annual broadleaf weeds in winter wheat in the Central Great Plains are pennycress, shepherd's-purse, and tansy mustard.

Summer annual weeds usually do not compete with winter wheat. In eastern Nebraska, green foxtail, common lambsquarters, kochia, Pennsylvania smartweed, and yellow woodsorrel are the major summer annual weeds. In western Nebraska, these are barnyardgrass, green foxtail, common sunflower, Russian thistle, and slimleaf lambsquarters. Although these weeds may not compete with winter wheat, they will mature and set an abundance of seed unless controlled after wheat harvest.

Field bindweed is the most widely distributed perennial weed in the Central Great Plains region. It is considered to be the most serious perennial weed in wheat in Colorado, Kansas, and Nebraska.

Pacific Northwest

This discussion of weeds and their control in the Pacific Northwest (Tables 39.1 and 39.2) will be separated into two parts (western Oregon and Washington and eastern Oregon and Washington). The two geographic regions are separated by the Cascade Mountain Range running north and south through the two states. The western region has an average annual precipitation of about 40 in., while the eastern region has an average annual precipitation of about 15 in.

Western Oregon and Washington

In western Oregon and Washington, wheat is normally planted from mid-October to early November. The wet, mild winters are favorable for winter wheat production, with yields among the highest in the United States, averaging more than 67 bu/A most years and not uncommonly yielding 135 bu/A in some fields. However, the mild winters favor luxuriant growth of a large number of fall-germinating weeds that can seriously reduce wheat yields. Winter wheat harvest begins in late July and is completed about mid-August.

In general, more than 100 species of broadleaf and 8 species of grass weeds infest winter wheat fields of western Oregon and Washington. Wild oat, Italian ryegrass, witchgrass, barnyardgrass, green foxtail, annual bluegrass, *Bromus* spp., and slender foxtail (*Alopecurus myosuroides*) are common annual grass problems. Nearly all serious brome infestations are in fields where wheat has been grown continuously for 2 or more years. Examples of annual broadleaf weeds are bedstraw spp., speedwell spp., common chickweed, mustard spp., vetch spp., among many others. The two major perennial grass weeds are quackgrass and bentgrass species. The major perennial broadleaf weeds are Canada thistle, field bindweed, and, occasionally, western wildcucumber.

Crop rotation may be the most satisfactory control measure for certain weeds, but the choice of rotational crops is limited and their use is often governed by market considerations. Common rotational crops include red clover or crimson clover for seed, oats, oats-vetch, or winter field peas. About 20% of the wheat grown in the area is not rotated, with wheat following wheat.

Diuron is the most widely used herbicide in western Oregon, with about 80% of the wheat acreage treated with it each year. Diuron occasionally causes wheat injury, but its benefits over the years have far outweighed any detrimental effects. It is less widely used in western Washington. Diuron controls a wide variety of winter annual broadleaf weeds and has been widely used since 1960. Diuron does not control perennial weeds, and some resistant annuals (such as species of speedwell and bedstraw) have increased as problem weeds.

The growth regulator–type herbicides (2,4-D, MCPA, clopyralid, dicamba), bromoxynil, chlorsulfuron, and chlorsulfuron + metsulfuron are used primarily in the spring for broadleaf weed control. Chlorsulfuron is highly effective against a broad spectrum of annual broadleaf weeds and the perennial Canada thistle. Wheat has a high degree of resistance to chlorsulfuron, but the herbicide poses a hazard to most rotational crops due to its soil persistence.

Frequent rains in western Oregon and Washington make herbicide application difficult during the winter, often at less than optimal timing. Depending on the weed species present, herbicides may be applied as many as four times in one season, beginning with a preemergence or early postemergence treatment to the wheat in the fall and extending into early April. For example, diuron may be applied early postemergence in the fall to control many annual broadleaf weeds and annual bluegrass, and to retard Italian ryegrass. If necessary, the diuron treatment may be combined with chlorsulfuron + metsulfuron to control speedwell and retard bedstraw, both resistant to diuron and phenoxy herbicides. In January, diclofop may be used to control Italian ryegrass and wild oat. In March, MCPA and dicamba may be applied to control Canada thistle, vetch, and spring-emerging broadleafs. If necessary, difenzoquat may be used in April to control late-emerging wild oat.

Eastern Oregon and Washington

In eastern Oregon and Washington, wheat is planted in the fall and harvested in July and August. It is grown in a 2-year rotation of wheat and summer fallow. Following the wheat harvest, the land is kept free of vegetation during the following crop season, primarily to conserve moisture for the next wheat crop. Commonly, fallow fields either are not tilled until the following spring or are tilled once in the fall for weed control or to help prevent runoff from frozen soil. In eastern Oregon, tillage is with large sweeps cutting at a depth of 4–5 in. to undercut the stubble and weeds and leave most of the crop residue on the soil surface to prevent erosion and trap winter snows. About 10% of the fallow land in eastern Oregon is fall-tilled. In eastern Washington, tillage is by disking, with disks set to cut 3–4 in. deep, or by chiseling, with chisels spaced about 12 in. apart and set to cut at a depth of 8–10 in. When weed control is the primary objective, tillage by disking is more common. When improved water infiltration is the main objective, chiseling is preferred.

The moldboard plow is still used, usually in the spring, followed by disking or other secondary tillage to break up large clods. The rodweeder is used through the summer, when needed, to eliminate weeds and to "set the moisture line" by creating a dust mulch to reduce evaporation. Although still used successfully, these methods leave the soil bare and susceptible to wind and water erosion. For this reason, the use of these methods has declined. In a fallow system termed "stubble mulch" or "trashy fallow," primary tillage is done in the spring with sweeps, chisels, and disks. The rodweeder is used as needed (two to six times) until planting time in the fall. A considerable amount of crop residue is left on the soil surface to reduce wind and water erosion. Using this system, the following crop generally has more weeds in it than when the moldboard plow is used, but the moldboard plow is not recommended because of soil conservation considerations.

The major annual grass weed in wheat in eastern Oregon and Washington is downy brome, a troublesome weed since the 1800s. Rye, a cereal, and jointed goatgrass are problem weeds in certain areas of this region. Jointed goatgrass has been spread in the area primarily by planting contaminated wheat seed. Jointed goatgrass is genetically close to cultivated wheat and, at present, there is no selective herbicide available to control it in wheat. Wild oat is found in the area but does not present a problem. Except for wild oat, all of these grass weeds usually germinate in the fall along with the wheat. Fall-germinated wild oat often fails to survive the winter, but may be replaced by spring-germinating wild oat. At harvest, the seeds of rye, jointed goatgrass, and wild oat are common contaminants in the crop grain, and they are major culprits contributing to dockage.

There are more than 30 broadleaf weed species common in wheat in eastern Oregon and Washington. The principal broadleaf weeds that germinate in the fall include the mustards (blue, tansy, and tumble) and coast fiddleneck.

Major weed species germinating in the early spring include Russian thistle and prostrate knotweed. Weeds germinating in late spring, such as common lambsquarters, kochia, and pigweed species, are generally not problem weeds in wheat unless the crop stand is sparse or retarded in growth. Where 2,4-D has been used alone year after year, populations of bur buttercup and corn gromwell have increased.

The major perennial weed is field bindweed, both in the crop and during the fallow year. Tillage equipment spreads it more each year.

The most common grass weeds in fallow are downy brome, rye, and volunteer wheat. The most common broadleaf weeds in fallow are coast fiddleneck, prostrate knotweed, prostrate pigweed, Russian thistle, and tumble mustard. Unless controlled, these weeds grow rapidly and deplete the soil of moisture.

Prevention as a means of weed control has not been high on growers' priority lists, partly due to the fact that most of the common weeds have been in the area for many years. Combines are seldom cleaned between fields, and wheat seed is usually obtained from the grower's own farm or from a neighbor's. The use of certified seed is common only when changing to a new wheat cultivar. Some growers reduce the spread of weed seed by burning wheat stubble, using nonselective herbicides, or establishing a perennial grass sod along fence rows. Untended fence rows surrounding wheat fields are a common source of weed infestations.

Over 90% of the wheat in eastern Oregon and Washington is treated with at least one herbicide. Paraquat and glyphosate may be used postemergence to control emerged weeds prior to wheat emergence. Trifluralin and diclofop may be applied preplant to control downy brome. Triallate may be applied and incorporated preplant or shortly after planting for wild oat control. Diclofop and difenzoquat may be used postemergence to control wild oats in the spring. Unlike downy brome, wild oat is most sensitive to diclofop when in the two- to four-leaf stage. Difenzoquat is most effective on wild oat in the three- to five-leaf stage.

Broadleaf herbicides such as bromoxynil and chlorsulfuron may be applied in the fall to wheat. The growth regulator–type herbicides (2,4-D, MCPA, dicamba, clopyralid, and picloram) may be applied after the wheat is well tillered and before the jointing stage. 2,4-D may also be applied after the wheat has reached the soft dough stage, a stage too late to reduce weed competition but possibly helpful as a harvest aid and in reducing weed seed production.

Of the herbicides used in wheat in eastern Oregon and Washington, only 2,4-D has been implicated in a significant number of cases involving damage from spray drift. The production of highly sensitive crops, such as fruits, vegetables, and especially grapes, to 2,4-D has increased on irrigated lands in northern Oregon and southern Washington; this has lead to restrictions on the use of 2,4-D esters. The use of high-volatile esters is prohibited in a zone surrounding the grape-growing areas, and low-volatile esters may not be used in this zone during the late spring and summer months.

Canadian Prairie Provinces

The major weed problems for wheat in the Canadian Prairie Provinces (Alberta, Saskatchewan, and Manitoba) are the introduced weed species. Species native to this region that have generally become minor problem weeds include tumble pigweed, tansy mustard, false ragweed, blue lettuce, greenleaf pepperweed, and Nuttall povertyweed.

Wild mustard was one of the first weeds introduced by early settlers, and by the early 1930s, it had reached epidemic proportions, threatening wheat production on the Regina Plains. Wild oat was introduced as a contaminant in seed grain, feed, and packing material. It was the first grass weed to become a serious problem in wheat. A survey reported in 1983 indicated that wild oat occurred in 61, 71, and 75% of the cereal and flax fields in Alberta, Saskatchewan, and Manitoba, respectively.

The widespread use of the herbicide 2,4-D in wheat in the Canadian Prairie Provinces, following its introduction in 1946, caused a major shift in the weed spectrum. Weeds highly susceptible to 2,4-D, such as field pennycress and wild mustard, were effectively removed, leaving an ecological niche that favored a rapid increase of cowcockle, a less competitive weed but one more resistant to 2,4-D. Similarly, pale smartweed and wild buckwheat, both tolerant to 2,4-D, increased in population and became major weeds in wheat. Later, dicamba and bromoxynil were introduced to control hard-to-kill weeds such as pale smartweed and wild buckwheat, leaving an ecological niche for erect knotweed, mouseear chickweed, hempnettle, roundleaved mallow, and narrowleaf hawksbeard. Narrowleaf hawksbeard is a winter annual, and the seed capsules of roundleaf mallow break up into varying sized pieces, making it difficult to separate them from seed grain regardless of the size of the crop seed. Following the more recent use of the sulfonylurea herbicides, shifts in the weed spectrum have again occurred, with a decrease in the density of hempnettle and roundleaf mallow and an increase in the populations of cutleaf nightshade.

The easily killed common lambsquarters, wild mustard, and field pennycress continue to be everpresent weed problems in wheat in the Canadian Provinces. Redroot pigweed is one of the top 10 weeds in all three Prairie Provinces. It is relatively resistant to recommended applications of 2,4-D, bromoxynil, and dicamba. Also, it is a later germinator, and its seedlings emerge after the wheat crop has been sprayed with these herbicides. Normally, this weed is not a strong competitor with wheat, but even small plants under the wheat canopy are prolific seed producers.

Russian thistle and tumble mustard have been reduced to comparatively insignificant problems by advances in agricultural technology. Although Russian thistle occurs throughout the Prairie Provinces, it continues to be a major weed only in the southwest corner of Saskatchewan and the southeast corner of Alberta. It ranks as the fifth most common weed in Saskatchewan. If not controlled by tillage soon after wheat harvest, it grows rapidly and produces an abundance of seed before killing frosts occur.

Wheat growers in western Canada list wild oat and green foxtail, in that order, as their worst weeds. Wild oat germinates earlier, grows taller, and competes better than green foxtail. However, the relative success in controlling wild oat has opened an ecological niche for the invasion of green foxtail.

Herbicides presently used in western Canada to control annual grasses in wheat include diclofop, difenzoquat, triallate, and trifluralin. There is no herbicide currently available that selectively controls green foxtail in wheat after the weed reaches the four-leaf stage. Green foxtail may only be 2–2.5 in. tall by the time it reaches the four-leaf stage and thus may not be recognized as a problem at this stage.

Canada thistle was introduced into Canada from the Mediterranean area as an impurity in crop seed and feeds. After its introduction, Canada thistle spread as rapidly as agriculture itself and became, in 1871, the first weed to be legislated against. In spite of the legislation, Canada thistle occurs in all three Prairie Provinces, with the highest frequency in Manitoba where it ranks as the fifth most abundant weed. Canada thistle readily establishes on newly disturbed areas where competition is limited. Once established, it becomes very competitive and has been reported to cause heavier yield losses than any other perennial weed.

Summer Fallow

Summer fallowing has been credited as the single most important development in sustaining agriculture in the semiarid region of the Prairie Provinces. The unpredictability of precipitation during the growing season was the major factor in its development. However, weeds grew unchecked in the fallowed land, robbing the soil of the very moisture that fallowing was intended to conserve. Thus to ensure the success of summer fallowing, a weed-control program must be in place.

Prior to the introduction of herbicides, cultivation was the only means of controlling weeds in fallow, but tillage during the summer and fall left the soil surface open to wind erosion. With the introduction of 2,4-D, the use of cultural practices decreased rapidly and "chemical fallow" became a possibility. The winter annual weeds field penny-cress and flixweed could be controlled in fallow by low rates of 2,4-D, and the wheat stubble could be left standing as an obstacle to wind erosion. Other herbicides are now available that control 2,4-D-resistant weeds such as cowcockle and wild buckwheat and that enhance the control of Canada thistle. A number of herbicides are registered for the control of wild oat and green foxtail. The introduction of the nonselective, nonresidual herbicides paraquat and glyphosate has made it possible to control all annual weeds and volunteer cereals and to slow the growth of perennial weeds in fallow land, and chemical fallow is now a reality.

■ SELECTED REFERENCES

Agricultural Chemical Usage: 1993 Field Crop Summary. United States Department of Agriculture, National Agricultural Statistics Service, Washington, D.C. pp. 82 and 84.

Agricultural Statistics, 1992. United States Department of Agriculture, Government Printing Office, Washington, D.C. p. 5.

Anonymous. 1992. Agricultural Statistics, USDA.

Anonymous. 1993. Agricultural Chemical Usage, 1992. Field Crop Summary. USDA Statistics.

Banks, P. A. 1990. The Southeast. In *Systems of weed control in wheat in North America,* W. W. Donald, ed. Monograph No. 6, pp. 182–190. Champaign, Ill.: Weed Science Society of America.

Brengle, K. G. 1982. Principles and methods of summer fallow. In *Principles and practices of dryland farming.* pp. 73–101. Boulder, Colo.: Colorado Associated University Press.

Bridges, D. C. 1992. *Crop losses due to weeds in the United States.* Champaign, Ill.: Weed Science Society of America.

Donald, W. W. 1990. Northern Great Plains. In *Systems of weed control in wheat in North America,* W. W. Donald, ed. Monograph No. 6, pp. 90–126. Champaign, Ill.: Weed Science Society of America.

Donald, W. W., and T. Prato. 1991. Profitable, effective herbicides for planting-time weed control in no-till spring wheat (*Triticum aestivum*). *Weed Sci.* 39:83–90.

Donald, W. W. 1991. Seed survival, germination ability, and emergence of jointed goatgrass (*Aegilops cylindrica*). *Weed Sci.* 39:210–216.

Donald, W. W., and A. G. Ogg, Jr. 1991. Biology and control of jointed goatgrass (*Aegilops cylindrica*), a review. *Weed Technol.* 5:3–17.

Fay, P. K. 1990. A brief overview of the biology and distribution of weeds of wheat. In *Systems of weed control in wheat in North America,* W. W.

Donald, ed. Monograph No. 6, pp. 33–50. Champaign, Ill.: Weed Science Society of America.

Holm, F. A., and K. J. Kirkland. 1986. Annual broadleaf weed control in wheat. In *Wheat production in Canada—a review,* A. E. Slinkard and D. B. Fowler, eds., pp. 375–390. Saskatoon, Canada: Univ. Saskatchewan.

Hunter, J. H., I. N. Morrison, and D. R. S. Rourke. 1990. The Canadian Prairie Provinces. In *Systems of weed control in wheat in North America,* W. W. Donald, ed., Monograph No. 6, pp. 51–89. Champaign, Ill.: Weed Science Society of America.

Jones, D. P., ed. 1976. *Wild oats in world agriculture.* London: Agric. Research Council.

Mitich, L. W., and G. B. Kyser. 1990. Wheat production and weed control in Arizona. In *Systems of weed control in wheat in North America,* W. W. Donald, ed., Monograph No. 6, pp. 257–265. Champaign, Ill.: Weed Science Society of America.

Mitich, L. W., and G. B. Kyser. 1990. Wheat production practices in relation to weed control in California. In *Systems of weed control in wheat in North America,* W. W. Donald, ed., Monograph No. 6, pp. 233–256. Champaign, Ill.: Weed Science Society of America.

McMullan, P. M. and J. D. Nalewaja. 1991. Triallate antidotes for wheat (*Triticum aestivum*). *Weed Sci.* 39:57–61.

Nalewaja, J. D. 1981. Integrated pest management for weed control in wheat. In *CRC handbook of pest management in agriculture,* Vol. 3, pp. 343–354. Boca Raton, Fla.: CRC Press.

Peeper, T. F., and A. F. Wiese. 1990. Southern Great Plains. In *Systems of weed control in wheat in North America,* W. W. Donald, ed., Monograph No. 6, pp. 158–181. Champaign, Ill.: Weed Science Society of America.

Peters, E. J. 1990. The Midwest and Northeast. In *Systems of weed control in wheat in North America,* W. W. Donald, ed., Monograph No. 6, pp. 191–199. Champaign, Ill.: Weed Science Society of America.

Staniforth, D. W., and A. F. Wiese. 1985. Weed biology and its relationship to weed control in limited-tillage systems. In *Weed control in limited-tillage systems,* A. F. Wiese, ed., Monograph No. 2, pp. 215–225. Champaign, Ill.: Weed Science Society of America.

Stobbe, E. H. 1976. Tillage practices on the Canadian Prairies. *Outlook Agric.* 10(1):21–26.

Swanton, C. J., K. N. Harker, and R. L. Anderson. 1993. Crop losses due to weeds in Canada. *Weed Technol.* 7:537–542.

Tottman, D. R. 1976. Spray timing and the identification of cereal growth stages. (Proceedings British Crop Protection Conference.) *Weed Sci.* 13:791–800.

Unger, P. W., and T. M. McCalla. 1981. Conservation tillage systems. *Adv. Agron.* 33:1–44.

Wicks, G. A. 1985. Weed control in conservation tillage systems—small grains. In *Weed control in limited-tillage systems,* A. F. Wiese, ed., Monograph No. 2, pp. 77–92. Champaign, Ill.: Weed Science Society of America.

Wicks, G. A., and D. E. Smika. 1990. Southern Great Plains. In *Systems of weed control in wheat in North America,* W. W. Donald, ed., Monograph No. 6, pp. 127–157. Champaign, Ill.: Weed Science Society of America.

Young, F. L. 1986. Russian thistle (*Salsola iberica*) growth and development in wheat (*Triticum aestivum*). *Weed Sci.* 34:901–905.

40 LETTUCE

■ INTRODUCTION

In 1991, a total of 221,900 A was planted to lettuce in the United States (Table 40.1). In that year, weeds caused an estimated loss of $29 million in lettuce, even though herbicides were used. Under the same conditions, but without herbicides, the estimated loss would have been $135 million. In 1992, 68% of the head lettuce acreage and 59% of the other lettuce acreage were treated with herbicides. In 1992, in eastern Canada, weeds caused an estimated reduction in the lettuce crop of 10%, a loss valued at $2.4 million.

Arizona and California produce about 90% of the lettuce grown in the United States. The principal weeds of lettuce grown in these states are given in Table 40.2. The principal herbicides used in lettuce are benefin, bensulide, and pronamide.

■ INFORMATION OF PRACTICAL IMPORTANCE

Benefin (Balan EC and DF) is a dinitroaniline herbicide (Chapter 17). It is used preplant-incorporated to control grass and broadleaf weeds. It does not control emerged weeds. In the western United States, incorporate benefin 2–3 in. deep within 4 hours of application to prevent herbicide loss by volatility; elsewhere, incorporate within 8 hours. Benefin is applied broadcast at 1.1–1.5 lb ai/A (3–4 qt Balan EC/A *or* 2–2.5 lb Balan DF/A), depending on soil type.

Bensulide (Prefar 4EC) is a nonfamily herbicide (Chapter 30). It is applied either preplant-incorporated or preemergence to control annual grass and certain broadleaf

TABLE 40.1 ■ Lettuce acreage in the United States in 1991.*

STATE	ACRES	STATE	ACRES
Arizona	49,000	New Jersey	2,500
California	152,000	New York	2,600
Colorado	4,700	New Mexico	2,100
Florida	5,700	Texas	2,100
Michigan	800	Washington	1,300
*Total acreage planted to lettuce in 1991 was 221,900 A.			

SOURCE: *Agricultural statistics 1992.* United States Department of Agriculture. Government Printing Office, Washington, D.C. p. 155.

TABLE 40.2 ■ Principal weeds in lettuce grown in Arizona and California.

ARIZONA	CALIFORNIA
Common Weeds*	**Common Weeds***
Common purslane	Burning nettle
Groundcherries	Common groundsel
Littleseed canarygrass	Common purslane
London rocket	Hairy nightshade
Nettleleaf goosefoot	Nettleleaf goosefoot
Prickly lettuce	Redroot pigweed
Purple nutsedge	Shepherd's-purse
Shepherd's-purse	
Silversheath knotweed	
Spiny sowthistle	
Troublesome Weeds**	**Troublesome Weeds****
Groundcherries	Common groundsel
Nettleleaf goosefoot	False chamomile
Prickly lettuce	Sowthistles
Shepherd's-purse	
Silversheath knotweed	
Spiny sowthistle	

*Common weeds are those that abundantly infest a significant portion of the lettuce acreage.
**Troublesome weeds are those that, despite weed-control efforts, are inadequately controlled and interfere with crop production and/or yield, crop quality, or harvest efficiency.

SOURCE: D. C. Bridges, ed. 1992. *Crop losses due to weeds in the United States.* Champaign, Ill.: Weed Science Society of America.

weeds. For best results, a preemergence treatment must be followed by irrigation or rainfall to move bensulide into the weed seed germination zone. Bensulide does not control emerged weeds. It is only effective on mineral soils (soils not of high organic matter content), due to its being strongly adsorbed to organic colloids. Bensulide is applied broadcast at 5–6 lb ai/A (5–6 qt product/A).

Pronamide (Kerb 50W) is an acid amide herbicide (Chapter 12) and a Restricted Use Pesticide because of its oncogenicity. Most lettuce varieties are highly tolerant to pronamide. It can be applied either preplant, preemergence, or postemergence to the lettuce. It is used preemergence and early postemergence to control winter annual and perennial grasses and chickweed (a broadleaf). To be effective when applied preemergence, pronamide must be incorporated by rainfall or overhead sprinkler irrigation to move it into the weed seed germination zone. It is not recommended for use on soils of high organic matter content, as it is strongly adsorbed to organic colloids. The broadcast rate of pronamide is 1–2 lb ai/A (2–4 lb product/A).

■ SELECTED REFERENCES

Agricultural chemical usage: Vegetables, 1992 summary. National Agricultural Statistics Service. United States Department of Agriculture. Washington, D.C. p. 140.

Agricultural statistics 1992. United States Department of Agriculture. Government Printing Office, Washington, D.C. pp. 155–156.

Bridges, D. C., ed. 1992. *Crop losses due to weeds in the United States.* Champaign, Ill.: Weed Science Society of America.

Herbicide product labels.

Meister, R. T., ed. 1994. *Weed control manual 1994.* Willoughby, Ohio: Meister Publishing. pp. 197–198.

41 ONIONS

■ INTRODUCTION

In 1991, 134,000 A of dry bulb onions were harvested in the United States (Table 41.1). In that year, weeds caused an estimated crop loss of $36.2 million in onions, even though herbicides were used. Under the same conditions, but without herbicides, the estimated loss would have been $184.4 million. In the United States in 1992, 86% of the dry bulb onion acreage was treated with herbicides.

In 1989 in eastern Canada, weeds caused an estimated onion crop reduction of 10%, a loss valued at $2.1 million.

The principal weeds in onions grown in the United States are given in Table 41.2. The principal herbicides registered in the United States for dry bulb onions in 1991, and their comparative use, are given in Table 41.3.

■ INFORMATION OF PRACTICAL IMPORTANCE

Bensulide (Prefar 4EC) is a nonfamily herbicide (Chapter 30). Bensulide is registered for use in onions only in Arizona and New Mexico. It is not effective in soils of moderate to high organic matter content. It can be applied either preplant-incorporated or preemergence to onions.

When applied preemergence, bensulide must be moved into the seed germination zone by irrigation or rainfall. Bensulide controls annual grasses and certain broadleaf weeds such as common lambsquarters and redroot pigweed. It does not control emerged weeds. It is applied at 3 lb ai/A (3 qt product/A).

Bromoxynil (Buctril 2EC) is a benzonitrile herbicide (Chapter 14) and a Restricted Use Pesticide because it can cause birth defects in laboratory animals. Bromoxynil is applied early postemergence to control annual broadleaf weeds. It does not control grasses and has no soil-residual activity. Bromoxynil can be applied preplant, preemergence, and postemergence to onions in the two- to five-true-leaf stage. It is applied at 0.25–0.38 lb ai/A (1–1.5 pt product/A), depending on weed susceptibility.

DCPA (Dacthal 75W) is a phthalic acid herbicide (Chapter 24). It is used preemergence to control annual grasses and certain broadleaf weeds. *It provides excellent control of common purslane on mineral soils.* It is not effective on muck soils, as it is strongly adsorbed to organic colloids. DCPA can be applied to onions at seeding, transplanting, and/or lay-by. At transplanting, DCPA may be applied directly over the onion transplants without injury. A lay-by application of DCPA can be made to onions either alone or

TABLE 41.1 ■ Dry bulb onion acreage in the United States in 1991.

STATE	ACRES	STATE	ACRES
Arizona	900	New York	11,800
California	36,600	Ohio	490
Colorado	12,700	Oregon	14,200
Georgia	6,000	Texas	15,800
Idaho	8,000	Utah	1,900
Michigan	7,300	Washington	8,600
Minnesota	980	Wisconsin	1,600
New Mexico	7,100		
*Total onion acreage harvested in 1991 was 133,970 A.			

SOURCE: *Agricultural Statistics 1992.* United States Department of Agriculture. Government Printing Office, Washington, D.C. p. 157.

TABLE 41.2 ■ Principal weeds in dry bulb onions grown in the United States in 1991.

COMMON WEEDS*	TROUBLESOME WEEDS**
Barnyardgrass	Barnyardgrass
Common lambsquarters	Common lambsquarters
Common purslane	Common purslane
Common sunflower	Common sunflower
Kochia	Hairy nightshade
Ladysthumb	Kochia
London rocket	Nutsedge, purple
Nightshades	Nutsedge, yellow
Nutsedge, yellow	Pigweeds
Pigweeds	*Polygonum* spp.
Spurges	Prostrate spurge
	Quackgrass
	Russian thistle
	Wild mustard

*Common weeds are those that abundantly infest a significant portion of the onion acreage.
**Troublesome weeds are those that, despite weed-control efforts, are inadequately controlled and interfere with crop production.

SOURCE: D. C. Bridges, ed. 1992. *Crop losses due to weeds in the United States.* Champaign, Ill.: Weed Science Society of America.

in addition to a preemergence application up to 14 weeks after planting. Before the lay-by application, the crop should be clean-tilled. DCPA is applied at 4.5–10.5 lb ai/A (6–14 lb product/A).

Fluazifop-P-butyl (Fusilade 2000, 1EC) is an aryloxyphenoxypropionate herbicide (Chapter 13). It is used postemergence to control annual and perennial grasses. It

TABLE 41.3 ■ Comparative use of the principal herbicides registered for dry bulb onions in 1991.

HERBICIDE	PERCENT*
Oxyfluorfen	59
DCPA	40
Bromoxynil	35
Fluazifop-P-butyl	20
Bensulide	10

*Percent of total acres of dry bulb onions. Not all herbicides applied in every state.

SOURCE: D. C. Bridges, ed. 1992. *Crop losses due to weeds in the United States.* Champaign, Ill.: Weed Science Society of America.

does not control broadleaf weeds or sedges. For best results, a nonionic surfactant or crop oil concentrate must be added to the spray mixture. *Do not tank-mix fluazifop-P-butyl with other herbicides, as its effectiveness may be reduced.* Fluazifop-P-butyl is applied at a rate of 0.09–0.38 lb ai/A (0.75–3 pts product/A), depending on weed susceptibility.

Oxyfluorfen (Goal 1.6E) is a diphenyl ether herbicide (Chapter 18). It can be used postemergence in direct-seeded or transplanted onions to control grass and broadleaf weeds. Transplanted onions are most tolerant to oxyfluorfen immediately after transplanting. Oxyfluorfen can cause injury, such as necrotic lesions, twisting, or stunting, to onion leaves. Injury is more severe if oxyfluorfen is applied during cool, wet weather and/or if applications are made prior to the development of three fully developed true leaves on onions grown in the Northeast, or two fully developed true leaves on onions grown in other states. The recommended rate of oxyfluorfen varies from 0.03 to 0.25 lb ai/A (2.4–20 oz product/A), depending on geographic location. Multiple applications of oxyfluorfen may be made as needed.

■ SELECTED REFERENCES

Agricultural chemical usage: Vegetables, 1992 summary. National Agricultural Statistics Service. United States Department of Agriculture. Washington, D.C. pp. 177–178.

Agricultural Statistics 1992. United States Department of Agriculture. Government Printing Office, Washington, D.C. p. 157.

Bridges, D. C., ed. 1992. *Crop losses due to weeds in the United States.* Champaign, Ill.: Weed Science Society of America.

Herbicide product labels.

Meister, R. T., ed. 1994. *Weed control manual 1994.* Willoughby, Ohio: Meister Publishing. pp. 199–200.

42 POTATOES

■ INTRODUCTION

In 1991, potatoes were harvested from 1.59 million A in the United States. Of this acreage, 6% was harvested in the spring, 6% in the summer, and 88% in the fall. In that year, weeds caused an estimated loss of $65.3 million in potatoes, even though herbicides were used. Under the same conditions, but without herbicides, the estimated loss due to weeds would have been $493 million.

In 1989 in the provinces of eastern Canada, weeds caused an estimated reduction in potato yields of 7%, a loss valued at $25 million. In the western provinces, the estimated yield reduction from weeds was 8%, a loss valued at $7.6 million.

In 1993, fall-harvested potatoes were grown on 1.12 million A in the 11 major potato-growing states of the United States (Table 42.1). Herbicides were applied to 83% of the potato acreage in 1993, with coverage ranging from 89% or more in Idaho, Maine, Michigan, New York, Washington, and Wisconsin to 53 and 64% in North Dakota and Minnesota, respectively.

■ WEEDS

Eastern black nightshade and hairy nightshade are the two most common nightshade species infesting potato fields in the United States. Eastern black nightshade is most prevalent in the midwestern and eastern United States. Hairy nightshade is more common in irrigated potatoes grown in Minnesota, Wisconsin, and the western United States. Nightshades harbor a number of pests that attack potatoes, such as insects, nematodes, and certain potato diseases (e.g., early blight, late blight, and verticillium wilt).

In the Red River Valley of North Dakota, the major weeds in potatoes include foxtails and redroot pigweed. In Idaho, the most prevalent broadleaf weeds in potatoes are hairy nightshade, redroot pigweed, common lambsquarters, and kochia. Of these, hairy nightshade is the most difficult to control. Season-long control of hairy nightshade is difficult to achieve because its seeds germinate throughout the growing season. The predominant weed species common to potatoes vary from region to region, and representative weeds are listed in Table 42.2.

TABLE 42.1 ■ Acreages planted to fall-harvested potatoes in 1993 in the 11 major potato-growing states.

STATE	ACRES (× 1000)
Colorado	73
Idaho	390
Maine	81
Michigan	40
Minnesota	70
New York	29
N. Dakota	143
Oregon	50
Pennsylvania	21
Washington	150
Wisconsin	72
Total	**1,119**

SOURCE: *Agricultural chemical usage: 1993 field crop summary.* United States Department of Agriculture, National Agricultural Statistics Service, Washington, D. C. pp. 42–43.

■ WEED CONTROL

Cultivation, hilling, and herbicides are commonly used to control weeds in potatoes. Weeds emerging between the potato rows after planting are controlled by the hilling process, but weeds emerging after hilling cannot be controlled by mechanical cultivation.

Weeds competing season-long with potatoes can reduce yields by about 55%, compared to a 16% reduction when the crop is kept weed-free for the first 3 weeks after emergence. Weeds competing with potatoes during the first 8 or 10 weeks after crop emergence, but not thereafter, reduced tuber yields 19 and 35%, respectively.

A weed-free period of 4–6 weeks after potato emergence is, in some cases, considered optimal for potatoes. However, where redroot pigweed was the predominant weed, a weed-free period of about 9 weeks was optimum. Where barnyardgrass was the predominant weed, a weed-free period of 2–4 weeks was sufficient to assure maximum tuber yield. Potato yields were reduced 19% when one redroot pigweed or one barnyardgrass plant per 3.3 ft of row competed with the crop for the entire season.

Early-germinating hairy nightshade can be controlled with preemergence herbicides such as EPTC, linuron, metolachlor, or a mixture of pendimethalin + metribuzin. However, nightshade species are resistant to trifluralin, with black nightshade being 100 times more resistant than barnyardgrass. Postemergence control of hairy nightshade and black nightshade can be achieved with bentazon ap-

TABLE 42.2 ■ Representative problem weeds in fall-harvested potatoes in the United States.

COMMON WEEDS	TROUBLESOME WEEDS
Grasses	*Grasses*
Barnyardgrass	Green foxtail
Foxtail spp.	Large crabgrass
Large crabgrass	Quackgrass
Quackgrass	Wild oat
Wild oat	Yellow nutsedge*
Yellow nutsedge*	
Broadleafs	*Broadleafs*
Canada thistle	Canada thistle
Common lambsquarters	Common ragweed
Common ragweed	False chamomile
Kochia	Nightshade, eastern black
Nightshade, eastern black	Nightshade, hairy
Nightshade, hairy	Redroot pigweed
Polygonum spp.	Russian thistle
Redroot pigweed	Texas blueweed
Russian thistle	
Texas blueweed	
Wild buckwheat	
A sedge, not a grass.	

SOURCE: Adapted from D. C. Bridges, ed. 1992. *Crop losses due to weeds in the United States.* Champaign, Ill.: Weed Science Society of America.

plied at 0.5 lb ai/A mixed with a spray adjuvant (e.g., nonionic surfactant, petroleum oil concentrate, methylated seed oil, or urea-ammonium nitrate) without a reduction in No. 1 grade market tubers. Bromoxynil applied as an early postemergence, directed spray at a rate of 0.25 lb ai/A effectively controlled hairy nightshade, as well as redroot pigweed, common lambsquarters, and kochia, without a reduction in No. 1 grade market tubers. Potatoes are susceptible to bromoxynil applied to their foliar canopy.

Metribuzin applied preemergence at a rate of 0.5 lb ai/A provided 100% control of prostrate pigweed, kochia, and Russian thistle. Tank mixes of metribuzin at 0.27 lb with pendimethalin at 1.0 lb or metolachlor at 2.0 lb ai/A applied preemergence provided 100% control of these same species.

Potato yields are greatest when preemergence herbicides are used. However, the effectiveness of preemergence herbicides can be reduced when coupled with early hilling, as the herbicide is tumbled, mixed, and diluted with untreated soil thrown over the row during hilling.

Metribuzin applied postemergence at 0.17 lb ai/A (0.19 kg ai/ha) after hilling effectively controlled late-germinating weeds. Sequential, split-rate herbicide applications, made preemergence to the potatoes and after hilling, provided effective weed control. Reduced rates of linuron, metribuzin, and oryzalin applied after hilling have also provided good weed control.

Residues of rotational crops with allelopathic properties such as rye or wheat can suppress early weed emergence and reduce the amount of herbicide needed for adequate control. Also, the use of allelopathic crop residues in combination with delayed low-rate herbicide applications after hilling can provide successful weed control in potatoes.

In 1993, metribuzin was applied to 64% of the fall-harvested potato acreage, and the percent of acreage treated with EPTC was 28; pendimethalin, 17; metolachlor, 13; linuron, 6; and trifluralin, 6.

■ INFORMATION OF PRACTICAL IMPORTANCE

Metribuzin (Lexone 75DF, Sencor 75DF) is a triazine herbicide (Chapter 27). It is applied preemergence after planting or after drag-off but before crop emergence. Do not soil-incorporate. Metribuzin can be applied postemergence directly over-the-top of potato plants; apply before weeds are 1 in. tall. Potato cultivars vary as to their tolerance to metribuzin applied postemergence; do not use on early-maturing white or red-skinned cultivars. Metribuzin is applied at 0.5–1.0 lb ai/A (0.67–1.33 lb product/A), depending on soil type.

EPTC (Eptam 7E) is a thiocarbamate herbicide (Chapter 26). It is applied preplant, after drag-off, and/or at lay-by. EPTC must be soil-incorporated at least 0.5 in. deep to prevent herbicide loss by volatility. When possible, application and incorporation should be done in the same operation. EPTC is applied at a rate of 3–6 lb ai/A (3.5–7 pt product/A), depending on soil type.

Pendimethalin (Prowl 3.3EC) is a dinitroaniline herbicide (Chapter 17). It can be applied either preemergence surface, preemergence-incorporated, or early postemergence-incorporated. It can be applied to the soil surface after planting or after drag-off, but before the potatoes and weeds emerge. It can be soil-incorporated mechanically, or by rainfall or irrigation within 7 days after application.

Pendimethalin can be applied early postemergence, from crop emergence to 6 in. in height. Effective weed control is dependent on rainfall or sprinkler irrigation occurring within 7 days after application to move herbicide into the upper soil surface. Pendimethalin does not control emerged weeds, and the soil must be clean-tilled before

application. Pendimethalin is applied at 0.75–1.5 lb ai/A (1.8–3.6 pt product/A).

Metolachlor (Dual 8E) is an acid amide herbicide (Chapter 12). It can be applied preplant or postplant-incorporated, preemergence, or after hilling, or lay-by. When applied preplant, soil-incorporate metolachlor 3 in. (7.6 cm) deep. Postplant-incorporate metolachlor 2 in. deep anytime after planting to drag-off, but before potato emergence; avoid damaging seed potatoes or their sprouts during incorporation. Metolachlor is applied at 1.5–3 lb ai/A (1.5–3 pt product/A), depending on soil type.

Linuron (Lorox 50DF) is a substituted urea herbicide (Chapter 29). It is applied preemergence after planting but before crop emergence. Plant seed potatoes at least 2 in. deep. Do not spray over-the-top of emerged potatoes or crop injury will result. Where drag-off or hilling is practiced, apply linuron after the final drag-off or hilling operation. Apply before grasses are 2 in. tall and before broadleaf weeds are 6 in. tall. If emerged weeds are present at the time of application, add 1 pt of nonionic surfactant for each 25 gal of spray mixture. Preferably, apply just before or as weed seedlings emerge. Linuron is applied at a rate of 0.5–2 lb ai/A (1–4 lb product/A), depending on soil type and geographic location.

Trifluralin (Treflan 4EC) is a dinitroaniline herbicide (Chapter 17). It can be applied and incorporated prior to crop emergence, immediately following drag-off, or after potato plants have fully emerged. Trifluralin must be soil-incorporated within 4 hours of application to avoid herbicide loss by volatility. Incorporation equipment should be set so that beds and furrows are uniformly covered with a layer of treated soil. If the herbicide is concentrated over the bed, potato emergence may be retarded and stem brittleness may occur. When incorporating after crop has emerged, do not completely cover the plants with the treated soil to avoid damage to the seed potatoes and their sprouts during incorporation. Trifluralin does not control emerged weeds, and the crop should be clean-tilled before using it. Trifluralin is applied at 0.5–1 lb ai/A (1 to 2 pts product/A), depending on soil type.

■ SELECTED REFERENCES

Agricultural chemical usage: 1992 field crop summary. United States Department of Agriculture, National Agricultural Statistics Service, Washington, D.C. pp. 45.

Agricultural chemical usage: 1993 field crop summary. United States Department of Agriculture, National Agricultural Statistics Service, Washington, D.C. pp. 42–43.

Agricultural statistics 1992. United States Department of Agriculture. Government Printing Office, Washington, D.C. pp. 159–160.

Bridges, D. C., ed. 1992. *Crop losses due to weeds in the United States.* Champaign, Ill.: Weed Science Society of America.

Eberlein, C. V., M. J. Guttieri, and W. C. Schaffers. 1992. Hairy nightshade (*Solanum sarrachoides*) control in potatoes (*Solanum tuberosum*) with bentazon plus additives. *Weed Technol.* 6:85–90.

Eberlein, C. V., M. J. Guttieri, and F. N. Fletcher. 1993. Broadleaf weed control in potatoes (*Solanum tuberosum*) with postemergence directed herbicides. *Weed Technol.* 7:298–303.

Herbicide product labels.

Lanfranconi, L. E., R. R. Bellinder, and R. W. Wallace. 1993. Grain rye residues and weed control strategies in reduced tillage potatoes. *Weed Technol.* 6:1021–1026.

Meister, R. T., ed. 1994 *Weed control manual* 1994 Willoughby, Ohio: Meister Publishing. pp. 204–207.

Murray, M. W., et al. 1994. Early broadleaf weed control in potato (*Solanum tuberosum*) with herbicides. *Weed Technol.* 8:165–167.

Nelson, D. C., and M. C. Thoreson. 1981. Competition between potatoes (*Solanum tuberosum*) and weeds. *Weed Sci.* 29:672–677.

Ogg, A. G., Jr., and B. S. Rogers. 1989. Taxonomy, distribution, biology, and control of black nightshade (*Solanum nigrum*) and related species in the United States and Canada. *Rev. Weed Sci.* 4:25–58.

Vangessel, M. J., and K. A. Renner. 1990. Redroot pigweed (*Amaranthus retroflexus*) and barnyardgrass (*Echinochloa crus-galli*) interference in potatoes (*Solanum tuberosum*). *Weed Sci.* 38:338–343.

43

SWEET CORN

■ INTRODUCTION

In 1991, sweet corn was grown on 719,000 A in the United States (Table 43.1). Sweet corn for processing was grown on 76% of this acreage, and that for the fresh market on the remainder. In 1991, weeds caused an estimated loss of $16.9 million in processed sweet corn and $17.5 million in the fresh-market crop, even though herbicides were used. Under the same conditions, but without herbicides, the estimated loss would have been $64.6 and $80.6 million dollars, respectively. In 1992, 92% of the processed sweet corn acreage was treated with herbicides, as was 75% of the fresh sweet corn acreage.

In 1992, weeds reduced the sweet corn crop in eastern Canada by an estimated 11%, a loss valued at $2.4 million.

The principal weeds in sweet corn grown in the United States are given in Table 43.2. Herbicides applied to sweet corn in 1991 and the acreage treated with the respective herbicides in each region is given in Table 43.3.

■ INFORMATION OF PRACTICAL IMPORTANCE

Alachlor (Lasso 4EC) is an acid amide herbicide (Chapter 12) and a Restricted Use Pesticide, due to its oncogenicity. Alachlor can be applied either preplant or preemergence to sweet corn to control annual grass and broadleaf weeds. It may be tank-mixed with atrazine and applied either preplant-incorporated or preemergence. It may be tank-mixed with cyanazine and applied preemergence. Alachlor is applied at a rate of 2–4 lb ai/A (2–4 qt product/A).

Atrazine (AAtrex 4L) is a triazine herbicide (Chapter 27) and a Restricted Use Pesticide because of ground and surface water concerns. Atrazine should be used in combination with a grass-control herbicide such as butylate or metolachlor, to broaden the spectrum of weeds controlled. Atrazine may be applied either preplant-surface, preplant-incorporated, preemergence, or postemergence to sweet corn. It is applied at 1.6–2 lb ai/A (1.6–2 qt product/A).

TABLE 43.1 ■ Sweet corn (fresh and processing) acreage in the United States in 1991.*

STATE	ACRES	STATE	ACRES
Northeast		*Corn Belt*	
Connecticut	3,700	Illinois	49,900
Delaware	6,700	Ohio	9,900
Maryland	7,100	**Regional total: 59,800 A**	
Massachusetts	7,100	*Lake States*	
New Jersey	9,000	Michigan	19,100
New York	57,600	Minnesota	141,400
Pennsylvania	18,400	Wisconsin	166,700
Regional total: 109,600 A		**Regional total: 327,200 A**	
Southeast		*Mountain*	
Alabama	3,200	Colorado	3,100
Florida	46,300	Idaho	21,600
N. Carolina	4,900	**Regional total: 24,700 A**	
Texas	1,600	*Pacific Coast*	
Virginia	2,100	California	16,300
Regional total: 58,100 A		Oregon	50,100
Other states		Washington	61,500
Total: 11,760 A		**Regional total: 127,900 A**	

*Total sweet corn acreage planted in 1991 was 719,060 A.

SOURCE: *Agricultural statistics 1992.* United States Department of Agriculture. Government Printing Office, Washington, D.C. p. 153.

TABLE 43.2 ■ Most common problem weeds in sweet corn grown in the United States.

COMMON WEEDS*	TROUBLESOME WEEDS**
Barnyardgrass	Barnyardgrass
Common lambsquarters	Canada thistle
Common ragweed	Common lambsquarters
Fall panicum	Common ragweed
Giant foxtail	Fall panicum
Giant ragweed	Giant foxtail
Large crabgrass	Morningglories
Nutsedge, yellow	Nutsedge, purple
Pennsylvania smartweed	Nutsedge, yellow
Quackgrass	Quackgrass
Redroot pigweed	Redroot pigweed
Smooth pigweed	Velvetleaf
Velvetleaf	Wild-proso millet

*Common weeds abundantly infest a significant portion of the sweet corn acreage.
**Troublesome weeds, despite weed-control efforts, are inadequately controlled and interfere with crop production.

SOURCE: D. C. Bridges, ed. 1992. *Crop losses due to weeds in the United States.* Champaign, Ill.: Weed Science Society of America.

Cyanazine (Bladex 4L and 90DF) is a triazine herbicide (Chapter 27) and a Restricted Use Pesticide because it causes serious maternal illness and birth defects in laboratory animals. Cyanazine can be applied either preplant-incorporated or preemergence to sweet corn to control annual grass and broadleaf weeds. To avoid crop injury, do not apply cyanazine postemergence to sweet corn. Rotary hoeing is recommended to incorporate preemergence applications of cyanazine if the herbicide is not moved into the weed seed germination zone (upper 2-in. soil layer) by rainfall or sprinkler irrigation within 10 days after application. Cyanazine is applied at 1.25–4.75 lb ai/A (1.25–4.75 qt Bladex 4L/A or 1.3–5.3 lb Bladex 90 DF/A), depending on soil type.

Metolachlor (Dual 8E) is an acid amide herbicide (Chapter 12). It is used to control annual grasses and certain broadleaf weeds and also to control or suppress yellow nutsedge. Metolachlor can be applied either preplant-surface, preplant-incorporated, or preemergence to sweet corn. For a preplant-surface treatment, apply

TABLE 43.3 ■ Principal herbicides used in sweet corn in 1991, expressed as the average percentage of the total acres grown in each region.

HERBICIDE**	REGIONS*					
	1	2	3	4	5	6
Alachlor	39	51	31	25	20	10
Atrazine	86	35	46	38	5	90
Butylate	18	8	0	3	0	30
Cyanazine	12	35	28	10	0	0
EPTC	7	2	5	10	20	0
Metolachlor	25	33	13	23	10	50

*Regions: 1 = Northeast; 2 = Corn Belt; 3 = Lake States; 4 = Pacific Coast; 5 = Mountain; 6 = Southeast.
**A specific herbicide may not be used in every state within a region.

SOURCE: D. C. Bridges, ed. 1992. *Crop losses due to weeds in the United States.* Champaign, Ill.: Weed Science Society of America.

two-thirds of the recommended rate 30–45 days before planting in a split application, with the remainder at planting. Metolachlor is applied at a broadcast rate of 1.5–3 lb ai/A (1.5–3 pt product/A).

■ SELECTED REFERENCES

Agricultural chemical usage: Vegetables, 1992 summary. United States Department of Agriculture, National Agricultural Statistics Service, Washington, D.C. pp. 87–109.

Agricultural statistics 1992. United States Department of Agriculture. Government Printing Office, Washington, D.C. p. 153.

Bridges, D. C., ed. 1992. *Crop losses due to weeds in the United States.* Champaign, Ill.: Weed Science Society of America.

Herbicide product labels.

Meister, R. T., ed. 1994. *Weed control manual 1994.* Willoughby, Ohio: Meister Publishing. pp. 209–217.

44 TOMATOES

■ INTRODUCTION

In 1991, 487,460 A of tomatoes were grown in the United States (Table 44.1). Of this acreage, 27 and 73% were grown for the fresh and processing markets, respectively. California and Florida had 30 and 35% of the fresh-market tomato acreage, while Ohio and California had 5 and 88% of the processing tomato acreage.

In the United States in 1991, weeds caused an estimated loss of $64 million to the fresh-market tomato crop and $66 million to the processing crop, even though herbicides were used. Under the same conditions, but without herbicides, the estimated loss would have been $336 and $190 million, respectively.

In 1992 in Quebec and Ontario, both eastern provinces of Canada, weeds caused a 6% reduction in the fresh-market tomato crop, a loss valued at $1.5 million, and a 5% reduction in processing tomatoes, a loss valued at $3.2 million.

The principal weeds in the fresh-market tomato crop grown in California and Ohio are given in Table 44.2. In 1992, herbicides were applied to 75% of the fresh-market acreage and 90% of the processing acreage. The principal herbicides used in fresh-market and processing tomatoes grown in the United States in 1992 are given in Table 44.3.

■ INFORMATION OF PRACTICAL IMPORTANCE

Glyphosate (Roundup 4WS) is a nonfamily herbicide (Chapter 30). It is a nonselective herbicide applied preemergence to direct-seeded tomatoes to control emerged weeds. Do not apply it to transplanted tomatoes, as the plants will be killed. Glyphosate has no soil-residual activity. The broadcast rate of glyphosate is 0.25–5 lb ai/A (0.25–5 qt product/A), depending on weed susceptibility.

Metribuzin (Lexone 75DF, Sencor 75DF) is a triazine herbicide (Chapter 27). It is used to control annual broadleaf weeds and certain annual grasses. Metribuzin can be used only on tomatoes grown east of the Rocky Mountains. It can be applied postemergence as an over-the-top or directed spray to established seeded or transplanted tomatoes in at least the five- to six-leaf stage and before weeds are 1 in. tall. One or more postemergence applications can be made per season, not to exceed 1.3 lb product/A in any 35-day period per crop season. Metribuzin is applied postemergence at a rate of 0.25–0.5 lb ai/A (0.33–0.66 lb product/A).

Metribuzin can be applied preplant-incorporated only to transplanted tomatoes, not to direct-seeded tomatoes. Metribuzin is applied preplant-incorporated at 0.25–0.5 lb

TABLE 44.1 ■ Tomato (fresh and processing) acreage in the United States in 1991.

STATE	ACRES Fresh	ACRES Processing*
Alabama	2,800	
California	40,000	312,000
Florida	46,200	
Georgia	2,600	
Indiana	1,400	8,800
Maryland	2,700	
Michigan	2,700	6,700
New Jersey	4,800	2,600
New York	2,700	
North Carolina	1,600	
Ohio	3,200	17,700
Pennsylvania	4,300	2,000
South Carolina	3,700	
Tennessee	4,700	
Texas	3,300	
Virginia	3,500	
Other states		5,980
Totals:	131,680	355,780
Grand total: 487,460 A		

*Blank spaces denote no commercial acreage.

SOURCE: *Agricultural statistics 1992.* United States Department of Agriculture, Government Printing Office, Washington, D.C. p. 165.

ai/A (0.33 to 0.66 lb product/A). When metribuzin is applied preplant-incorporated, trifluralin can be added as a tank mix at 0.5–1 lb ai/A for added annual grass control; soil-incorporate 2–4 in. deep after application.

Napropamide (Devrinol 2E or 50DF) is an acid amide herbicide (Chapter 12). It is used preemergence to control annual grass and broadleaf weeds. It does not control emerged weeds. To be effective, napropamide must be incorporated into the weed seed germination zone, 1–2 in. deep. It may be applied preplant-incorporated for seeded or transplanted tomatoes. Napropamide is applied at 1–2 lb ai/A (2–4 qt Devrinol 2E/A or 2–4 lb Devrinol 50-DF/A). Devrinol may be tank-mixed with Tillam 6E (pebulate) to broaden the spectrum of weeds controlled.

Paraquat (Gramoxone Extra 2.5EC) is a bipyridinium herbicide (Chapter 15) and a Restricted Use Pesticide because it is fatal to humans if swallowed, inhaled, or skin-absorbed. It is a nonselective, contact-type, postemergence herbicide. It is not active through the soil. It is used to control annual grass and broadleaf weeds up to about

TABLE 44.2 ■ Principal weeds in fresh-market tomatoes grown in California and Ohio.

CALIFORNIA	OHIO
Common Weeds*	**Common Weeds***
Barnyardgrass	*Amaranthus* spp.
Burning nettle	Common lambsquarters
Common groundsel	Common purslane
Common purslane	Fall panicum
Hairy nightshade	*Galinsoga* spp.
Little mallow	Large crabgrass
Redroot pigweed	Ragweeds
Shepherd's-purse	Venice mallow
Troublesome Weeds*	Velvetleaf
Little mallow	**Troublesome Weeds***
Hairy nightshade	Each of the above common
Yellow nutsedge	weeds

*Common weeds abundantly infest a significant portion of the crop acreage.
**Troublesome weeds, despite control efforts, are inadequately controlled and interfere with crop production.

SOURCE: D. C. Bridges, ed. 1992. *Crop losses due to weeds in the United States.* Champaign, Ill.: Weed Science Society of America.

TABLE 44.3 ■ Herbicides used in fresh and processing tomatoes in the United States in 1992, expressed as a percentage of the acreage planted.

HERBICIDE	TYPE OF TOMATO CROP Fresh*	TYPE OF TOMATO CROP Processing*
Glyphosate	9	52
Metribuzin	33	6
Napropamide	5	29
Paraquat	45	5
Pebulate	4	17
Trifluralin	13	35

*Based on a crop of 105,100 A grown in eight principal fresh-market states.
**Based on a crop of 252,300 A with 242,000 A grown in California and an additional 6200 A grown in four other states.

SOURCE: *Agricultural chemical usage: Vegetables, 1992 summary.* United States Department of Agriculture, National Agricultural Statistics Service, Washington, D.C. pp. 231–249.

4 in. tall, and it can suppress the growth of certain perennials. It can be applied preplant or preemergence to seeded tomatoes to control emerged weeds. It may be applied postemergence to tomatoes using shields to avoid spraying the crop plants. The broadcast rate of paraquat is 0.63–0.94 lb ai/A (2–3 pt product/A).

Pebulate (Tillam 6E) is a thiocarbamate herbicide (Chapter 26). It is used to control annual grass and broadleaf weeds prior to emergence. It provides control of purple nutsedge and yellow nutsedge if existing stands are thoroughly chopped up prior to treatment to encourage sprouting of the tubers. Pebulate must be soil-incorporated immediately after application to avoid loss by volatility. Incorporation also moves the herbicide into the weed seed germination zone. It can be used preplant-incorporated or incorporated at lay-by in seeded or transplanted tomatoes. Pebulate is applied at 4–6 lb ai/A (2.6–4 qt product/A), depending on soil type. Pebulate can be tank-mixed with napropamide to broaden the spectrum of weeds controlled.

Trifluralin (Treflan 4EC) is a dinitroaniline herbicide (Chapter 17). It is used preemergence to control annual grass and broadleaf weeds. It does not control emerged weeds. For direct-seeded tomatoes, apply trifluralin at blocking or thinning as a directed spray to the soil between rows and beneath plants, and incorporate 1–2 in. deep to avoid herbicide loss by volatility. Incorporation also moves the herbicide into the weed seed germination zone. For transplant tomatoes, apply and incorporate trifluralin 1–2 in. deep before transplanting. Do not apply after transplanting. The broadcast rate of trifluralin is 0.5–1 lb ai/A (1–2 pt Treflan EC/A), depending on soil type.

■ SELECTED REFERENCES

Agricultural chemical usage: Vegetables, 1992 summary. United States Department of Agriculture, National Agricultural Statistics Service, Washington, D.C. pp. 231–249.

Agricultural statistics 1992. United States Department of Agriculture. Government Printing Office, Washington, D.C. p. 165.

Bridges, D. C., ed. 1992. *Crop losses due to weeds in the United States.* Champaign, Ill.: Weed Science Society of America.

Herbicide product labels.

Meister, R. T., ed. 1994. *Weed control manual 1994.* Willoughby, Ohio: Meister Publishing. pp. 219–221.

TURFGRASS

■ INTRODUCTION

There is an estimated 17 million A of turfgrass in the United States. For convenience, turfgrass species are categorized as either cool- or warm-season grasses. The cool-season species are more common to Canada and the northern United States. They include creeping bentgrass, Kentucky bluegrass, fine fescues, tall fescue, and perennial ryegrass. The warm-season species are common to the southern and southwestern regions of the United States. The principal warm-season grasses are bahiagrass, bermudagrass, buffalograss, carpetgrass, centipedegrass, kikuyugrass, St. Augustinegrass, and zoysiagrass.

■ WEEDS

Weeds are usually minor problems in established, well-managed turfgrass. A good turf-management program utilizes proper watering, fertilizing, mowing, aerating or verticutting, adequate drainage, and effective insect and disease control. Proper fertilization encourages the development of a dense turf that, in turn, tends to inhibit weed seed germination and seedling growth. However, even in well-managed turfgrass, weeds will occur due to the continual introduction of weed seeds by wind, birds, dogs, flooding waters, shoes and clothing, maintenance and mowing equipment, and other means.

Weed species common to turfgrass vary with geographic regions, but many are common to more than one region. Weeds common to turfgrass in the North Central region of the United States are given in Table 45.1 on the following page. Many of these weeds are also problems in other regions of the United States and Canada.

In the southern United States, large crabgrass and dallisgrass are the most prevalent summer annual grass weeds in turfgrasses, and annual bluegrass is the most serious winter annual grass. Goosegrass is the most difficult summer grass weed to control. In the Southwest, annual bluegrass and rescuegrass are the predominant winter annual grass weeds. Purple nutsedge and, to a lesser extent, yellow nutsedge are problem turf weeds in the South and Southwest. Weeds common to turfgrass in the southern United States are given in Table 45.2.

■ WEED CONTROL

Hand-weeding (pulling and digging) is the oldest method of controlling weeds. It is often the most practical method for

TABLE 45.1 ■ Problem weeds in turfgrass in the North Central region of the United States (common names).

Broadleaf	Grass
Bedstraw, catchweed	Barnyardgrass
Bellflower, creeping	Brome, smooth
Bindweed, field	Crabgrass, large
Carpetweed	Crabgrass, smooth
Chickweed, common	Fescue, tall
Chickweed, mouseear	Foxtail
Clover, white	Goosegrass
Dandelion	Quackgrass
Deadnettle	Sandbur, longspine
Henbit	
Ivy, ground	**Nutsedge**
Knotweed, prostrate	Nutsedge, yellow
Kochia	
Mallow, common	**Others**
Medic, black	Moss
Nimblewill	Wild garlic
Pigweed, prostrate	
Plantain, blackseed	
Plantain, buckhorn	
Puncturevine	
Purslane, common	
Shepherd's-purse	
Sorrel, red	
Speedwell	
Spurge, prostrate	
Thistle, Canada	
Thistle, musk	
Vervain, prostrate	
Violets	
Woodsorrel, yellow	
Yarrow, common	

SOURCES: B. N. Stougaard. 1969. *Lawn weeds and their control* (North Central Regional Publication No. 26). Lincoln, Nebr.: University of Nebraska. pp. 1–18.
L. M. Wax et al. 1981. *Weeds of the North Central States* (North Central Regional Research Publ. No. 281, Bulletin 772). Urbana, Ill.: University of Illinois at Urbana-Champaign.

TABLE 45.2 ■ Turfgrass weeds in the southern United States (common names).

Broadleaf	Grass
Betony, Florida	Annual bluegrass
Buttercup	Bahiagrass
Carpetweed	Crabgrass, large
Chickweed, common	Crabgrass, smooth
Chickweed, mouseear	Dallisgrass
Clover, white	Goosegrass
Dandelion, common	
Dichondra	**Nutsedge**
Ground ivy	Purple nutsedge
Henbit	Yellow nutsedge
Knotweed, prostrate	
Pennywort	**Others**
Plantain, broadleaf	Wild garlic
Plantain, buckhorn	Wild onion
Pusley, Florida	
Shepherd's-purse	
Speedwell	
Spurge, prostrate	
Spurge, spotted	
Spurweed	
Virginia pepperweed	
Virginia buttonwood	

SOURCES: Anonymous. 1992. *Weed control guidelines for Mississippi, 1992.* Mississippi State, Miss.: Mississippi State University. p. 119.
T. R. Murphy. 1993. *Weeds of southern turfgrasses.* Athens, Ga.: University of Georgia, Cooperative Extension Service.

plants were taller. When necessary, there are herbicides for the selective control of most problem weeds in turfgrass lawns.

■ HERBICIDES

Herbicides registered for use in established turfgrasses are applied preemergence and postemergence to the weeds. They may effectively control grass and/or broadleaf weeds.

A typical weed-control program includes a preemergence herbicide applied in either the fall or spring for annual grass control, followed by application of postemergence herbicides on an as-needed basis. In dormant, warm-season turfgrasses, a nonselective herbicide such as glyphosate or diquat may be applied postemergence to the weeds. Bermudagrass is the dominant warm-season turfgrass species. In general, herbicides that can be safely used on bermudagrass can also be used on zoysiagrass. However, special attention should be paid to the directions on herbicide labels regarding applications to centipedegrass and St. Augustinegrass, which are similar to one another in their tolerance

controlling weeds in small turf areas and individual plants in large areas. It is effective against annual and biennial weeds when care is taken to remove the crown and taproots. Perennial weeds are easiest to remove while in the seedling stage, as their underground reproductive parts (roots, rhizomes, tubers) make control difficult. Even a small portion of a dandelion's taproot, a quackgrass's rhizome, or a nutsedge's tuber left in the soil can resprout. Mowing, timed before weeds flower, prevents weeds from producing seeds. Repeated mowings may weaken and eventually kill some types of weeds. However, some weeds may flower and set seedheads below the height of the mower cutterbar following an earlier mowing when the to herbicides. Bentgrass is susceptible to many herbicides, and care should be taken to avoid using any herbicide not specifically recommended for use on this turfgrass.

Water

Rainfall or irrigation can influence the effectiveness of soil-applied turf herbicides. Weed control is enhanced by 0.5 in. of rainfall or irrigation following soon after applications of preemergence herbicides; this moves them into the zone of weed seed germination and seedling growth. Foliar-applied herbicides are often so quickly absorbed that they are not adversely affected by rainfall or irrigation occurring 1 hour after application. However, there are a few foliar-applied herbicides that may be washed from the foliage before being absorbed in toxic amounts. Water may leach some herbicides too deeply into the soil, presenting hazards to ornamental plantings with root systems extending under the turfgrass or posing a hazard to groundwater. Dicamba is an example of a herbicide that leaches readily in soils and is absorbed by the roots of trees and shrubs. Atrazine is a Restricted Use Pesticide because of its ability to leach deeply in soils and contaminate groundwater.

Preemergence

To be effective, preemergence herbicides must be applied before weed seeds germinate. Thus it is vital to know when problem weed germination will occur. For example, crabgrass seed will germinate in sandy soil at a soil temperature of 49–51°F (9–11°C), in loam soil at 50–52°F, and in heavy wet clay at 53°F, with soil temperatures taken at a soil depth of 2 in. between 7 and 8 A.M. Goosegrass germinates at a higher soil temperature, 60–65°F (15–18°C). *It is the soil temperature, not the air temperature, that is important.*

Preemergence herbicides do not control established weeds, and such weeds should be removed by hand or with postemergence herbicides. A thick layer of thatch will reduce the effectiveness of preemergence herbicides by preventing herbicide contact with the soil, as will leaves and other debris overlying the turfgrass. Be prepared for weed escapes, and hand-pull or spot-treat with a postemergence herbicide. Split applications of most preemergence herbicides usually provide better weed control than a single application. Apply one-half of the maximum recommended rate on the labeled date followed in about 60 days with the remaining one-half. Split applications tend to maintain a phytotoxic level of the herbicide(s) in the soil for control of germinating weeds over a broader span of time.

Examples of problem weeds controlled by preemergence herbicides in turfgrasses include: *Large crabgrass and dallisgrass* controlled with benefin, benefin + oryzalin, bensulide, DCPA, or pendimethalin; *annual bluegrass* controlled with benefin, benefin + oryzalin, DCPA, oxadiazon, pendimethalin, pronamide, or simazine; *goosegrass* controlled with benefin, benefin + trifluralin, dithiopyr, oryzalin, or oxadiazon; *rescuegrass* and *ryegrass* controlled with endothall; and *yellow nutsedge* controlled in warm-season turfgrass with metolachlor.

Postemergence

To be effective, most postemergence herbicides should be applied to young, actively growing weeds 1–2 in. high. The turfgrass should not be mowed 3 to 4 days prior to the postemergence application, nor 3 to 4 days after the application. It is vital that sufficient weed foliage is present at the time of application to intercept the applied herbicide, and that sufficient time is allowed for foliar absorption and translocation (if a systemic herbicide is used). In general, turfgrasses are most tolerant to postemergence herbicides when they are fully dormant or actively growing, not stressed for water, and the air temperature is less than 90°F. Rainfall or irrigation immediately after application may wash the herbicides from the treated foliage, resulting in poor weed control. However, most herbicides are foliar absorbed within about the first 30 minutes.

Examples of weeds controlled by postemergence herbicides include: *Large crabgrass, smooth crabgrass, and dallisgrass* and other grass weeds controlled in bermudagrass and zoysiagrass with MSMA (repeat applications usually necessary); *crabgrasses* controlled by dithiopyr; and *annual bluegrass* controlled by pronamide in bermudagrass. Fenoxaprop controls annual grass weeds, but not broadleaf weeds, in cool-season turfgrass.

■ INFORMATION OF PRACTICAL IMPORTANCE

Preemergence Herbicides

Atrazine (AAtrex Nine-0) is a triazine herbicide (Chapter 27) and a Restricted Use Pesticide due to ground and surface water concerns. It is used in warm-season turfgrasses to control annual grass and broadleaf weeds. It is applied at 1–2 lb ai/A (1.1–2.2 lb product/A) or 0.4 oz product/1000 ft^2.

Benefin (Balan 2.5G) is a dinitroaniline herbicide (Chapter 17). It is used in cool- and warm-season turfgrasses to control annual grass weeds. Use only on well-established turf. Benefin should not be used on bentgrass. It is applied at 1.5–3 lb ai/A (1.4–2.8 lb product/1000 ft^2). Two applications spaced 6–8 weeks apart are usually required for season-long control.

Bensulide (Bensumec 4LF) is a nonfamily herbicide (Chapter 30). It is used in warm-season turfgrasses to control annual grass (especially crabgrass) and certain broadleaf weeds in established turfgrass. For crabgrass control, apply 6.5 fl oz product/1000 ft^2, and for control of annual bluegrass and other weeds, apply 9.4 fl oz product/1000 ft^2. Do not attempt to reseed until at least 4 months after application.

DCPA (Dacthal W75) is a phthalic acid herbicide (Chapter 24). It is used in cool- and warm-season turfgrasses to control many annual broadleaf and grass weeds in established turfgrass. *DCPA may be applied postemergence for control of creeping speedwell. (Note: This postemergence treatment is a variant to DCPA's usual preemergence use.)* Do not use on dichondra or putting greens. DCPA is applied at 10.5–15 lb ai/A (14–20 lb product/A). For crabgrass control, apply 0.3 lb product per 1000 ft^2 in early spring. If necessary, make a second application at one-half this rate 2 months after first application. For annual bluegrass control, apply 0.5 lb product/1000 ft^2 before seeds germinate (late summer or early fall in northern states).

Dithiopyr (Dimension 1EC) is a nonfamily herbicide (Chapter 30). It is used in cool- and warm-season turfgrasses to control crabgrass and other susceptible annual grass and broadleaf weeds. Dithiopyr is applied at 0.25–0.5 lb ai/A (2–4 pt product/A), or 22–44 ml product/1000 ft^2. Do not make repeat or split applications during the same growing season.

Endothall (Endothal Turf Herbicide 1.46WS) is a phthalic acid herbicide (Chapter 24). It is used in cool- and warm-season turfgrasses to control a broad spectrum of annual and perennial weeds and certain winter annual grass weeds. Endothall is applied at 0.17 lb ae/A, or 1 pt product/10,000 ft^2 or 47 ml product/1000 ft^2. Do not use on putting greens or red fescue. Temporary browning of turfgrasses may occur at temperatures over 80°F (26°C).

Ethofumesate (Progress 1.5EC) is a nonfamily herbicide (Chapter 30). It is used in cool-season turfgrasses to control annual grass and broadleaf weeds. Ethofumesate is applied at 0.5–2 lb ai/A (0.33–1.33 gal product/A).

Isoxaben (Gallery 75DF) is a nonfamily herbicide (Chapter 30). It is used in cool- and warm-season turf-grasses to control a very broad spectrum of broadleaf weeds (46 species), and it also suppresses the growth of about 10 grass species in established turfgrasses. Do not use on putting greens. Isoxaben is applied at 0.5–1 lb ai/A (0.67–1.33 lb product/A) or 0.32 oz–0.64 oz product/1000 ft^2. Apply in early spring, late summer, or early fall.

Metolachlor (Pennant Liquid 8E) is an acid amide herbicide (Chapter 12). It is in used warm-season turf-grasses to control yellow nutsedge prior to emergence. Do not use on golf greens, tees, or aprons. Do not apply more than once each year. Metolachlor is applied at 4 lb ai/A (4 pt product/A), or 43 ml product/1000 ft^2.

Oxadiazon (Chipco Ronstar 2G) is a oxydiazolin herbicide. It is used on warm-season turfgrasses to control annual grass weeds. It can be used on newly sprigged bermudagrass. Oxadiazon is applied at 2–4 lb ai/A (100–200 lb product/A) or 2.3–4.6 lb product/1000 ft^2. Do not use on putting greens or tees.

Oryzalin (Surflan 4A.S.) is a dinitroaniline herbicide (Chapter 17). It is used in warm-season turfgrasses to control annual grasses and certain broadleaf weeds. Oryzalin is applied at 1.5–2 lb ai/A (1.5–2 qt product/A) or 1–1.5 fl oz/1000 ft^2.

Pendimethalin (Lesco Pre-M 3.3EC) is a dinitroaniline herbicide (Chapter 17). It is used in cool- and warm-season turfgrasses to control annual grasses and chickweeds, henbit, prostrate spurge, and yellow woodsorrel (oxalis). Pendimethalin is applied at 1.5–3 lb ai/A or 39–78 ml product/1000 ft^2. For season-long control, make a second application 6–8 weeks after first application. This product is also available as a 60% water-dispersible granule.

Prodiamine (Barricade 65WG) is a dinitroaniline herbicide (Chapter 17). It is used in warm-season turfgrasses. Prodiamine provides residual control for many annual grass and broadleaf weeds. Do not apply to golf putting greens, tees, or to areas where dichondra, colonial or velvet bentgrasses, and annual bluegrass are desired species. Prodiamine is applied at 0.5–1.5 lb ai/A (0.77–2.3 lb product/A) or 0.3– 0.9 oz product/1000 ft^2.

Pronamide (Kerb 50W) is an acid amide herbicide (Chapter 12) and a Restricted Use Pesticide because it produces tumors in laboratory animals. Pronamide is used on bermudagrass and warm-season turfgrass. *It provides excellent control of annual bluegrass.* It can be used on actively growing or dormant bermudagrass (common, hybrids, improved strains). Pronamide is applied at 0.5–1.0 lb ai/A (1–2 lb product/A) or 0.2–0.38 lb product/1000 ft^2. Use the higher rate for longer residual control.

Siduron (Tupersan 50WP) is a substituted urea herbicide (Chapter 29). It is used on cool-season turfgrasses to control annual grasses such as barnyardgrass, foxtails, and large and smooth crabgrasses. *Siduron is unique in that it does not interfere with the germination of newly seeded cool-season perennial grasses, including bluegrass.* Siduron is applied at 8–12 lb ai/A (16–32 lb product/A) or 0.36–0.7 lb product/1000 ft².

Benefin and *oryzalin* are formulated as a premix (Dow-Elanco's XL 2G) that contains 1% w/w of each active ingredient. Both herbicides are dinitroanilines (Chapter 17). XL 2G is applied at 2–3 lb ai/A (100–150 lb product/A) or 2.5–3.5 lb product/1000 ft² in the spring for control of summer annual grass and broadleaf weeds and in the fall for control of annual bluegrass and winter annual broadleaf weeds in established warm-season turfgrasses and tall fescue. Do not use on dichondra or bentgrass.

Postemergence Herbicides

Asulam (Asulox 3.34 WS) is a nonfamily herbicide (Chapter 30). It is used in warm-season turfgrasses (bermudagrass and St. Augustinegrass) to control annual grass weeds such as bullgrass, crabgrass, goosegrass, and sandbur. Asulam is applied at 2 lb ai/A, or 54 ml product/1000 ft². Do not add surfactant to the spray mixture.

Bentazon (Basagran 4WS) is a nonfamily herbicide (Chapter 30). It is used in cool- and warm-season turfgrasses to control annual broadleaf weeds and yellow nutsedge. It does not control grasses or purple nutsedge. Follow-up applications may be necessary for complete yellow nutsedge control. Bentazon is applied at 1–2 lb ai/A (2–4 pt product/A) or 22 ml product/1000 ft².

Dicamba (Banvel 4WS) is a growth regulator–type herbicide (Chapter 19). It is used in cool- and warm-season turfgrasses to control a broad spectrum of annual and perennial broadleaf weeds. It does not control grasses or sedges. Dicamba is applied at 0.25–1 lb ai/A (0.5–2 pt product/A, or 5–20 ml product/1000 ft²). Rates greater than 1 pt product/A may cause noticeable stunting or discoloration of sensitive grasses such as bentgrass, buffalograss, carpetgrass, and St. Augustinegrass. Dicamba is commonly used in mixtures with the phenoxy herbicides 2,4-D, 2,4-DP, MCPA, and mecoprop (MCPP).

Diclofop (Illoxan 3EC) is an aryloxyphenoxypropionate herbicide (Chapter 13) and a Restricted Use Pesticide because it is toxic to fish. It is used on established bermudagrass to control emerged annual grass weeds. Diclofop provides excellent control of small, emerged goosegrass. It does not control broadleaf weeds. Diclofop is registered for use on golf courses in the southern United States. Diclofop is applied at 0.75–1.5 lb ai/A, or 22–44 ml product/1000 ft². Refer to the product label for state restrictions.

2,4-D, 2,4-DP, MCPA, and *mecoprop* (MCPP) are growth regulator–type herbicides. They are applied in cool- and warm-season turfgrasses. Applied alone and in combinations with one another and/or with dicamba, they control a broad spectrum of broadleaf weeds. They do not control grass weeds. Refer to the labels on such products as Formula 40, Weedar 64, Chipco Weedone DPC Amine, Trimec (several other premix products), Turflon D and Turflon II Amine for specific application rates. When using these herbicides, be very careful to avoid spray drift to desired broadleaf plants, and use products formulated as amine salts, rather than esters, to minimize herbicide vapor drift. Their application rate varies from about 1 to 1.5 lb ai/A.

DSMA (Drexel DSMA Liquid 3.6L) is a methanearsonate herbicide (Chapter 21). It is used on cool- and warm-season turfgrasses to control annual grasses and certain broadleaf weeds. Bluegrass, bermudagrass, and zoysiagrass are quite tolerant to DSMA. Do not apply DSMA to carpetgrass, centipedegrass, or St. Augustinegrass. DSMA is applied at 3.6 lb ai/A (1 gal product/A).

Ethofumesate (Progress 1.5EC) is a nonfamily herbicide (Chapter 30). It is used in warm-season turfgrasses to control annual grass and broadleaf weeds and purple nutsedge and yellow nutsedge. Ethofumesate is applied at 0.5–2 lb ai/A (0.33–1.33 gal product/A).

Fenoxaprop (Acclaim 1EC) is an aryloxyphenoxypropionate herbicide (Chapter 13). It is used in cool-season turfgrasses to control annual grass weeds and common bermudagrass. It does not control broadleaf weeds or nutsedges. Fenoxaprop is applied at 0.08–0.35 lb ai/A or 7–30 ml product/1000 ft². It can also be used on zoysiagrass, a warm-season turfgrass.

Glyphosate (Jury, Rattler, Roundup 4L) is a nonfamily herbicide (Chapter 30). It is a broad-spectrum, nonselective herbicide. Glyphosate is applied in dormant bermudagrass to control emerged annual and perennial grass and broadleaf weeds. Glyphosate is applied at 0.38 lb ai/A (0.75 pt product/A) or 8 ml product/1000 ft². Apply to dormant bermudagrass prior to spring greenup to control winter annual weeds and tall fescue.

Imazaquin (Image 1.5LC) is an imidazolinone herbicide (Chapter 20). It is used on warm-season turfgrasses to control chickweeds, henbit, and certain other broadleaf weeds. Imazaquin also controls field sandbur, purple nutsedge, yellow nutsedge, wild garlic, and wild onion. Imazaquin is applied at 0.25–0.5 lb ai/A or 15–30 mL product/1000 ft². Add 1 qt of a nonionic surfactant to each 100 gal of spray mixture.

MSMA (Bueno 6 + surfactant, others) is a methanearsonate herbicide (Chapter 21). It is used on established bermudagrass to control many small, actively growing grass and broadleaf weeds. It is effective on crabgrass, dallisgrass, sandbur, bahiagrass, chickweed, woodsorrel, and others. MSMA controls emerged purple and yellow nutsedges, but repeat applications are required when regrowth reaches about 4 in. high. MSMA is applied at 2 lb ai/A or 1 fl oz product/1000 ft^2 or 30 ml product/1000 ft^2. Apply to bermudagrass and zoysiagrass, but not to carpetgrass, centipedegrass, St. Augustinegrass, or dichondra.

Pronamide (Kerb 50WP) is an acid amide herbicide (Chapter 12) and a Restricted Use Pesticide because it produces tumors in laboratory animals. Pronamide is used on cool- and warm-season turfgrasses. *It provides excellent control of annual bluegrass at any stage of growth, from the seedling stage through tillering, heading, and seed formation.* Pronamide acts slowly on annual bluegrass; at first, the plants become dark green and then turn yellow and die over a period of 3–5 weeks. Pronamide is applied at 0.75–1.0 lb ai/A (1.5–2 lb product/A) or 0.6–0.75 oz product/1000 ft^2 to annual bluegrass in the early tillering to heading stage. When annual bluegrass is in the seed-forming stage, pronamide is applied at 1.0–1.5 lb ai/A (2–3 lb product/A) or 0.75–1.1 oz product/1000 ft^2.

Sethoxydim (Vantage) is a cyclohexanedione herbicide (Chapter 16). It is used postemergence to control actively growing grass weeds, such as bahiagrass, goosegrass, large crabgrass, and smooth crabgrass. It is used on centipedegrass (newly planted or established) and fine fescues (seedling or established) such as Chewings, creeping red, and hard fescue. Do not use on tall fescue or other turfgrass species. *Note:* The use of sethoxydim for selective grass control in centipedegrass and fine fescues is in contrast to its principal use: the control of annual and perennial grasses in broadleaf crops. Sethoxydim is applied in turfgrass at 3 to 4.5 oz ai/A (1.5–2.25 pt product/A or 16.3–24.5 ml product/1000 ft^2).

Triclopyr and *clopyralid* are formulated as a premix (Confront) that contains a total of 3 lb ae/gal. Both active ingredients are growth regulator–type herbicides (Chapter 19). Confront is just one of a number of premix formulations available for use on turfgrasses. Confront is used in cool- and warm-season turfgrasses. It controls a broad spectrum of annual and perennial broadleaf weeds. It does not control grasses. The premix is applied at 0.38–0.75 lb ae/A (1–2 pt product/A) or 11–22 ml product/1000 ft^2.

■ SELECTED REFERENCES

Anonymous. 1981. *Lawn weed control.* Pullman, Wash.: Washington State University, Cooperative Extension Service.

Anonymous. 1992. *Weed control guidelines for Mississippi, 1992.* Mississippi State, Miss.: Mississippi State University, Cooperative Extension Service.

Bingham, S. W. 1994. Weed control in cool-season turfgrass. *Landscape Management* 33(3):36, 38.

Herbicide product labels.

Landry, G., and T. Murphy. 1992. Weed control strategies for sports fields. *Landscape Management* 31(12):21–22.

McCarty, B. 1993. Winter weed control. *Landscape Management* 32(12):26, 35, 38.

Murphy, T. R. 1993. *Weeds of southern turfgrasses.* Athens, Ga.: University of Georgia, Cooperative Extension Service.

Murphy, T. R. 1994. Postemergence weed control in warm-season turfgrasses. *Landscape Management* 33(3):42, 43, 46.

Murphy, T. R. 1994. Turfgrass. In *Weed control manual 1994.* R. T. Meister, ed. Willoughby, Ohio: Meister Publishing. Pages 252–261.

Stougaard, B. N. 1989. *Lawn weeds and their control.* (North Central Regional Publ. No. 26.). Lincoln, Nebr.: University of Nebraska.

Wax, L. M., R. S. Fawcett, and D. Isely. 1981. *Weeds of the North Central States* (North Central Regional Research Publ. No. 281, Bulletin 772). Urbana, Ill.: University of Illinois at Urbana-Champaign.

46 PASTURES AND RANGELANDS

■ INTRODUCTION

There are approximately 988 million A of land devoted to pastures and range in the United States. *Grasses are the most important forage of pastures and rangelands. Range forbs* are broadleaved (dicot), flowering plants that do not develop woody stems aboveground. Palatable forbs are high in nutritive value, and the young leaves and flowers are especially favored by sheep and deer. The majority of forbs may not be eaten by livestock because they have spines, barbs, hairs, or secretions of undesirable odor or flavor. For this reason, forbs often increase in abundance on heavily grazed ranges, and they are also common on burned-over lands where competition from taller grasses, shrubs and trees has been removed. Control of weedy forbs becomes a major objective on some ranges. *Woody plants* are the shrubs and trees. Some species of woody plants provide feed and cover for wildlife and livestock. Their palatable twigs, leaves, flowers, and fruits are known as *browse*.

The objectives of weed control in pastures and rangelands are the control of plant species that (1) are poisonous to livestock, especially cattle and sheep; (2) reduce forage production; and (3) inhibit livestock movement. Methods of weed control in pastures and rangelands include proven management practices (grazing programs, cultivation, fertilizing, seeding, water management), mechanical (grubbing, bulldozing, chaining), controlled burning, and chemical, used alone or in various combinations.

■ POISONOUS PLANTS

Poisonous plants are largely a problem in the western rangelands of the United States and Canada, where natural vegetation has not been disturbed by cultivation. However, some poisonous plants continue to cause livestock losses in the eastern United States. With few exceptions, poisonous plants are not climax species, but invading species that increase with heavy grazing. An increase in poisonous plant populations is indicative of overgrazing and poor range conditions. *Grazing abuse on rangelands is the most significant cause of livestock death losses from poisonous plants.*

Species, breeds, and individual grazing animals differ in susceptibility to plant poisoning. Larkspur is a major threat to cattle but not to sheep. Conversely, losses of sheep grazing on halogeton-infested rangelands are sometimes very great, while cattle often use such rangelands without

TABLE 46.1 ■ Selected list of plants poisonous to livestock.

COMMON NAME	SCIENTIFIC NAME
Arrowgrass	*Triglochin maritima*
Brachenfern	*Pteridium aquilinum* var. *pubescens*
Chokecherry, black	*Prunus virginiana* var. *melanocarpa*
Chokecherry, western	*Prunus virginiana* var. *demissa*
Copperweed	*Oxytenia acerosa*
Death camas	*Zigadenus* spp.
False-hellebore, California	*Veratrum californicum*
Greasewood	*Sarcobatus vermiculatus*
Groundsel, Riddell	*Senecio riddellii*
Groundsel, threeleaf	*Senecio longilobus*
Halogeton	*Halogeton glomeratus*
Hemlock, poison	*Conium maculatum*
Hemlock, water	*Cicuta douglasii*
Hemp, dogbane	*Apocynum cannabinum*
Larkspur, low	*Delphinium nuttallianum*
Larkspur, tall	*Delphinium barbeyi*
Locoweed	*Astragalus* spp.
Locoweed, white	*Oxytropis lambertii*
Lupine	*Lupinus* spp.
Milkvetch, timber	*Astragalus miser*
Milkweed, labriform	*Asclepias labriformis*
Milkweed, western whorled	*Asclepias subverticillata*
Oak, scrub live	*Quercus geminata*
Oak, shinnery	*Quercus harvardii*
Parsley, spring	*Cymopterus watsonii*
Pinque	*Hymenoxys richardsonii* var. *floribunda*
Ragwort, tansy	*Senecio jacobaea*
Rubberweed, bitter	*Hymenoxys odorata*
Snakeweed, broom	*Gutierrezia sarothrae*
Sneezeweed	*Helenium* spp.
Sneezeweed, orange	*Helenium hoopesii*
St. Johnswort, common	*Hypericum perforatum*

apparent ill effects even though they too can be poisoned by consuming it.

A poisonous plant is one that contains naturally produced toxic compounds (chemicals) that, when consumed, cause adverse biochemical and physiological changes in livestock. More than 200 range plants indigenous to the grazing lands of the United States produce compounds toxic to livestock. In addition to indigenous poisonous plants, there are numerous introduced plants that are

poisonous to livestock. A list of some plants poisonous to livestock in the western United States is given in Table 46.1.

Poisonous plants, most of which are forbs, are one of the most important economic impediments to profitable livestock production. They take their toll through death of livestock, abortions, birth defects, photosensitization, chronic illnesses, and general debilitation. Each year, poisonous plants kill 3–5% of the cattle, sheep, and horses grazing on rangelands. In addition, their presence requires additional fences, altered grazing programs, decreased forage production and utilization, and in some cases, supplemental feeding.

Poisonous plants can be found in most plant communities on both private and public grazing lands. Some plant species are poisonous when ingested at any time of the year, while others are poisonous only during certain seasons. Some plant species contain toxic chemicals in all their parts, while others contain poisonous compounds only in certain plant parts. Unusual environmental conditions may initiate the production of poisons in some plant species. Some poisonous plants will cause signs of poisoning and/or death within minutes after ingestion (e.g., water hemlock and timber milkvetch). In other cases, the poisoning may not be apparent for 3 or 4 months.

Normally, livestock do not eat poisonous plants. *The most common reason animals eat poisonous plants is the lack of palatable forage.* Change in both the palatability and nutrients of grasses and grasslike plants can occur after they reach maturity, and this can drive animals to new, poisonous forage. Hungry livestock grazing in very early spring or on depleted range are most likely to feed on poisonous plants. Livestock losses from poisonous plants are particularly heavy if animals are driven or trailed through areas infested with poisonous plants, if they are not watered regularly, and if they graze infested areas at times when poisonous plants are most dangerous.

In general, poisonous plants must be consumed in relatively large quantities to be toxic. This enables the development of vegetation- and livestock-management programs that make productive use of rangelands that support poisonous plant species.

■ CHEMICAL WEED CONTROL

Herbicides are widely used in pasture and range management to suppress and kill growth of unwanted vegetation, thereby opening the way for improved growth of desired forage species. However, care must be taken to remove livestock from herbicide-treated pastures or rangelands for

TABLE 46.2 ■ Herbicides registered in 1994 for use in pastures and rangelands.

Clopyralid	Metsulfuron
2,4-D	Paraquat
Dicamba	Picloram
Dicamba + 2,4-D	Tebuthiuron
Diuron	Triclopyr
Glyphosate	Triclopyr + 2,4-D
MCPA ester	

SOURCE: R. T. Meister, ed. 1994. *Weed control manual 1994.* Willoughby, Ohio: Meister Publishing. pp. 281–284.

at least 30 days after treatment, as some herbicides (e.g., 2,4-D), are known to alter the chemical composition of some plants or the palatability of other plants not normally eaten, which , of course, poses a potential poisoning hazard to livestock if ingested. In 1984, the EPA withdrew the registrations for 2,4,5-T and silvex, two highly effective herbicides that controlled many weedy species common to pastures and rangelands, due to the presence of dioxin (a carcinogen) in their formulations. Herbicides registered for use in pastures and rangelands in 1994 are shown in Table 46.2. Examples of chemical control of selected pasture and rangeland weeds follow.

■ BROOM SNAKEWEED

Broom snakeweed (*Gutierrezia sarothrae*) infests an estimated 156 million A of rangeland in 11 states of the western United States, with 40 million A of this infested acreage in New Mexico. The economic impact of broom snakeweed in eastern New Mexico and west Texas has been estimated at approximately $40 million. Broom snakeweed poisoning usually occurs during winter and late spring when available forage is low, forcing livestock to consume large amounts of the plant. Broom snakeweed poisoning can result in livestock abortions and death.

Range-burning kills broom snakeweed, but it is considered a high-risk practice in the southwestern United States because of the low and erratic precipitation. Broom snakeweed is temporarily suppressed by shredding or mowing. In New Mexico, herbicides are more widely used to control this weed than other methods.

Applied as a broadcast foliar spray in October or November, metsulfuron applied at 0.25 lb ai/A, or picloram at 0.5 lb ae/A, killed 95% of more of the broom snakeweed plants. Applied in April, these same treatments controlled less than 40% of the plants.

■ COMMON GOLDENWEED

Common goldenweed (*Isocoma coronopifolia*) occurs only in the Rio Grande Plains of southern Texas and northeastern New Mexico. Infestations are especially severe in areas that have been mechanically cleared of brush and seeded to grasses, especially buffelgrass (*Pennisetum ciliare*).

Applied in March, a broadcast foliar spray of tebuthiuron at 2 lb ai/A, or a 5–10% pelleted formulation of picloram at 2 lb ae/A, provided excellent control of common goldenweed, with 97% control after 20 months and 85% control after 30 months.

Burning in February killed 33% of the mature common goldenweed plants, but burning in February *plus* tebuthiuron at 1 lb ai/A or pelleted picloram at 1 lb ae/A applied in March, several weeks after burning, provided excellent control of common goldenweed, equal to that with either herbicide applied alone at twice the rate but without the burning pretreatment.

An application of 2,4-D granules at 2 lb ae/A was not effective without the burning pretreatment. However, applied in March several weeks after a burning pretreatment, this same rate of granular 2,4-D reduced the live canopy cover of common goldenweed by 87% 30 months after treatment. Broadcast sprays of 2,4-D, picloram, and triclopyr have been effective at rates as low as 0.5 lb ae/A during years with above-average spring precipitation.

■ LARKSPURS

Larkspur species kill more cattle on mountain rangelands in the western United States and Canada than any other plant or disease. Larkspur species are palatable to cattle and are readily grazed even in the presence of desirable forage plants. Losses occur when cattle graze in dense patches of larkspur and consume large amounts in a short time.

Western larkspur species have been categorized into two broad groups: the "low larkspurs" and the "tall larkspurs." The low larkspurs generally have a maximum height of 3 ft, whereas the tall larkspurs are generally 5–7 ft tall. The terms also serve to describe the elevations where they generally cause problems for stockmen.

Most losses to low larkspurs occur at lower elevations where moisture is severely limited during the summer months, usually in the sagebrush zone on the foothill ranges used for spring grazing. The low larkspurs are frequently among the first plants to produce green herbage in the spring, sometimes appearing before the snow has melted. It is during this period, when livestock are seeking new green forage, that losses to low larkspurs are most likely to occur.

Losses to tall larkspurs occur at high elevations, from the low montane zone up into the subalpine areas. Tall larkspurs grow at high elevations on deep soils where moisture is readily available over most of the growing season. They are frequently the dominant forb, whether under dense tree canopies, in open meadows, or in the various transitional communities. They are climax species and tend to increase with improving range conditions. They remain green and palatable to livestock throughout the growing season.

Tall larkspur (*Delphinium barbeyi*) is the dominant larkspur species under aspen and in the subalpine vegetation zone in the southern Rocky Mountains from Colorado and central Utah southward. Duncecap larkspur (*Delphinium occidentale*) is the dominant species in the tall forb community in deep soils of the mountain big sagebrush and aspen zones in the Great Basin and northern Rocky Mountains.

Duncecap larkspur is twice as susceptible to glyphosate, compared to tall larkspur. Glyphosate, applied as a broadcast foliar spray at 1 lb ae/A, killed more than 95% of the duncecap larkspur plants, while a rate of 2 lb ae/A was required to kill 95% of the tall larkspur plants. The glyphosate treatment was nonselective, and all other vegetation in the treated area was also killed. Unacceptable levels of bare soil were still present 3 years after the glyphosate treatment.

To avoid total vegetation kill, glyphosate may be applied as a spot treatment to individual plants, using a spray (2% aqueous/product solution), or hand-held wiper (33% aqueous/product solution). Both methods were equally effective, but the sprayer treatment was more rapid and easier to apply. To ensure kill, glyphosate must be applied to the entire plant. When applied to only half of the larkspur stems, glyphosate killed only 40% of the treated plants. Glyphosate was most effective when applied prior to the flower stage.

Applied as a broadcast foliar spray, picloram was equally effective in controlling duncecap larkspur over all growth stages and tall larkspur in the bud and flower stage when applied at 1 or 2 lb ae/A. The lower rate killed about 75% of the plants, the higher rate 95%.

The two larkspur species responded differently to a broadcast foliar spray of triclopyr applied at 4 lb ae/A. Duncecap larkspur was most susceptible in the vegetative stage, with more than 85% of the plants killed. Tall larkspur was most susceptible in the flower stage, with about the same degree of control.

Applied as a broadcast foliar spray, metsulfuron was most effective when applied in the vegetative stage, rather than in the bud or flower stage, to either duncecap larkspur or tall larkspur. However, duncecap larkspur was much more susceptible to metsulfuron than was tall larkspur, with 95% of the duncecap larkspur plants killed by 0.56 oz ai/A, while 2.2 oz ai/A was needed to kill the same percentage of tall larkspur.

■ LEAFY SPURGE

Leafy spurge is a widely established perennial weed in the northwestern United States and western Canada. It is a particularly serious problem due to the speed with which it spreads and the difficulty of controlling it with available herbicides. Leafy spurge infestations can reduce livestock carrying capacity by as much as 75%. Cattle avoid grazing in areas with even a 10% infestation.

In North Dakota in 1962, an estimated 200,000 A were infested with leafy spurge; in 1973, 423,000 A; in 1982, 860,000 A; and in 1987, 1.2 million A. From 1985 to 1987, the total expenditure for leafy spurge control in North Dakota exceeded $1 million per year. In 1992, leafy spurge infestations reduced the annual net income of North Dakota ranchers by nearly $9 million, and the regional impacts were about $75 million in reduced business activity for all sectors.

Picloram applied as a single broadcast foliar spray at 2 lb ae/A resulted in 95–100% control of leafy spurge, and it maintained 90% control for a period of 3–4 years before retreatment was necessary. However, the cost of the picloram treatment was more than $80/A, far exceeding the value of the increased production of forage. Although reduced rates of picloram applied alone are less expensive, control of leafy spurge is greatly reduced. Single applications of picloram at 0.25, 0.5, 0.8, and 1.0 lb ae/A respectively averaged 0, 5, 22, and 61% control of leafy spurge 3 years after treatment. A minimum of 1.5 lb ae/A of picloram was required to maintain effective leafy spurge control and canopy reductions for 3 years without repetitive treatments. The use of low rates of picloram in tank mix with 2,4-D applied annually have been economical.

■ LOCOWEEDS

Locoweeds are found all over the western United States, and locoweed poisoning of livestock is one of the most widespread poisonous plant problems of the western United States. Locoweeds are common to the western foothills and desert regions. The more poisonous species on western rangelands include the locoweeds named white, purple, blue, Bigbend, and western. Locoweed flowers resemble those of sweetpea, and the blossoms may be white, purple, blue, or yellow.

White locoweed (*Oxytropis lambertii*) grows from Montana and North Dakota south to Arizona, New Mexico, and Texas. Purple locoweed, or "woolly loco," (*Astragalus mollissimus*) is found from southwestern South Dakota south to New Mexico and Texas. It is one of the most common species and most harmful to animals. Blue locoweed grows from eastern Washington to California, east to Colorado. Bigbend locoweed is native to the Big Bend region of Texas, and it also occurs in southern New Mexico. Western locoweed grows in Arizona and in the Big Bend region of Texas.

Locoweeds are poisonous in all their parts and at all stages of growth. They are dangerous throughout the year, even when matured and dried. They lose very little toxicity, even after 3 years of storage. The poisonous substance in locoweeds damages the optic nerve of animals and, in cows and ewes with acute poisoning, abortion and skeletal malformations frequently occur.

Grubbing was the most common and effective method of control before development of the phenoxy herbicides. With the herbicides 2,4,5-T and silvex now unavailable, clopyralid is the most effective herbicide for control of white locoweed. Clopyralid at 0.25 or 0.5 lb ae/A, dicamba at 0.5 lb ae/A, and mixtures of 2,4-D at 1.0 lb ae/A + clopyralid *or* picloram, each at 0.25 lb ae/A, killed all white locoweed plants.

■ MESQUITE

Honey mesquite (*Prosopis glandulosa*) is a spiny deciduous shrub or small tree. It is a native plant that is more commonly found on dry rangelands in the southwestern United States. It is considered noxious on rangeland because of its successful competitiveness for soil moisture and its reproductive capacity. It resprouts if complete root-kill is not achieved following control practices. It competes with desirable forage species and hinders efficient livestock handling. Honey mesquite infests about 62 million A of rangeland in Texas.

Honey mesquite has generally been controlled by mechanical means, such as bulldozing or chaining, and with herbicides. However, honey mesquite populations continue to increase even on areas once "successfully treated" either by bulldozing or with herbicides. Two areas treated in southern New Mexico in 1959–1960 with either bulldozing or the herbicide fenuron, and with original kills of nearly 100%, had an average density of 153 honey mesquite plants/A by 1988. *Clopyralid is reportedly the most effective herbicide available for honey mesquite control.* It is reported to be more effective in controlling honey mes-

quite than 2,4,5-T, triclopyr, or tank mixes of 2,4,5-T + picloram, or triclopyr + picloram. Clopyralid, applied as a broadcast foliar spray in late May to mid-June at 1 lb ae/A, killed 87% of the honey mesquite plants 3–6 ft tall, evaluated 1 year after treatment. Tank mixes of clopyralid at 0.5 lb ae/A + picloram *or* triclopyr, each at 0.5 lb ae/A, were as effective as clopyralid alone at 1 lb ae/A.

Clopyralid is less effective in the control of some other weed and brush species. Therefore, it may be appropriate to control mixed stands of range weeds with combinations of clopyralid + picloram, triclopyr, or dicamba. Applied as a broadcast foliar spray, a tank mix of clopyralid + triclopyr, each at 0.3 lb ae/A, was synergistic, killing 87% of the honey mesquite plants. However, unfavorable environmental conditions and plant growth prior to treatment can reduce the effectiveness of the treatment.

■ SAGEBRUSH

Big sagebrush (*Artemisia tridentata*) is an aromatic evergreen growing on an estimated 143 million A in the western United States. Of this acreage, 5.4 million A are in New Mexico. Big sagebrush begins root growth about 1 month before vegetative growth begins in the spring and before growth of associated herbaceous species. Early fibrous root growth helps the plant to cope with a limited soil-water supply and enables the efficient extraction of soil moisture. Big sagebrush stands frequently become dense enough to reduce forage production and inhibit livestock movement.

Methods for controlling big sagebrush include controlled burning, mechanical control (bulldozing and chaining), and herbicide treatments. Big sagebrush is not difficult to control as the plant does not resprout from crown or root buds after control measures have been applied.

Ground application of pelleted tebuthiuron at 0.5 lb ai/A in January or May reduced big sagebrush density by an average of 84 and 92%, respectively, evaluated 24 months after treatment. A 1 lb ai/A treatment of tebuthiuron, similarly applied, killed 94 and 97% of the big sagebrush plants, respectively. Desirable forage grasses, such as galleta, blue grama, and crested wheatgrass, generally increased in the tebuthiuron-treated areas, compared to untreated areas.

Clopyralid, applied as a foliar spray in June at 1 and 2 lb ae/A, killed 81 and 92% of the plants, respectively. Clopyralid and 2,4-D, each applied at 2 lb ae/A, were equally effective in control of big sagebrush. However, when used to control big sagebrush in a mixed stand of antelope bitterbrush and Saskatoon serviceberry, both desired browse plants, the 2 lb ae/A 2,4-D treatment killed 84 and

96% of the desired species, respectively. In comparison, the 2 lb ae/A clopyralid treatment killed less than 10% of the desired species, making clopyralid an effective alternative to 2,4-D for control of big sagebrush when antelope bitterbrush and Saskatoon serviceberry are present.

■ YAUPON

Yaupon (*Ilex vomitoria*) is a root-sprouting evergreen plant found throughout the southeastern United States. It infests about 2.7 million A of Texas rangelands. Yaupon often forms impenetrable thickets that reduce forage production and hinder movement of livestock and equipment. Mechanical methods of control, such as mowing, bulldozing, or chaining, and controlled burning are ineffective in killing yaupon.

Foliar sprays of picloram at 4 lb ae/A, ground application of pelleted tebuthiuron at 4 lb ai/A, or boluses (large pellets) of hexazinone at 4 lb ai/A reduced the yaupon canopy by 87–99% and killed 68–93% of the plants 1 year after treatment. The treatments reduced yaupon canopy sufficiently to allow an increase in understory grass and forb vegetation.

■ WINTER FORBS

The most cost effective herbicide for controlling winter forbs is 2,4-D. Dicamba or picloram, each at 0.25 lb ae/A, are usually effective, but they may not be economical.

■ SELECTED REFERENCES

Anonymous. 1968. *Twenty-two plants poisonous to livestock in the western United States* (Agriculture Information Bulletin No. 327). Washington, D.C.: U.S. Dept. of Agriculture.

Bailey, E. M., Jr. 1978. Physiologic responses of livestock to toxic plants. *J. Range Manage.* 31:343–347.

Bovey, R. W., and R. E. Meyer. 1985. Herbicide mixtures for control of honey mesquite (*Prosopis glandulosa*). *Weed Sci.* 33:349–352.

Bovey, R. W., et al. 1987. Influence of adjuvants on the deposition, absorption, and translocation of clopyralid in honey mesquite (*Prosopis glandulosa*). *Weed Sci.* 35:253–258.

Bovey, R. W., and S. G. Whisenant. 1992. Honey mesquite (*Prosopis glandulosa*) control by synergistic action of clopyralid:triclopyr mixtures. *Weed Sci.* 40:563–567.

Cronin, E. H., and D. B. Nielsen. 1981. Larkspurs and livestock on the rangelands of western North America. *Down to Earth* 37(3):11–16.

Cronin, E. H., D. B. Nielsen, and N. Madson. 1976. Cattle losses, tall larkspur, and their control. *J. Range Manage.* 29:364–367.

Dahl, B. E., et al. 1989. Winter forb control for increased grass yield on sandy rangeland. *J. Range Manage.* 42:400–403.

Gibbens, R. F., et al. 1992. Recent rates of mesquite establishment in the northern Chihuahuan Desert. *J. Range Manage.* 45:585–588.

Hein, D. G., and S. D. Miller. 1991. Leafy spurge (*Euphorbia esula*) response to single and repetitive picloram treatments. *Weed Technol.* 5:881–883.

Herbicide product labels.

Hickman, M. V., C. G. Messersmith, and R. G. Lym. 1990. Picloram release from leafy spurge roots. *J. Range Manage.* 43:442–445.

Jacoby, P. W., C. H. Meadors, and R. J. Ansley. 1990. Control of honey mesquite with herbicides: Influence of plant height. *J. Range Manage.* 43:33–35.

James, L. F., and A. E. Johnson. Some major plant toxicities of the western United States. *J. Range Manage.* 29:355–362.

James, L. F., D. B. Nielsen, and K. E. Panter. 1992. Impact of poisonous plants on the livestock industry. *J. Range Manage.* 45:3–8.

Johnson, H. F., and H. S. Mayeux. 1992. Viewpoint: A view on species additions and deletions and the balance of nature. *J. Range Manage.* 45:322–333.

Leistritz, F. L., F. Thompson, and J. A. Leitch. 1992. Economic impact of leafy spurge (*Euphorbia esula*) in North Dakota. *Weed Sci.* 40:275–280.

Lym, R. G., and C. G. Messersmith. 1985. Leafy spurge control with herbicides in North Dakota.: 20-year summary. *J. Range Manage.* 38:149–154.

Marten, G. C., and R. N. Andersen. 1975. Forage nutritive value and palatability of 12 common annual weeds. *Crop Sci.* 15:821–827.

Maxwell, J. F., et al. 1992. Effect of grazing, spraying, and seeding on knapweed in British Columbia. *J. Range Manage.* 45:180–182.

McDaniel, K. C. 1984. *Snakeweed control with herbicides* (Bulletin No. 706). Las Cruces, N.M.: New Mexico State University, Agriculture Experimental Station.

McDaniel, K. C., et al. 1984. *Taxonomy and ecology of perennial snakeweeds in New Mexico* (Bulletin 711). Las Cruces, N.M.: New Mexico State University, Agriculture Experimental Station.

Merrill, L. B., and J. L. Schuster. 1978. Grazing management practices affect livestock losses from poisonous plants. *J. Range Manage.* 31:351–354.

Meyer, R. E., and R. W. Bovey. 1985. Response of honey mesquite (*Prosopis glandulosa*) and understory vegetation to herbicides. *Weed Sci.* 33:537–543.

Mickelson, L. V., et al. 1990. Herbicidal control of duncecap larkspur (*Delphinium occidentale*). *Weed Sci.* 38:153–157.

Ralphs, M. H., et al. 1990. Herbicides for control of tall larkspur (*Delphinium barbeyi*). *Weed Sci.* 38:573–577.

Ralphs, M. H., et al. 1991. Selective application of glyphosate for control of larkspurs (*Delphinium* spp.). *Weed Technol.* 5:229–231.

Ralphs, M. H., J. O. Evans, and S. A. Dewey. 1992. Timing of herbicide applications for control of larkspurs (*Delphinium* spp.). *Weed Sci.* 40:264–269.

Stechman, J. V. 1986. *Common western range plants,* 3rd ed. San Luis Obispo, Calif.: California Polytechnic State University.

Valentine, K. A., and J. B. Gerard. 1968. Life-history characteristics of the creosotebush, *Larrea tridentata.* (Bulletin no. 526). Las Cruces, N.M.: New Mexico State University, Agriculture Experimental Station.

Williams, M. C. 1980. Purposefully introduced plants that have become noxious or poisonous weeds. *Weed Sci.* 28:300–305.

APPENDIX

Glossary

Herbicide Names

Conversion Tables

List of Weeds

GLOSSARY

Absorption   The surface penetration (uptake) of ions and molecules by or into any substance or organ, such as the uptake of herbicides and nutrients by plant roots, or water by a sponge.

Acid equivalent (ae).   The theoretical yield of parent acid from the active ingredient of a formulation present as a salt or ester of the parent acid.

Active ingredient (ai).   The chemical(s) in a formulated product principally responsible for the herbicidal effects, shown as the active ingredient(s) on the product's label.

Adjuvant.   Any substance in an herbicide formulation, or added to the spray mixture, that improves application characteristics or herbicidal activity. An adjuvant is generally a form of surfactant.

Adsorption.   The process whereby ions and molecules are held to the surface of soil colloids through the electrical attraction between themselves and the colloidal particles. It is the most important means by which herbicides are held in soils.

Anions.   Negatively charged ions.

Antagonism.   When the resulting effect of the interaction of two or more chemicals is less than the predicted effect based on the activity of each chemical applied separately.

Antagonism, biochemical.   When one chemical decreases the amount of herbicide that reaches the site of action by either reducing absorption or translocation or by enhancing metabolic inactivation or sequestering.

Antagonism, competitive.   When the antagonist binds at the site of action and thus prevents the herbicide from binding.

Antagonism, physiological.   When two herbicides have opposite biological effects and counteract one another.

Allelopathy.   The biochemical interaction between plants (or microorganisms). It is the direct or indirect, harmful or stimulatory, effect of one plant (including microorganisms) on another through the production of chemical compounds that escape into the environment.

At-cracking.   Herbicide applied when the soil surface begins to crack from the pressure exerted by the emerging crop seedling, such as with peanuts or soybeans.

At-emergence.   Herbicide treatment applied when the crop or specified weeds are emerging from the soil.

Band treatment.   Herbicide applied in a broadcast band of specified width located over or between the seed row.

Basal treatment.   Herbicide applied so as to encircle the stem of a plant above or at ground level to enhance absorption and minimize foliar contact. The term is mostly used to describe treatment of woody plants.

Bed.   (1) A ridge of soil formed for planting crops above adjacent furrows. (2) An area in which seedlings or plants are grown for later transplanting.

Bioassay.   Determination of the presence or concentration of an herbicide (or metabolite) by use of a susceptible indicator plant or other biological organism.

Blind cultivation.   Cultivation before the planted crop emerges.

Broadcast treatment.   Herbicide applied uniformly over the entire treated area.

Broadleaf (broadleaved) plants.   Plant species classified botanically as dicotyledous plants having two cotyledonary leaves in the seedling stage. Their true leaves have netlike or reticulate veins.

Brush control.   Control of woody plants such as brambles, sprout clumps, shrubs, trees, and vines.

Carcinogen.   A substance or agent that produces or incites cancerous growth in animals.

Carrier.   A gas, liquid, or solid substance used to dilute, propel, or suspend an herbicide during application.

Cations.   Positively charged ions.

Chemical name.   The scientific name of the active ingredient of a pesticide as derived from its chemical structure.

Chlorosis.   Loss of green color in foliage followed by yellowing or whitening of the tissue.

Common weeds.   Weeds that abundantly infest a significant portion of the crop acreage.

Compatible.   Mixable in a formulation or spray solution for application in the same carrier without undesirably altering the separate effects of the components.

Competition.   Competition is the disproportionate acquisition of one or more growth factors, such as light, water, nutrients, or space, by one plant that proves detrimental to another's growth.

Concentration. The amount of active ingredient or herbicide equivalent in a given amount of diluent (formulation) expressed as percent, pounds per gallon, milliliters per liter, and so forth.

Conservation Reserve Program (CRP). An erosion-control program designed to help owners and operators of highly erodible cropland, or other eligible cropland that poses a threat to degradation of water quality, to conserve and improve the soil and water resources of their farms or ranches. It is a volunteer program entered into for a minimum of 5–10 years. The participant is subsidized during the period of participation. The program is administered by the Natural Resources Conservation Service (formerly the Soil Conservation Service), USDA.

Contact herbicide. An herbicide that is phytotoxic on contact with plant tissue rather than as the result of translocation. Implies localized toxicity.

Conventional tillage. Moldboard plow used for primary tillage followed by other tillage implements, such as disks and other harrows, for seedbed preparation. Weed control is by cultivation and/or herbicides.

CRP. Conservation Reserve Program.

Critical weed-control period. That period during crop production in which weed-control efforts must be maintained to prevent loss in crop yield.

Critical weed-free period. That period during crop production in which weeds are most likely to reduce crop growth and yield.

Cross resistance. Refers to a weed biotype that has developed resistance (after repeated use) to one herbicide, or to members of a chemical family of herbicides with the same mode of action, and that then exhibits resistance to other chemically dissimilar herbicides with different modes of action.

Defoliant. Any substance or mixture of substances whose primary use is to cause the leaves to drop from the treated plant.

Desiccant. Any substance or mixture of substances used to accelerate the drying of plant tissue.

Diluent. Any gas, liquid, or solid material used to reduce the concentration of an active ingredient in a formulation.

Directed application. Precise application to a specified area or plant organ such as to a row or bed, to plant stems, or onto the soil surface or weeds located below the foliar canopy.

Dormancy. State of inhibited germination of seeds or growth of plant organs. A state of suspended development.

Drop-nozzle. Spray nozzle located at the lower end of an adjustable extended arm that allows the spray to be applied below the crop canopy.

Double-crop. Two crops harvested in one growing season.

Early postemergence. Herbicide applied just after crop or weed seedlings emerge from the soil.

Ecofallow. The intervening period between the harvest of one crop and the planting of the next crop.

Ecological succession. A term used to describe sequential changes in species composition of a plant population in a defined habitat.

Economic weed threshold. The point at which weed-control measures should be employed to prevent economic crop loss due to weed population density or duration of weed competition.

Emersed plant. A rooted or anchored aquatic plant adapted to grow with most of its vegetative tissue above the water surface and that does not rise or lower in response to changes in the water level.

Emulsifier. A surface-active substance that promotes the suspension of one liquid in another.

Emulsifiable concentrate (EC). A concentrated herbicide formulation containing organic solvent and adjuvants to facilitate emulsification with water.

Emulsion. The suspension of one liquid as minute globules in another liquid (e.g., oil dispersed in water).

Encapsulated formulation. Herbicide enclosed in tiny capsules of thin polyvinyl or other material intended to control the release of the herbicide.

Environment. Summation of all living (biotic) and nonliving (abiotic) factors that surround and potentially affect an organism.

EPA. Environmental Protection Agency.

Epinasty. Plant response to an applied chemical (plant growth regulator) that results in disturbance of the normal growth pattern and that is accompanied by curvature, twisting, cupping, or curling of the affected plant part (usually the petiole and/or leaves).

Fitness. A relative measure of the evolutionary advantage of a plant species that is based on the average number of seed produced by individuals of that species, subsequent seed viability, and the fertility of subsequent plants. Fitness is one of the most important factors influencing the appearance and persistance of an herbicide-resistant biotype, eventually affecting the rate at which the biotype's genes flow through a population.

Floating plant. A free-floating or anchored aquatic plant that grows with most of its vegetative tissue at or above the water surface and that rises or lowers with changes in the water level.

Foliar application. Herbicide applied directly to the leaves of the target plant(s).

Formulation. An herbicidal preparation designed for practical use, usually supplied by a manufacturer.

Gene flow. The processes that influence the maintenance of a particular genotype in a population. Dormant seed within the soil seed bank conserve genes within a plant population. An integral part of seed bank dynamics is seed longevity in the soil. Seed viability influences the length of time genes of an herbicide-resistant or herbicide-susceptible biotype can remain in the population.

Granule or granular. A dry formulation of the herbicide and other components in discrete particles generally less than 10 mm^3 in size.

Grass. Botanically, any plant in the *Gramineae* family. Grass plants are characterized by fibrous roots, rather than taproots; stems that are round, hollow (sometimes solid or semisolid), with solid, often swollen nodes; leaves that consist of blade, sheath, ligule (usually), and auricles (occasionally); flowers that are inconspicuous and borne in spikelets. The spikelet is considered "the basic unit of grass classification."

Growth regulator. A substance used to control or modify plant growth processes without appreciable phytotoxic effects at the rate applied.

Growth stages of cereal crops. (1) *Tiller* or *tillering*—when additional shoots are developing from the crown. (2) *Joint* or *jointing*—when stem internodes begin elongating. (3) *Flag leaf*—when the spike or panicle begins to enlarge within the leaf sheath of the uppermost leaf on the stem. The leaf blade is usually at right angles to the stem at this stage. (4) *Boot* or *booting*—when the upper leaf sheath swells because of the growth of the developing spike or panicle within (the flag leaf and boot stages are generally one and the same). (5) *Heading*—when spike or panicle emerges from leaf sheath. (6) *Milk stage*—when endosperm of developing seed is in a liquid (milky) state. (7) *Dough stage*—when the endosperm of the developing seed has a doughlike consistency.

Herbaceous plant. A vascular plan that does not develop persistent woody tissue aboveground.

Herbicide. A chemical used for killing or severely inhibiting normal plant growth processes.

Herbicide equivalent. The possible or theoretical yield of herbicidal compound from the active ingredient in a formulation (especially compounds not derived from acids).

Herbicide residue. The amount of herbicide persisting in the soil after its mission has been accomplished, even though it may not be present in phytotoxic concentrations or be immediately available for plant absorption. Decomposition by-products of the herbicide may be considered residues of the herbicide, and they may or may not pose a problem to subsequent land use.

Humus. The amorphous (noncrystalline) organic colloidal portion of soils derived from the breakdown of organic soil matter by microorganisms.

Incompatible formulations. Two or more pesticide formulations (products) that are incapable of being mixed or used together.

Inorganic soil colloids. Of the mineral (inorganic) entities of soils, only the clay particles are colloidal in size, and even some clay particles are too large to fit this classification. In contrast to the amorphous structure of organic colloids, the inorganic colloids are crystalline, having definite size and shape.

Interference. An all-inclusive term that denotes all the effects that one plant might impose upon another, such as competition, allelopathy, parasitism, and indirect (usually unknown) effects, without referring to any one effect in particular.

Invert emulsion. The suspension of minute water droplets in a continuous oil phase.

Ions. An atom or group of atoms that carries a positive (cation) or negative (anion) electric charge as a result of having lost or gained one or more electrons.

Late postemergence. Applied after the specified crop or weeds are well established.

Lay-by. Last normal cultivation.

Lay-by application. Herbicide applied with or after the last cultivation of the crop.

LD$_{50}$. Abbreviation for median lethal dose of toxicant that produces a 50% kill of the organism being tested. It is usually expressed as milligrams per kilogram (mg/kg) of body weight.

Leaching. Downward movement of an herbicide in the soil as influenced by percolating water.

Limited-tillage. Encompasses a wide range of practices from no-till to one postseeding cultivation in a conventional clean-tillage system. Its significant impact has been substitution of the chisel plow and sweeps for the moldboard plow.

Mulch-till. The soil is worked prior to planting using tillage tools such as chisels, cultivators, disks, and sweeps. The stubble of the preceding crop is only partially buried.

Mutagen. A compound having the property to induce mutations.

Natural Resources Conservation Service. Formerly, Soil Conservation Service.

Necrosis. Localized death of living tissue (e.g., following desiccation, browning, or loss of function).

Nonselective herbicide. A chemical that is generally toxic to plants without regard to species (may be a function of applied rate, method of application, stage of growth, etc.).

No-till. The soil is left undisturbed from harvest to planting, except for nutrient injection. Planting or drilling is accomplished in a narrow seedbed with coulters, row cleaners, disk openers, in-row chisels, or rototillers. Weed control is achieved primarily with herbicides, with cultivation for emergency control. In a modified no-till system, herbicides may be applied in a band over the row, thereby reducing the total amount applied per acre, accompanied by interrow cultivation.

No-tillage. Residues of previous crop and weeds remain on the soil surface as a mulch to reduce wind and water erosion.

Noxious weed. A weed specified by law as being especially undesirable, troublesome, and difficult to control. Definition will vary according to legal interpretations.

NRCS. Natural Resources Conservation Service (formerly Soil Conservation Service, or SCS).

Oncogenicity. Capacity to induce tumors in animals.

Organic soil. A soil containing more than 30% organic matter and greater than 15.7 in. (40 cm) in thickness.

Organic soil colloids. The organic colloidal portion of soils is called *humus*. It is derived from microbial decomposition of organic matter in the soil, and it is amorphous (noncrystalline) in structure.

Over-the-top application (OTA). Herbicide applied over the foliar canopy of transplanted or established plants such as by airplane or raised spray boom of ground equipment.

Parasitism. The process in which one plant lives on another and obtains food from the host while contributing nothing to the host's survival.

Pegging. The formation of the peanut runners and the setting of the underground nut. It denotes the downward growth (elongation) of the ovary from the peanut flower and its subsequent penetration 1–3 in. deep in the soil below the plant. The tip of the "peg" (elongated ovary) becomes the peanut.

Pellet. A dry formulation of herbicide and other components in discrete particles usually larger than 10 mm^3 in size.

Population. A collection of live individuals of a species, including seeds, in a given field. Geneticists define population as a set of plants that exchange genetic information.

Persistent herbicide. An herbicide that, when applied at the recommended rate, remains active in the soil longer than desired and injures susceptible rotational crop plants or interferes with regrowth of native vegetation in noncrop areas.

Phytotoxic. Injurious or toxic to plants.

Plant ecophysiology or physiological plant ecology. The study of the functioning of plants (e.g., photosynthesis) in relation to the environment.

Postemergence (POST). Herbicide applied after the specified crop or weeds emerge from the soil.

Preemergence (PRE). Herbicide applied to soil surface (not incorporated) before the specified crop or weeds emerge.

Preemergence-incorporated. Herbicide applied to the soil surface after planting and soil-incorporated above the seed.

Premix. A manufacturer-prepared formulation containing two or more pesticides (e.g., herbicide/herbicide, herbicide/insecticide).

Preplant (PP). Herbicide applied to soil surface (not incorporated) or to emerged weeds before planting.

Preplant-incorporated (PPI). Herbicide applied prior to planting or transplanting and soil-incorporated.

Product. The formulated herbicide available for use under one or more trade names.

Rate. The amount of herbicide applied per unit area, usually expressed in the English system as pounds of active ingredient or acid equivalent per acre, or as pounds or liquid volume (pints, quarts, gallons) of product (formulation) per acre. In the metric system, the rate is expressed as kilograms of active ingredient or acid equivalent per hectare or as kilograms or liters of product per hectare.

Reduced-tillage. Implies that tillage operations have been reduced to a minimum without adversely affecting crop yields.

Residue. The amount of herbicide remaining in or on the soil, plant parts, or animal tissue at the time of analysis.

Residual herbicide. An herbicide that persists in the soil at phytotoxic levels over an extended period of time as opposed to one that is inactive when applied to the soil or that is rapidly degraded in the soil.

Resistance. Herbicide resistance is the ability of a weed biotype to survive treatment of a given herbicide to which the weed species is normally susceptible. The mechanism of herbicide resistance takes place at the site of herbicide activity. Differential absorption, translocation, and metabolism are not significant factors endowing herbicide resistance.

Ridge-till. Ridge-tillage is characterized by ridges that are left intact throughout the year. There is usually no preplant tillage or tillage between successive crops. At planting time, the top of the ridge is cleaned and an interrow cultivation is performed. Annual weeds are controlled with herbicides applied in a band over the row, along with one or two interrow cultivations.

Ruderal plant. An introduced plant growing as a weed where the vegetative cover has been disturbed by man, as in fields and roadsides. Ruderal plants grow best in low stress but high disturbance habitats. A ruderal plant that colonizes successfully may be categorized as a competitive ruderal species.

SCS. Soil Conservation Service (now called Natural Resources Conservation Service, or NRCS).

Selectivity. Herbicide selectivity is the ability of a given herbicide to kill certain plant species without significant injury to others (may be a function of applied rate, mode of application, stage of growth, etc.).

Sleeping sod. Maintaining a living cover crop while growing a crop. Growth of the cover crop is suppressed or partially killed with a sublethal application of a herbicide (e.g., atrazine).

Soil application. Herbicide applied to the soil surface rather than to vegetation.

Soil colloids. Inorganic (clay) and organic (humus) soil particles with a maximum size of 0.001 mm (1 μ). They are the most chemically active portion of the soil.

Soil fumigant. A herbicide or other pesticide that, when added to the soil, vaporizes and forms a toxic gas.

Soil-incorporate. To mix or blend a herbicide into the soil by mechanical tillage or by leaching.

Soil injection. Placement of the herbicide beneath the soil surface with a minimum of mixing or stirring of the soil (often with an injection blade, knife, or tine).

Soil-layered. Placement of herbicide in a discrete horizontal zone under a lifter or tilled layer of soil.

Soil persistence. The length of time that a herbicide remains active in the soil.

Soil sterilant. A soil-applied herbicide that is intended to kill all green vegetation over an extended period of time.

Soluble solid. A dry herbicide formulation that is soluble in the carrier liquid, which is usually water.

Spray drift. Movement of airborne spray droplets from the intended area of application.

Stale seedbed. No soil preparation is done within 2–3 months prior to spring planting. Weed control is with a "burndown" herbicide applied 30–45 days ahead of planting.

Stubble mulch. The stubble of the preceding crop is left on the soil surface with or without soil disturbance, as with mulch-tillage.

Submersed plant. An aquatic plant adapted to grow with all or most of its vegetative tissue below the water surface.

Surfactant. A material that favors or improves the emulsifying, dispersing, spreading, wetting, or other surface-modifying properties of liquids.

Susceptibility. The degree of plant response of a given species to an applied herbicide. Susceptibility is a positive response.

Suspension. Finely divided solid particles dispersed in a solid, liquid, or gas.

Synergism. Complementary action of different chemicals such that the total effect is greater than the sum of the independent effects.

Tank mix. Two or more pesticide formulations mixed in the same spray mixture by the applicator prior to application as opposed to pesticide mixtures offered in one formulation (pre-mixes) by manufacturers.

Teratogen. A compound that can cause congenital malformations in the fetus (birth defects).

Tolerance. Herbicide tolerance is the capacity of a crop plant to withstand herbicide treatment without marked deviation from normal growth or function. Tolerance is a negative response. It is largely attributed to differential herbicide absorption, translocation, and/or detoxification.

Topical application. Treatment of a localized surface site such as a single leaf blade, petiole, or growing point.

Trade name. The registered brand name, number, or mark used by a company to identify and separate its product from others on the market.

Translocated herbicide. An herbicide that moves from one location to another in a plant. Translocated herbicides may be either phloem or xylem mobile or both.

Troublesome weeds. Weeds that, despite control measures, are inadequately controlled and interfere with crop production and/or yield, crop quality, or harvest efficiency.

USDA. United States Department of Agriculture.

Vapor drift. The movement of herbicide vapors carried by air movement from one area of herbicide application to others.

Volatile herbicide. An herbicide with sufficiently high vapor pressure that it vaporizes when applied at normal rates and temperatures. Such vapors pose a hazard to nearby susceptible plants. It is often difficult to distinguish between spray- and vapor-drift plant injuries.

Water-dispersible slurry. A two-phase concentrated formulation containing a solid herbicide suspended in a liquid that is readily suspended in water, which is the carrier.

Weed. Any plant growing where it is not wanted. Plants are considered weeds when they interfere with people's activities and welfare.

Weed control. The process of limiting weed infestations or of killing weeds for aesthetic, economic, public health, or other reasons.

Weed eradication. The elimination of all viable plant propagules and plant parts of one or more specified weeds from a designated area.

Weed seed bank. The seed bank consists of the viable (dormant and nondormant) weed seeds in the soil, usually considered to the plowshare depth (about 8 in. or 20 cm).

Wetting agent. A substance that serves to reduce interfacial tensions between a liquid and solid (e.g., spray solution and leaf surface) resulting in the spray mixture making better contact with the treated surface. A wetting agent is a surfactant.

Wettable powder (WP). A finely divided, dry herbicide formulation that can be readily suspended in water.

Xenobiotics. Molecules foreign to an organism such as, but not limited to, those synthesized by humankind.

HERBICIDE NAMES

TABLE A.1 ■ Herbicides listed by common name: general information.

A note on mammalian toxicity: Categories of herbicide mammalian toxicity are based on hazards to skin and eyes and LD_{50}. The smaller the category number, the greater the hazard. Exceptions may occur between different formulations of the same active ingredient.

LD_{50} denotes a 50% kill of the population of the designated test animal (usually mice, rats, rabbits) at a given dosage. LD_{50} is expressed as weight of applied chemical per unit body weight of test animal (mg/kg).

Category I: *Skin*, corrosive; *eyes*, corrosive with corneal opacity not reversible within 7 days.
LD_{50}: *Oral*, up to 50 mg/kg; *dermal*, up to 200 mg/kg; *inhalation*, up to 0.2 mg/L.

Category II: *Skin*, irritation severe at 72 hours; *eyes*, corrosive with corneal opacity reversible within 7 days, but irritation persisting for 7 days.
LD_{50}: *Oral*, from 50 to 500 mg/kg; *dermal*, from 200 to 2000 mg/kg; *inhalation*, from 0.2 to 2 mg/L.

Category III: *Skin*, irritation moderate at 72 hours; *eyes*, no corneal opacity and irritation reversible within 7 days.
LD_{50}: *Oral*, from 500 to 5000 mg/kg; *dermal*, from 2000 to 20,000 mg/kg; *inhalation*, from 2.0 to 20 mg/L.

Category IV: *Skin*, irritation mild or slight at 72 hours; *eyes*, no irritation.
LD_{50}: *Oral*, greater than 5000 mg/kg; *dermal*, greater than 20,000 mg/kg; *inhalation*, greater than 20 mg/L.

| COMMON NAME | TRADE NAME | MAMMALIAN TOXICITY | | MANUFACTURER |
		Category	Signal Word	
Acetochlor	Harness Plus Surpass	I	Warning	Monsanto Zeneca
Acifluorfen	Blazer	III	Danger (eyes)	BASF
Alachlor	Cropstar, Lasso, Micro-Tech Stall Judge	II	Warning	Monsanto Monsanto Platte Riverside
Ametryn	Evik	III	Caution	Ciba
Amitrole	Amitrol T, Amizol	III	Caution	Rhone-Poulenc
Asulam	Asulox	IV	Caution	Rhone-Poulenc
Atrazine	AAtrex Drexel Atrazine Riverside Atrazine	III	Caution	Ciba Drexel Riverside
Benefin	Balan	IV	Varies with formulation	DowElanco
Bensulfuron	Londax	IV	Caution	DuPont
Bensulide	Prefar, Betasan Bensumec	II	Caution	Zeneca PBI/Gordon
Bentazon	Basagran Pledge	III	Caution	BASF Setre
Bromacil	Hyvar	II	Warning	DuPont
Bromoxynil	Buctril	II	Warning	Rhone-Poulenc
Butylate	Sutan	III	Caution	Zeneca

Continued

TABLE A.1 ■ Herbicides listed by common name: general information, *continued*.

COMMON NAME	TRADE NAME	MAMMALIAN TOXICITY*		MANUFACTURER
		Category	Signal Word	
Cacodylic acid	Montar	III	Caution	Monterey
Chlorimuron	Classic	III	Caution	DuPont
Chlorsulfuron	Glean, Telar	IV	Caution	DuPont
Cinmethylin	Cinch	III	Caution	DuPont
Clethodim	Select	II	Warning	Valent
Clomazone	Command	III	Caution	FMC
Clopyralid	Reclaim, Stinger Transline	I	Danger (eyes)	DowElanco DowElanco
Chloridazon	Pyramin FL	III	Caution	BASF
Cyanazine	Bladex	II	Warning	DuPont
Cycloate	Ro-Neet	III	Caution	Zeneca
2,4-D	Weedar, Weedone, others	Varies with formulation		Rhone-Poulenc others
2,4-DB	Butoxone, Butyrac	III	Caution	Cedar Rhone-Poulenc
Dazomet	Basamid Granular	III	Caution	BASF
DCPA	Dacthal	IV	Caution	ISK Biotech
Desmedipham	Betanex	III	Caution	Nor-Am
Dicamba	Banvel	III	Warning	Sandoz
Diclobenil	Casoron, Dyclomec, Norosac	III	Caution	Uniroyal PBI/ Gordon
Dichlorprop (2,4-DP)	Weedone 2,4-DP	II	Warning	Rhone-Poulenc
Diclofop	Hoelon, Illoxan	II	Warning	Hoechst-Roussel
Diethatyl	Antor	III	Caution	Nor-Am
Difenzoquat	Avenge	II	Danger	American Cyanamid
Diglycolamine salt of dicamba	Clarity, Vanquish	III	Caution	Sandoz
Diquat	Reward Weedtrine-D	II	Warning	Zeneca Applied Biochemicals
Dithiopyr	Dimension	II	Warning	Monsanto
Diuron	Karmex Direx Drexel Diuron Riverside Diuron	III	Warning	DuPont Griffin Drexel Riverside
DSMA	DSMA Liquid Liquid DSMA Methar 30 Drexel DSMA	III	Caution	ISK Biotech Setre Cleary Drexel
Endothall	Endothal Turf Herbicide, Aquathol, Herbicide 273, Hydrothol 191	II	Warning	Atochem
EPTC	Eptam, Eradicane	III	Caution	Zeneca
Ethalfluralin	Sonalan Curbit	II	Warning	DowElanco Platte
Ethofumesate	Nortron, Prograss	I	Danger	Nor-Am
Fenoxaprop	Acclaim, Bugle, Horizon, Option, Whip	III	Caution	Hoechst-Roussel Hoechst-Roussel

Continued

TABLE A.1 ■ Herbicides listed by common name: general information, *continued*.

| COMMON NAME | TRADE NAME | MAMMALIAN TOXICITY* | | MANUFACTURER |
		Category	Signal Word	
Fluazifop-P-butyl	Fusilade 2000 Ornamec Take-Away	II	Warning (irritant)	Zeneca PBI/ Gordon Lesco
Fluometuron	Cotoran Meturon Riverside Fluometuron	II	Warning	Ciba Griffin Riverside
Fluridone	Sonar	IV	Caution	DowElanco
Fomesafen	Flex, Flexstar, Reflex	III	Warning	Zeneca
Fosamine	Krenite	III	Warning	DuPont
Glufosinate	Finale, Ignite	II	Warning	Hoechst-Roussel
Glyphosate	Accord, Roundup, Rodeo Jury Rattler	III	Caution	Monsanto Riverside Setre
Halosulfuron	Battalion, Manage, Permit	II	Warning	Monsanto Monsanto
Haloxyfop	Verdict	II	Warning	DowElanco
Hexazinone	Velpar	IV	Warning (eyes)	DuPont
Imazamethabenz	Assert	I	Danger	American Cyanamid
Imazapyr	Arsenal	IV	Caution	American Cyanamid
Imazaquin	Image,Scepter	III	Caution	American Cyanamid
Imazethapyr	Pursuit	III	Caution	American Cyanamid
Isoxaben	Gallery	IV	Caution	DowElanco
Lactofen	Cobra	I (Skin and dye irritant)	Danger	Valent
Linuron	Drexel Linuron Linex Lorox	III	Caution	Drexel Griffin DuPont
MCPA amine	Gordon MCPA Amine Rhomene	I	Danger	PBI/Gordon Rhone-Poulenc
MCPA ester	Rhonox	I	Danger	Rhone-Poulenc
MCPB	Thistrol	II	Caution	Rhone-Poulenc
Mecoprop (MCPP)	Cleary's MCPP Mecomec	II	Warning	Cleary PBI/Gordon
Metham	Vapam	II	Warning	Zeneca
Metolachlor	Dual,Pennant	III	Caution	Ciba
Metribuzin	Lexone Sencor	III	Caution	DuPont Bayer
Metsulfuron	Ally, Escort	IV	Caution	DuPont
Molinate	Ordram	II	Varies with formulation	Zeneca
MSMA	Ansar, Bueno, Daconate Drexel MSMA Setre MSMA	III	Caution	ISK Biotech Drexel Setre
Napropamide	Devrinol Ornamental Herbicide 5G	III	Caution	Zeneca Lesco
Naptalam	Alanap	II	Caution	Uniroyal
Nicosulfuron	Accent	IV	Caution	DuPont

Continued

TABLE A.1 ■ **Herbicides listed by common name: general information,** *continued.*

| COMMON NAME | TRADE NAME | MAMMALIAN TOXICITY* | | MANUFACTURER |
		Category	Signal Word	
Norflurazon	Evital, Predict, Solicam, Zorial	IV	Caution	Sandoz Sandoz
Oryzalin	Surflan	IV	Caution	DowElanco
Oxadiazon	Chipco Ronstar	II	Warning	Rhone-Poulenc
Oxyfluorfen	Goal	IV	Warning	Rohm & Haas
Paraquat	Cyclone, Starfire, Gramoxone Extra	I	Danger-Poison	Zeneca Zeneca
Pebulate	Tillam	III	Caution	Zeneca
Pendimethalin	Pre-M Prowl, Stomp Weedgrass Control	III	Caution	Lesco American Cyanamid O.M. Scott
Phenmedipham	Spin-Aid	II	Warning	Nor-Am
Picloram	Tordon K	IV	Warning	DowElanco
Primisulfuron	Beacon	IV	Caution	Ciba
Prodiamine	Barricade	IV	Caution	Sandoz
Prometon	Pramitol 25E	I	Danger (corrosion)	Ciba
Prometryn	Caparol Cotton-Pro Prometryne 4L Gowan Prometryne	II, III	Varies with formulation	Ciba Griffin Riverside Gowan
Pronamide	Kerb	IV	Caution	Rohm & Haas
Propachlor	Ramrod	III	Warning	Monsanto
Propanil	Cedar Propanil, Wham DF Stam, Stampede	II	Warning	Cedar Rohm & Haas
Pyridate	Tough	III	Caution	Cedar
Quinclorac	Facet	III	Caution	BASF
Quizalofop	Assure	III	Caution	DuPont
Sethoxydim	Poast, Vantage	III	Caution	BASF
Siduron	Tupersan	IV	Caution	DuPont
Simazine	Drexel Simazine Princep Riverside Simazine	IV	Caution	Drexel Ciba Riverside
Sulfometuron	Oust	IV	Caution	DuPont
Sulfosate	Touchdown	III	Caution	Zeneca
Tebuthiuron	Spike	III	Caution	DowElanco
Terbacil	Sinbar	IV	Caution	DuPont
Thifensulfuron	Pinnacle	IV	Caution	DuPont
Thiobencarb	Bolero	III	Caution	Valent
Triallate	Far-Go	III	Caution	Monsanto
Triasulfuron	Amber	IV	Caution	Ciba
Tribenuron	Express	IV	Caution	DuPont
Triclopyr	Redeem, Remedy	I	Danger	DowElanco
Trifluralin	Treflan Tri-4 Trifluralin 4EC Trific	Varies with formulation		DowElanco American Cyanamid Riverside Terra
Vernolate	Vernam	II	Caution	Drexel

TABLE A.2 ■ Herbicides listed by trade name, active ingredient, and manufacturer.

TRADE NAME	ACTIVE INGREDIENT	MANUFACTURER
AAtrex	atrazine	Ciba
Accent	nicosulfuron	DuPont
Acclaim	fenoxaprop	Hoechst-Roussel
Accord	glyphosate	Monsanto
Alanap	naptalam	Uniroyal
Ally	metsulfuron	DuPont
Amber	triasulfuron	Ciba
Amitrol T	amitrole	Rhone-Poulenc
Amizol	amitrole	Rhone-Poulenc
Ansar	MSMA	ISK Biotech
Antor	diethatyl	Nor-Am
Aquathol	endothall	Atochem
Arsenal	imazapyr	American Cyanamid
Assert	imazamethabenz	American Cyanamid
Assure	quizalofop	DuPont
Asulox	asulam	Rhone-Poulenc
Avenge	difenzoquat	American Cyanamid
Balan	benefin	DowElanco
Banvel	dicamba	Sandoz
Barricade	prodiamine	Sandoz
Basagran	bentazon	BASF
Basamid Granular	dazomet	BASF
Battalion	halosulfuron	Monsanto
Beacon	primisulfuron	Ciba
Bensumec	benefin	PBI/Gordon
Betanex	desmedipham	Nor-Am
Betasan	bensulide	Zeneca
Bladex	cyanazine	DuPont
Blazer	acifluorfen	BASF
Bolero	thiobencarb	Valent
Buctril	bromoxynil	Rhone-Poulenc
Bueno	MSMA	ISK Biotech
Bugle	fenoxaprop	Hoechst-Roussel
Butoxone	2,4-DB	Cedar
Butyrac	2,4-DB	Rhone-Poulenc
Caparol	prometryn	Ciba
Casoron	dichlobenil	Uniroyal
Cedar Propanil	propanil	Cedar
Chipco Ronstar	oxadiazon	Rhone-Poulenc
Cinch	cinmethylin	DuPont
Clarity	dicamba	Sandoz
Classic	chlorimuron	DuPont
Cleary's MCPP	MCPP	Cleary
Cobra	lactofen	Valent
Command	clomazone	FMC
Cotoran	fluometuron	Ciba

Continued

TABLE A.2 ■ Herbicides listed by trade name, active ingredient, and manufacturer, *continued.*

TRADE NAME	ACTIVE INGREDIENT	MANUFACTURER
Cotton-Pro	prometryn	Griffin
Cropstar	alachlor	Monsanto
Curbit	ethalfluralin	Platte
Cyclone	paraquat	Zeneca
Daconate	MSMA	ISK Biotech
Dacthal	DCPA	ISK Biotech
Devrinol	napropamide	Zeneca
Dimension	dithiopyr	Monsanto
Direx	diuron	Griffin
Drexel Atrazine	atrazine	Drexel
Drexel Diuron	diuron	Drexel
Drexel DSMA	DSMA	Drexel
Drexel Linuron	linuron	Drexel
Drexel MSMA	MSMA	Drexel
Drexel Simazine	simazine	Drexel
DSMA Liquid	DSMA	ISK Biotech
Dual	metolachlor	Ciba
Dyclomec	dichlobenil	PBI/Gordon
Endothal Turf Herb.	endothall	Atochem
Eptam	EPTC	Zeneca
Eradicane	EPTC + safener	Zeneca
Escort	metsulfuron	DuPont
Evik	ametryn	Ciba
Evital	norflurazon	Sandoz
Express	tribenuron	DuPont
Facet	quinclorac	BASF
Far-Go	triallate	Monsanto
Finale	glufosinate	Hoechst-Roussel
Flex	fomesafen	Zeneca
Flexstar	fomesafen	Zeneca
Fusilade 2000	fluazifop-P-butyl	Zeneca
Gallery	isoxaben	DowElanco
Glean	chlorsulfuron	DuPont
Goal	oxyfluorfen	Rohm & Haas
Gordon MCPA Amine	MCPA amine	PBI/Gordon
Gowan Prometryne	prometryn	Gowan
Gowan Trifluralin	trifluralin	Gowan
Gramoxone Extra	paraquat	Zeneca
Harness Plus	acetochlor	Monsanto
Herbicide 273	endothall	Atochem
Hoelon	diclofop	Hoechst-Roussel
Horizon	fenoxaprop	Hoechst-Roussel
Hydrothol 191	endothall	Atochem
Hyvar	bromacil	DuPont
Ignite	glufosinate	Hoechst-Roussel
Illoxan	diclofop	Hoechst-Roussel
		Continued

TABLE A.2 ■ Herbicides listed by trade name, active ingredient, and manufacturer, *continued.*

TRADE NAME	ACTIVE INGREDIENT	MANUFACTURER
Image	imazaquin	American Cyanamid
Judge	alachlor	Riverside
Jury	glyphosate	Riverside
Karmex	diuron	DuPont
Kerb	pronamide	Rohm & Haas
Krenite	fosamine	DuPont
Lasso	alachlor	Monsanto
Lexone	metribuzin	DuPont
Linex	linuron	Griffin
Liquid DSMA	DSMA	Setre
Londax	bensulfuron	DuPont
Lorox	linuron	DuPont
Manage	halosulfuron	Monsanto
Mecomec	MCPP	PBI/Gordon
Methar 30	DSMA	Cleary
Meturon	fluometuron	Griffin
Micro-Tech	alachlor	Monsanto
Montar	cacodylic acid	Monterey
Norosac	diclobenil	PBI/Gordon
Nortron	ethofumesate	Nor-Am
Option	fenoxaprop	Hoechst-Roussel
Ordram	molinate	Zeneca
Ornamec	fluazifop-P-butyl	PBI/Gordon
Ornamental Herbicide 5G	napropamide	Lesco
Oust	sulfometuron	DuPont
Pennant	metolachlor	Ciba
Permit	halosulfuron	Monsanto
Pinnacle	thifensulfuron	DuPont
Pledge	bentazon	Setre
Poast	sethoxydim	BASF
Predict	norflurazon	Sandoz
Pramitol	prometon	Ciba
Prefar	bensulide	Zeneca
Pre-M	pendimethalin	Lesco
Princep	simazine	Ciba
Prograss	ethofumesate	Nor-Am
Prometryne 4L	prometryn	Riverside
Prowl	pendimethalin	American Cyanamid
Pursuit	imazethapyr	American Cyanamid
Pyramin	chloridazon	BASF
Ramrod	propachlor	Monsanto
Rattler	glyphosate	Setre
Reclaim	clopyralid	DowElanco
Redeem	triclopyr	DowElanco
Reflex	fomesafen	Zeneca
Remedy	triclopyr	DowElanco

Continued

TABLE A.2 ■ Herbicides listed by trade name, active ingredient, and manufacturer, *continued.*

TRADE NAME	ACTIVE INGREDIENT	MANUFACTURER
Reward	diquat	Zeneca
Rhomene	MCPA amine	Rhone-Poulenc
Rhonox	MCPA ester	Rhone-Poulenc
Riverside Atrazine	atrazine	Riverside
Riverside Diuron	diuron	Riverside
Riverside Fluometuron	fluometuron	Riverside
Riverside Simazine	simazine	Riverside
Rodeo	glyphosate	Monsanto
Ro-Neet	cycloate	Zeneca
Roundup	glyphosate	Monsanto
Scepter	imazaquin	American Cyanamid
Select	clethodim	Valent
Sencor	metribuzin	Bayer
Setre MSMA	MSMA	Setre
Sinbar	terbacil	DuPont
Solicam	norflurazon	Sandoz
Sonalan	ethalfluralin	DowElanco
Sonar	fluridone	DowElanco
Spike	tebuthiuron	DowElanco
Spin-Aid	phenmedipham	Nor-Am
Stall	alachlor	Platte
Stam	propanil	Rohm & Haas
Stampede	propanil	Rohm & Haas
Starfire	paraquat	Zeneca
Stinger	clopyralid	DowElanco
Stomp	pendimethalin	American Cyanamid
Surflan	oryzalin	DowElanco
Surpass	acetochlor	Zeneca
Take-Away	fluazifop-P-butyl	Lesco
Telar	chlorsulfuron	DuPont
Thistrol	MCPB	Rhone-Poulenc
Tillam	pebulate	Zeneca
Touchdown	sulfosate	Zeneca
Tough	pyridate	Cedar
Tordon K	picloram	DowElanco
Transline	clopyralid	DowElanco
Treflan	trifluralin	DowElanco
Tri-4	trifluralin	American Cyanamid
Trific	trifluralin	Terra
Trifluralin 4EC	trifluralin	Riverside
Tupersan	siduron	DuPont
Vanquish	dicamba	Sandoz
Vantage	sethoxydim	BASF
Vapam	metham	Zeneca
Verdict	haloxyfop	DowElanco
Velpar	hexazinone	DuPont

Continued

TABLE A.2 ■ **Herbicides listed by trade name, active ingredient, and manufacturer,** *continued.*

TRADE NAME	ACTIVE INGREDIENT	MANUFACTURER
Vernam	vernolate	Drexel
Weedar 64	2,4-D amine	Rhone-Poulenc
Weedone 2,4-DP	2,4-DP ester	Rhone-Poulenc
Weedone LV4	2,4-D ester	Rhone-Poulenc
Weedone 638	2,4-D ester + acid	Rhone-Poulenc
Weed Rhap 2,4-D	2,4-D amine	Setre
Weed Rhap LV6	2,4-D ester	Setre
Weedgrass Control	pendimethalin	O.M. Scott
Weedgrass Preventer	bensulide	O.M. Scott
Weedtrine-D	diquat	Applied Biochemists
Wham DF	propanil	Cedar
Whip	fenoxaprop	Hoechst-Roussel
Zorial	norflurazon	Sandoz

TABLE A.3 ■ Herbicide *premixes* listed by trade name, active ingredients, and manufacturer.

TRADE NAME	ACTIVE INGREDIENTS	MANUFACTURER
Access	picloram + triclopyr	DowElanco
Arrosolo	molinate + propanil	Zeneca
Betamix	desmedipham + phenmedipham	Nor-Am
Bicep	atrazine + metolachlor	Ciba
Bronate	bromoxynil + MCPA	Rhone-Poulenc
Bronco	alachlor + glyphosate	Monsanto
Buckle	triallate + trifluralin	Monsanto
Buctril + Atrazine	bromoxynil + atrazine	Rhone-Poulenc
Bullet	alachlor + atrazine	Monsanto
Canopy	metribuzin + chlorimuron	DuPont
Chipco Weedone DPC Amine	2,4-D + 2,4-DP amine	Rhone-Poulenc
Chipco Weedone DPC Ester	2,4-D + 2,4-DP ester	Rhone-Poulenc
Cleary's MCPP-2,4-D	mecoprop + 2,4-D	Cleary
Cleary's Weedone DPC	2,4-D ester + 2,4-DP ester	Cleary
Commence	clomazone + trifluralin	FMC
Concert	thifensulfuron + chlorimuron	DuPont
Confront	triclopyr + clopyralid	DowElanco
Crossbow	triclopyr + 2,4-D	DowElanco
Curtail	clopyralid + 2,4-D	DowElanco
Cycle	metolachlor + cyanazine	Ciba
Dakota	fenoxaprop + MCPA ester	Hoechst-Roussel
Derby	metolachlor + simazine	Ciba
Dissolve	2,4-D + 2,4-DP + MCPP	Riverdale
Envert 171	2,4-D + dichlorprop	Rhone-Poulenc
Extrazine II	cyanazine + atrazine	DuPont
Fallow Master	glyphosate + dicamba	Monsanto
Finesse	chlorsulfuron + metsulfuron	DuPont
Fluid Broadleaf Weed Control	2,4-D + 2,4-DP amine	O.M. Scott
Freedom	alachlor + trifluralin	Monsanto
Fusion	fluazifop-P-butyl + fenoxaprop	Zeneca
Galaxy	bentazon + acifluorfen	BASF
Gemini	linuron + chlorimuron	DuPont
Harmony Extra	thifensulfuron + tribenuron	DuPont
Krovar	bromacil + diuron	DuPont
Laddock	bentazon + atrazine	BASF
Landmaster BW	glyphosate + 2,4-D	Monsanto
Lariat	alachlor + atrazine	Monsanto
Marksman	dicamba + atrazine	Sandoz
Moncide	MSMA + cacodylic acid	Monterey
New Lorox Plus	linuron + chlorimuron	DuPont
Ornamental Herbicide II	oxyfluorfen + pendimethalin	O.M. Scott
Passport	imazethapyr + trifluralin	American Cyanamid

Continued

TABLE A.3 ■ Herbicide *premixes* listed by trade name, active ingredients, and manufacturer, *continued.*

TRADE NAME	ACTIVE INGREDIENTS	MANUFACTURER
Pathway	picloram + 2,4-D	DowElanco
Pramitol 5PS	prometon + simazine + sodium chlorate + sodium metaborate	Ciba
Preview	metribuzin + chlorimuron	DuPont
Prompt	bentazon + atrazine	BASF
Pursuit Plus	pendimethalin + imazethapyr	American Cyanamid
Quadmec Trimec Plus	MSMA + 2,4-D + MCPP + dicamba	PBI/Gordon
Ramrod/Atrazine	propachlor + atrazine	Monsanto
Rout	oxyfluorfen + oryzalin	Grace-Sierra
Salute	trifluralin + metribuzin	Bayer
Scepter O.T.	imazaquin + acifluorfen	American Cyanamid
Simazat	simazine + atrazine	Drexel
Snapshot	isoxaben + oryzalin	DowElanco
Squadron	pendimethalin + imazaquin	American Cyanamid
Stocktrine	mixed copper complexes	Applied Biochemists
Storm	bentazon + acifluorfen	BASF
Super Trimec	2,4-D + dichlorprop + dicamba	PBI/Gordon
Surpass 100	acetochlor + atrazine	Zeneca
Sutazine+	atrazine + butylate	Zeneca
Team 2G	benefin + trifluralin	DowElanco
Three-Way	2,4-D + MCPP + dicamba	Lesco
Tiller	fenoxaprop + 2,4-D + MCPA	Hoechst-Roussel
Topsite	imazapyr + diuron	American Cyanamid
Tordon 101 Mixture	picloram + 2,4-D	DowElanco
Tornado	fluazifop-P-butyl + fomesafen	Zeneca
Total	2% bromacil + 2% diuron + 40% sodium chlorate + 50% sodium metaborate	Riverside
Trex-San	2,4-D + 2,4-DP + dicamba	Grace-Sierra
Trimec Classic	2,4-D + mecoprop + dicamba	PBI/Gordon
Trimec Encore	MCPA + mecoprop + dicamba	PBI/Gordon
Trimec Southern	2,4-D + mecoprop + dicamba	PBI/Gordon
Tri-Power	MCPA + MCPP + dicamba	Riverdale
Tri-Scept	imazaquin + trifluralin	American Cyanamid
Turbo	metolachlor + metribuzin	Bayer
Turflon D	triclopyr ester + 2,4-D ester	DowElanco
Turflon II	triclopyr amine + 2,4-D amine	DowElanco
2 Plus 2	MCPP + 2,4-D amine	ISK Biotech
Vegemec Vegetation Killer	prometon + 2,4-D + aromatic oil	PBI/Gordon
Weed Blast 4G	bromacil + diuron	Setre
Weedmaster	dicamba + 2,4-D	Sandoz
Weedone CB	2, 4-D + 2,4-DP ester	Rhone-Poulenc
Weedone 170	2, 4-D + 2,4-DP ester	Rhone-Poulenc
XL 2G	benefin + oryzalin	DowElanco

CONVERSION TABLES

TABLE A.4 ■ Converting broadcast rates to band rates

HERBICIDE BAND WIDTH (INCHES)	CONVERSION FACTORS FOR ROW SPACINGS (INCHES)							
	20	24	28	30	32	36	38	40
6	.3	.25	.21	.20	.19	.17	.16	.15
8	.4	.33	.29	.27	.25	.22	.21	.20
10	.5	.42	.36	.33	.31	.28	.26	.25
12	.6	.50	.43	.40	.37	.33	.31	.30
14	.7	.58	.50	.47	.44	.39	.37	.35
16	.8	.67	.57	.53	.50	.44	.42	.40
18	.9	.75	.64	.60	.56	.50	.47	.45
20	1.0	.83	.71	.67	.62	.56	.53	.50

To convert: Multiply the broadcast rate by the factor for row spacing and band width.

Example 1: The broadcast rate is 1.0 lb/A, band width is 14 in., and row spacing is 24 in. Multiply 1.0 by .58 to get 0.58 lb/A.

Example 2: The broadcast rate is 1.5 pt/A, band width is 8 in., and row spacing is 36 in. Multiply 1.5 by .22 to get 0.33 pt/A.

TABLE A.5 ■ **Milliliters of liquid formulations* required per 1000 ft² to be equivalent to pounds of active ingredient per acre (lb ai/A).**

LB ai/A	ACTIVE INGREDIENT (POUNDS PER GALLON OF FORMULATION)															
	0.25	0.33	0.50	0.67	0.75	1.00	1.33	1.50	1.67	2.00	2.50	3.00	3.33	4.00	5.00	6.00
0.25	87	66	44	32	29	22	16.2	14.4	13.0	10.9	8.7	7.2	6.5	5.4	4.4	3.6
0.33	115	87	57	43	38	29	22	19.1	17.2	14.3	11.5	9.6	8.6	7.2	5.7	4.8
0.50	174	132	87	65	58	44	33	29	26	22	17.4	14.5	13.0	10.9	8.7	7.2
0.67	233	176	116	87	78	58	44	39	35	29	23	19.4	17.5	14.6	11.6	9.7
0.75	261	198	130	97	87	65	49	43	39	33	26	22	19.5	16.3	13.0	10.9
1.00	348	263	174	130	116	87	65	58	52	44	35	29	26	22	17.4	14.5
2.00	695	527	348	259	232	174	130	116	104	87	69	58	52	43	35	29
3.00	1,043	790	521	389	348	261	195	174	156	130	104	87	78	65	52	43
4.00	1,390	1,053	695	518	463	348	260	231	208	174	139	116	104	87	70	58
5.00	1,738	1,317	869	648	579	435	325	289	260	217	174	145	130	109	87	72

*Soluble, emulsfiable, flowable (slurry).

Example: To apply the equivalent of 1.5 lb/A active ingredient to an area of 1000 ft² using an emulsifiable-concentrate formulation containing 4.0 lb ai/gal, apply 32.9 ml (10.9 ml + 22 ml) of the formulation uniformly over the 1000-ft² area.

TABLE A.6 ■ Grams of dry formulations* required per 1000 ft^2 to be equivalent to pounds of active ingredient per acre (lb ai/A).

LB ai/A	FORMULATION—PERCENT (%) ACTIVE INGREDIENT									
	0.5	1	2	3	4	5	50	75	80	100
0.25	520	260	130	87	65	52	5.2	3.5	3.3	2.6
0.33	688	344	172	115	86	69	6.9	4.6	4.3	3.4
0.50	1,042	521	261	174	130	104	10.4	7.0	6.5	5.2
0.67	1,394	697	349	232	174	139	13.9	9.3	8.7	7.0
0.75	1,562	781	391	260	195	156	15.6	10.4	9.8	7.8
1.00	2,082	1,041	521	347	260	208	21	13.9	13.0	10.4
2.00	4,166	2,083	1,042	694	521	417	42	28	26	21
3.00	6,248	3,124	1,562	1,041	781	625	63	42	39	31
4.00	8,330	4,165	2,083	1,388	1,051	833	83	56	52	42
5.00	10,404	5,202	2,604	1,736	1,302	1,041	104	69	65	52

*Wettable powders, dry flowables, granules, pellets.
Example: To apply the equivalent of 1.5 lb/A active ingredient to an area of 1000 ft^2 using a wettable-powder formulation that contains 80% ai, apply 19.5 g (6.5 g + 13.0 g) of the formulation uniformly over the 1000-ft^2 area.

TABLE A.7 ■ Conversion tables: Metric and English systems.*

ENGLISH TO METRIC							METRIC TO ENGLISH		
lb/A	equals	lb/ha	equals	kg/ha	equals	g/m²	kg/ha	equals	lb/A
0.25		0.62		0.28		0.028	0.25		0.22
0.33		0.82		0.35		0.035	0.50		0.44
0.50		1.24		0.56		0.056	0.75		0.67
0.67		1.63		0.69		0.069	1.00		0.89
0.75		1.85		0.84		0.084	1.50		1.33
1.00		2.47		1.12		0.112	1.75		1.55
2.00		4.95		2.24		0.224	2.00		1.78
3.00		7.42		3.37		0.337	2.50		2.23
4.00		9.89		4.49		0.448	3.00		2.67
5.00		12.36		5.61		0.561	3.50		3.11
6.00		14.84		6.73		0.673	4.00		3.56
7.00		17.31		7.85		0.785	4.50		4.01
8.00		19.78		8.98		0.898	5.00		4.45
9.00		22.26		10.10		1.010	5.50		4.90
10.00		24.70		11.22		1.122	6.00		5.34
15.00		37.06		16.83		1.683	6.50		5.79
20.00		49.50		22.44		2.244	7.00		6.23
25.00		61.86		28.05		2.805	7.50		6.68
							8.00		7.12
							8.50		7.57
							9.00		8.01
							9.50		8.46
							10.00		8.90
							15.00		13.36
							20.00		17.80
							25.00		22.26
							30.00		26.72
							35.00		31.15
							40.00		35.60
							45.00		40.05
							50.00		44.50
							60.00		53.44
							70.00		62.30
							80.00		71.20
							90.00		80.10
							100.00		89.00

*To convert lb/A to kg/ha, divide lb/A by 0.893.
To convert kg/ha to lb/A, multiply kg/ha by 0.893.

TABLE A.8 ■ Convenient areas, weights, and liquid measures.

Area

1 acre (A) = 43,560 square feet (ft^2)

1 acre = 0.407 hectare (ha)

1 square mile = 640 acres = 1 section

1 square mile = 250 hectares

1 hectare = 2.47 acres

1 hectare = 10,000 square meters (m^2)

1 square meter = 10.764 square feet

1 square kilometer = 100 hectares

Liquid Measure

1 U.S. gallon = 3785 milliliters (ml)

1 U.S. gallon = 0.833 Imperial gallon

1 Imperial gallon = 1.20 U.S. gallons

1 Imperial gallon = 4545 ml

1 U.S. gallon = 4 quarts = 8 pints

1 quart = 2 pints = 946 ml

1 pint = 2 cups = 473 ml

1 pint = 16 fluid ounces (fl oz)

1 cup = 8 fl oz

1 cup = 236.56 ml

1 tablespoon = 15 ml

1 teaspoon = 5 ml

1 fl oz = 2 tablespoons

1 fl oz = 29.563 ml

Weight

lb/A = pounds per acre

lb/ha = pounds/hectare

kg/ha = kilograms/hectare

g/m^2 = grams per square meter

1 pound = 16 ounces (oz)

1 ounce = 28.35 grams

1 pound = 453.59 grams

1 pound = 0.4536 kilograms

1 kilogram = 2.205 pounds

1 kilogram = 35.28 ounces

1 kilogram = 1000 grams

1 gram = 1000 milligrams (mg)

1 mg = 1000 micrograms (μg)

1 gram = 1 million μg

TABLE A.9 ■ Approximate conversions—Metric and English systems.

YOU CAN FIND—	IF YOU MULTIPLY—	BY—
Length		
millimeters	inches	25.4
centimeters	feet	30.48
meters	yards	0.914
kilometers	miles	1.609
inches	millimeters	0.03937
inches	centimeters	0.3937
yards	meters	1.094
miles	kilometers	0.6214
feet	meters	3.28
Area		
square centimeters	square inches	6.452
square meters	square feet	0.093
square meters	square yards	0.836
square kilometers	square miles	2.589
square hectares (hectometers)	acres	0.404
square inches	square centimeters	0.1550
square yards	square meters	1.196
square miles	square kilometers	0.347
acres	square hectares (hectometers)	2.471
Mass		
grams	ounces	28.35
kilograms	pounds	0.4536
metric tons (megagrams)	short tons	0.9072
ounces	grams	0.0353
pounds	kilograms	2.2046
short tons	metric tons (megagrams)	1.102
1 short ton = 2000 pounds; 1 metric ton = 2204.622 pounds		
Liquid Volume		
milliliters	ounces	29.573
liters	pints	0.473
liters	quarts	0.946
liters	gallons (U.S.)	3.785
ounces	milliliters	0.0338
pints	liters	2.11
quarts	liters	1.057
gallons (U.S.)	liters	0.2642
Imperial liquid measure × 0.8327 = U.S. liquid measure		
Temperature		
degrees Celsius (centigrade)	degrees Fahrenheit	0.556 (after subtracting 32)
degrees Fahrenheit	degrees Celsius (centigrade)	1.8 (then add 32)

LIST OF WEEDS

TABLE A.10 ■ Common and scientific names of weeds mentioned in text.

COMMON NAME	SCIENTIFIC NAME
Alligatorweed	*Alternanthera philoxeroides*
Amaranth, Palmer	*Amaranthus palmeri*
Amaranth, Powell	*Amaranthus powellii*
Amaranth, spiny	*Amaranthus spinosus*
Ammi, greater	*Ammi majus*
Anoda, spurred	*Anoda cristata*
Artichoke thistle	*Cynara cordunculus*
Aster	*Aster* spp.
Aster, spiny	*Aster spinosus*
Bahiagrass	*Paspalum notatum*
Balloonvine	*Cardiospermum halicacabum*
Barberry, European	*Berberis vulgaris*
Barley, volunteer	*Hordeum vulgare*
Barley, foxtail	*Hordeum jubatum*
Barley, hare	*Hordeum leporinum*
Barley, wall	*Hordeum glaucum*
Barnyardgrass	*Echinochloa crus-galli*
Bassia, fivehook	*Bassia hyssopifolia*
Bean, field	*Phaseolus vulgaris* var. Red Mexican
Bedstraw	*Galium* spp.
Bedstraw, catchweed	*Galium aparine*
Beggarticks, bur	*Bidens tripartita*
Beggarticks, tall	*Bidens vulgata*
Bellflower, creeping	*Campanula rapunculoides*
Bentgrass colonial	*Agrostis tenuis*
Bentgrass, creeping	*Agrostis stolonifera*
Bermudagrass	*Cynodon dactylon*
Bindweed, field	*Convolvulus arvensis*
Bindweed, hedge	*Calystegia sepium*
Bluegrass, annual	*Poa annua*
Bluegrass, Kentucky	*Poa pratensis*
Bluestem, big	*Andropogon gerardi*
Bluestem, little	*Andropogon scoparius*
Blueweed, Texas	*Helianthus ciliaris*
	Continued

TABLE A.10 ■ Common and scientific names of weeds mentioned in text, *continued.*

COMMON NAME	SCIENTIFIC NAME
Brackenfern	*Pteridium aquilinum*
Brome, downy	*Bromus tectorum*
Brome, false	*Brachypodium distachyon*
Brome, smooth	*Bromus inermis*
Broomrape	*Orobanche* spp.
Broomweed, common	*Gutierrezia dracunculoides*
Buckwheat, wild	*Polygonum convolvulus*
Buffalobur	*Solanum rostratum*
Buffelgrass	*Pennisetum ciliaris*
Burcucumber	*Sicyos angulatus*
Burdock, common	*Arctium minus*
Bursage, annual	*Ambrosia acanthicarpa*
Burweed, lawn (spurweed)	*Soliva pterosperma*
Buttercup, bur	*Ranunculus testiculatus*
Buttercup, creeping	*Ranunculus repens*
Buttonweed, Virginia	*Diodia virginiana*
Canarygrass	*Phalaris canariensis*
Canarygrass, Carolina	*Phalaris caroliniana*
Canarygrass, hood	*Phalaris paradoxa*
Canarygrass, littleseed	*Phalaris minor*
Canarygrass, reed	*Phalaris arundinacea*
Capeweed	*Arctotheca calendula*
Carelessweed	*Amaranthus palmeri*
Carpetweed	*Mollugo verticillata*
Carrot, wild (Queen Anne's lace)	*Dauca carota*
Castorbean	*Ricinus communis*
Cattail, common	*Typha latifolia*
Centipedegrass	*Eremochloa ophiuroides*
Chamomile, false	*Matricaria maritima*
Cheat	*Bromus secalinus*
Chess, hairy	*Bromus commutatus*
Chickweed, common	*Stellaria media*
Chickweed, mouseear	*Cerastium vulgatum*
Clover, white	*Trifolium repens*
Cocklebur, common	*Xanthium strumarium*
Corn, volunteer	*Zea mays*
Crabgrass, large	*Digitaria sanguinalis*
Crabgrass, smooth	*Digitaria ischaemum*
Crazyweed, Lambert	*Oxytropis Lambertii*
Cress, hoary	*Cardaria draba*
Crotalaria, showy	*Crotalaria spectabilis*
Croton, tropic	*Croton glandulosus*
Cucumber, squirting	*Ecballium elaterium*
Cucumber, western bur	*Marah oreganus*

Continued

TABLE A.10 ■ **Common and scientific names of weeds mentioned in text,** *continued.*

COMMON NAME	SCIENTIFIC NAME
Dallisgrass	*Paspalum dilatatum*
Dandelion	*Taraxacum officinale*
Darnel, Persian	*Lolium persicum*
Datura, sacred	*Datura innoxia*
Datura, small	*Datura discolor*
Deadnettle, purple	*Lamium purpureum*
Deadnettle, spotted	*Lamium maculatum*
Dichondra	*Dichondra repens*
Dock, curly	*Rumex crispus*
Dodder, field	*Cuscuta campestris*
Dogbane, hemp	*Apocynum cannabinum*
Ducksalad	*Heteranthera limosa*
Duckweed, common	*Lemna minor*
Dumbcane	*Diffenbachia* spp.
Eclipta	*Eclipta prostrata*
Elodea, common	*Elodea canadensis*
Evening primrose, common	*Oenothera biennis*
Everlasting, small	*Antennaria microphylla*
Falsedandelion, Carolina	*Pyrrhopappus carolinianus*
Falsehellebore, California	*Veratrum californicum*
Fennel	*Foeniculum vulgare*
Fescue, tall	*Festuca arundinacea*
Fiddleneck, coast	*Amsinckia intermedia*
Fieldcress, yellow	*Rorippa sylvestris*
Filaree, broadleaf	*Erodium botrys*
Filaree, redstem	*Erodium cicutarium*
Filaree, whitestem	*Erodium moschatum*
Fleabane, hairy	*Conyza bonariensis*
Fleabane, Philadelphia	*Erigeron philadelphicus*
Flixweed	*Descurainia sophia*
Foxtail, bristly	*Setaria verticillata*
Foxtail, giant	*Setaria faberi*
Foxtail, green	*Setaria viridis*
Foxtail, slender (rigid)	*Alopecurus myosuroides*
Foxtail, yellow	*Setaria glauca*
Galinsoga, hairy	*Galinsoga ciliata*
Galinsoga, smallflower	*Galinsoga parviflora*
Garlic, wild	*Allium vineale*
Goatgrass, jointed	*Aegilops cylindrica*
Goldenrod	*Solidago* spp.
Goldenrod, gray	*Solidago nemoralis*
Goldenweed, common	*Isocoma coronopifolia*
Goosefoot, Jeruselum-oak	*Chenopodium botrys*
Goosefoot, nettleleaf	*Chenopodium murale*
Goosefoot, pigweed	*Chenopodium paganum*
	Continued

TABLE A.10 ■ **Common and scientific names of weeds mentioned in text,** *continued.*

COMMON NAME	SCIENTIFIC NAME
Goosegrass	*Eleusine indica*
Gooseweed	*Sphenoclea zeylandica*
Gourd, buffalo	*Cucurbita foetidissima*
Grama, sixweeks	*Bouteloua barbata*
Groundcherry, lanceleaf	*Physalis lanceifolia*
Groundcherry, longleaf	*Physalis longifolia*
Groundcherry, Virginia	*Physalis virginiana*
Groundcherry, Wright	*Physalis wrightii*
Groundsel, common	*Senecio vulgaris*
Halogeton	*Halogeton glomeratus*
Hatico	*Ixophorus unisetus*
Hawksbeard, Asiatic	*Youngia japonica*
Hawksbeard, narrowleaf	*Crepis tectorum*
Healall	*Prunella vulgaris*
Hellebore, western false	*Veratrum californicum*
Henbit	*Lamium amplexicaule*
Hogpotato	*Hoffmanseggia glauca*
Honeysuckle, Japanese	*Lonicera japonica*
Horsenettle	*Solanum carolinense*
Horsetail, field	*Equisetum arvense*
Horseweed	*Conyza canadensis*
Hydrilla	*Hydrilla verticillata*
Indiangrass	*Sorghastrum nutans*
Ivy, ground	*Glechoma hederacea*
Jimsonweed	*Datura stramonium*
Johnsongrass	*Sorghum halepense*
Jointvetch, northern	*Aeschynomene virginica*
Junglerice	*Echinochloa colona*
Knapweed, diffuse	*Centaurea diffusa*
Knapweed, Russian	*Centaurea repens*
Knotweed, Douglas	*Polygonum douglasii*
Knotweed, prostrate	*Polygonum aviculare*
Knotweed, silversheath	*Polygonum argyrocoleon*
Kochia	*Kochia scoparia*
Ladysthumb	*Polygonum persicaria*
Lambsquarters, common	*Chenopodium album*
Larkspur, duncecap	*Delphinium occidentale*
Larkspur, tall	*Delphinium barbeyi*
Lettuce, blue	*Lactuca pulchella*
Lettuce, prickly	*Lactuca serriola*
Licorice, wild	*Glycyrrhiza lepidota*
Locoweeds	*Astragalus* spp.
Locoweed, purple (wooly)	*Astragulus mollissimus*
Locoweed, white (Crazyweed, Lambert)	*Oxytropis lambertii*

Continued

TABLE A.10 ■ Common and scientific names of weeds mentioned in text, *continued.*

COMMON NAME	SCIENTIFIC NAME
Mallow, common (roundleaf)	*Malva neglecta*
Mallow, little	*Malva parviflora*
Mallow, Venice	*Hibiscus trionum*
Maple	*Acer* spp.
Marestail	*Hippuris vulgaris*
Medic, black	*Medicago lupulina*
Mesquite	*Prosopis juliflora*
Milkvetch, twogrooved	*Astragalus bisulcatus*
Milkweed, climbing	*Sarcostemma cynanchoides*
Milkweed, common	*Asclepias syriaca*
Millet, wild-proso	*Panicum miliaceum*
Mistletoe, small	*Arceuthobium* spp.
Morningglory, ivyleaf	*Ipomoea hederacea*
Morningglory, pitted	*Ipomoea lacunosa*
Morningglory, tall	*Ipomoea purpurea*
Mullein	*Verbascum* spp.
Mullein, moth	*Verbascum blattaria*
Mustard, birdsrape	*Brassica rapa*
Mustard, blue	*Chorispora tenella*
Mustard, tumble	*Sisymbrium altissimum*
Mustard, wild	*Sinapis arvensis*
Nettle, burning	*Urtica urens*
Nettle, stinging	*Urtica dioica*
Nightshade, black	*Solanum nigrum*
Nightshade, cutleaf	*Solanum triflorum*
Nightshade, eastern black	*Solanum ptycanthum*
Nightshade, hairy	*Solanum sarrachoides*
Nightshade, silverleaf	*Solanum elaeagnifolium*
Nimblewill	*Muhlenbergia schreberi*
Nutsedge, purple	*Cyperus rotundus*
Nutsedge, yellow	*Cyperus esculentus*
Oat, wild	*Avena fatua*
Oat, winter wild	*Avena ludoviciana*
Oats, volunteer	*Avena sativa*
Oharechinoqiku	*Erigeron sumatrensis*
Oleander	*Nerium oleander*
Onion, wild	*Allium canadense*
Orach, spreading	*Atriplex patula*
Orchardgrass	*Dactylis glomerata*
Panicum, fall	*Panicum dichotomiflorum*
Panicum, Texas	*Panicum texanum*
Parsley, spring	*Cymopterus watsonni*
Parsnip, wild	*Pastinaca sativa*
Pennycress, field	*Thlaspi arvense*
Pennywort	*Hydrocotyl* spp.

Continued

TABLE A.10 ■ **Common and scientific names of weeds mentioned in text,** *continued.*

COMMON NAME	SCIENTIFIC NAME
Pepperweed, Virginia	*Lepidium virginicum*
Pigweed, fringed	*Amaranthus fimbriatus*
Pigweed, prostrate	*Amaranthus blitoides*
Pigweed, redroot	*Amaranthus retroflexus*
Pigweed, smooth	*Amaranthus hybridus*
Pigweed, tumble	*Amaranthus albus*
Pine	*Pinus* spp.
Pineappleweed	*Matricaria matricariodes*
Plantain, blackseed	*Plantago rugelii*
Plantain, broadleaf	*Plantago major*
Plantain, buckhorn	*Plantago lanceolata*
Poinsettia, wild	*Euphorbia heterophylla*
Poison hemlock	*Conium maculatum*
Poison ivy	*Rhus radicans*
Poison oak	*Rhus toxicodendron*
Poison oak, Pacific	*Rhus diversiloba*
Poison sumac	*Rhus vernix*
Pokeweed, common	*Phytolacca americana*
Polypogon, rabbitfoot	*Polypogon monspeliensis*
Povertyweed, Nuttall	*Monolepis nuttalliana*
Pricklypear, common	*Opuntia inermis*
Pricklypear, spiny	*Opuntia stricta*
Puncturevine	*Tribulus terrestris*
Purslane, common	*Portulaca oleracea*
Purslane, horse	*Trienthema portulacastrum*
Pusley, Florida	*Richardia scabra*
Quackgrass	*Agropyron repens*
Queen Anne's lace (wild carrot)	*Dauca carota*
Radish, wild	*Raphanus raphanistrum*
Ragweed, common	*Ambrosia artemisiifolia*
Ragweed, giant	*Ambrosia trifida*
Ragwort, tansy	*Senecio jacobaea*
Redvine	*Brunnichia ovata*
Reed, giant	*Arundo donax*
Rescuegrass	*Bromus catharticus*
Rhubarb	*Rhuem rhaponticum*
Rice (including red rice)	*Oryza sativa*
Rocket, London	*Sisymbrium irio*
Rocket, yellow	*Barbarea vulgaris*
Rose, Macartney	*Rosa bracteata*
Rose, multiflora	*Rosa multiflora*
Rye, cereal	*Secale cereale*
Ryegrass, Italian	*Lolium multiflorum*
Ryegrass, perennial	*Lolium perenne*

Continued

TABLE A.10 ■ Common and scientific names of weeds mentioned in text,
continued.

COMMON NAME	SCIENTIFIC NAME
Ryegrass, rigid	*Lolium rigidum*
Sage, whiteleaf	*Salvia leucophylla*
Sagebrush, big	*Artemesia tridentata*
St. Augustinegrass	*Stenotaphrum secundatum*
St. Johnswort, common	*Hypericum perforatum*
Salsify, common	*Tragopogon porrifolius*
Salsify, meadow	*Tragopogon pratensis*
Salsify, western	*Tragopogon dubius*
Saltbush	*Atriplex* spp.
Salvia	*Salvia splendens*
Sandbur, field	*Cenchrus incertus*
Sandbur, southern	*Cenchrus echinatus*
Sandwort, thymeleaf	*Arenaria serpyllifolia*
Sesbania, hemp	*Sesbania exaltata*
Shattercane	*Sorghum bicolor*
Shepherd's-purse	*Capsella bursa-pastoris*
Sicklepod	*Cassia obtusifolia*
Sida, prickly	*Sida spinosa*
Signalgrass, broadleaf	*Brachiaria platyphylla*
Smartweed, Pennsylvania	*Polygonum pensylvanicum*
Snakeweed, broom	*Gutierrezia sarothrae*
Snakeweed, threadleaf	*Gutierrezia microcephala*
Sorghum-almum	*Sorghum almum*
Sorrel, red	*Rumex acetosella*
Sowthistle, annual	*Sonchus oleraceus*
Sowthistle, perennial	*Sonchus arvensis*
Sowthistle, spiny	*Sonchus asper*
Spanishneedles	*Bidens bipinnata*
Speedwell	*Veronica* spp.
Speedwell, bilobed	*Veronica biloba*
Speedwell, common	*Veronica officinalis*
Speedwell, corn	*Veronica arvensis*
Speedwell, creeping	*Veronica filiformis*
Speedwell, Persian	*Veronica persica*
Speedwell, purslane	*Veronica peregrina*
Sprangletop, red	*Leptochloa filiformis*
Spruce	*Picea* spp.
Spurge, ground	*Euphorbia prostrata*
Spurge, leafy	*Euphorbia esula*
Spurge, prostrate	*Euphorbia humistrata*
Spurge, spotted	*Euphorbia maculata*
Spurweed (lawn burweed)	*Soliva pterosperma*
Starbur, bristly	*Acanthospermum hispidum*
Starwort, water	*Myosoton aquaticum*
Sudangrass	*Sorghum sudanese*

Continued

TABLE A.10 ■ Common and scientific names of weeds mentioned in text, *continued.*

COMMON NAME	SCIENTIFIC NAME
Sumpweed, poverty	*Iva axillaris*
Sunflower, common	*Helianthus annuus*
Sunflower, prairie	*Helianthus petiolaris*
Sweetclover, white	*Melilotus alba*
Sweetclover, yellow	*Melilotus officinalis*
Switchgrass	*Panicum virgatum*
Tansymustard, pinnate	*Descurainia pinnata*
Tarweed, cluster	*Madia glomerata*
Tarweed, coast	*Madia sativa*
Thistle, barbwire Russian	*Salsola paulsenii*
Thistle, Canada	*Cirsium arvense*
Thistle, musk	*Carduus nutans*
Thistle, Russian	*Salsola iberica*
Threeawn, prairie	*Aristida oligantha*
Toadflax, Dalmatian	*Linaria genistifolia*
Trumpetcreeper	*Campsis radicans*
Velvetleaf	*Abutilon theophrasti*
Vervain, prostrate	*Verbena bracteata*
Violet, common blue	*Viola papilionacea*
Violet, field	*Viola arvensis*
Walnut, black	*Juglans nigra*
Watergrass, early	*Echinochloa oryzoides*
Waterhemlock, western	*Cicuta douglasii*
Waterhemp, common	*Amaranthus rudis*
Waterhyacinth	*Eichornia crassipes*
Waterhyacinth, anchored	*Eichornia azurea*
Waterlettuce	*Pistia stratiotes*
Wheat, volunteer	*Triticum aestivum*
Wheatgrass, intermediate	*Agropyron intermedium*
Wildbuckwheat, longleaf	*Eriogonum longifolium*
Wildrice, annual	*Zizania aquatica*
Willowweed	*Epilobium* spp.
Willowweed, American	*Epilobium adenocaulon*
Witchgrass	*Panicum capillare*
Witchweed	*Striga asiatica*
Woodsorrel, creeping	*Oxalis corniculata*
Woodsorrel, yellow	*Oxalis stricta*
Wormwood, biennial	*Artemisia biennis*
Yarrow, common	*Achillea millefolium*
Yarrow, western	*Achillea lanulosa*
Yaupon	*Ilex vomitoria*
Yucca	*Yucca* spp.

SOURCE: *Composite list of weeds.* 1989. Champaign, Ill.: Weed Society of America.

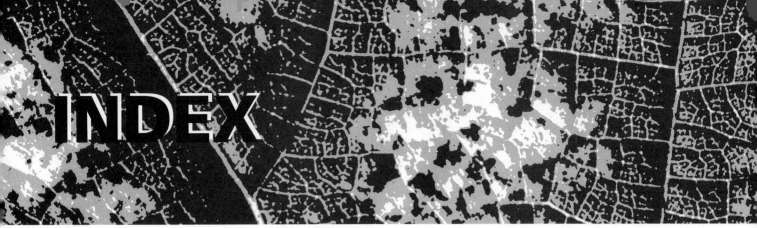

INDEX